Lecture Notes in Computer Science 10386

Commenced Publication in 1973
Founding and Former Series Editors:
Gerhard Goos, Juris Hartmanis, and Jan van Leeuwen

Editorial Board

David Hutchison
  Lancaster University, Lancaster, UK
Takeo Kanade
  Carnegie Mellon University, Pittsburgh, PA, USA
Josef Kittler
  University of Surrey, Guildford, UK
Jon M. Kleinberg
  Cornell University, Ithaca, NY, USA
Friedemann Mattern
  ETH Zurich, Zurich, Switzerland
John C. Mitchell
  Stanford University, Stanford, CA, USA
Moni Naor
  Weizmann Institute of Science, Rehovot, Israel
C. Pandu Rangan
  Indian Institute of Technology, Madras, India
Bernhard Steffen
  TU Dortmund University, Dortmund, Germany
Demetri Terzopoulos
  University of California, Los Angeles, CA, USA
Doug Tygar
  University of California, Berkeley, CA, USA
Gerhard Weikum
  Max Planck Institute for Informatics, Saarbrücken, Germany

More information about this series at http://www.springer.com/series/7407

Ying Tan · Hideyuki Takagi
Yuhui Shi · Ben Niu (Eds.)

Advances in Swarm Intelligence

8th International Conference, ICSI 2017
Fukuoka, Japan, July 27 – August 1, 2017
Proceedings, Part II

 Springer

Editors
Ying Tan
Peking University
Beijing
China

Hideyuki Takagi
Kyushu University
Fukuoka
Japan

Yuhui Shi
Southern University of Science
and Technology
Shenzhen
China

Ben Niu
Shenzhen University
Shenzhen
China

ISSN 0302-9743   ISSN 1611-3349   (electronic)
Lecture Notes in Computer Science
ISBN 978-3-319-61832-6   ISBN 978-3-319-61833-3   (eBook)
DOI 10.1007/978-3-319-61833-3

Library of Congress Control Number: 2017944216

LNCS Sublibrary: SL1 – Theoretical Computer Science and General Issues

© Springer International Publishing AG 2017
This work is subject to copyright. All rights are reserved by the Publisher, whether the whole or part of the material is concerned, specifically the rights of translation, reprinting, reuse of illustrations, recitation, broadcasting, reproduction on microfilms or in any other physical way, and transmission or information storage and retrieval, electronic adaptation, computer software, or by similar or dissimilar methodology now known or hereafter developed.
The use of general descriptive names, registered names, trademarks, service marks, etc. in this publication does not imply, even in the absence of a specific statement, that such names are exempt from the relevant protective laws and regulations and therefore free for general use.
The publisher, the authors and the editors are safe to assume that the advice and information in this book are believed to be true and accurate at the date of publication. Neither the publisher nor the authors or the editors give a warranty, express or implied, with respect to the material contained herein or for any errors or omissions that may have been made. The publisher remains neutral with regard to jurisdictional claims in published maps and institutional affiliations.

Printed on acid-free paper

This Springer imprint is published by Springer Nature
The registered company is Springer International Publishing AG
The registered company address is: Gewerbestrasse 11, 6330 Cham, Switzerland

Preface

This book and its companion volumes, LNCS vols. 10385 and 10386, constitute the proceedings of the 8th International Conference on Swarm Intelligence (ICSI 2017) held from July 27 to August 1, 2017, in Fukuoka, Japan.

The theme of ICSI 2017 was "Serving Life with Intelligence and Data Science." ICSI 2017 provided an excellent opportunity and/or an academic forum for academics and practitioners to present and discuss the latest scientific results and methods, innovative ideas, and advantages in theories, technologies, and applications in swarm intelligence. The technical program covered all aspects of swarm intelligence and related areas.

ICSI 2017 was the eighth international gathering in the world for researchers working on all aspects of swarm intelligence, following successful events in Bali (ICSI 2016), Beijing (ICSI-CCI 2015), Hefei (ICSI 2014), Harbin (ICSI 2013), Shenzhen (ICSI 2012), Chongqing (ICSI 2011), and Beijing (ICSI 2010), which provided a high-level academic forum for participants to disseminate their new research findings and discuss emerging areas of research. It also created a stimulating environment for participants to interact and exchange information on future challenges and opportunities in the field of swarm intelligence research. ICSI 2017 was held in conjunction with the Second International Conference on Data Mining and Big Data (DMBD 2017) at Fukuoka, Japan, to share common mutual ideas, promote transverse fusion, and stimulate innovation.

ICSI 2017 was held in the center of the Fukuoka City. Fukuoka is the fifth largest city in Japan with 1.55 million inhabitants and is the seventh most liveable city in the world according to the 2016 Quality of Life Survey by *Monocle*. Fukuoka is in the northern end of the Kyushu Island and is the economic and cultural center of Kyushu Island. Because of its closeness to the Asian mainland, Fukuoka has been an important harbor city for many centuries. Today's Fukuoka is the product of the fusion of two cities in the year 1889, when the port city of Hakata and the former castle town of Fukuoka were united into one city called Fukuoka. The participants of ICSI 2017 enjoyed traditional Japanese dances, the local cuisine, beautiful landscapes, and the hospitality of the Japanese people in modern Fukuoka, whose sites are part of UNESCO's World Heritage.

ICSI 2017 received 267 submissions from about 512 authors in 35 countries and regions (Algeria, Australia, Austria, Brazil, Brunei Darussalam, Canada, China, Colombia, Germany, Hong Kong SAR China, India, Indonesia, Iran, Italy, Japan, Lebanon, Malaysia, Mexico, New Zealand, Nigeria, Norway, Romania, Russia, Serbia, Singapore, South Africa, South Korea, Spain, Sweden, Chinese Taiwan, Thailand, Tunisia, Turkey, UK, USA) across six continents (Asia, Europe, North America, South America, Africa, and Oceania). Each submission was reviewed by at least two, and on average 2.9 reviewers. Based on rigorous reviews by the Program Committee members and reviewers, 133 high-quality papers were selected for publication in this proceedings volume, with an acceptance rate of 49.81%. The papers are organized in 24 cohesive sections covering all major topics of swarm intelligence and computational intelligence research and development.

On behalf of the Organizing Committee of ICSI 2017, we would like to express our sincere thanks to the Research Center for Applied Perceptual Science of Kyushu University and the Computational Intelligence Laboratory of Peking University for their sponsorship, to the IEEE Computational Intelligence Society for its technical sponsorship, to a Japan Chapter of IEEE Systems, Man and Cybernetics Society for its technical cosponsorship, as well as to our supporters the International Neural Network Society, World Federation on Soft Computing, IEEE Beijing Section, and Beijing Xinghui Hi-Tech Co. and Springer. We would also like to thank the members of the Advisory Committee for their guidance, the members of the international Program Committee and additional reviewers for reviewing the papers, and the members of the Publications Committee for checking the accepted papers in a short period of time. We are particularly grateful to Springer for publishing the proceedings in the prestigious series *Lecture Notes in Computer Science*. Moreover, we wish to express our heartfelt appreciation to the plenary speakers, session chairs, and student helpers. In addition, there are still many more colleagues, associates, friends, and supporters who helped us in immeasurable ways; we express our sincere gratitude to them all. Last but not the least, we would like to thank all the speakers, authors, and participants for their great contributions that made ICSI 2017 successful and all the hard work worthwhile.

May 2017
                                                                    Ying Tan
                                                              Hideyuki Takagi
                                                                   Yuhui Shi
                                                                    Ben Niu

Organization

General Co-chairs

Ying Tan — Peking University, China
Hideyuki Takagi — Kyushu University, Japan

Program Committee Chair

Yuhui Shi — Southern University of Science and Technology, China

Advisory Committee Co-chairs

Russell C. Eberhart — IUPUI, USA
Gary G. Yen — Oklahoma State University, USA
Hisao Ishibuchi — Osaka Prefecture University, Japan

Technical Committee Co-chairs

Kay Chen Tan — National University of Singapore, Singapore
Xiaodong Li — RMIT University, Australia
Nikola Kasabov — Auckland University of Technology, New Zealand
Ponnuthurai N. Suganthan — Nanyang Technological University, Singapore

Plenary Session Co-chairs

Mengjie Zhang — Victoria University of Wellington, New Zealand
Andreas Engelbrecht — University of Pretoria, South Africa

Invited Session Co-chairs

Yan Pei — University of Aizu, Japan
Chaomin Luo — University of Detroit Mercy, USA

Special Sessions Co-chairs

Shangce Gao — University of Toyama, Japan
Ben Niu — Shenzhen University, China
Qirong Tang — Tongji University, China

Tutorial Co-chairs

Milan Tuba   John Naisbitt University, Serbia
Andreas Janecek   University of Vienna, Austria

Publications Co-chairs

Swagatam Das   Indian Statistical Institute, India
Xinshe Yang   Middlesex University, UK

Publicity Co-chairs

Yew-Soon Ong   Nanyang Technological University, Singapore
Carlos Coello   CINVESTAV-IPN, Mexico
Yaochu Jin   University of Surrey, UK
Shi Cheng   Shanxi Normal University, China
Bin Xue   Victoria University of Wellington, New Zealand

Finance and Registration Chairs

Chao Deng   Peking University, China
Suicheng Gu   Google Corporation, USA

Local Arrangements Chair

Ryohei Funaki   Kyushu University, Japan

Conference Secretariat

Jie Lee   Peking University, China

International Program Committee

Mohd Helmy Abd Wahab   Universiti Tun Hussein Onn Malaysia, Malaysia
Harshavardhan Achrekar   University of Massachusetts Lowell, USA
Miltos Alamaniotis   Purdue University, USA
Peter Andras   Keele University, UK
Esther Andrés   INTA, Spain
Helio Barbosa   Laboratório Nacional de Computação Científica, Brazil
Carmelo J.A. Bastos Filho   University of Pernambuco, Brazil
David Camacho   Universidad Autonoma de Madrid, Spain
Bin Cao   Tsinghua University, China
Jinde Cao   Southeast University, China
Kit Yan Chan   DEBII, Australia
Shi Cheng   Shaanxi Normal University, China
Manuel Chica   European Centre for Soft Computing, Spain

Carlos Coello Coello	CINVESTAV-IPN, Mexico
Jose Alfredo Ferreira Costa	Federal University, Brazil
Prithviraj Dasgupta	University of Nebraska, USA
Mingcong Deng	Tokyo University of Agriculture and Technology, Japan
Ke Ding	Tencent Corporation, China
Yongsheng Dong	Xi'an Institute of Optics and Precision Mechanics of CAS, China
Mark Embrechts	RPI, USA
Andries Engelbrecht	University of Pretoria, South Africa
Jianwu Fang	Xi'an Institute of Optics and Precision Mechanics of CAS, China
Shangce Gao	University of Toyama, Japan
Beatriz Aurora Garro Licon	IIMAS-UNAM, Mexico
Maoguo Gong	Xidian University, China
Yinan Guo	China University of Mining and Technology, China
J. Michael Herrmann	University of Edinburgh, UK
Lu Hongtao	Shanghai Jiao Tong University, China
Andreas Janecek	University of Vienna, Austria
Yunyi Jia	Clemson University, USA
Changan Jiang	Ritsumeikan University, Japan
Mingyan Jiang	Shandong University, China
Chen Junfeng	Hohai University, China
Liangjun Ke	Xian Jiaotong University, China
Thanatchai Kulworawanichpong	Suranaree University of Technology, Thailand
Rajesh Kumar	MNIT, India
Hung La	University of Nevada, USA
Germano Lambert-Torres	PS Solutions, Brazil
Xiujuan Lei	Shanxi Normal University, China
Bin Li	University of Science and Technology of China
Xiaodong Li	RMIT University, Australia
Xuelong Li	Chinese Academy of Sciences, China
Andrei Lihu	Politehnica University of Timisoara, Romania
Fernando B. De Lima Neto	University of Pernambuco, Brazil
Ju Liu	Shandong University, China
Qun Liu	Chongqing University of Posts and Communications, China
Wenlian Lu	Fudan University, China
Wenjian Luo	University of Science and Technology of China
Mohamed Arezki Mellal	M'Hamed Bougara University, Algeria
Sanaz Mostaghim	Institute IWS, Germany
Ben Niu	Arizona State University, USA
Quan-Ke Pan	Huazhong University of Science and Technology, China
Shahram Payandeh	Simon Fraser University, Canada

Yan Pei   The University of Aizu, Japan
Somnuk Phon-Amnuaisuk   Universiti Teknologi, Brunei
Radu-Emil Precup   Politehnica University of Timisoara, Romania
Kai Qin   RMIT University, Australia
Quande Qin   Shenzhen University, China
Boyang Qu   Zhongyuan University of Technology, China
Robert Reynolds   Wayne State University, USA
Luneque Silva Junior   Federal University of Rio de Janeiro, Brazil
Pramod Kumar Singh   ABV-IIITM Gwalior, India
Ponnuthurai Suganthan   Nanyang Technological University, Singapore
Hideyuki Takagi   Kyushu University, Japan
Ying Tan   Peking University, China
Qian Tang   Xidian University, China
Qirong Tang   Tongji University, China
Milan Tuba   John Naisbitt University, Serbia
Mario Ventresca   Purdue University, USA
Gai-Ge Wang   Jiangsu Normal University, China
Guoyin Wang   Chongqing University of Posts and
                               Telecommunications, China
Lei Wang   Tongji University, China
Ka-Chun Wong   City University of Hong Kong, SAR China
Shunren Xia   Zhejiang University, China
Bo Xing   University of Johannesburg, South Africa
Benlian Xu   Changshu Institute of Technology, China
Yingjie Yang   De Montfort University, UK
Wei-Chang Yeh   National Tsing Hua University, Taiwan
Kiwon Yeom   NASA Ames Research Center, USA
Peng-Yeng Yin   National Chi Nan University, Taiwan
Zhuhong You   Shenzhen University, China
Zhigang Zeng   Huazhong University of Science and Technology,
                               China
Zhi-Hui Zhan   Sun Yat-sen University, China
Jie Zhang   Newcastle University, UK
Jun Zhang   Waseda University, Japan
Junqi Zhang   Tongji University, China
Lifeng Zhang   Renmin University of China
Mengjie Zhang   Victoria University of Wellington, New Zealand
Qieshi Zhang   Shaanxi Normal University, China
Qiangfu Zhao   The University of Aizu, Japan
Yujun Zheng   Zhejiang University of Technology, China
Cui Zhihua   Complex System and Computational Intelligence
                               Laboratory, China
Guokang Zhu   Shanghai University of Electric Power, China
Xingquan Zuo   Beijing University of Posts and Telecommunications,
                               China

Additional Reviewers

Abdul Rahman, Shuzlina
Ambar, Radzi
Chen, Zonggan
Farias, Felipe
Feng, Jinwang
Figueiredo, Elliackin
Haji Mohd, Mohd Norzali
Huang, Yu-An
Jiao, Yuechao
Joyce, Thomas
Joyce, Thomes
Junyi, Chen
Li, Liping
Li, Xiangtao
Liu, Xiaofang
Liu, Zhenbao
Lu, Quanmao
Lyu, Yueming
Mohamad Mohsin, Mohamad Farhan
Oliveira, Marcos

Qiujie, Wu
Saharan, Sabariah
Shang, Ke
Siqueira, Hugo
Vázquez Espinoza De Los Monteros,
    Roberto Antonio
Wang, Lei
Wang, Lingyu
Wang, Tianyu
Wang, Yanbin
Wang, Zi-Jia
Yahya, Zainor Ridzuan
Yan, Shankai
Yang, Le
Yu, Jun
Zhang, Hao
Zhang, Jianhua
Zhang, Jiao
Zhang, Qinchuan
Zheng, Lintao

Contents – Part II

Community Detection

Multi-agent Systems and Swarm Robotics

Hybrid Optimization Algorithms and Applications

Fuzzy and Swarm Approach

Clustering and Forecast

Classification and Detection

Planning and Routing Problems

Dialog System Applications

Robotic Control

Other Applications

Contents – Part I

Particle Swarm Optimization

Applications of Particle Swarm Optimization

Ant Colony Optimization

Artificial Bee Colony Algorithms

Genetic Algorithms

Differential Evolution

Fireworks Algorithm

Brain Storm Optimization Algorithm

Cuckoo Search

Firefly Algorithm

Multi-objective Optimization

A Parametric Study of Crossover Operators in Pareto-Based Multiobjective Evolutionary Algorithm

Shohei Maruyama and Tomoaki Tatsukawa[✉]

Tokyo University of Science, Tokyo 125-8585, Japan
4416631@ed.tus.ac.jp , tatsukawa@rs.tus.ac.jp

Abstract. The parameter setting in Multi-Objective Evolutionary Algorithms (MOEA) often affects the performance of optimization. Besides, the optimal values of parameters often depend on each optimization problem. However, it is difficult to decide the appropriate parameter setting for each problem in advance. In this study, the effect of parameter difference of the major crossover operators: Simulated Binary crossover (SBX), Differentioal Evolution operator (DE), Simplex crossover (SPX), Parent Centric crossover (PCX), and Unimodal Normal Distribution crossover (UNDX), is investigated by using 10 benchmark problems. DTLZ, WFG, and ZDT benchmark problems were considered. Non-dominated sorting genetic algorithm-II (NSGA-II) were used as a Pareto-based MOEA. The number of objectives was set to three. The experimental results on benchmark problems show that the effect of parameter variation is relatively small for SBX and SPX. On the other hand, there are optimum parameters in each benchmark problem as other crossover operators. This indicates that the choice of the crossover operator is significantly important in MOEA for achieving the good performance.

Keywords: Crossover operator · NSGA-II · SBX · DE · SPX · PCX · UNDX

1 Introduction

Multi-objective optimization problems (MOP) exist in various real-world applications such as engineering, financial, and scientific applications. In MOP, each objective function is often in a trade-off relationship, and in such a case, it is impossible to obtain one optimal solution. Therefore, the purpose of MOP is to get a set of solutions called Pareto-optimal (or non-dominated) solutions. Multi-objective optimization requires not only convergence but also the diversity of solutions. In recent years, many kinds of multi-objective evolutionary algorithms (MOEA) have been proposed and shown to demonstrate good performance in real-world optimization problems.

In MOEA, it is known that the performance is changed in different crossover operators. Each crossover operator has some parameters. For example, Simulated Binary crossover(SBX) [1] has a parameter η, and Parent Centric crossover

© Springer International Publishing AG 2017
Y. Tan et al. (Eds.): ICSI 2017, Part II, LNCS 10386, pp. 3–14, 2017.
DOI: 10.1007/978-3-319-61833-3_1

(PCX) [10] has two parameters η and ζ. When we run MOEA, we have to properly determine the values of these parameters in advance. However, in most cases, we use the default values set in each MOEA as they are. Table 1 shows the default crossover operator and the default value of the crossover operator in some major MOEA. The default crossover operator as shown in Table 1 is almost SBX, and η is set to between 15 and 30. It is known that the optimal parameter value is changed by the optimization problem [7,15]. In [7], the effect of Differential Evolution(DE) operator [13] is investigated with 4 benchmark problems. However, the effect of various crossover operators has not yet been investigated.

In this study, we consider 5 crossover operators such as SBX, DE, simplex crossover (SPX) [14], PCX, and unimodal normal distribution crossover (UNDX) [8] as well as 10 benchmark problems to widely investigate the effects of crossover operators. For the number of parameters in each crossover operator, there is one in SBX and SPX, and two in DE, PCX, and UNDX. The objective of this study is to investigate the effect of changing parameters of each crossover operator. NSGA-II is used as a MOEA. DTLZ2,3,4 [3], WFG1,2,6,8,9 [9] and ZDT1,4 [17] were used as benchmark problems.

Table 1. Default values of crossover operator in MOEA

MOEA	Crossover method	Parameter value
NSGA-II [2]	SBX	$\eta = 20$
ϵ-MOEA [12]	SBX	$\eta = 15$
NSGA-III [11]	SBX	$\eta = 30$
MOEA/D	SBX	$\eta = 20$
MOEA/D-DE [16]	DE	$[cr, F] = [1.0, 0.5]$
IBEA [5]	SBX	$\eta = 20$
SPEA2 [4]	SBX	$\eta = 20$

2   Background

2.1   Related Work

Yuan et al. compared the performance when changing parameter values in the DE operator using several test problems. In Yuan's paper, cr is fixed to 0.1, and the performance is compared when F is varied. The performance of F is good at about 0.5 [15]. Besides, F is fixed to 0.5, and the performance is compared when cr is varied. The performance of cr is good at about 0.1. Hadka et al. examined the correlation of various parameters in the Borg MOEA, which has adaptive operator selection (AOS). However, in Hadka's paper, performance on each parameter value of crossover operator has not been reported [6].

Thus, a comparison of performance when fixing one of the parameter values in DE and a comparison of correlation with various parameters in AOS has already been done. However, the case of not fixing one of the parameter values and the case of changing to various parameter values in crossover operators excluding DE has not been compared yet.

In this study, we compare the performance when various parameter values are changed variously in SBX, DE, SPX, PCX, and UNDX. Although DE, PCX, and UNDX each have 2 parameters, in this study, we examine the performance without fixing one of these parameters. In addition, to gain insight into the parameters of each operator for each property of the problem, we used 10 kinds of benchmark problems with different features.

2.2   Crossover Operator

In MOEA, offspring solutions are generated from parent solutions by crossover and mutation in each generation. Crossover and mutation operators are different depending on whether the design variables are handled as binary or real numbers. In this research, we focus on cases where design variables are treated as real numbers. Five crossover operators are used in this research.

SBX imitates one point crossover of a binary number with a real number. The algorithm of SBX is shown below.

$$\beta(u) = \begin{cases} (2u)^{\frac{1}{\eta}}, & if \quad u(0,1) \leq 0.5 \\ \frac{1}{2(1-u)^{\frac{1}{\eta+1}}}, & else \end{cases} \quad (1)$$

$$y^{(1)} = \frac{1}{2}\{(1-\beta)x^{(1)} + (1+\beta)x^{(2)}\}. \quad (2)$$

$$y^{(2)} = \frac{1}{2}\{(1+\beta)x^{(1)} + (1-\beta)x^{(2)}\}. \quad (3)$$

where $x^{(1)}, x^{(2)}$ denote parent solutions, and $y^{(1)}, y^{(2)}$ denote offspring solutions. β denotes a distribution function, and η denotes a SBX parameter.

DE is a crossover operator featuring stochastic linear search. DE randomly selects 3 solutions from the parent population and calculates a vector from the first 2 solutions. It then generates an offspring solution from at last a parent solution to the vector direction. The algorithm of DE is shown below.

$$x'_i = x_{p1} + F(x_{p2} - x_{p3}). \quad (4)$$

$$y_{i,j} = \begin{cases} x'_{j,i}, & if \quad u(0,1) \leq cr \\ x_{j,i}, & else \end{cases} \quad (5)$$

Here, $\{x_{p1}, x_{p2}, x_{p3}\}$ denote parent solutions, y denotes offspring solution, and j denotes a design variable. Terms cr and F denote DE parameters.

SPX is a crossover operator that generates a child solution from the part that extends past the range surrounded by the centroid. These are parent individuals of multiple parent solutions. The SPX algorithm wich three parents is shown below.

$$O = \frac{1}{3}\sum_{i=1}^{3} x^{(i)}. \quad (6)$$

$$r_i = u^{\frac{1}{i}}, \ i = 1, 2, 3. \tag{7}$$

$$v^{(i)} = O + \epsilon(x^{(i)} - O). \tag{8}$$

$$c^{(i)} = \begin{cases} 0, \ i = 1 \\ r_{i-1}(y^{(i-1)} - v^{(i)} + c^{(i-1)}), \ i = 2 \end{cases} \tag{9}$$

$$y = v^{(3)} + c^{(3)}. \tag{10}$$

Here, $\{x^i, i = 1, 2, 3\}$ denotes parent solutions, y denotes offspring solutions, and O denotes centroid parent solutions. Term ϵ denotes the SPX parameter.

PCX is a crossover operator that generates offspring solutions around the parent solution via the centroid of multiple parent solutions. The algorithm of PCX with three parents is shown below.

$$O = \frac{1}{3} \sum_{i=1}^{3} x^{(i)}. \tag{11}$$

$$d^{(p)} = x^{(p)} - O \quad p = 1, 2 \ or \ 3. \tag{12}$$

$$\overline{D} = \frac{1}{2} \sum_{i=1, i \neq p}^{3} \frac{\sqrt{d(x^{(i)})}}{||d||}. \tag{13}$$

$$y = x^{(p)} + \omega_\zeta |d^{(p)}| + \sum_{i=1, i \neq p}^{3} \omega_\eta \overline{D} e^{(i)}. \tag{14}$$

Here, $\{x^i, i = 1, 2, 3\}$ denotes parent solutions, y denotes offspring solution, and O denotes the centroid of the parent solutions. Term $e^{(i)}$ is the orthogonal basis vector defined by $d^{(p)}$, and D denotes the distance of the perpendicular vectors of the remaining 2 individuals and the $d^{(p)}$. Terms η and ζ are parameters of PCX. Term ω_η denotes a random number generated from a Gaussian distribution with mean 0, and variance σ_η^2. Term ω_ζ denotes a random number generated from a Gaussian distribution with mean 0 and variance σ_ζ^2.

UNDX is a crossover operator creating a centroid from parent solutions. It generates offspring solutions using a normal distribution around the centroid. The algorithm of UNDX with three parents is shown below.

$$O = \sum_{i=1}^{2} x^{(i)}. \tag{15}$$

$$d = x^{(2)} - x^{(1)}. \tag{16}$$

$$D = \sqrt{|x^{(3)} - x^{(1)}|^2 - \frac{[d(|x^{(3)} - x^{(1)})]^2}{|d|^2}}. \tag{17}$$

$$y = O + \sigma_\zeta d + D \sum_{i=1}^{n-1} \sigma_{\eta_i} e_i. \tag{18}$$

$$\sigma_\zeta \sim N(0, \zeta^2), \qquad \sigma_\eta \sim N(0, \eta^2). \tag{19}$$

Here, $\{x^i, i = 1, 2, 3\}$ denotes parent solutions, y denotes offspring solutions, and O denotes centroid of 2 parent solutions. Term $e^{(i)}$ is the orthogonal basis vector defined by d. Terms η and ζ denote parameters of UNDX.

3 Computational Condition

In this study, we use NSGA-II as MOEA. The crossover operator in NSGA-II is calculated for each of the 5 operators described above. We also use polynomial mutation for mutation operator. Table 2 shows the parameter values of each crossover operator in this study.

Table 2. Parameter values of crossover operators

Crossover	Parameter	Range of parameter values
SBX	η	$\{5, 15, 25, 35, 45\}$
DE	cr	$\{0.1, 0.3, 0.5, 0.7, 0.9\}$
	F	$\{0.1, 0.3, 0.5, 0.7, 0.9\}$
SPX	ϵ	$\{1, 3, 5, 7, 9\}$
PCX	η	$\{0.1, 0.3, 0.5, 0.7, 0.9\}$
	ζ	$\{0.1, 0.3, 0.5, 0.7, 0.9\}$
UNDX	η	$\{0.1, 0.3, 0.5, 0.7, 0.9\}$
	ζ	$\{0.1, 0.3, 0.5, 0.7, 0.9\}$

Therefore, we adopt 10 benchmark problems with different properties (DTLZ2, 3, 4, WFG1, 2, 6, 8, 9, and ZDT1, 4). DTLZ and WFG problems can be scaled with any number of objectives and design variables. Here, we consider that the number of objectives in DTLZ and WFG is 3. The number of the objective functions in ZDT problems is set to 2. For DTLZ and WFG problems, the number of design variables is 12. In the ZDT1 problem, the number of design variables is 30, and in the ZDT4 problem, the number of design variables is 10. Therefore, none of the benchmark problems have a constraint function. Table 3 summarizes the properties of the benchmark problems from the view of separability of design variables, modality, shape of Pareto front, and whether each problem has bias or not. In all problems, both the population size and the number of generations are set to 100. Furthermore, the number of trials is 10.

To evaluate the performance in each pattern, the hypervolume (HV) [18] is used as the performance metric. The HV can measure both convergence and diversity of a optimal solution set and a higher value means better performance. The calculation formula of HV is shown below.

$$HV(P) = volume(\bigcup_{y \in S} [y_1, y_1^*] \times \dots \times [y_m, y_m^*]). \tag{20}$$

Table 3. Summary of the problems of the test problems

Problem	Separability	Modality	Bias	Shape
DTLZ2	S	U	×	concave
DTLZ3	S	M	×	concave
DTLZ4	S	U	○	concave
WFG1	S	U	○	convex
WFG2	N	M	×	convex
WFG6	N	U	×	concave
WFG8	N	U	○	concave
WFG9	N	M	○	concave
ZDT1	S	U	×	convex
ZDT4	S	M	×	convex

For separability, "S" denotes that design variables are separable, and "N" indicates that the design variables are non-separable. For each modality, "U" indicates that the problem is uni-modal, and "M" denotes that the problem is multi-modal. For bias, we indicate whether the problem has bias or not. Shape shows the shape of the Pareto front.

Here, m denotes the number of objectives, and S denotes the Pareto set. In this study, y^* is set to 1.5 times the maximum value of each objective function value in the true Pareto front except for DTLZ3. In DTLZ3, y^* is set to $\{150, 150, 150\}$. Furthermore, to avoid changing the value for each problem, divide by the volume surrounded by the origin point and y^* to convert it to the proportion of the HV value surrounded by the origin point and y^*. Pareto set in all search solutions is used for calculation of HV.

4   Results and Discussion

In all parameters mentioned above in Table 2, the trials are run using different 10 random seeds. The HV calculates the average of 10 trials and compares it to each parameter value. Figures 1, 2, 3, 4 and 5 show the average values of HV wich 10 trials in each crossover operator. A bluer color is preferred. In Figs. 1, 2, 3, 4 and 5, the HV value is converted from 0 to 1 for each problem, and in that range, each cell is colored in 50 hierarchies. The conversion formula is shown below.

$$HV_{new} = \frac{HV_{old} - min(HV_{old})}{max(HV_{old}) - min(HV_{old})}. \tag{21}$$

For each cell in Figs. 1, 2, 3, 4 and 5, each cell shows the proportion of the HV value surrounded by the origin point and y^*.

Figure 1 shows the result of SBX - the number of parameters is only one. From Fig. 1, it is clear when the value of η is large, each cell is bluer, but there

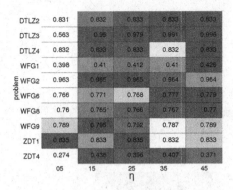

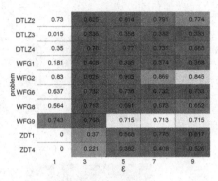

Fig. 1. Result of SBX   **Fig. 2.** Result of SPX

is little difference in the HV value for each parameter except DTLZ3 and ZDT4. We think that η does not change the performance in most test problems. In the case of DTLZ3 and ZDT4, there is a large difference in the $\eta = 5$ case and other cases. Performance is not good with $\eta = 5$.

Figure 2 shows the result of SPX. The number of parameters is only one. Figure 2 shows that the value of $\epsilon = 3, 5, 7$, each cell is bluer. Thus, the performance is considered good with $\epsilon = 3, 5, 7$ except WFG9. With WFG9, there is almost no difference in the HV value for each parameter, and the value of ϵ does not change the performance in WFG9.

Figure 3 shows the result of DE - the number of parameters is two. In DE, PCX, and UNDX, there are two parameters: the horizontal axis and the vertical axis, and these are set as the respective parameters. The performance is compared with each problem. Therefore, in DE, PCX, and UNDX, the performance is compared to 10 figures. Thus, Fig. 3(a)–(j) show the result of DE in each test problem. From Fig. 3, it is understood when the value of cr and F are small, each cell is bluer in most test problems. However, for ZDT4, it is understood that each cell is bluer when the value of cr and F are small. Thus, in most test problems, the performance is good when the value of cr and F are small. For ZDT4, the performance is good when the value of cr and F are large.

Figure 4 shows the result of PCX, and Fig. 4(a)–(j) show the results in each problem. From Fig. 4, the color tone differs for each problem, and the optimal parameter value may be different depending on the problem in PCX. Especially for WFG2, 6, 8, 9, each cell is bluer when the value of η is large. A common characteristic of WFG2, 6, 8, 9 is that design variables are non-separable. Thus in PCX, if the design variables are non-separable, then the performance is good when the value of η in PCX is large.

Figure 5 shows the result of UNDX, and Fig. 5(a)–(j) show the results in each problem. Figure 5 shaws that when the value of η is large, each cell is bluer in most test problems. Thus, the performance is good when the value of η is large. However, in the case of DTLZ3 and WFG1, there is little difference in the HV value for each parameter. Thus, the value of η likely does not change

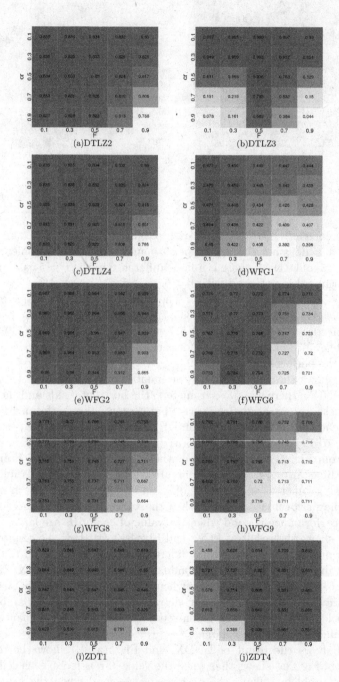

Fig. 3. Result of DE

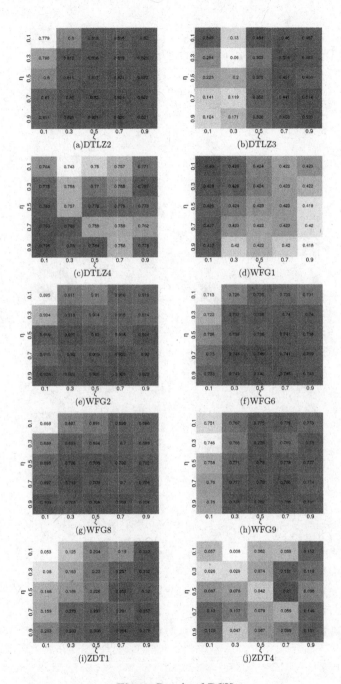

Fig. 4. Result of PCX

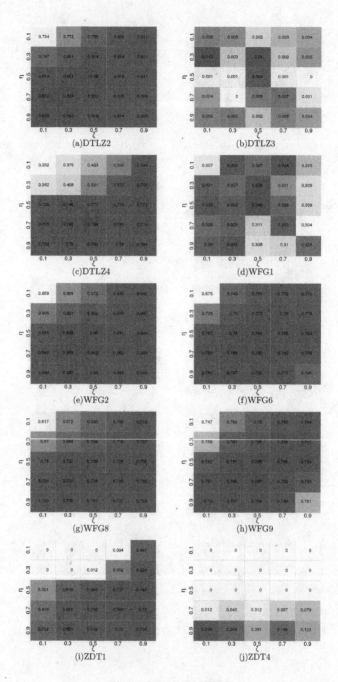

Fig. 5. Result of UNDX

the performance in DTLZ3 and WFG1. On the other hand, in the case of ζ, it becomes bluer from 0.3 to 0.5 in each problem. Thus, the performance is good when $\zeta = 0.3, 0.5$.roblem. Thus, the performance is good when $\zeta = 0.3, 0.5$.

5 Conclusion

In this study, we consider five crossover operators such as SBX, DE, SPX, PCX, and UNDX as well as 10 benchmark problems to widely investigate the effects of crossover operators. For the number of parameters in each crossover operator, there is one in SBX and SPX, and there are two in DE, PCX, and UNDX. The objective of this study is to investigate the effect of changing the parameters in each crossover operator. NSGA-II is used as a MOEA. DTLZ2,3,4 [3], WFG1,2,6,8,9 [9] and ZDT1,4 [17] is used as benchmark problems.

In SBX, a large value of η is preferred. In SPX, values of 3 to 7 are preferred for ϵ. However, there is little difference in the performance for η and ϵ. These results indicate that the effect of parameter variation is relatively small.

In DE, small values are preferred for both cr and F. In PCX, appropriate values of η and ζ depend on the problem. When the design variable is non-separable, a large value is preferred for η. In UNDX, a large value of η is preferred. For ζ, values about 0.3 to 0.5 are preferred. These results indicate that as for DE, PCX, and UNDX, there are optimum parameters in each benchmark problem. This also indicates that the choice of crossover operator is significantly important in MOEA for achieving good performance.

In future work, we will investigate not only the effect when the number of objectives is more than four but also the effect when the different mutation is considered. In addition, it is also necessary to consider other types of MOEA such as MOEA/D and IBEA.

Acknowledgment. This research was partially supported by MEXT as "Priority Issue 8 (Sub-issue A) on Post-K computer" (Development of innovative design and production processes that lead the way for the manufacturing industry in the near future, sub-issue A: Research and development of multi-objective design exploration and high-performance computing technologies for design innovation).

References

1. Deb, K., Agrawal, R.B.: Simulated binary crossover for continuous search space. Complex Syst. **50**(9), 115–148 (1994)
2. Deb, K., Pratap, A., Agarwal, S., Meyarivan, T.: A fast and elitist multiobjective genetic algorithm: NSGA-II. IEEE Trans. Evol. Comput. **6**(2), 182–197 (2002)
3. Deb, K., Thiele, L., Laumanns, M., Zitzler, E.: Scalable multi-objective optimization test problems. In: Proceedings of the Congress Evolutionary Computation CEC 2002, vol. 1, pp. 825–830, May 2002
4. Zitzler, E., Laumanns, M., Thiele, L.: Spea 2: improving the strength pareto evolutionary algorithm for multiobjective optimization. International Center for Numerical Methods in Engineering (CIMNE) (2002)

5. Zitzler, E., Künzli, S.: Indicator-based selection in multiobjective search. In: Yao, X., et al. (eds.) PPSN 2004. LNCS, vol. 3242, pp. 832–842. Springer, Heidelberg (2004). doi:10.1007/978-3-540-30217-9_84

6. Hadka, D., Reed, P.M., Simpson, T.W.: Diagnostic assessment of the borg moea for many-objective product family design problems. In: Proceedings of IEEE Congress on Evolutionary Computation, pp. 1–10, June 2012

7. Hitomi, N., Selva, D.: A classification and comparison of credit assignment strategies in multiobjective adaptive operator selection. IEEE Trans. Evol. Comput. (2016) (in press)

8. Kita, H., Ono, I., Kobayashi, S.: Multi-parental extension of the unimodal normal distri- bution crossover for real-coded genetic algorithms. In: Proceedings of Congress on Evolutionary Computation, pp. 581–1588 (1999)

9. Huband, S., Hingston, P., Barone, L., While, R.L.: A review of multiobjective test problems and a scalable test problem toolkit. IEEE Trans. Evol. Comput. 10(5), 477–506 (2006). http://dx.doi.org/10.1109/TEVC.2005.861417

10. Deb, K., Joshi, D., Anand, A.: Real-coded evolutionary algorithms with parent-centric recombination. In: Proceedings of the World Congress on Computational Intelligence, pp. 61–66 (2002)

11. Deb, K., Jain, H.: An evolutionary many-objective optimization algorithm using reference-point-based nondominated sorting approach, part I: Solving problems with box constraints. IEEE Trans. Evol. Comput. 18(4), 577–601 (2014)

12. Deb, K., Mohan, M., Mishra, S.: A fast multi-objective evolutionary algorithm for finding well-spread pareto-optimal solutions. KanGAL report 2003002, pp. 1–18 (2003)

13. Storn, R., Price, K.: Differential evolution - a simple and e cient heuristic for global optimization over continuous spaces. J. Global Optim. 11(4), 341–359 (1997)

14. Tsutsui, S., Yamamura, M., Higuchi, T.: Multi-parent recombination with simplex crossover in real coded genetic algorithms. In: Proceedings of the Genetic and Evolutionary Computation Conference (GECCO 1999), pp. 657–664 (1999)

15. Yuan, Y., Xu, H., Wang, B.: An experimental investigation of variation operators in reference-point based many-objective optimization. In: Proceedings of the 2015 Annual Conference on Genetic and Evolutionary Computation, pp. 775–782. ACM (2015)

16. Zhang, Q., Liu, W., Li, H.: The performance of a new version of MOEA/D on CEC09 unconstrained MOP test instances. In: IEEE CEC, pp. 203–208 (2009). http://dx.doi.org/10.1109/CEC.2009.4982949

17. Zitzler, E., Deb, K., Thiele, L.: Comparison of multiobjective evolutionary algorithms: empirical results. Evol. Comput. 8(2), 173–195 (2000). http://dx.doi.org/10.1162/106365600568202

18. Zitzler, E., Thiele, L.: Multiobjective evolutionary algorithms: a comparative case study and the strength pareto approach. IEEE Trans. Evol. Comput. 3(4), 257–271 (1999). http://dx.doi.org/10.1109/4235.797969

Non-dominated Sorting and Crowding Distance Based Multi-objective Chaotic Evolution

Yan Pei[1](✉) and Jia Hao[2]

[1] Computer Science Division, University of Aizu,
Aizu-wakamatsu, Fukushima 965-8580, Japan
peiyan@u-aizu.ac.jp
[2] School of Mechanical Engineering, Beijing Institute of Technology,
Beijing 100081, China
haojia632@bit.edu.cn
http://web-ext.u-aizu.ac.jp/~peiyan/

Abstract. We propose a new evolutionary multi-objective optimization (EMO) algorithm based on chaotic evolution optimization framework, which is called as multi-objective chaotic evolution (MOCE). It extends the optimization application of chaotic evolution algorithm to multi-objective optimization field. The non-dominated sorting and tournament selection using crowding distance are two techniques to ensure Pareto dominance and solution diversity in EMO algorithm. However, the search capability of multi-objective optimization algorithm is a serious issue for its practical application. Chaotic evolution algorithm presents a strong search capability for single objective optimization due to the ergodicity of chaotic system. Proposed algorithm is a promising multi-objective optimization algorithm that composes a search algorithm with strong search capability, dominant sort for keeping Pareto dominance, and tournament selection using crowding distance for increasing the solution diversity. We evaluate our proposed MOCE by comparing with NSGA-II and an algorithm using the basic framework of chaotic evolution but different mutation strategy. From the evaluation results, the MOCE presents a strong optimization performance for multi-objective optimization problems, especially in the condition of higher dimensional problems. We also analyse, discuss, and present some research subjects, open topics, and future works on the MOCE.

Keywords: Chaotic evolution · Multi-objective chaotic evolution · Evolutionary multi-objective optimization · Chaos theory · Chaotic optimization

1 Introduction

In the conventional mathematical optimization field, the optimization problems are categorized into continuous optimization and discrete optimization according to the domain of optimized variables. The canonical form of objective function for continuous optimization is defined as a single objective function with

© Springer International Publishing AG 2017
Y. Tan et al. (Eds.): ICSI 2017, Part II, LNCS 10386, pp. 15–22, 2017.
DOI: 10.1007/978-3-319-61833-3_2

some equality or inequality constraint functions. The mathematical programming methods, such as Karush-Kuhn-Tucker conditions [3], are efficient algorithms that can solve this set of problems when the objective functions and constraints have some perfect characteristics, such as continuous, differentiable, etc. However, most of the real world problem's objectives are not singleton. They have multiple objectives, i.e., multiple objective (multi-objective) functions. If the objectives are more than four, they also are referred as to many objectives problem [2]. This leads us to think whether the single objective problem or singleton concept only exist in our human brain. The multi-objective optimization problem is formally defined as in Eq. (1), where it defines a m $(m \geq 2)$ objectives optimization function, and the optimized variable is vector $\mathbf{x}$ in a search space Ω.

$$\begin{aligned} \underset{\mathbf{x}}{\text{minimize}} \quad & F(\mathbf{x}) = (f_1(\mathbf{x}), f_2(\mathbf{x}), ..., f_m(\mathbf{x})) \\ \text{subject to} \quad & \mathbf{x} \in \Omega. \end{aligned} \quad (1)$$

Evolutionary multi-objective optimization (EMO) algorithms are efficient and effective to handle the multi-objective optimization problem. From the methodological viewpoint, the class of EMO algorithms belongs to stochastic optimization, whose theoretical explanation lies on probability theory. The EMO keeps the multiple objective functions independently and uses Pareto-based ranking schemes to maintain the feasible solution, rather than the determinative programming methods use the scalarization method that needs to transfer multiple objectives into one objective. Most of the state-of-the-art studies in EMO concentrate on Pareto dominance handling in objective space, which ties to generate solutions to approximate the Pareto frontiers [4]. However, the main disadvantage of EMO are its worse optimization capability and non guarantee of Pareto optimality, which cannot be perfectly solved by Pareto dominance studies in EMO.

In this paper, we propose a non-dominated sorting and tournament selection using crowding distance based chaotic evolution (CE) algorithm for the multi-objective optimization problem. This work extends the fundamental optimization framework of CE in the EMO field. The non-dominated sorting method handles the Pareto dominance issue, and tries to keep the non-dominated solution into the next generation, and tournament selection using crowding distance technique maintains the diversity of solution. The CE algorithm implements the search function in optimization process. We do not only handle the Pareto dominance and solution diversity issues, but also compose an efficient optimization algorithm in proposed algorithm. Because the CE has better optimization capability in a variety of problems, the proposed non-dominated sorting and tournament selection using crowding distance based CE can therefore appeal its optimization capability in multi-objective optimization problem. It presents one of originalities and contributions in this work.

Following this introductory section, we briefly make an overview of the CE and NSGA-II in Sect. 2. In Sect. 3, we present the proposed multi-objective CE algorithm and its framework. We explain the optimization principle of the proposal and its optimization framework. The evaluation and discussion of the proposed method are presented in Sects. 4 and 5. We report the performance of proposed algorithm, and address how and why the proposed algorithm has a better optimization performance by comparing with NSGA-II. Finally, we conclude the whole work, and discuss the future works, open topics in Sect. 6.

2 An Overview on Chaotic Evolution and NSGA-II

2.1 Chaotic Evolution

The fundamental of chaotic evolution (CE) simulates the chaotic motion when generating new search points, which is a critical implementation determining the exploration and exploitation capabilities [5]. For example, if we have k individuals (vectors) in the $n-1th$ generation, we can directly generate k mutant vectors in nth generation using different chaotic systems. This idea is motivated by some natural phenomenons that can be well modeled and explained with chaotic systems while deterministic or stochastic methods fail. Therefore, we believe that the evolution of a set of individuals can be modeled by chaotic systems. The feasibility and performance of CE are guaranteed by the ergodicity property of chaotic systems, which means iterative variables of the systems can approach to any location in a search space with an arbitrary accuracy [6].

2.2 NSGA-II

Non-dominant sorting genetic algorithm II (NSGA-II) was firstly introduced in reference [1], and it is an improved version of NSGA [7]. NSGA is one of the EMO algorithms, which have the capability of finding multiple Pareto optimal solutions with a single simulation running [1]. Although NSGA contributed to the EMO community, it also suffered from several criticisms, including high computational cost, lack of elitism and requirement for the setting of sharing parameter. The NSGA-II succeed in solving all the above three issues at once by introducing fast non-dominated sorting and tournament selection using crowding distance.

3 Multi-objective Chaotic Evolution Using Non-dominated Sorting and Tournament Selection with Crowding Distance

There are two research issues and subjects in Pareto-based EMO. The one is Pareto dominance problem, most of studies try to define new dominant concepts to apply in the selection and survive processing for maintaining performance of Pareto dominance. The other is diversity issue that EMO should support a variety of Pareto solutions to be selected. This research issue also leads to another EMO related subject that how to select one Pareto solution from Pareto frontier for a real world application in practice. The canonical approaches to cope with these two issues are non-dominated sorting and tournament selection using crowding distance, respectively. However, the optimization performance of EMO does not only lie on the solutions of Pareto dominance and solution diversity, but also depends on the optimization capability of search algorithm. The design of EMO algorithm should consider three aspects of issue, i.e., optimization capability of search algorithm, Pareto dominance, and solution diversity. This paper contributes a new EMO algorithm that implements a better optimization capability in EMO, i.e., multi-objective chaotic evolution (MOCE).

The proposed MOCE composes three design elements in its optimization framework, i.e., optimization capability of search algorithm, Pareto dominance, and solution diversity. CE algorithm implements the basic search function that ensures a strong optimization capability, non-dominant sorting solves the Pareto dominance issue, and tournament selection using crowing distance keeps the solution diversity when selecting the dominant solutions. We can design a variety of MOCE algorithms by implementing these three techniques. From the single objective optimization results of CE, the basic performance of search and optimization can be maintained [5]. The non-dominant sorting and crowding distance selection are two canonical methods that can solve the issues of Pareto dominance and solution diversity [1]. The proposed non-dominant sorting and tournament selection using crowding distance based MOCE, therefore, be considered as an efficient and effective implementation of EMO algorithm. It is one of the implementations of MOCE algorithms as well.

Table 1. Abbreviations of algorithms in evaluation.

Abbreviations	Meaning
Multi-CE	MOCE based on non-dominated sort and crowding distance tournament selection
NSGA-II	non-dominant sort genetic algorithm
Multi-Rand	MOCE implemented by a uniform distribution parameter generator based on non-dominated sort and crowding distance tournament selection

4   Numerical Evaluations

We use five multi-objective benchmark problems to evaluate our proposed algorithm. The abbreviations of investigated algorithms are in Table 1. The population size of all evaluated algorithms is set to 100, we evaluate each algorithm using 1000 generations. 30 running trails are conducted to be applied in statistical test. For each running trail, each evaluated algorithm is set with the same initialization for a comparison. The benchmark problems [9], dimensional setting, search range, and Pareto frontier are listed in Table 2. The NSGA-II is also applied to these problems for a comparison. The crossover rate and mutation

Table 2. Multi-objective benchmark function used in evaluation. All of the Pareto frontier are $g(x) = 1$.

Func	Dim	Definition	Search Range
ZDT1	30	$f_1(x) = x_1$	$x_i \in [0, 1]$
		$f_2(x) = g(x)[1 - \sqrt{\frac{x_1}{g(x)}}]$	
		$g(x) = 1 + 9\frac{\sum_{i=2}^{n} x_i}{n-1}$	
ZDT2	30	$f_1(x) = x_1$	$x_i \in [0, 1]$
		$f_2(x) = g(x)[1 - (\frac{x_1}{g(x)})^2]$	
		$g(x) = 1 + 9\frac{\sum_{i=2}^{n} x_i}{n-1}$	
ZDT3	30	$f_1(x) = x_1$	$x_i \in [0, 1]$
		$f_2(x) = g(x)[1 - \sqrt{\frac{x_1}{g(x)}} - \frac{x_1}{g(x)}\sin(10\pi x_1)]$	
		$g(x) = 1 + 9\frac{\sum_{i=2}^{n} x_i}{n-1}$	
ZDT4	10	$f_1(x) = x_1$	$x_1 \in [0, 1]$, $x_i \in [-5, 5]$, $i = 2, 3, ..., 10$
		$f_2(x) = g(x)[1 - \sqrt{\frac{x_1}{g(x)}}]$	
		$g(x) = 1 + 10(n - 1) + \sum_(i = 2)^n [x_i^2 - 10\cos(4\pi x_i)]$	
ZDT6	10	$f_1(x) = 1 - \exp(-4\pi x_1)\sin^6(6\pi x_1)$	$x_i \in [0, 1]$
		$f_2(x) = g(x)[1 - (\frac{x_1}{g(x)})^2]$	
		$g(x) = 1 + 9[\frac{\sum_{i=2}^{n} x_i}{n-1}]^{0.25}$	

rate of NSGA-II are set to 0.9 and 0.1, respectively. The chaotic system used in CE is the logistic map (Eq. (2)) with the $\mu = 4$, and the initial number of that is $random[0, 1]$ not equal to $\{0, 0.25, 0.5, 0.75, 1\}$. We also implement a random parameter based search algorithm that uses a uniform distribution parameter generator in CE algorithm to investigate the performance of chaotic parameter implementation with a comparison. Concretely, we use $CP_{i,j} = rand(0, 1)$ to replace $CP_{i,j} = ChaoticSystem(CP_{i,j})$ in CE algorithm to implement this compared algorithm. Figure 1 shows the average number of Pareto frontier solution of 30 running trails. Based on these evaluation results, we apply some statistical tests on these data, and discuss and analyse proposed MOCE algorithm.

$$x(i + 1) = \mu x(i)(1 - x(i)) \tag{2}$$

5   Discussions and Analyses

5.1   Discussion on the Number of Pareto Frontier Solution

Pareto dominance and solution diversity are two evaluation metrics to compare EMO algorithms. Because most EMO algorithms use Pareto-based ranking

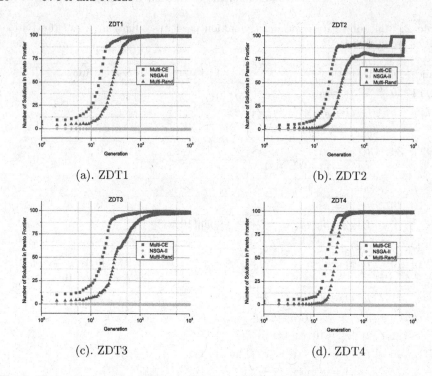

(a). ZDT1   (b). ZDT2

(c). ZDT3   (d). ZDT4

Fig. 1. The average number of Pareto frontier solution in each generation, the Wilcoxon signed rank test results present that there is a significant difference between Multi-CE and Multi-Rand.

schemes, the number and the distribution of solutions in Pareto frontier can present the optimization performance of EMO algorithms. From the Fig. 1, we can conclude that the number of Pareto frontier of Multi-CE and Multi-Rand are more than that of NSGA-II. We collect and calculate the number of Pareto frontier solution in 500 generation from 30 running trails, and apply a Friedman test to rank these numbers, and a Bonferroni-Dunn tests in the significant level of $\alpha = 0.05$ for a statistical evaluation. Figure 2 is a result of critical difference calculation. From Fig. 2, we observe that there is a significant difference between NSGA-II and Multi-CE/Multi-Rand in the aspect of the number of Pareto frontier solution, i.e., the number of Pareto frontier solution from our proposed Multi-CE is significantly more than that from NSGA-II.

5.2   Discussion on Comparison of Chaotic and Random Generators

The difference of Multi-CE and Multi-Rand lies on the mutation generator. For comparing generators from a chaotic system and a uniform distribution random system, we use two generators to conduct a comparison study. One is from a chaotic system (the logistic map in this paper), the other is from a uniform distribution. Whether the optimization performance is significantly influenced by

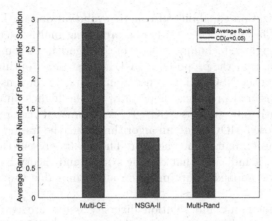

Fig. 2. Bonferroni-Dunn test in the significant level of $\alpha = 0.05$, we observe that there is a significant difference of the number of Pareto frontier solution between NSGA-II and Multi-CE/Multi-Rand. Critical difference (CD) used in Bonferroni-Dunn test is $CD = q_\alpha \sqrt{\frac{k(k+1)}{6*N}}$, and $k = 3$ and $N = 5$, q is equal to $q_\alpha(0.05) = 2.242$ from Appendix Table B.16 of [8].

mutation generator? We investigate this issue using a statistical test. The number of Pareto frontier solution in each generation in Multi-CE and Multi-Rand is pairwise related, and we do not know the normality of the data distribution, so we apply a Wilcoxon signed-rank test for five benchmark functions with the number of Pareto frontier solution in 500th generation. The five p-values from the tests are all less than 0.05, and we check Fig. 1, which indicates that the number of Pareto frontier solution from Multi-CE is significantly more than that from Multi-Rand, i.e., the optimization performance is significantly influenced by the mutation generator, and chaotic generator presents a better optimization performance than that of the random generator.

The primary difference of chaotic generator and random generator lies on the distribution of the generated numbers from the statistical viewpoint. The random generator has the uniform distribution of generated numbers among the whole range of (0,1). It have been investigated that the distribution characteristic of the logistic map is that most generated numbers are in the ranges of (0,0.1) and (0.9,1) [6], which presents the exploitation and exploration search capabilities of CE algorithm. The above analysis proves that this characteristic also influences the optimization results of the MOCE. This is one of discoveries in this work.

6   Conclusion

In this paper, we proposed a new EMO algorithm using the optimization framework of CE and two canonical techniques from NSGA-II. The proposed algorithm uses non-dominated sort to keep the Pareto dominance, applies the tournament selection using crowding distance to increasing solution diversity, and composes the strong optimization capability of chaotic evolution due to the ergodicity of chaotic system. The proposed MOCE is a promising multi-objective optimization because of the three technique aspects of the algorithm. From the evaluation results, it indicates that the proposed MOCE has a stronger optimization performance than that of the NSGA-II. It is useful in both lower dimensional problems and higher dimensional problems. Especially for the higher dimensional problems, it significantly outperforms NSGA-II. We compared the optimization performance of proposed MOCE and an algorithm using the framework of CE and a uniform distribution mutation generator. The results present the advantage of chaotic generator, it indicates that chaotic system and theory have the promising potential to be applied in optimization algorithms due to characteristic of ergodicity.

The Pareto dominant and solution diversity issues are two research subjects in EMO field. In the future, we will implement MOCE algorithm using other techniques to solve these two problems. The chaotic generator presents an advantage for evolution optimization. Why the chaotic generator can results in the better optimization performance but the uniform random distribution generator cannot, we need to investigate this study subject in both aspects of theory and application. These two research subjects will be involved in our future work. Another research subject is the application of the proposed algorithm. We will also apply it in a variety of applications to benefit our society.

Acknowledgements. The author, Jia Hao, would like to thank the strong support provided by Beijing Natural Science Foundation (BJNSF 3172028).

References

1. Deb, K., Pratap, A., Agarwal, S., Meyarivan, T.: A fast and elitist multiobjective genetic algorithm: NSGA-II. IEEE Trans. Evol. Comput. **6**(2), 182–197 (2002)
2. Farina, M., Amato, P.: On the optimal solution definition for many-criteria optimization problems. In: Proceedings of the NAFIPS-FLINT international conference, pp. 233–238 (2002)
3. Karush, W.: Minima of functions of several variables with inequalities as side constraints. Master's thesis, Department of Mathematics, University of Chicago (1939)
4. Li, B., Li, J., Tang, K., Yao, X.: Many-objective evolutionary algorithms: a survey. ACM Comput. Surv. (CSUR) **48**(1), 13 (2015)
5. Pei, Y.: Chaotic evolution: fusion of chaotic ergodicity and evolutionary iteration for optimization. Nat. Comput. **13**(1), 79–96 (2014)
6. Pei, Y.: From determinism and probability to chaos: chaotic evolution towards philosophy and methodology of chaotic optimization. Sci. World J., Article ID 704587 (2015)
7. Srinvas, N., Deb, K.: Multi-objective function optimization using non-dominated sorting genetic algorithms. Evol. Comput. **2**(3), 221–248 (1994)
8. Zar, J.H.: Biostatistical Analysis. Pearson Education, India (1999)
9. Zitzler, E., Deb, K., Thiele, L.: Comparison of multiobjective evolutionary algorithms: empirical results. Evol. Comput. **8**(2), 173–195 (2000)

On Performance Improvement Based on Restart Meta-Heuristic Implementation for Solving Multi-objective Optimization Problems

Christina Brester[(✉)], Ivan Ryzhikov, and Eugene Semenkin

Siberian State Aerospace University, Krasnoyarsk, Russia
christina.brester@gmail.com,
{ryzhikov-88, eugenesemenkin}@yandex.ru

Abstract. One of the possible goals of multi-objective optimization is finding a set of non-dominated solutions or, in other words, a Pareto set approximation. Population-based algorithms, in particular, genetic algorithms, are widely used for this purpose because they deal with a set of alternative solutions, which might be helpful when a number of trade-off points should be obtained. To get a representative approximation, various regions of a search space should be explored. However, during the algorithm execution a search might be stuck in some areas. Therefore, in this article we present a new restart operator for multi-objective genetic algorithms which can prevent a search from stagnating, help to explore new regions and, as a result, improve the algorithm performance significantly. In our proposal we answer the two crucial questions of a restarting concept which are *when* to restart an algorithm and *how* to use previously found solutions. We introduce the algorithm independent restart operator, even though in this work we investigate it in combination with a certain MOGA. The experimental results prove the high effectiveness of the modified MOGA with the incorporated restart operator in comparison with the conventional one.

Keywords: Multi-objective genetic algorithm · Restart operator · Performance · Benchmark problems

1 Introduction

The balanced combination of exploration and exploitation is vitally important in an evolutionary search. Both of these strategies are embodied in a genetic algorithm (GA) as well-known genetic operators: selection, crossover and mutation. However, for complex multi-modal problems it might be impossible to prevent a GA from stagnating or from converging prematurely by just applying these operators. Therefore, to leave a local optimum, it is reasonable to interrupt the current run and launch the algorithm again (it refers to exploration) but with the use of valuable information about the search space gathered during the previous work of the algorithm (it relates to exploitation).

To incorporate the described mechanism into GAs, a restarting technique has been proposed [1, 2] and implemented as an additional restart operator [3, 4]. Contrary to the

© Springer International Publishing AG 2017
Y. Tan et al. (Eds.): ICSI 2017, Part II, LNCS 10386, pp. 23–30, 2017.
DOI: 10.1007/978-3-319-61833-3_3

conventional genetic operators, it is applied not at each generation but according to a predefined restart schedule or when a specified criterion is satisfied. The question "when" to restart the algorithm is the first open question which should be properly solved while implementing this operator. Frequent use of the restart may divert GAs from the exact value of a global optimum, whereas rare use may eliminate its positive effects. The second crucial question is about how to involve the obtained information about the search space in the next runs. On the one hand, good solutions found should be included in the new population, but on the other hand, attraction basins of local optima should be penalized so that GAs could not be trapped again.

While useful restart properties are obvious for one-criterion optimization and some effort has been made to improve the performance of one-criterion GAs with the use of this operator [1, 2, 5, 6], there are almost no examples of restarting in multi-objective optimization [7]. The outcome of any multi-objective genetic algorithm (MOGA) is a set of non-dominated points, i.e. a Pareto set approximation. To find a representative approximation, diversity preservation techniques are engaged in MOGAs, which helps to keep points uniformly distributed along a true Pareto set or front. However, MOGAs might, firstly, be subjected to stagnation or, secondly, the population might concentrate on some areas of the search space and as a result, we get a partly approximated Pareto set. Therefore, in this study we propose the use of restart meta-heuristic to increase the MOGA effectiveness. We introduce an algorithm independent scheme for a restart which might be involved in any MOGA. As a case in point, we investigate the effectiveness of the restart operator in a preference-inspired co-evolutionary algorithm using goal vectors (PICEA-g) [8]. This MOGA has proved its high performance on a set of different benchmark problems. Moreover, it is capable of solving many-objective problems (when the number of criteria excesses three). Thus, in this paper we also show that the restart operator enables the efficiency of the algorithm to be increased significantly even though its conventional version is rather effective too.

The rest of the paper is organized as follows: Sect. 2 contains a brief description of PICEA-g and all the details about the restart operator. The test problems used, the experiments conducted, the results obtained and the main inferences are included in Sect. 3. The conclusion and future work are presented in Sect. 4.

2   PICEA-g and Restart Meta-Heuristic for Multi-objective Optimization Algorithms

2.1   Preference-Inspired Co-evolutionary Algorithm with Goal Vectors

The preference-inspired co-evolutionary algorithm using goal vectors (PICEA-g) was proposed by Wang in 2013 [8]. This algorithm relates to a class of preference-inspired co-evolutionary algorithms (PICEAs) which are based on the concept of co-evolving the population with decision-maker preferences.

PICEA-g includes the following steps:

1. Generate an initial population and evaluate objective values for individuals. Find non-dominated candidate solutions in the population and copy them into the *archive*. Determine the set of *goal vectors* as a number of targets randomly generated within the goal vector bounds.
2. Produce the offspring solutions with *genetic operators*: selection, crossover and mutation. Evaluate objective values for new generated individuals.
3. Pool together parents and children; compile the set of objective values.
4. Append to the set of goal vectors the additional targets generated within the determined bounds.
5. Assign fitness values for goal vectors and for individuals in the united population.
6. Form the new population and the set of goal vectors based on their fitness.
7. Update the archive with new non-dominated solutions.
8. Check the stopping criterion: if it is satisfied, then finish the search with the archive set, otherwise proceed with the second step.

In Steps 1 and 4 decision-maker preferences are incorporated into the algorithm by using goal vectors. They represent points generated in the criteria search space within the bounds which are determined according to the rule:

$$
\begin{aligned}
g_i^{\min} &= \min(BestF_i), \\
\Delta F_i &= \max(BestF_i) - \min(BestF_i), \\
g_i^{\max} &= \min(BestF_i) + \alpha \times \Delta F_i,
\end{aligned}
\tag{1}
$$

where $g_i^{\min}$ is the lower bound and $g_i^{\max}$ is the upper one for the i-th goal vector component, $BestF_i$ is the best value of the i-th objective function amid solutions in the archive, $i = \overline{1, M}$, M is the number of criteria. The recommended value of the α parameter is 1.2.

2.2 Restarting Operator Meta-Heuristic

In this study we consider a black-box multi-objective optimization problem:

$$
\begin{aligned}
C(a) &: A \rightarrow C_A \subset R^m, \ \dim(A) = n, \\
C(a) &= (C_1(a) \ldots C_m(a)) \rightarrow \underset{a \in A}{extrem},
\end{aligned}
\tag{2}
$$

where A is a space of alternatives with dimension n, C_A is a subspace of some Euclidean vector space R^m, $i = \overline{1, m} : C_i(\cdot) : A \rightarrow C_A^i \subset R, \prod_j C_j(A) = C_A$ are the unknown mappings. We assume that there is a bijection between the alternatives and the binary strings, so every alternative would be represented in such a way.

Since we consider a population-based optimization tool, let the population in the i-th generation be noted as P_i. Each population consists of different solutions – a set of alternatives – and our aim is not to find the non-dominated set, but the set which maps into a good approximation of the whole Pareto front. In this case one can face a contradiction between the need for a search in depth to improve the current solutions and for a search in breadth to approximate the whole front.

To resolve this contradiction we put forward a hypothesis of implementing the population algorithm independent restarting operator meta-heuristic. The following operator acts like a simple evaluation estimating if the condition that causes the algorithm restart is met. We propose a restart condition that uses the observations of the distances between the Pareto front estimations in each generation. If the distance does not change for some period, the algorithm restarts. A more detailed explanation is given below.

Let $S_i = \left\{ a_j \in A : \nexists k < i, j(k) \leq |S_k| : a_j \overset{c}{p} a_{j(k)}, a_{j(k)} \in S_k \right\}$ be the Pareto set and $F_i = \left\{ C(a_j), a_j \in S_i \right\}$ be the Pareto front estimations in the i-th generation. It is easy to see that $\forall i \ S_i \subset S_{i-1} \bigcup P_i$, so the distance $\rho(F_i, F_{i-1})$ between two different sets F_i and F_{i-1} is performed by the non-dominated solutions found in the current generation. Let F be a set of any limited cardinality $F = \left\{ f_i \in R^m, i = \overline{1, |F|} \right\}$, then

$$\rho(F_a, F_b) : \ F \times F \to R^+ \bigcup \{0\},$$
$$\rho(F_a, F_b) = \frac{1}{|F_a|} \cdot \sum_{i=1}^{|F_a|} \min_{j \leq |F_b|} \left(\left\| (F_a)_i - (F_b)_j \right\|_{R^m} \right), \tag{3}$$

where $\|\cdot\|_{R^m} : \ R^m \to R^+ \bigcup \{0\}$ is a norm on the R^m vector space.

To estimate if there is a need for a restart we use a specific variable, which is a queue that consists of metrics (3) values of the previous l_{tail} iterations,

$$Tail_i(l_{tail}) = \left\{ \rho(F_j, F_{j-1}) : \ i - l_{tail} < j \leq i \right\}, \tag{4}$$

and the meta-heuristic performs a restart if the following condition is met

$$\max_{j < l_{tail}} \{ Tail_i(j) \} - \min_{j < l_{tail}} \{ Tail_i(j) \} \leq \delta_{tail}. \tag{5}$$

In Eqs. (4) and (5) the settings of two different operators are presented: the tail length l_{tail} controls the size of the observation period and δ_{tail} is the threshold level. Now, if the restart takes place, we save the current algorithm's run data into the sets, which is a representation of memory. Since all these features are significant for forming the final solution and performing the next algorithm run, the following sets are used: $Memory_S = Memory_S \bigcup \{S_i\}$, $Memory_F = Memory_F \bigcup \{F_i\}$, $Memory_P = Memory_P \bigcup \{P_i\}$ and $Memory_C = Memory_C \bigcup \{\tilde{C}_i\}$, where $\tilde{C}_i = \left\{ F(c_j) : c_j = (P_i)_j, ; j = \overline{1, |P_i|} \right\}$.

First of all, let us describe a possible way in which the new starting population may perform. The generation of the initial population is controlled by two parameters: the probability of each individual in the initial population being randomly generated - α, and the probability of each gene being changed to the opposite - β, in the case of the individual being a mutant of a randomly chosen previously found solution. So, each j-th individual's k-th gene in the initial population is generated in one of the following ways:

$$\left((P_0)_j\right)_k = r_{j,k}, \; P(r_{j,k} = 0) = P(r_{j,k} = 1), \tag{6}$$

with probability α and with probability $1 - \alpha$:

$$\left((P_0)_j\right)_k = f_c\left(\left((Memory_S)_{r_j^1}\right)_{r_j^2}, r_{j,k}^3\right), \tag{7}$$

where k is the number of gene, r_j^1, r_j^2, $r_{j,k}^3$ are the random values: $P\left(r_j^1 = 1\right) = \ldots = P\left(r_j^1 = |Memory_S|\right)$, $P\left(r_j^2 = 1\right) = \ldots = P\left(r_j^2 = (Memory_S)_{r_j^1}\right)$, $P\left(r_{j,k}^3 = 0\right) = 1 - P\left(r_{j,k}^3 = 1\right) = \beta$, and a special function $f_c(v,p) = \begin{cases} v, \; p = 0 \\ \neg v, p = 1 \end{cases}$.

Varying the parameters α and β, we control the initial population generation process. If we want the initial population to be completely randomized, we set α to 1, and if we want it to be in some sense near to the previously estimated Pareto set, we set it closer to 0 and β closer to 0 too, where β represents the closeness of the new individual to a found one.

Memory sets are also used to perform the final IGD metric (8) and estimate the algorithm efficiency.

3   Performance Assessment

3.1   Test Multi-objective Problems

To investigate the effectiveness of PICEA-g with the restart operator in comparison with its conventional version, we have engaged a set of high-dimensional test problems designed by the international scientific community to compare the effectiveness of developed algorithmic schemes (the CEC 2009 competition [9]). There are problems with discrete and continuous, convex and non-convex Pareto Sets and Fronts. In this study we use a number of these test instances which are unconstrained two- and three-objective optimization problems with real variables.

In the CEC 2009 competition the metric IGD was used to estimate the quality of obtained Pareto Front approximations:

$$IGD(A, P^*) = \frac{\sum_{v \in P^*} d(v, A)}{|P^*|} \tag{8}$$

where P^* is a set of uniformly distributed points along the Pareto Front (in the objective space), A is an approximate set to the Pareto Front, $d(v, A)$ is the minimum Euclidean distance between v and the points in A. In short, the $IGD(A, P^*)$ value reflects the average distance from P^* to A.

The next section provides a description of the experiments conducted, the results obtained and a brief discussion of them.

3.2 Experimental Results

Firstly, the conventional PICEA-g algorithm was applied to solve the problems introduced. PICEA-g was provided with the following amount of resources: according to the rules of the CEC 2009 competition, the maximal number of function evaluations was equal to 300,000. The maximal number of solutions in the approximate set produced by each algorithm for computing the IGD metric was 100 and 150 for two-objective and three-objective problems respectively. For all of the test instances, IGD values were averaged over 25 runs for each algorithm.

In the experiments conducted the following settings were defined: binary tournament selection, uniform recombination and the mutation probability $p_m = 1/L$, where L is the length of the chromosome. PICEA-g operated with binary strings and therefore, we used standard binary coding to obtain the real values of variables.

The results obtained in this experiment are presented in Table 1 (columns 'PICEA-g'). For each test problem we give the average ('Mean'), minimum ('Min') and maximum ('Max') values of the IGD metric.

Table 1. Experimental results

Test problem	PICEA-g			PICEA-g, the restart with "average" settings		
	Mean	Min	Max	Mean	Min	Max
UF1	0.10296	0.07429	0.18265	**0.03626**	**0.02528**	**0.04426**
UF2	0.05824	0.04780	0.08146	**0.03723**	**0.03116**	**0.04455**
UF3	0.22356	0.16639	0.33290	**0.19150**	**0.14439**	**0.27185**
UF4	0.05763	0.05269	0.06790	0.06668	0.06203	0.07522
UF5	0.52541	0.37200	0.74903	**0.22604**	**0.16217**	**0.29384**
UF6	0.37426	0.21349	0.64676	**0.20703**	**0.08549**	**0.26497**
UF7	0.10919	0.03991	0.42055	**0.03011**	**0.02465**	**0.04095**
UF8	0.19015	0.16446	0.20097	0.19875	0.16788	0.21412
UF9	0.27406	0.21015	0.38429	**0.22292**	**0.16236**	**0.30561**
UF10	0.41424	0.22110	0.85904	**0.30983**	**0.21222**	**0.42998**

Test problem	PICEA-g, the restart with the best settings						
	Mean	Min	Max	α	β	l_{tail}	δ_{tail}
UF1	**0.03407**	**0.02428**	**0.04935**	0.9	0.5	5	0.001
UF2	**0.03723**	**0.03116**	**0.04455**	0.9	0.7	5	0.0005
UF3	**0.16350**	**0.12724**	**0.21743**	0.1	0.7	15	0.0005
UF4	**0.05358**	**0.04836**	**0.05673**	0.1	0.7	10	5.00E−05
UF5	**0.15967**	**0.12315**	**0.21449**	0.9	0.5	5	0.01
UF6	**0.18777**	**0.06162**	**0.32109**	0.1	0.7	10	0.001
UF7	**0.02823**	**0.02400**	**0.03937**	0.9	0.5	5	0.0005
UF8	**0.18377**	**0.16396**	**0.19064**	0.1	0.05	5	0.0001
UF9	**0.20374**	**0.14696**	**0.28560**	0.3	0.5	5	0.001
UF10	**0.26515**	**0.19683**	**0.44773**	0.1	0.7	5	0.001

Then, a series of similar experiments was conducted for the PICEA-g with the restart operator. We varied its parameters: $\alpha = 0, 0.1, 0.3, 0.5, 0.9, 1$; $\beta = 0.05, 0.2, 0.5, 0.7$; $l_{tail} = 5, 10, 12, 15$; $\delta_{tail} = 0.01, 0.005, 0.001, 0.0005, 0.0001, 0.00005$ to find their best combination for the set of test problems on average. After the normalization of IGD values for all the problems we found that $\alpha = 0.9$, $\beta = 0.7$, $l_{tail} = 5$, $\delta_{tail} = 0.0005$ are the best on average. The experimental results of the PICEA-g algorithm with the restart operator which has the stated settings are shown in Table 1, columns 'PICEA-g, restart with 'average' settings'. For almost all the problems (except UF4 and UF8) we obtained better results in comparison with the IGD values (mean, min, max) provided by the conventional PICEA-g algorithm. It means that to avoid multiple numerical experiments with various values for the restart parameters, these recommended values might be used.

In Table 1 for each test problem we also demonstrate the best results of the PICEA-g algorithm with the restart operator and settings given to them (columns 'PICEA-g, restart with the best settings'). It can be noticed that for all the test problems, the average values of the IGD metric are better than the corresponding statistics for the conventional PICEA-g.

4 Conclusion

In this paper, we have proposed a restart operator for MOGAs to improve their performance. We consider that the good qualities of this operator are useful not only for one-criterion GAs when it helps to avoid stagnation or premature convergence to a local optimum, but also for multi-objective algorithms. To find a representative and uniformly distributed approximation of the Pareto set or front, we should explore various areas of the search space properly. When the population is concentrated in a limited part of it and does not demonstrate any improvement within a number of generations, the restart operator should be applied. We also need to take advantage of the information about the search space gathered for previous iterations. Thus, these key points underlie our restarting meta-heuristic.

The experimental results have revealed that owing to the use of the restart operator, it is possible to improve the MOGA performance tremendously. We have developed the algorithm independent restarting technique in such a way that any MOGA could use it. Taking into account the lack of existing methodology about the use of the restart in multi-objective optimization, this study may serve as a fundamental basis for further research.

Moreover, in this article we give some recommendations about the appropriate values for the restart operator parameters.

Future work should be devoted to the thorough investigation of the restart operator on a set of MOGAs based on different heuristics. Besides, we also intend to design some mechanisms to adjust the restart parameters during the algorithm execution and make them adaptive.

Acknowledgements. This research is supported by the Russian Foundation for Basic Research within project No. 16-01-00767.

References

1. Fukunaga, A.S.: Restart scheduling for genetic algorithms. In: Eiben, A.E., Bäck, T., Schoenauer, M., Schwefel, H.-P. (eds.) PPSN 1998. LNCS, vol. 1498, pp. 357–366. Springer, Heidelberg (1998). doi:10.1007/BFb0056878
2. Beligiannis, G.N., Tsirogiannis, G.A., Pintelas, P.E.: Restartings: a technique to improve classic genetic algorithms' performance. Int. J. Comput. Intell. **1**, 112–115 (2004)
3. Ryzhikov, I., Semenkin, E.: Restart operator meta-heuristics for a problem-oriented evolutionary strategies algorithm in inverse mathematical MISO modelling problem solving. In: IOP Conference Series: Materials Science and Engineering, vol. 173 (2017). doi:10.1088/1757-899X/173/1/012015
4. Ryzhikov, I., Semenkin, E., Sopov, E.: A meta-heuristic for improving the performance of an evolutionary optimization algorithm applied to the dynamic system identification problem. In: IJCCI (ECTA), pp. 178–185 (2016)
5. Dao, S.D., Abhary, K., Mariam. R.: An adaptive restarting genetic algorithm for global optimization. In: Proceedings of the World Congress on Engineering and Computer Science, WCES 2015, 21–23 October, San Francisco, USA (2015)
6. Mohamed, A.W.: RDEL: restart differential evolution algorithm with local search mutation for global numerical optimization. Egypt. Inform. J. **15**(3), 175–188 (2014)
7. Gacto, M.J., Alcala, R., Herrera, F.: An improved multi-objective genetic algorithm for tuning linguistic fuzzy system. In: Proceedings of 2008 International Conference on Information Processing and Management of Uncertainty in Knowledge-Based Systems (IPMU 2008), pp. 1121–1128 (2008)
8. Wang, R.: Preference-inspired co-evolutionary algorithms. a thesis submitted in partial fulfillment for the degree of the Doctor of Philosophy, p. 231. University of Sheffield (2013)
9. Zhang, Q., Zhou, A., Zhao, S., Suganthan, P.N., Liu, W., Tiwari, S.: Multi-objective optimization test instances for the CEC 2009 special session and competition. University of Essex and Nanyang Technological University, Technical report. CES-487 (2008)

Using Multi-objective Evolutionary Algorithm to Solve Dynamic Environment and Economic Dispatch with EVs

Boyang Qu[⊠], Baihao Qiao, Yongsheng Zhu, Yuechao Jiao,
Junming Xiao, and Xiaolei Wang

School of Electrical Engineering, Zhongyuan University of Technology,
Zhengzhou 450001, China
qbyl984@hotmail.com

Abstract. In order to cope with the challenges brought by the large-scale Electric Vehicles (EVs) application to the power system dispatch, a dynamic economic emission dispatch model with the EVs is established. The vehicle to grid (V2G) power and conventional generator outputs of each dispatch period are set as the decision variables. The main optimization objectives are minimizing the total fuel cost and the pollution emission, so that the charging and discharging behavior of EVs is dynamically managed in the premise of meeting the demands of system energy and user travel. In this paper, the nondominated sorting genetic algorithm-II (NSGA-II) is used to solve such a model.

Keywords: Power system · Electric Vehicles · V2G · NSGA-II

1 Introduction

Electric vehicles have evident advantages compared with traditional cars, such as reduced pollution, low noise and high energy efficiency. Governments have made series of policies to encourage the development of electric vehicles. Studies demonstrate that, under the development of medium speed, electric vehicles' market share in USA will reach 35% in 2020, 51% in 2030 and 62% in 2050 [1]. Economic dispatch (ED) is one key point of the power system optimal operation problems. With the large-scale development of electric vehicles, the demand in traveling of the user, the net number of vehicles, and the performance of the battery device, etc., have a great impact on the charging load of the electric vehicles. Meanwhile, it brings a typical random uncertainty as the increase of the burden of power grid. Especially if the on-board energy storage device interactive Vehicle to Grid (V2G) is connected to the electricity grid [2], the frequent charge and discharge behavior makes the scheduling problem become a dynamic coupling system at different time stages. That will bring a new challenge to dynamic economic dispatch (DED) [3].

In electrical power systems, environment factor is one of the major concerns because of the fossil fuel fired electric generators causing global warming. At the same time, continuous attention to environmental problems is paid around the world. Therefore, the environmental factors in the power system dispatching becomes more

© Springer International Publishing AG 2017
Y. Tan et al. (Eds.): ICSI 2017, Part II, LNCS 10386, pp. 31–39, 2017.
DOI: 10.1007/978-3-319-61833-3_4

and more important, which conforms to the demand of the electric vehicle energy saving and environmental protection [4]. If the environmental dispatch objective is considered, the DED becomes dynamic environmental economic dispatch (DEED). The single optimal objective of the traditional ED problem is minimizing the cost in a dispatch period. In contrast, DEED is a typical multi-objectives, many time stages, high dimension and strong constraint nonlinear optimization problem. Particularly, if the electric vehicles driving and charging/discharging requirements in the term of time and energy are considered, modelling and solving the dispatch problem will be vitally complex.

At present, there are few researches in DEED with electric vehicles. In [5, 6], a single objective dispatch model was established by including economic and environmental factors, and was solved by using the PSO algorithm. But it only considers the registered electric vehicles and the basic battery capacity rather than the user travel demand and battery charging and discharging characteristics. In [7], it is pointed out that with the increased number of EVs the computational complexity increases and the decentralized approach could be potentially suitable. Some studies have focused on the benefits of V2G-enabled electric vehicles for aiding the integration of renewable resources into the electric grid [8–10]. Many researches established the model by considering either the power demand or the user demand. But few studies consider the environmental factors as a condition of constraint or an objective function [11–15].

In this paper, a multi-objective DEED model of electric vehicles dispatch V2G in accessing mode is established. This model takes the electric vehicles power and of the units output as the decision variables, and takes the total fuel cost and the pollutant emission as the optimization objective [16, 17]. Besides conventional power constraints, the user travel demand and on-board battery charging and discharging characteristics etc., are considered and the corresponding model is solved by nondominated sorting genetic algorithm-II (NSGA-II) algorithm.

2 DEED of Power System with Electric Vehicles Modeling

2.1 Objective Functions

Since electric vehicles participate in power system dispatch, the objective functions become dynamic compared with classical EED problems [4, 5].

Fuel cost objective:

$$\text{Minimize } FC = \sum_{t=1}^{T} \sum_{i=1}^{N} (a_i + b_i P_{i,t} + c_i P_{i,t}^2) \tag{1}$$

where a_i, b_i, c_i are the cost coefficients of ith thermal power generator. T is the total scheduling period, $P_{i,t}$ is output power of ith unit at time t, F_C is the total fuel cost of the system while N identifies the number of thermal units.

Pollution emission objective:

$$\text{Minimize } E_M = \sum_{t=1}^{T} \sum_{i=1}^{N} [(\alpha_i + \beta_i P_{i.t} + \gamma_i P_{i.t}^2 + \zeta_i \exp(P_{i.t})] \tag{2}$$

where α_i, β_i, γ_i, ζ_i and φ_i are the emission coefficients of the ith power generator and E_M is the total pollution emission of the system.

2.2 System Constrains

(1) System power balance
Considering the electric vehicles charging and discharging, the system power balance equation can be express as

$$\sum_{t=1}^{N} P_{i.t} + P_{Dch.t} = P_{D.t} + P_{L.t} + P_{ch.t} \qquad (t = 1, 2, \ldots, T) \tag{3}$$

where $P_{ch.t}$ is the electric vehicles charging load in period t. $P_{Dch.t}$ is the discharging load at time stage t. $P_{L.t}$ is the network losses at time stage t, which can be calculated by using the B-coefficients method.

$$P_{L.t} = \sum_{i=1}^{N} \sum_{j=1}^{N} P_{i.t} B_{ij} P_{j.t} + \sum_{i=1}^{N} P_{i.t} B_{i0} + B_{00} \tag{4}$$

where $B_{i,j}$, $B_{i,0}$ and $B_{0,0}$ are the network loss coefficients.
(2) Battery remaining power constraint
The electric vehicle on-board battery remaining power St at time stage t is defined as follow.

$$S_t = S_{t-1} + \eta_C P_{ch.t} \Delta t - \frac{1}{\eta_D} P_{Dch.t} \Delta t - S_{Trip.t} \tag{5}$$

where η_C and η_D are the coefficients of the charging and discharging efficiency, respectively. Δt is the scheduling interval. $S_{Trip.t}$ is the electric vehicles consumption in the process of driving power, which can be calculated as follow.

$$S_{Trip.t} = \Delta SL \tag{6}$$

where ΔS is the average distance unit consumption of power. L is driving distance. To ensure the safety of the operation and service lifespan of the battery, remaining power S_t is constrained by lower and upper limits.

$$S_{\min} \le S_t \le S_{\max} \tag{7}$$

(3) Electric vehicles charging and discharging power constraint
Generally, the charging and discharging power of the electric vehicles is no more than the rated power.

$$\begin{cases} P_{ch.t} \leq P_{NCh} \\ P_{Dch.t} \leq P_{NDch} \end{cases} \tag{8}$$

(4) User travel constraint
The basic function of the electric vehicles is to meet the user's travel requirements. Suppose that a charge and discharge cycle is completed in a scheduling cycle, the following formula should be satisfied.

$$\sum_{t=1}^{T} S_{Trip.t} = \sum_{t=1}^{T} \eta_C P_{ch.t} \Delta t - \sum_{t=1}^{T} \frac{1}{\eta_D} P_{Dch.t} \Delta t \tag{9}$$

(5) Generation capacity constraint
The real output power of each generator is constrained by lower and upper limits. These are also the bounds in the optimization formulation.

$$P_{i \cdot \min} \leq P_{i \cdot t} \leq P_{i \cdot \max} \tag{10}$$

(6) Conventional unit ramp rate constraint

$$\begin{cases} P_{i.t} - P_{i.t-1} - U_{Rt} \Delta t \leq 0 \\ P_{i.t-1} - P_{i.t} - D_{Rt} \Delta t \leq 0 \end{cases} \tag{11}$$

where U_{Rt} and D_{Rt} are the up and down rate of the ith unit.
(7) Spinning reserve
To maintain system reliability, adequate spinning reserves are required.

$$\sum_{t=1}^{N} P_{i.max} + P_{Dch.t} \geq P_{D.t} + P_{L.t} + P_{ch.t} + S_{R.t} \tag{12}$$

where $S_{R.t}$ is spinning reserve capacity requirements at time stage t.

3 Constraints Handing Method and Populations Set

3.1 Constraints Handling Method

To solve the constrained optimization problem, a constraint handling technique is required. In this paper, a penalty function method using the objective functions and constraints conditions is adopted to construct the unconstrained augmented objective function. In this case, the nonlinear constrained programming problem becomes

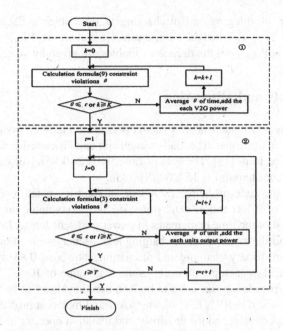

Fig. 1. Flowchart for equality constraints processing.

unconstrained programming problem. Figure 1 presents the flowchart for equality constraints processing.

For inequality constraints, the following methods are employed.

(1) The electric vehicles charging and discharging power constraints, the output of unit constraints, climbing constraints are added the inequality in processing.
(2) Infeasible solutions violation after equality constraint adjustment, battery remaining power constraint violation and spinning reserve constraint violation are set as the total of constraint violations.

3.2 Populations Set

In this paper, the dispatch model, decision variables including generator power outputs and electric vehicles charging and discharging power constitute the individual x_i of NSGA-II population, as follow.

$$x^i = \begin{Bmatrix} P_{1,1} & P_{1,2} & \cdots & P_{1,T} \\ P_{2,1} & P_{2,2} & \cdots & P_{2,T} \\ \vdots & \vdots & \vdots & \vdots \\ P_{N,1} & P_{N,2} & \cdots & P_{N,T} \\ P_{ev,T} & P_{ev,T} & \cdots & P_{ev,T} \end{Bmatrix} \tag{13}$$

where $P_{ev.t}$ is the of charging and discharging at time stage t. Each individual of dimension is $(N + 1) \times T$. When the electric vehicles is charging, $P_{ev.t} = P_{ch.t}$; while for discharging, $P_{ev.t} = P_{Dch.t}$. All the decision variables are given by vectors $\{x_1,..., x_N\}$.

4 Experiments and Discussion

A simulation study of an independent system operator of a ten-unit system with 50000 EVs is carried out in this paper. The load demand and unit characteristics of the ten-unit system are collected from [18]. The battery capacity is 24 kW·h (for example: Nissan Leaf) and power consumption is 15 kW·h/100 km.

Suppose that the state of charge (SOC) of an electric vehicle every morning is 100%, beginning of 07:00 and 17:00 1 h driving in the commute on the road (total 50 km) and the rest of the time participates in power grid scheduling. In dispatch cycle minimum SOC and the rated power of charging and discharging are limited to 20% of the rating, on-board battery charging and discharging power of 0.85, system of spinning reserve demand in each period is set as the load value of 10%.

For NSGA-II, population size is set to be 200, iterations is set to be 50000, penalty coefficient is 100, and $\varepsilon = 10^{-6}$, $K = 10$, crossover and mutation probability are taken as 0.9 and 0.2, respectively, and the crossover and mutation operator distribution index is 20. The optimal result of Pareto Fronts (PFs) are shown in Fig. 2 and the optimal solutions are displayed in Table 1.

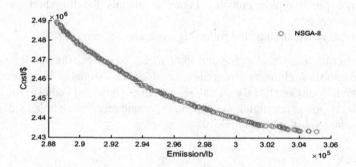

Fig. 2. Comparison of Pareto optimal fronts.

Table 1. comparison of extreme solutions and best compromise solutions.

	Objective	Emission/lb	Cost/$
NSGA-II	Best economic	3.0508×10^5	2.4333×10^6
	Best environmental	2.8851×10^5	2.4894×10^6
	Best compromise	2.9460×10^5	2.4544×10^6

The optimization result shows that the multi-objective problem does not have the best solutions which satisfy two objectives at the same time, instead, the Pareto best solutions. If decision makers only consider economy optimal and then choose the optimal scheme, it will be adverse to the environmental protection. Contrarily, if decision makers only minimum pollution emission and then select the optimal scheme environment, the fuel costs will increase. Therefore, it is necessary to find Pareto optimal solution set, especially for DEED problem. At the same time, in Fig. 2, NSGA-II get the optimal frontier that is wider and more uniform distribution, so it can provide a more abundant scheduling information for DEED problem with EVs.

To validate the scheduling scheme, in Table 2, details are given about the best compromise solutions. In Fig. 3, the electric vehicle charging and discharging power under different schemes (in figure, 1 represent 01:00–02:00, and so on) are illustrate.

Table 2. Best compromise solutions.

Time (H)	Unit1 (MW)	Unit2 (MW)	Unit3 (MW)	Unit4 (MW)	Uni5 (MW)	Uni6 (MW)	Unit7 (MW)	Unit8 (MW)	Unit9 (MW)	Unit10 (MW)	V2G (MW)	Pi (MW)	Pb (MW)
1	150.12	135.51	107.8	163.57	194.91	160	129.64	119.32	72.87	53.31	222.49	28.54	1036
2	152.72	138.27	163.9	169.61	203.95	159.29	129.09	119.02	78.84	54.15	226.71	32.13	1110
3	150.06	135.29	199.43	193.88	239.88	159.81	129.82	119.81	79.82	54.8	168.07	36.53	1258
4	156.14	137.87	230.14	208.03	243	160	130	120	80	55	74.7	39.49	1406
5	158.22	135.01	219.97	221.31	242.88	159.88	129.86	119.9	79.88	54.88	2.25	39.55	1480
6	165.54	213.52	298.49	269.82	192.88	158.51	128.51	118.51	78.51	53.51	1.03	48.77	1628
7	197.65	285.14	244.24	242.31	242.92	159.72	129.52	119.92	79.92	54.9	–	54.24	1702
8	169.06	272.09	311.7	291.77	243	160	130	120	80	55	-1.97	58.59	1776
9	220.56	292.01	339.98	299.98	242.98	159.98	129.98	119.97	79.98	54.98	-49.88	66.29	1924
10	299.16	364.28	339.94	299.93	242.93	159.94	129.94	119.93	79.93	54.94	-9.71	78.63	2022
11	298.49	350.76	340	300	243	160	130	120	80	55	-106.22	77.43	2106
12	277.36	323.49	339.99	299.99	242.99	159.99	129.99	119.99	79.99	54.99	-194.61	73.35	2150
13	296.26	323.35	339.83	299.82	242.81	159.8	129.81	119.79	79.81	54.81	-100.73	74.82	2072
14	218.69	313.11	333.94	299.97	242.97	159.97	129.97	119.97	79.97	54.97	-37.9	67.4	1924
15	176.59	281.25	305.33	299.99	243	160	129.99	119.99	80	54.99	15.21	59.91	1776
16	152.45	201.25	253.36	277.3	242.63	159.64	129.7	119.68	79.67	54.72	68.41	48.01	1554
17	162.45	135	206.6	230.52	242.62	159.45	129.52	119.37	79.52	54.42	–	39.47	1480
18	150.92	213.83	261.58	264.25	243	160	130	120	80	55	2.05	48.56	1627.97
19	184.48	258.83	303.41	299.96	242.96	159.96	129.96	119.96	79.96	54.96	-0.24	58.66	1776
20	258.65	338.31	340	300	243	160	130	120	80	55	-20.07	73.06	1971.97
21	219.5	313.25	338.65	300	243	160	130	120	80	55	-32.37	67.81	1923.95
22	150.08	233.25	258.65	250.95	240.77	159.65	129.79	119.79	79.73	54.79	0.76	48.68	1628
23	158.75	171.56	222.74	222.82	242.81	159.81	129.81	119.81	79.81	54.81	188.82	41.91	1332
24	152.23	136.09	186.59	197.98	240.95	159.78	129.68	119.67	79.61	54.76	237.05	36.3	1184

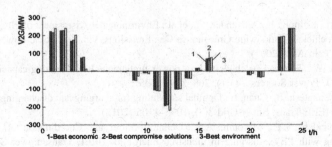

Fig. 3. comparisons of EV charge/discharge power among different solutions.

Figure 3 shows that, the extreme solution and best compromise solution corresponding to the electric vehicles charging and discharging rule is similar, which is just different on the specific power. It changes the load distribution among the conventional units, eventually, leads to make a great difference in fuel costs and pollution emissions.

As it can be seen from the above results, in order to ensure that the electric vehicles travel demand, in the 22:00 to the next day 06:00 in the charging status, and SOC reaches 100% in 07:00. Driving on the road in 07:00 to 08:00, on-board battery discharge, leads to SOC falling. In 08:00 to 15:00 peak load, up to 2150 MW, the lowest reaches 1776 MW, electric vehicles in the discharge state, alleviate the pressure of the conventional thermal power units. Due to the owner of the driving needs in 17:00 to 18:00, the vehicle is charging in 16:00. In 20:00 and 21:00 peak load at night, the electric vehicles continue to discharge to SOC reaching the lower limit. Then electric vehicles charge at night valley load until the next day trip.

5 Conclusion

In this paper, under the background of the electric vehicles on large scale accessing to grid, a multi-objective DEED model of electric vehicles scheduling V2G access mode is established. In this model the electric vehicles energy and traffic characteristics, and the system of network loss etc. are considered. This model can bring both economic and environmental benefits, meanwhile, reflect the influence of electric vehicle access on power system dispatch.

Acknowledgments. This research is partially supported by National Natural Science Foundation of China (61673404 and 61473266), Scientific and Technological Projects of Henan (132102210521, 152102210153, 172102210601), Innovative Talents in Universities Support Project of Henan (16HASTIT033) and Key Scientific and Technological Research Projects of Education Department of Henan (17A470006).

References

1. Duvall, M., Knippig, E., Alexander, M., et al.: Environmental Assessment of Plug-in Hybrid Electric Vehicles. Nationwide Greenhouse Gas Emissions, vol. 1. Electric Power Research Institute, Palo Alto (2007)
2. Kempton, W., Tomic, J.: Vehicle-to-grid power fundamentals: calculating capacity and net revenue. J. Power Sources **144**(1), 268–279 (2005)
3. He, Y., Venkatesh, B., Guan, L.: Optimal scheduling for charging and discharging of electric vehicles. IEEE Trans. Smart Grid **3**(3), 1095–1105 (2012)
4. Derakshuandeh, S.Y., Masoum, A.S., Deilami, S., et al.: Coordination of generation scheduling with PEVs charging in industrial microgrids. IEEE Trans. Power Syst. **28**(3), 3451–3461 (2013)
5. Saber, A.Y., Venayagamoorthy, G.K.: Plug-in vehicles and renewable energy sources for cost and emission reductions. IEEE Trans. Indus. Electron. **58**(4), 1229–1238 (2011)

6. Gholami, A., Ansari, J., Jamei, M., et al.: Environmental/economic dispatch incorporating renewable energy sources and plug-in vehicles. IET Gener. Transm. Distr. **8**(12), 2183–2198 (2014)

7. Gan, L., Topcu, U., Low, S.H.: Optimal decentralized protocol for electric vehicle charging. IEEE Trans. Power Syst. **28**(2), 940–951 (2013)

8. Haddadian, G., et al.: Security-constrained power generation scheduling with thermal generating units, variable energy resources, and electric vehicle storage for V2G deployment. Int. J. Electr. Power Energ. Syst. **73**, 498–507 (2015)

9. Haque, A.N.M.M., Ibn Saigf, A.U.N., Nguyen, P.H.: Exploration of dispatch model integrating wind generators and electric vehicles. Appl. Energ. **183**, 1441–1451 (2016)

10. Garcia-Villalobos, J., Zamora, I., San Martin, J.I., Asensio, F.J., Aperribay, V.: Plug-in electric vehicles in electric distribution networks: a review of smart charging approaches. Renew. Sustain. Energ. Rev. **38**, 717–731 (2014)

11. Zhou, B., Yao, F., Litter, T., Zhang, H.: An electric vehicle dispatch module for demand-side energy participation. Appl. Energ. **177**, 464–474 (2016)

12. Andersson, S.L., Elofsson, A.K., Galus, M.D., et al.: Plug-in hybrid electric vehicles as regulating power providers: case studies of Sweden and Germany. Energ. Policy **38**(6), 2751–2762 (2010)

13. De Los Rios, A., Goentzel, J., Nordstrom, K.E., et al.: Economic analysis of vehicle-to-grid (V2G)-enabled fleets participating in the regulation service market. In: IEEE PES Innovative Smart Grid Technologies, Washington D.C., USA, pp. 16–24 (2012)

14. Han, S., Han, S.: Economic feasibility of V2G. Energies **6**(2), 748–765 (2013)

15. Zhao, Y., Noori, M., Tatari, O.: Vehicle to grid regulation services of electric delivery trucks: economic and environmental benefit analysis. Appl. Energy **170**, 161–175 (2016)

16. Qu, B.Y., Suganthan, P.N., Pandi, V.R., et al.: Multi objective evolutionary programming to solve environmental economic dispatch problem. In: 11th International Conference on Control Automation Robotics & Vision (ICARCV), pp. 1673–1679. IEEE, Singapore (2010)

17. Qu, B.Y., Liang, J.J., Zhu, Y.S.: Economic emission dispatch problems with stochastic wind power using summation based multi-objective evolutionary algorithm. Inf. Sci. **351**, 48–66 (2016)

18. Basu, M.: Dynamic economic emission dispatch using nondominated sorting genetic algorithm-II. Int. J. Electr. Power Energ. Syst. **30**, 140–149 (2008)

Improved Interval Multi-objective Evolutionary Optimization Algorithm Based on Directed Graph

Xiaoyan Sun, Pengfei Zhang$^{(\boxtimes)}$, Yang Chen, and Yong Zhang

School of Information and Control Engineering,
China University of Mining and Technology, Xuzhou 221116, China
zhangpongfy@sina.com

Abstract. Multi-objective evolutionary algorithm for optimizing objectives with interval parameters is becoming more and more important in practice. The efficient comparison metrics on interval values and the associated offspring generations are critical. We first present a neighboring dominance metric for interval numbers comparisons. Then, the potential dominant solutions are predicted by constructing a directed graph with the neighboring dominance. We design a directed graph using those competitive solutions sorted with NSGA-II, and predict the possible evolutionary paths of next generation. A PSO mechanic is applied to generate the potential outstanding solutions in the paths, and these solutions are further used to improve the crossover efficiency. The experimental results demonstrate the performance of the proposed algorithm in improving the convergence of interval multi-objective evolutionary optimization.

Keywords: Interval multi-objective · Evolutionary optimization · Directed graph

1 Introduction

Multi-objective evolutionary algorithms (MOEAs) based on the concept of Pareto dominance such as vector evaluation genetic algorithm [1], non-dominated sort genetic algorithm (NSGA) [2], improved NSGA-II [3] and multi-objective evolutionary optimization algorithm based on decomposition [4] are powerful for solving multi-objective problems [5]. In practice, numerous multi-objective optimization problems encounter uncertain parameters.

When the parameters are intervals, the values of the objectives are intervals too, which are called interval multi-objective optimization problems. Aiming at solving the multi-objective optimization problems with the objective functions having noise, Limbourgetc. used the interval to represent the uncertain objectives, defined the interval vector partial order relationship and the hyper-volume to measure the distribution and approximation of the optimized solutions, and then proposed the evolutionary optimization method, i.e., Imprecision Propagating Multi-Objective Evolutionary Algorithm(IP-MOEA) [6]. Eskandarietc. proposed a stochastic Pareto genetic algorithm (SPGA) to solve the above similar problems [7]. In addition, Gong et al. proposed

© Springer International Publishing AG 2017
Y. Tan et al. (Eds.): ICSI 2017, Part II, LNCS 10386, pp. 40–48, 2017.
DOI: 10.1007/978-3-319-61833-3_5

a multi-objective evolutionary optimization method of interval parameters based on the crowding distance and interval Pareto dominance [8]. Zhang Yong et al. proposed a multi-objective micro-particle swarm optimization algorithm based on probability dominance via defining the probability dominance relationship and comparing the quality of the solutions [9]. These algorithms only emphasize the dominance metrics without sufficiently using the knowledge of the evolutionary population to promote the search.

It is critical to speed up the search process of the evolutionary population, and get a solution set with better convergence in the Pareto front. Intuitively, the entire performance of the MOEAs will be greatly improved if the convergence direction or trend is known in prior. Unfortunately, the most MOEAs for interval multi-objective problems have not mining the potential knowledge of the evolutionary population and the elitist individuals. Motivated by this, we present a directed graph based MOEA for interval multi-objective problems by predicting the possible convergence direction and generating potential super solutions. To this end, we here focus on three critical issues in the following subsection: the construction of the directed graph, the prediction of the convergence direction and the outstanding offspring, the usage of the predicted individuals.

2   Multi-objective Evolutionary Algorithm with Directed Graph and Individual Prediction

2.1   Framework

The minimization problem with interval parameters is considered and shown in Eq. (1):

$$\begin{cases} \min \tilde{F}(X, C) = (\tilde{f}_1(X, C), \tilde{f}_2(X, C), \ldots, \tilde{f}_m(X, C)) \\ s.t.\, X \in S \subset R^n \\ C = (c_1, c_2, \ldots, c_l)^T, c_k = [\underline{c}_k, \bar{c}_k], k = 1, 2, \ldots, l \end{cases} \tag{1}$$

Where $X = (x_1, x_2, \ldots, x_n)^T \in S \subset R^n$ is an n-dimensional decision vector. S is the n-dimension decision space. C is internal vector parameter with l elements. c_k is the kth component of C. $\underline{c}_k$ and $\bar{c}_k$ are lower limit and upper limit of c_k, respectively. The ith objective $\tilde{f}_i(X, C) = [\underline{f}_i(X, C), \bar{f}_i(X, C)]$ is in an interval form due to the interval property of the parametric vector C.

The pseudocode of our algorithm is first illustrated in Fig. 1. The framework of NSGA-II is adopted as the base, but in which, the dominance metric and the generation of the offspring are different from the traditional NSGA-II. The main contributions of our algorithm are bolded as in Steps 4, 9,10 and 11.

The main procedures of our algorithm are as follows. Based on our $\mu \oplus P$ dominance metric, the NSGA-II is first conducted to get the sorted individuals. And then with those best individuals who are at the first layer of the Pareto front are archived and used to construct/update the directed graph. With the directed graph, the possible convergence directions or paths are obtained. The potential outstanding solutions are

1.Determine the optimization parameters such as population scale N

2.Initialize the population $P(0)$

3. Initialize the archive set A, namely put the initialized population $P(0)$ into A

4.Calculate the $\mu \oplus P$ index, and conduct non-dominated ranking

5. According to the non-dominated relationship, and select $N/2$ dominant individuals by tournament

6. Perform genetic operations (Based on the crossing and variation of the predicted individuals), and generate the population $Q(t)$ with the size of N

7.Merge $P(t)$ and $Q(t)$ into $R(t)$

8. According to the $\mu \oplus P$ Pareto dominance non-dominated ranking, select the top N dominant population to generate $P(t+1)$

9.Put the top N non-dominant individuals into archive set A to replace the original individuals in A

10. According to the $\mu \oplus P$ Pareto dominated relationship and neighbor dominance relationship, apply the individuals in archive set A to construct or update the directed graph

11.Using PSO Strategy to predict potential better individuals

12. Does it meet the terminal conditions?

No, turn to step 5

Yes, consider the individuals at the end of the directed graph as the obtained Pareto front, and output them

Fig. 1. Pseudocode of our algorithm

predicted according to the particle update of PSO. These potential solutions are further used to make a crossover for guiding the search to the potential convergence direction. As for the $\mu \oplus P$ dominance metric please refer to our previous work [10].

2.2   Construction of the Directed Graph

An example of the directed graph of the solutions is shown in Fig. 2, where each vertex corresponds to an individual selected from the sorted individuals, and the directed edge corresponds to the dominant relationships among these individuals. We here select *num* solutions from the lower levels of the sorted individuals, and the directed graph together with the prediction are conveyed based on these individuals.

For simply getting the dominant relationships of these selected individuals, we here present a neighboring dominance metric to measure or cluster the archived individuals for obtaining the directed edges among all these individuals.

Definition 1 Neighboring Dominance. If two solutions $(x_1, f(x_1))$ and $(x_2, f(x_2))$ meet the conditions of $|x_1 - x_2| < \varepsilon$ and $x_1 \prec x_2$, namely x_2 is dominated by x_1, it is called that the individual x_2 is neighbor dominated by the individual x_1. The symbol $\prec$ is used to represent the general occupation relationship.

With the neighboring dominant relationship, only those individuals whose Euclidean distances less than ε are compared, which reduces the comparisons of dominant relationships among all the individuals. An example of constructed directed graph is given in Fig. 2, and the direction from the left to the right is the direction from the inferior to the superior. On the path 1 shown in the figure, the directed edge from the vertex C to the vertex A indicates that the individuals corresponding to the vertex C is neighboring dominant to the individuals corresponding to A. Three paths along the neighboring relationship are depicted in the figure, namely path 1 (A-C-D-F), path 2 (B-C-D-F) and path 3 (B-C-E-H). Taking the path 1 for example, the in-degree of point F is 0, which is the optimal individual in A. After the determination of the point F, the $(num - 1)$ points before the point F and point F constitute a path. The individual performance along the direction A-C-D-F is getting better and better. If the decision space positions corresponding to A, C, D, and F are regarded as the coordinates of four moments, A-C-D-F becomes the trajectory with the direction from the inferior to the superior, which paves the way for further prediction.

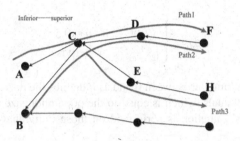

Fig. 2. Directed graph

2.3 Individual Prediction Based on PSO

After the directed graph is constructed, it is necessary to predict the individuals on each path. In this paper, Based on the PSO algorithm [11], the individual prediction strategy based on the PSO algorithm is proposed. The formulas are shown in Eqs. (2) and (3)

$$v_d^{k+1} = \omega * v_d^k + c_1 * Rand() * (pbest_{id} - x_{id}^k) + c_2 * Rand() * (gbest^k - x_{id}^k) \quad (2)$$

$$x_{id}^{k+1} = x_{id}^k + v_{id}^{k+1} \quad (3)$$

The parameters of these equations are the same as they are in PSO. Here, we use the individual in evolutionary population with the highest rank as the *gbest* and the individuals in the end of the directed graph as the *pbest*. Rand () is a random number that changes in the range of [0,1].

2.4 Simulation Binary Crossover (SBX) Strategy Based on the Predicted Individuals

Crossover operator is an important strategy to guide the direction of individual evolution. In this paper, it is very important to rationally use the new individuals predicted using the method in the previous section. In the traditional NSGA-II evolutionary algorithm, the simulation binary crossover operator [12] is often considered as the crossover strategy. Based on the simulation of the binary crossover strategy, the individuals predicted in Sect. 2.3 are involved in the crossover process, which aims at guiding the direction of population evolution in the evolutionary algorithm. The design of the crossover operator in this paper is shown in Eq. (4):

$$x_{1,d} = \frac{1}{2}[(1 - \beta_d) \bullet x_d^{k+1} + (1 + \beta_d) \bullet x_{r,d}]$$
$$x_{2,d} = \frac{1}{2}[(1 + \beta_d) \bullet x_d^{k+1} + (1 - \beta_d) \bullet x_{r,d}]$$
$$(4)$$

where x^{k+1} is the predicted individuals using the PSO strategy. $x_{r,d}$ is the dth dimension value of the rth individual in the original population. β_d is calculated according to the Eq. (5):

$$\beta_d = \begin{cases} (2u)^{\frac{1}{\eta+1}}, & \text{if } u \leq 0.5 \\ [\frac{1}{2(1-u)}]^{\frac{1}{\eta+1}}, & \text{otherwise} \end{cases}$$
$$(5)$$

where u is the random number between 0 and 1. η defines the distribution index of the newly generated individuals, which is equal to the population size in the method used in this paper. Besides, mutation is performed on those individuals without matching predicted individuals.

3   Experiments

3.1   Experimental Settings

Since the IP-MOEA and SPGA are two competitive methods for solving interval multi-objective problems, they are compared with our algorithm to illustrate the performance.

The standard optimization problems DTLZ1, DTLZ2, DTLZ3 and DTLZ5 [13] are selected, and the range of the decision variables for each optimization problems is [0,1], where the dimension of the decision variable of the DTLZ series function is 7. In the experiments, the number of objective functions of DTLZ1, DTLZ2 and DTLZ3 is 2 and 3, respectively. Given that the objectives of all the above optimization problems do not contain the internal parameters and in order to test the performance of different methods, the disturbance factor δ_i defined in Reference [6] is introduced to make the objective value be an interval as $\tilde{f}_i(X) = [\underline{f}_i(X), \overline{f}_i(X)]$. The optimized interval minimum multi-objective optimization problems are denoted as DTLZ1′, DTLZ2′, DTLZ3′. and DTLZ5′.

First, the parameter of the neighboring dominance is set. In view of the choice of ε to evaluate the Euclidean distance among the individuals in the decision space, the parameter in the experiments is defined as follows:

$$\varepsilon = \sqrt{\sum_{i=1}^{n}[(E_i \bullet \frac{1}{10})^2]}$$
$$(6)$$

Where n is the number of the variables, and $E_i = x_{i_{max}} - x_{i_{min}}$, i.e., the varied range of the ith decision variable.

Thirty groups of experiments were performed, and the average values were recorded and compared. All the experiments parameters were displayed as below in Table 1.

Table 1. Values of experiments indicators

Parameters	Size of crossover and mutation	Crossover probability	Mutation probability	Evolutionary generation	Parameters in Eq. (2)		
					ϖ	c_1	c_2
Values	2	0.9	0.01	100	11	6	4

In this paper, the performance indicators of the best super-volume BH [6] and the uncertainty I [6] were used to compare the proposed interval order relationship with the performance of the other methods.

3.2 Experimental Results and Analysis

3.2.1 Dominant Proportion of the Newly Generated Individuals
In view of the crossover strategy adopted in Sect. 2.4 for the predicted individuals, the comparison of individual dominance before and after the crossover was designed. Higher dominant proportion indicates more efficient individuals' generations of our algorithm, and the results are shown in Fig. 3.

It can be found that the proportion of the individuals after the crossover dominated by the individuals before the crossover continuously decreases. This proportion significantly increases at the early evolutionary stage, which indicates that the individuals obtained via directed graph prediction at the early evolutionary stage have a good guiding effect, and enhance the overall convergence via crossover operators. At the later evolutionary stage, there is a case where the dominance ratio is 0. The reason is that there are no effective paths in the directed graph at the later evolutionary stage when most individuals are non-dominated. In such cases, only the mutation works.

3.2.2 Analysis on Convergence
The algorithm in this paper and the other two algorithms are denoted as "G-NSGA", "IP-MOEA" and "SPGA", respectively. The results are listed in Table 2, and the † indicates that the comparison method and our method has significant differences (Mann-Whitney U test, confidence level is 0.05). It can be seen from Table 1 that for the DTLZ1' series test functions, when the evolutionary generation is set as 100, the population of each algorithm converges. When the test function is three-dimension, the convergence and uncertainty of the algorithm in this paper are optimal. In the case of 2-dimension and 4-dimension, the convergence of SPGA is optimal, while IP-MOEA

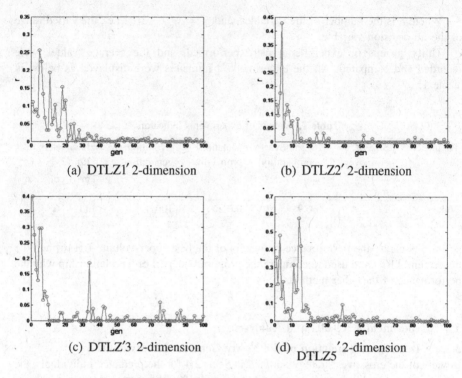

(a) DTLZ1′ 2-dimension   (b) DTLZ2′ 2-dimension

(c) DTLZ′3 2-dimension   (d) DTLZ5 ′2-dimension

Fig. 3. Analysis of the dominance proportion

is optimal in the aspect of uncertainty. When the evolutionary generation is 20, the convergence of our method is optimal in all dimensions. The reason is that the directed graph strategy improves the search ability of the algorithm, strengthens the convergence speed, and thus achieves good convergence at the early evolutionary stage. In contrast, the convergence of IP-MOEA is quite different from that of this paper. The reason is that $\mu \oplus P$ individual comparison strategy adopted in this paper can achieve the elaborative comparison among the intervals, while the accuracy of IP-MOEA interval comparison strategy is low, and thus the convergence speed is poor. Similar conclusions can be obtained for the other problems.

In summary, the proposed multi-objective evolutionary algorithm based on the directed graph theory can obtain solutions with less uncertainty and better convergence. In addition, it can be seen from the first two experiments that the guiding effect of the directed graph model on the population at the early evolutionary stage is relatively obvious, which indicates that the advantages of the proposed method are reflected in the aspect of the convergence speed.

Table 2. Experimentalresults on convergence

Functions	Dimension	Performance indicator	100 generations			
			G-NSGA	IP-MOEA	SPGA	NSGA-II $(\mu \oplus P)$
DTLZ1'	2	BH	9.91e + 12	4.36e + 12 †	1.09e + 13 †	**1.18e + 13** †
		I	0.4290	**0.3637** †	0.4897	0.4479
	3	BH	3.63e + 22	3.55e + 22	2.61e + 22	**4.29e + 22**
		I	0.5702	0.5796	0.6112	**0.5523**
	4	BH	5.82e + 17	1.13e + 17 †	6.05e + 17	**6.63e + 17**
		I	0.2989	**0.2913**	0.3476 †	0.3015
DTLZ2'	2	BH	35.7565	8.8694	35.0136	**36.11**
		I	0.5535	**0.3533** †	0.5920	0.4988
	3	BH	**75.8153**	15.2716 †	69.9141	74.1929
		I	0.6030	0.6100	0.6458	**0.5948**
	4	BH	**150.7496**	94.2002 †	136.4655	147.7313
		I	0.3458	0.4480	0.4082	**0.3274**
DTLZ3'	2	BH	3.57e + 14	5.07e + 12 †	3.57e + 14	**4.11e + 14**
		I	0.4450	**0.3952**	0.4930	0.4355
	3	BH	7.04e + 31	1.88e + 31 †	7.17e + 31	**8.18e + 31**
		I	0.5569	**0.4952**	0.6384	0.591
	4	BH	1.55e + 17	7.26e + 15 †	1.29e + 17	**1.74e + 17**
		I	**0.3105**	0.3538	0.3866	0.3604
DTLZ5'	2	BH	**9.7389**	4.0144 †	9.3205	6.8243
		I	0.5327	**0.3584** †	0.5858	0.6204
	3	BH	**12.0439**	10.1114 †	9.1990 †	8.7680 †
		I	0.6438	**0.6121**	0.6685	0.6793
	4	BH	**237.1157**	208.8099 †	210.8112	156.69 †
		I	**0.4209**	0.4392	0.4540	0.4801

4 Conclusion

An MOEA for solving interval multi-objective problems is improved by using directed graph to predict the potential evolutionary direction and get outstanding individuals. We present the construction method of the directed graph by using sorted individuals in the Pareto front and a relaxed dominance metric, i.e., neighboring dominance. And then along with the paths of the directed graph, the potential outstanding individuals are predicted and applied to guide the crossover. The experimental results illustrate the efficiency of our algorithm.

Acknowledgments. This paper is supported by the National Natural Science Foundation of China with granted No. 61473298 and 61473299.

References

1. Schaffer, J.D.: Multiple objective optimization with vector evaluated genetic algorithms. In: International Conference on Genetic Algorithms, pp. 93–100 (1985)
2. Srinivas, N., Deb, K.: Muiltiobjective optimization using nondominated sorting in genetic algorithms. Evol. Comput. **2**, 221–248 (1994)
3. Deb, K., Pratap, A., Agarwal, S., Meyarivan, T.: A fast and elitist multiobjective genetic algorithm: NSGA-II. IEEE Trans. Evol. Comput. **6**, 182–197 (2002)
4. Zhang, Q., Li, H.: MOEA/D: a multiobjective evolutionary algorithm based on decomposition. IEEE Trans. Evol. Comput. **11**, 712–731 (2007)
5. MaoGuo, G., LiCheng, J., DongDong, Y., WenPing, M.: Research on evolutionary multi-objective optimization algorithms. J. Software **20**, 271–289 (2009)
6. Limbourg, P., Aponte, D.E.S.: An optimization algorithm for imprecise multi-objective problem functions. In: The 2005 IEEE Congress on Evolutionary Computation 2005, vol. 451, pp. 459–466 (2005)
7. Eskandari, H., Geiger, C.D., Bird, R.: Handling uncertainty in evolutionary multiobjective optimization: SPGA. In: IEEE Congress on Evolutionary Computation, pp. 4130–4137 (2007)
8. Gong, D.W., Qin, N.N., Sun, X.Y.: Evolutionary algorithms for multi-objective optimization problems with interval parameters. In: IEEE Fifth International Conference on Bio-Inspired Computing: Theories and Applications, pp. 411–420 (2010)
9. Zhang, Y., Gong, D., Hao, G., Jiang, Y.: Particle swarm optimization for multi-objective systems with interval parameters. Acta Automatica Sinica **34**, 921–928 (2008)
10. Sun, X., Zhang, P., Chen, Y., Shi, L.: Interval multi-objective evolutionary algorithm with hybrid rankings and application in RFID location of underground mine. Control Decis. **32**, 31–38 (2017)
11. Kennedy, J., Eberhart, R.: Particle swarm optimization. In: Proceedings IEEE International Conference on Neural Networks 1995, vol. 1944, pp. 1942–1948 (1995)
12. Kittler, J., Hatef, M., Duin, R.P.W., Matas, J.: On combining classifiers. IEEE Trans. Pattern Anal. Mach. Intell. **20**, 226–239 (1998)
13. Deb, K., Thiele, L., Laumanns, M., Zitzler, E.: Scalable Test Problems for Evolutionary Multiobjective Optimization. Springer, London (2005)

A Novel Linear Time Invariant Systems Order Reduction Approach Based on a Cooperative Multi-objective Genetic Algorithm

Ivan Ryzhikov[✉], Christina Brester, and Eugene Semenkin

Siberian State Aerospace University, Krasnoyarsk, Russia
{ryzhikov-88, eugenesemenkin}@yandex.ru,
christina.brester@gmail.com

Abstract. Cooperative multi-objective optimization tool is proposed for solving the order reduction problem for linear time invariant systems. Normally, the adequacy of an order reduction problem solution is estimated using two different criteria, but only one of them identifies the model. In this study, it was suggested to identify the parameters using both of the criteria, and since the criteria are complex and multi-extremum there is a need for a powerful optimization algorithm to be used. The proposed approach is based on the cooperation of heterogeneous algorithms implemented in the islands scheme and it has proved its efficiency in solving various multi-objective optimization problems. It allows us to receive a set of lower order models, which are non-dominated solutions for the given criteria and an estimation of the Pareto set. The results of this study are compared to the results of solving the same problems using various approaches and heuristic optimization tools and it is demonstrated that the set of solutions not only outperforms these approaches by the main criterion, but also provides good solutions with another criterion and a combination of them using the same computational resources.

Keywords: Cooperation meta-heuristics · Multi-objective optimization · Multi-Objective Genetic Algorithm · Linear time invariant system · Parameters identification

1 Introduction

This study is focused on solving multi-objective optimization (MO) problems related to the parameter identification problem for linear time invariant (LTI) systems. The initial problem is important in identification theory and system theory and is currently applied for dynamical systems. We consider both single input single output (SISO) and multiple input multiple output (MIMO) systems and our aim is to identify the parameters of low order models. Various approaches of solving the order reduction problem exist, but all of them are based on a single criterion, which is a numerical representation of model adequacy. Model adequacy is calculated as the distance between the model output and desired output, where outputs are reactions on the unit-step function, so the identification problem is reduced to the complex multi-extremum optimization problem of minimizing the sum of the error.

© Springer International Publishing AG 2017
Y. Tan et al. (Eds.): ICSI 2017, Part II, LNCS 10386, pp. 49–56, 2017.
DOI: 10.1007/978-3-319-61833-3_6

Recent studies have shown that heuristic stochastic optimization tools (commonly nature-inspired or evolution-based) are the most efficient in solving the LTI order reduction problem. These complex tools are used because of the impossibility of representing the criteria in symbolic form, and thus one faces a black-box optimization problem. In the study [1] the genetic algorithm (GA) is used to solve the reduced problem. In studies [2] and [3] the Big Bang Big Crunch and Cuckoo Search Optimization algorithm were used. Many of these solvers are a combination of an optimization tool and a stability equation [4]. This combination was first met in the paper [5], where a stability equation was combined with a genetic algorithm. In this paper we are using the approach that is given in the study [6], but with two different criteria to optimize.

The representative characteristic of the reduced order model is also the distance between its output and initial model output as a reaction on the Dirac function, but actually, these criteria are different. This fact can be seen in the studies [2, 3] and others. This is why it was proposed to estimate not the parameters of the model that bring the minimum to unit-step function reaction distances, but the set of the solutions, which is an estimation of the Pareto front for these two criteria. In summary, it is clear that solving the identification problem in a generalized form requires an MO evolution-based or nature-inspired optimization tool.

2 Cooperative Multi-objective Genetic Algorithm

The common scheme of any Multi-Objective Genetic Algorithm(MOGA) includes the same steps as any conventional one-criterion GA. However, in contrast to one-criterion GAs, the outcome of MOGAs is the set of non-dominated points which form the Pareto set approximation. To avoid a choice of the most effective search strategy, in this study we apply a cooperation of several GAs based on various heuristic mechanisms [7], which are the Non-Sorting Genetic Algorithm II (NSGA-II) [8], the Preference-Inspired Co-Evolutionary Algorithm with goal vectors (PICEA-g) [9], and the Strength Pareto Evolutionary Algorithm 2 (SPEA2) [10]. The multi-agent model is expected to preserve a higher level of genetic diversity. The benefits of the particular algorithm could be advantageous in different stages of optimization.

In all these algorithms fitness assignment strategies are based on Pareto-dominance, but they are implemented in different ways: in NSGA-II a fitness function value of a candidate solution depends on the niche it belongs to and a crowding distance (the greater this distance, the better); in PICEA-g fitness function values are calculated with the use of special goal vectors; in SPEA2 a niching mechanism in combination with the density estimation (the distance to the k-th nearest neighbour in the objective space) is applied. To preserve diversity among solutions and obtain a representative approximation of the Pareto front a crowding distance metric is used in NSGA-II, whereas a nearest neighbour technique is implemented in PICEA-g as well as in SPEA2. In order not to lose good solutions found during the algorithm execution, elitism is applied: in NSGA-II parent and offspring are merged and the best solutions are chosen for the next generation; in SPEA2 a special archive set is used to save non-dominated points found; in PICEA-g both of these strategies are applied.

In this study, the chosen MOGAs are combined based on an island model. Generally speaking, the island model [11] of a GA implies the parallel work of several algorithms. A parallel implementation of GAs has shown not just an ability to preserve genetic diversity, since each island can potentially follow a different search trajectory, but also could be applied to separable problems. The initial number of individuals M is spread across L subpopulations: $M_i = M/L, i = 1, \ldots, L$. At each T-th generation algorithms exchange the best solutions (migration): these chromosomes substitute the worst ones in each population. There are two parameters: migration size, the number of candidates for migration, and migration interval, the number of generations between migrations. Moreover, it is necessary to define the island model topology, in other words, the scheme of migration. We use fully connected topology, which means that each algorithm shares its best solutions with all other algorithms included in the island model.

This algorithm was examined on different benchmark functions from [12] and complex optimization problems. With the use of the algorithm in cardiovascular predictive modelling [13] it became possible to reduce the number of features approximately from four hundred down to forty with no detriment to the model performance. For this problem, the result obtained is rather valuable because it means a decrease in the amount of clinical trials for patients. Moreover, in the speech-based emotion recognition problem the presented MOGA was successfully used not only for feature selection (the dimensionality of the feature vector was reduced from three hundred by a factor of two) [14], but also for the automated design of neural network models [15]. In both cases, the model performance was improved significantly. Thus, numerical results confirmed that the proposed MOGA with cooperation meta-heuristics implemented is a reliable and high performance multi-objective optimization tool. The next step is to perform the initial problem numerical criteria and then to apply the proposed optimization tool.

3 Order Reduction Problem for Linear Time Invariant Systems

Let the SISO LTI system model be determined by the following Laplace transformation of the linear differential equation

$$G(s) = \sum_{j=0}^{m} b_j \cdot s^j \cdot \left(\sum_{i=0}^{n} a_i \cdot s^i \right)^{-1} \tag{1}$$

where $a_i \in R, i = \overline{1, n}$ and $b_i \in R, i = \overline{1, m}$ are the coefficients, $n > m$ is the order. The initial point is $x(0) = 0$, where $x(t)$ is the inverse Laplace transformation for (1).

We know that the system (1) is stable and its output asymptote for the unit step function $H(t)$ can be determined with the following limit

$$a^s = \lim_{t \to +\infty} x(t) = a_0/b_0, \tag{2}$$

and since we know the initial model (1) to be reduced, we can evaluate the asymptote (2) and provide the model being convergent to the same asymptote by determination of its parameters. In this case, the model of the 2nd order can be represented with the transfer function

$$G_m(s,p) = \frac{p_2 \cdot s + a^s \cdot p_1}{s^2 + p_0 \cdot s + p_1}. \tag{3}$$

The same can be stated for the MIMO system (currently a two-input two-output system)

$$W_s(s) = \begin{pmatrix} W^{1,1}(s) & W^{1,2}(s) \\ W^{2,1}(s) & W^{2,2}(s) \end{pmatrix}, \tag{4}$$

where $W^{i,j}(s) = D^{i,j}(s)/N^{i,j}(s)$, $D^{i,j}$, $N^{i,j}$, $i,j = \overline{1,2}$ and its reduced order model

$$G_m(s,p) = \frac{1}{D_m} \cdot \begin{pmatrix} N_m^{1,1} & N_m^{1,2} \\ N_m^{2,1} & N_m^{2,2} \end{pmatrix},$$

$$D_m(s,p) = s^2 + p_0 \cdot s + p_1, \ N_m^{i,j}(s,p) = p_{1+2\cdot(i-1)+j} \cdot s + a_{2\cdot(i-1)+j}^s \cdot p_1. \tag{5}$$

Now, to provide lower order model stability, we demand the inequality $p_0 > 0$ to be satisfied, which can be added to the main criteria in the form of a penalty function. Using the inverse Laplace transformations of (1) and (3) or (4) and (5) the model adequacy for the parameters p and for the input function $u(t)$ can be estimated by the criterion

$$C_{SISO}^{u(t)}(p) = \sum_{i=0}^{N} (\hat{x}_u(t_i) - x_u(t_i, p))^2 + c_p \cdot P(p_0) \to \min_{p \in R^3}, \tag{6}$$

where t_i are the uniformly distributed points on $[0, 10]$, $N = 100$, $\hat{x}_H(t)$ and $\hat{x}_u(t_i)$ are the initial model output on the $u(t)$, c_p is a penalty function coefficient and $P(p) = \begin{cases} \|p\|, \ p < 0 \\ 0, \ p \geq 0 \end{cases}$ is a penalty function. The criterion for MIMO system is a sum of the proposed criterion for each output.

One can estimate the reduced model adequacy for some identified parameters by using the integral square error

$$I_1 = \int_{0}^{+\infty} (x_H(t) - \hat{x}_H(t))^2 dt, \tag{7}$$

$$I_2 = \int\limits_{0}^{+\infty} \left(x_H(t) - \hat{x}_H(t)\right)^2 dt \Big/ \int\limits_{0}^{+\infty} \left(x_H(\infty) - \hat{x}_H(t)\right)^2 dt, \tag{8}$$

$$I_3 = \int\limits_{0}^{+\infty} (x_\delta(t) - \hat{x}_\delta(t))^2 dt \Big/ \int\limits_{0}^{+\infty} x_\delta(t)^2 dt. \tag{9}$$

Criteria (7)–(9) were used to compare the results and to estimate the performance of the algorithm. The scheme of fitness function evaluation is given in the study [6].

4 Experimental Results

Since we propose a multi-objective optimization problem and using the proposed approach results in receiving an estimation of the Pareto set and not a single solution, for each problem we would present the best solution by the first criterion and the best by the second criterion. As in similar investigations, we make the limit for the maximum number of fitness function evaluations equal to 2500 and 25 launches. Migration size is equal to 5 and migration interval is 5 generations.

The first problem we consider is the SISO system

$$G(s) = \frac{s^3 + 7 \cdot s^2 + 24 \cdot s + 24}{s^4 + 10 \cdot s^3 + 35 \cdot s^2 + 50 \cdot s + 24}, \tag{11}$$

for which we received models $G^*_{H(t)}(s) = \frac{0.765927 \cdot s + 1.679831}{s^2 + 2.582943 \cdot s + 1.679831}$ and $G^*_{\delta(t)}(s) = \frac{0.858125 \cdot s + 0.866629}{s^2 + 1.776652 \cdot s + 0.866629}$. The numeric adequacy estimation is given in Table 1, where the results of the proposed approach are compared with results received in different studies using other approaches and optimization tools.

Table 1. Experimental results. SISO problem

Approach	Results		
	I_1	I_2	I_3
Proposed + MOGA, $G^*_{H(t)}$	$7.442 \cdot 10^{-5}$	$1.305 \cdot 10^{-4}$	$6.698 \cdot 10^{-3}$
Proposed + MOGA, $G^*_{\delta(t)}$	$3.054 \cdot 10^{-4}$	$5.356 \cdot 10^{-4}$	$6.084 \cdot 10^{-3}$
Proposed + COBRA, [6]	$7.458 \cdot 10^{-5}$	$1.308 \cdot 10^{-4}$	$6.901 \cdot 10^{-3}$
Stability + Big Bang Big Crunch, [2]	$2.841 \cdot 10^{-4}$	$4.982 \cdot 10^{-4}$	$5.236 \cdot 10^{-3}$
Stability + GA, [5]	$2.394 \cdot 10^{-4}$	$4.197 \cdot 10^{-4}$	0.018
Stability + CSO, [3]	$1.986 \cdot 10^{-3}$	$3.483 \cdot 10^{-3}$	$7.612 \cdot 10^{-3}$

A similar problem was solved for the MIMO system problem

$$H(s) = \begin{pmatrix} \frac{2\cdot(s+5)}{(s+1)\cdot(s+10)} & \frac{s+4}{(s+2)\cdot(s+5)} \\ \frac{s+10}{(s+1)\cdot(s+20)} & \frac{s+6}{(s+2)\cdot(s+3)} \end{pmatrix}. \tag{12}$$

for the same computational resources and algorithm runs we received a set of models and the two with the highest criteria values are given in Table 2.

Table 2. Solutions found. MIMO problem

Models	
$G^*_{H(t)}(s)$	$G^*_{\delta(t)}(s)$
$D^*_m(s) = s^2 + 4.254626 \cdot s + 3.410166$	$D^*_m(s) = s^2 + 4.975596 \cdot s + 4.326484$
$N^{*1,1}_m(s) = 1.323005 \cdot s + 3.410166,$	$N^{*1,1}_m(s) = 1.737371 \cdot s + 4.326484,$
$N^{*1,2}_m(s) = 1.074861 \cdot s + 1.364067,$	$N^{*1,2}_m(s) = 1.083585 \cdot s + 1.730593,$
$N^{*2,1}_m(s) = 0.562128 \cdot s + 1.705083,$	$N^{*2,1}_m(s) = 0.850862 \cdot s + 2.163242,$
$N^{*2,2}_m(s) = 1.710807 \cdot s + 3.410166,$	$N^{*2,2}_m(s) = 1.147347 \cdot s + 4.326484,$

Similar experimental results are compared in Table 3, where the criteria are summarized by all the model components.

Table 3. Experimental results. MIMO problem

Approaches	Results		
	$\sum I_1$	$\sum I_2$	$\sum I_2$
Proposed + MOGA, $G^*_{H(t)}$	$7.167 \cdot 10^{-3}$	0.023	0.132
Proposed + MOGA, $G^*_{\delta(t)}$	0.026	0.096	0.105
Proposed + COBRA, [6]	$3.323 \cdot 10^{-3}$	0.027	0.136
Stability + Big Bang Big Crunch, [2]	0.02	0.325	0.218
Stability + CSO, [3]	0.045	0.372	0.409

The approximations of the Pareto, which were made on every algorithm launch and randomly chosen single Pareto front estimation, are given in Fig. 1. As one can see, there is no one solution that would bring the minimum to the both of these criteria at the same time.

To summarize, all the figures and examination results prove that the proposed approach and optimization algorithm is a reliable technique for solving order reduction problems.

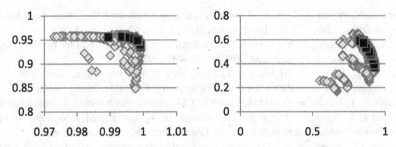

Fig. 1. The Pareto front estimations (light grey) and estimation made in a single run (black) for SISO order reduction problem (left) and MIMOorder reduction problem (right).

5 Conclusion

It is widely known that solving the order reduction problem for LTI systems requires a powerful and reliable global optimization tool. Many researchers are using heuristic optimization techniques, which allow them to achieve satisfactory results. However, for some problems there is an aim not just to identify the parameters by some criterion, but identify the parameters, which would fit two or more criteria. In this case, as was shown in this study, a MOGA can be used.

If it is required to solve the multi-objective problem, there is a necessity in using a multi-objective optimization algorithm, because the Pareto front is not just a single point in a vector space. Figures prove this hypothesis, since it can be seen that the Pareto front is a curve. By the single run results marked on these figures in black, we can see that the algorithm in the run gives an acceptable approximation of the Pareto front. This is one more study that proves the efficiency and high performance of the proposed MOGA optimization algorithm, based on the islands model and algorithm cooperation. The results of this work demonstrate that this algorithm is good not only for Pareto front estimation, but can also provide a good solution, close to the best solutions found by the single criterion optimization tools using the same resources.

Further work would be focused on improving the quality of estimation of the Pareto front in the case of more criteria numbers and developing meta-heuristics to improve the proposed MOGA algorithm performance. The other aspect of further work is related tousing the modified MOGA optimization tool to solve MIMO order reduction problems in which each output is characterized by two criteria.

Acknowledgements. This research is supported by the Russian Foundation for Basic Research within project No 16-01-00767.

References

1. Ramesh, K., Nirmalkumar, A., Gurusamy, G.: Order reduction of LTIV continuous MIMO system using stability preserving approximation method. Int. J. Comput. Appl. **36**, 1–8 (2011)
2. Desai, S., Prasad, R.: A novel order diminution of LTI systems using Big Bang Big Crunch optimization and Routh approximation. Appl. Math. Model. **37**, 8016–8028 (2013)

3. Narwal, A., Prasad, B.R.: A novel order reduction approach for LTI Systems using Cuckoo Search optimization and stability equation. IETE J. Res. **62**(2), 154–163 (2016)
4. Chen, T.C., Chang, C.Y., Han, K.W.: Reduction of transfer functions by the stability equation method. J. Franklin Inst. **308**(4), 389–404 (1979)
5. Parmar, G., Prasad, R., Mukherjee, S.: Order reduction of linear dynamic systems using stability equation method and GA. Int. J. comput. Inf. Eng. **1**(1), 26–32 (2007)
6. Ryzhikov, I., Semenkin, E., Akhmedova, Sh.: LTI system order reduction approach based on asymptotical equivalence and the co-operation of biology-related algorithms. In: IOP Conference Series: Materials Science and Engineering, vol. 173, Conference 1
7. Brester, C., Semenkin, E.: Cooperative multi-objective genetic algorithm with parallel implementation. In: Tan, Y., Shi, Y., Buarque, F., Gelbukh, A., Das, S., Engelbrecht, A. (eds.) ICSI 2015. LNCS, vol. 9140, pp. 471–478. Springer, Cham (2015). doi:10.1007/978-3-319-20466-6_49
8. Deb, K., Pratap, A., Agarwal, S., Meyarivan, T.: A fast and elitist multiobjective genetic algorithm: NSGA-II. IEEE Trans. Evol. Comput. **6**(2), 182–197 (2002)
9. Wang, R.: Preference-Inspired Co-evolutionary Algorithms. A thesis submitted in partial fulfillment for the degree of the Doctor of Philosophy. p. 231, University of Sheffield (2013)
10. Zitzler, E., Laumanns, M., Thiele, L.: SPEA2: Improving the strength pareto evolutionary algorithm for multiobjective optimization. In: Evolutionary Methods for Design Optimisation and Control with Application to Industrial Problems EUROGEN 2001 vol.3242, No. 103, pp. 95–100 (2002)
11. Whitley, D., Rana, S., Heckendorn, R.B.: Island model genetic algorithms and linearly separable problems. In: Corne, D., Shapiro, J.L. (eds.) AISB EC 1997. LNCS, vol. 1305, pp. 109–125. Springer, Heidelberg (1997). doi:10.1007/BFb0027170
12. Zhang, Q., Zhou, A., Zhao, S., Suganthan, P. N., Liu, W., Tiwari, S.: Multi-objective optimization test instances for the CEC 2009 special session and competition. University of Essex and Nanyang Technological University, Technical. report. CES-487 (2008)
13. Brester, C., Kauhanen, J., Tuomainen, T.P., Semenkin, E., Kolehmainen, M.: Comparison of two-criterion evolutionary filtering techniques in cardiovascular predictive modelling. In: Proceedings of the 13th International Conference on Informatics in Control, Automation and Robotics (ICINCO 2016), Lisbon, Portugal. vol. 1, pp. 140–145. (2016)
14. Brester, C., Semenkin, E., Sidorov, M., Kovalev, I., Zelenkov, P. Evolutionary feature selection for emotion recognition in multilingual speech analysis. In: Proceedings of IEEE Congress on Evolutionary Computation (CEC 2015), Sendai, Japan, pp. 2406–2411. (2015)
15. Brester, C., Semenkin, E., Sidorov, M., Semenkina, O. Multicriteria neural network design in the speech-based emotion recognition problem. In: Proceedings of the 12th International Conference on Informatics in Control, Automation and Robotics (ICINCO 2015), Colmar, France, vol. 1, pp. 621–628. (2015)

Solving Constrained Multi-objective Optimization Problems with Evolutionary Algorithms

Frikkie Snyman and Mardé Helbig[(✉)]

University of Pretoria, Pretoria, South Africa
u13028741@tuks.co.za, mhelbig@cs.up.ac.za

Abstract. Most optimization problems in real-life have multiple constraints. Constrained optimization problems with more than one objective, with at least two objectives in conflict with one another, are referred to as constrained multi-objective optimization problems (CMOPs). Two main approaches to solve constrained problems are to add a penalty to each objective function and then optimizing the new adapted objective function, or to adapt the Pareto-dominance principle that are used to compare two solutions in such a way that constraint violations are taken into consideration. This paper investigates how these two approaches affect the performance of the steady-state non-dominated sorting genetic algorithm II (SNSGAII), the Pareto-archived evolution strategy (PAES), the multi-objective evolutionary algorithm based on decomposition (MOEA/D) and a cultural algorithm (CA) when solving CMOPs. The results indicate that there is no statistical significant difference in performance between these two approaches. However, depending on the multi-objective evolutionary algorithm (MOEA) one approach does provide slightly better solutions than the other approach.

1 Introduction

Many optimization problems have more than one objective, with at least two objectives in conflict with one another. Therefore, improving on one objective leads to a weakening of another objective. This kind of problems are referred to as multi-objective optimization problems (MOPs). Evolutionary algorithms (EAs) are computational intelligence algorithms that are based on the natural model of evolution and are commonly used to solve MOPs [3]. Constrained multi-objective optimization problems CMOP refer to MOPs with inequality and/or equality constraints, in addition to boundary constraints of unconstrained MOPs.

Constrained MOPs are more difficult to solve, since the constraints limit the number of valid solutions in the search space, i.e. the feasible space. Two approaches that are frequently used to solve CMOPs with a (MOEA) are:

- Adding a penalty term (function) to each objective function and then optimizing the adapted objective functions [21]

© Springer International Publishing AG 2017
Y. Tan et al. (Eds.): ICSI 2017, Part II, LNCS 10386, pp. 57–66, 2017.
DOI: 10.1007/978-3-319-61833-3_7

– Adapting the Pareto-dominance [8,16] equations to take into account the
violation of constraints [5]

The problem with the penalty appraoch is that you are changing the objective function by adding an additional term, the penalty term, to the objective
function. Since the various objective functions of the CMOP can have different orders of magnitude, the order of magnitude of the penalty term should
also differ. However, the value selected for the penalty term will also influence
the quality of the solutions that are found, since the penalty term balances the
influence of constraint violation on the fitness of a specific solution. A too small
contribution by the penalty term to the fitness value may lead to a set of infeasible solutions being found. In contrast, a too high penalty term may lead to a
poor distribution of found solutions [5].

This paper investigates the influence of these two constraint-handling
approaches on the steady-state non-dominated sorting genetic algorithm II (SNS-
GAII), the Pareto-archived evolution strategy (PAES), the multi-objective evolutionary algorithm based on decomposition (MOEA/D) and the cultural algorithm (CA) when solving CMOPs with various characteristics.

The rest of the paper's layout is as follows: Sect. 2 provides an overview of
CMOP, the algorithms and the constraint-handling approaches. The experimental setup is discussed in Sect. 3. Section 4 presents the results of the study and
conclusions are highlighted in Sect. 5.

2   Background

This section discusses CMOP, as well as the algorithms and constraint-handling
approaches that were used in the study.

2.1   Constrained Multi-objective Optimization

A CMOP can be defined as follows:

$$
\begin{aligned}
minimize &: f_k(\mathbf{x}), \ k = 1, 2, \ldots, n_k \\
subject\ to &: g_m(\mathbf{x}) \geq 0, \ m = 1, \ldots, n_g \\
& h_m(\mathbf{x}) = 0, \ m = n_g + 1, \ldots, n_g + n_h \\
& x_i^L \leq x_i \geq x_i^U, \ i = 1, 2, \ldots, n_x
\end{aligned}
\tag{1}
$$

where f_k represents the k-th objective function, and g_m and h_m represent the m-th
inequality and equality constraint respectively. The i-th component of the upper
boundary value of $\mathbf{x}$ is indicated by $\mathbf{x}_i^U$ and the lower boundary value by $\mathbf{x}_i^L$.

Due to the inequality and equality constraints in Eq. 1 not all solutions in the
search space will be feasible. Various approaches exist to deal with CMOP [5,
11,18] and to guide the search of the algorithm towards feasible solutions. Two
approaches that are simple to implement are the following:

- Adding a penalty term to the objective function
- Modifying the Pareto-dominance equations to consider constraint violations

Assuming that you minimize all objective functions (a maximizing objective function can be managed by converting it to a minimization function using the duality principle), all constraints are first normalized before the constraint violation is calculated. Let the normalized constraint functions be represented by $g_m^*(\mathbf{x}_i) \geq 0$ for $m = 1, 2, \ldots, n_g$. Then for each solution $\mathbf{x}_i$ the constraint violation is calculated as follows [5]:

$$\omega_m(\mathbf{x}_i) = \begin{cases} |g^*(\mathbf{x}_i)|, & \text{if } g^*(\mathbf{x}_i) < 0 \\ 0, & \text{otherwise} \end{cases}$$

All constraint violations are then added together:

$$\Omega(\mathbf{x}_i) = \sum_{j=1}^{n_g} \omega_m(\mathbf{x}_i)$$

The combined constrained violation value (Ω) is multiplied with a penalty parameter (P_k) and this product is then added to each of the objective function values as follows:

$$f_k^*(\mathbf{x}_i) = f_k(\mathbf{x}_i) + P_k \Omega(\mathbf{x}_i) \tag{2}$$

For a feasible solution the Ω term in Eq. (2) will be zero and $f_k^* = f_k$.

The second approach is to keep the objective function (f_k) unchanged, but to modify the Pareto-dominance comparison of vectors. When solving a MOP a single fitness value does not exit. Therefore, a new approach is required to compare two solutions' quality. Typically, Pareto-dominance is used:

Definition 1. *Assuming minimization, a solution $\mathbf{x}_i$ is better in quality than another solution $\mathbf{x}_j$ if:*

$$f(x_i^a) < f(x_j^a) \quad \forall a = 1, \ldots, A \tag{3}$$
$$\exists\, x_i^b : f(x_i^b) \leq f(x_j^b) \quad \text{with } b \in [1, n_x] \tag{4}$$

If $\mathbf{x}_i$ is better than $\mathbf{x}_j$, $\mathbf{x}_i$ dominates $\mathbf{x}_j$, written as $\mathbf{x}_i \prec \mathbf{x}_j$. If not one of the solutions are better in quality than the other, the solutions are non-dominated. The set of non-dominated solutions in the objective space is referred to as the Pareto-optimal front (POF).

Definition 1 does not take constraint violations into consideration. Therefore, when comparing solutions of a CMOP Definition 1 has to be adapted as follows:

Definition 2. *Assuming minimization, a solution $\mathbf{x}_i$ constrain-dominates another solution $\mathbf{x}_j$ if:*

1. $\mathbf{x}_i$ *is feasible and* $\mathbf{x}_j$ *is infeasible*
2. $\mathbf{x}_i$ *and* $\mathbf{x}_j$ *are both infeasible, but* $\mathbf{x}_i$ *has a smaller constraint violation*
3. $\mathbf{x}_i$ *and* $\mathbf{x}_j$ *are both feasible and* $\mathbf{x}_i \prec \mathbf{x}_j$ *according to Definition 1.*

2.2   Algorithms

The non-dominated sorting genetic algorithm II (NSGA-II) is a multi-objective genetic algorithm (MOGA) modelled on genotypic behaviour and is still the standard benchmark algorithm in multi-objective optimization (MOO) studies [7,14]. The algorithm improves on the $O(MN^3)$ complexity of previous non-dominated sorting genetic algorithm (GA) [21], as well as adds elitism whilst removing the need for a sharing parameter [7].

PAES is a MOEA that uses an $(1+1)$-ES that uses only mutation on a single parent to create a single offspring. Therefore, PAES is a local search strategy [13]. It uses an archive to store the non-dominated solutions found so far and it maintains diversity through a crowding procedure that recursively divides the objective space into a grid.

MOEA/D is an EA based on the concept of decomposition [22]. It decomposes a MOP into a number of sub-problems and then optimizes the sub-problems simultaneously. Each sub-problem is optimized using only information from its neighbouring sub-problems. Each population contains the best solution for each of the sub-problems.

EAs mainly focus on genetic concepts and natural selection. However, CAs model cultural evolution and are based on theories in sociology and archeology [19]. Individuals are described through behavioural traits and can generate generalized descriptions of their experience. CA was extended to solve MOPs by Coello and Becerra [4].

3   Experimental Setup

This section discusses the experimental setup used for this study. The algorithms are discussed in Sect. 3.1. Sections 3.2 and 3.3 present the benchmarks and performance measures that were used to evaluate the performance of the algorithms on CMOPs. The statistical analyses conducted on the obtained results are discussed in Sect. 3.4.

3.1   Algorithms

The following algorithms were used in this study: SNSGA-II [7], PAES [13], MOEA/D [22] and CA [4]. For each of these algorithms two configurations were used in the study, namely one configuration that implemented the penalty function approach and one configuration that implemented the modified Pareto-dominance approach (refer to Sect. 2.1).

Each algorithm was run 30 times for a maximum of 25 000 fitness evaluations. Each algorithm had a population size of 100 and PAES used 3 bi-sections. The crossover probability was set to 0.75 and the mutation probability to 0.1, since these values promote a balance between exploration and exploitation of the search space [9]. Simulated binary crossover (SBX) [6] and polynomial mutation [5] were used to generate the offspring, where applicable. For MOEA/D, the neighbour size was set to 20, the neigbour selection probability to 0.9 and the maximum number of replaced solutions to 2.

3.2 Benchmark Functions

The following benchmark functions were used to evaluate the performance of the various algorithms:

- Binh2 with 2 decision variables and 2 inequality constraints [1], referred to as f_1
- Osyczka2 that has 6 decision variables and 6 inequality constraints [15], referred to as f_2
- Srinivas with 2 decision variables and 2 inequality constraints [21], referred to as f_3
- Two Bar Truss with 3 decision variables, 4 inequality constraints and a boundary constraint [12], referred to as f_4
- Welded Beam that has 4 decision variables and 4 constraints [17], referred to as f_5

3.3 Performance Measures

This section discusses the performance measures that were used to evaluate the performance of the MOEAs.

Inverted Generational Distance (IGD). Inverted generational distance (IGD) measures the distance between each solution in the optimal POF and the closest solution in the approximated POF [20]. A small value indicates that an algorithm performed well with regards to both convergence and diversity.

Hypervolume (HV). The hypervolume (HV) or S-metric measures how much of the objective space is dominated by a non-dominated set [23,24]. The reference vector or reference point that was used in the HV calculation was the vector that consisted of the worst value for each objective of the union of all non-dominated solutions of all approximated POFs that were compared against each other. A high HV value indicates good performance.

ϵ Metric (ϵ_m). Zitzler *et al.* presented the ϵ-metric (ϵ_m) to compare approximated sets [25]. It measures the factor by which an approximation set is worse than another approximation set with respect to all objectives, i.e. it provides the factor ϵ where for any solution in set B there is at least one solution in set A that is not worse by a factor of ϵ in all objectives. A small EP value indicates good performance.

3.4 Statistical Analysis

For each performance measure a Friedmann Test was performed to determine whether there was a statistical significant differences between the entire set of results. If the Friedmann Test indicated a statistical significant difference, pairwise Mann-Whitney U tests were conducted between the various algorithm configurations. If the Mann-Whitney U test indicated a statistical significant difference, the algorithm with the best average for the specific measure was awarded a win and the other algorithm a loss. All statistical tests were conducted with a confidence level of 95%.

4   Results

This section discusses the results that were obtained from the experiments.

4.1   Inverted Generational Distance

This section discusses the IGD values that were obtained by the various algorithms. The IGD mean and standard deviation values are presented in Table 1. In Table 1 D refers to the modified Pareto-dominance approach and P refers to the penalty function approach.

Table 1. IGD mean and standard deviation values

	SSNSGA-II(D)	SSNSGA-II(P)	PAES(D)	PAES(P)
f_1	$1.17E-04\pm2.17E-06$	$1.18E-04\pm1.68E-06$	$1.19E-02\pm1.86E-03$	$1.16E-02\pm1.93E-03$
f_2	$4.09E-03\pm1.19E-04$	$4.06E-03\pm1.53E-04$	$2.63E-02\pm1.91E-02$	$3.08E-02\pm1.67E-02$
f_3	$8.14E-04\pm3.55E-06$	$7.02E-04\pm3.74E-06$	$1.31E-02\pm1.01E-02$	$1.77E-02\pm8.08E-03$
f_4	$7.05E-04\pm6.20E-06$	$5.98E-04\pm1.28E-05$	$\mathbf{1.48E-02\pm7.02E-03}$	$1.65E-02\pm7.02E-03$
f_5	$2.05E-03\pm1.06E-06$	$2.03E-03\pm3.47E-05$	$2.23E-02\pm9.88E-03$	$1.85E-02\pm1.06E-02$
	MOEA/D(D)	MOAE/D(P)	CA(D)	CA(P)
f_1	$5.08E-04\pm3.72E-06$	$5.18E-04\pm7.91E-06$	$\mathbf{1.07E-02\pm1.37E-03}$	$1.19E-02\pm1.79E-03$
f_2	$4.84E-03\pm3.41E-04$	$\mathbf{4.48E-03\pm2.55E-04}$	$5.40E-02\pm2.79E-02$	$4.95E-02\pm3.14E-02$
f_3	$8.19E-04\pm8.87E-07$	$7.20E-04\pm1.85E-06$	$1.48E-02\pm8.48E-03$	$2.00E-02\pm7.86E-03$
f_4	$1.02E-02\pm1.72E-04$	$\mathbf{1.01E-02\pm1.62E-04}$	$1.28E-02\pm3.56E-03$	$1.32E-02\pm2.60E-03$
f_5	$1.27E-02\pm4.07E-04$	$\mathbf{1.10E-02\pm2.93E-04}$	$\mathbf{7.66E-02\pm8.49E-02}$	$2.65E-01\pm4.79E-01$

From Table 1 it can be seen that for SNSGA-II both constraint-dealing approaches performed equally well. Not one of the approaches obtained a better average and standard deviation value for any of the functions. For MOEA/D the penalty function approach obtained the best IGD average and standard deviation values for 3 of the functions (indicated in bold) and the modified Pareto-dominance approach for none of the functions. In contrast, for the CA and PAES, the modified Pareto-dominance approach obtained the best IGD average and standard deviation values for 2 and 1 of the 5 functions respectively, and the penalty function approach for none of the functions.

Applying a Friedman Test on the means of the IGD values revealed that the null hypothesis had to be rejected, and hence, there was a statistical significant difference in the performance of the algorithms. The results of the Mann-Whitney U tests are presented in Table 2. In Table 2, $<$ indicates that the left hand algorithm performed statistical significantly better than the right hand algorithm, $>$ indicates that the left hand algorithm performed statistical significantly worse than the right hand algorithm, and $=$ indicates equivalent performance. The

Table 2. Comparison of algorithms' performance using Mann-Whitney U tests

	SSNSGA-II(D)	SSNSGA-II(P)	PAES(D)	PAES(P)	MOEA/D(D)	MOAE/D(P)	CA(D)	CA(P)
SSNSGA-II(D)	N/A	=	<	<	=	=	<	<
SSNSGA-II(P)	=	N/A	<	<	=	=	<	<
PAES(D)	>	>	N/A	=	>	>	=	=
PAES(P)	>	>	=	N/A	=	>	=	=
MOEA/D(D)	=	=	<	=	N/A	=	<	<
MOAE/D(P)	=	=	<	<	=	N/A	<	<
CA(D)	>	>	=	=	>	>	N/A	=
CA(P)	>	>	=	=	>	>	=	N/A

results indicate that for each of the MOEAs there was not a statistical significant difference in performance between the configurations that incorporated
the two constraint-dealing approaches. However, there was a statistical significant difference in performance when the various MOEAs were compared against
one another. Both SNSGA-II configurations outperformed both configurations
of PAES and CA. MOEA/D using the penalty function also outperformed both
configurations of PAES and CA. The modified Pareto-dominance MOEA/D outperformed both configurations of CA and only the modified Pareto-dominance
approach of PAES. The modified Pareto-dominance MOEA/D experienced no
statistical significant difference in performance with the penalty function configuration of PAES. There was also no statistical significant difference in performance
between the SNSGA-II and MOEA/D configurations.

The percentage of wins obtained by each of these two constraint-dealing
approaches for IGD is presented in Table 3. The results indicate that the penalty
function approaches marginally outperformed the modified Pareto-dominance
approaches.

Table 3. Wins obtained by the two constraint-dealing approaches

	D	P
Win percentage	12.5%	19.64%

4.2 Hypervolume and ϵ-metric

Friedman Tests on the mean HV values and the mean ϵ_m values indicated that
there was no statistical significant difference in performance of the various algorithms. The HV and ϵ_m values are presented in Tables 4 and 5.

The penalty-based SNSGA-II outperformed the modified Pareto-dominace
SNSGA-II by obtaining the better HV mean and deviation for 2 functions.

Table 4. HV Mean and standard deviation values

	SSNSGA-II(D)	SSNSGA-II(P)	PAES(D)	PAES(P)
f_1	**8.12E−01 ± 2.76E−05**	8.12E−01 ± 4.38E−05	4.52E−01 ± 7.61E−02	4.64E−01 ± 9.86E−02
f_2	1.19E−01 ± 1.20E−03	1.20E−01 ± 1.75E−03	1.20E−01 ± 8.57E−02	**1.63E−01 ± 7.88E−02**
f_3	3.92E−01 ± 1.93E−04	**4.90E−01 ± 1.01E−04**	2.10E−01 ± 1.13E−01	1.46E−01 ± 1.13E−01
f_4	9.19E−01 ± 4.32E−04	9.09E−01 ± 2.42E−04	4.49E−01 ± 2.12E−01	4.98E−01 ± 2.41E−01
f_5	9.46E−01 ± 1.74E−03	**9.56E−01 ± 1.50E−03**	7.90E−01 ± 2.84E−01	6.57E−01 ± 2.92E−01
	MOEA/D(D)	MOAE/D(P)	CA(D)	CA(P)
f_1	**8.08E−01 ± 2.70E−05**	8.08E−01 ± 6.32E−05	**4.53E−01 ± 7.85E−02**	4.09E−01 ± 8.72E−02
f_2	1.00E−01 ± 6.96E−03	9.76E−02 ± 6.60E−03	1.49E−02 ± 3.39E−02	**1.87E−02 ± 3.22E−02**
f_3	3.91E−01 ± 1.09E−04	4.87E−01 ± 1.82E−04	**1.28E−01 ± 8.13E−02**	9.84E−02 ± 1.08E−01
f_4	1.00E−01 ± 1.12E−02	**1.08E−01 ± 1.10E−02**	2.35E−01 ± 1.90E−01	4.41E−01 ± 1.99E−01
f_5	9.48E−01 ± 7.17E−06	**9.50E−01 ± 6.61E−04**	**2.49E−01 ± 2.86E−01**	1.63E−01 ± 1.85E−01

Table 5. ϵ_m mean and standard deviation values

	SSNSGA-II(D)	SSNSGA-II(P)	PAES(D)	PAES(P)
f_1	**5.67E−03 ± 4.22E−04**	5.90E−03 ± 9.40E−04	4.68E−01 ± 1.23E−01	4.43E−01 ± 1.38E−01
f_2	**8.62E−01 ± 2.99E−03**	8.62E−01 ± 3.99E−03	8.08E−01 ± 1.59E−01	**7.60E−01 ± 1.19E−01**
f_3	5.36E−01 ± 2.94E−04	**3.69E−01 ± 1.96E−04**	6.94E−01 ± 1.99E−01	8.08E−01 ± 1.81E−01
f_4	7.20E−02 ± 4.66E−04	**6.18E−02 ± 3.22E−04**	5.47E−01 ± 2.15E−01	4.73E−01 ± 2.66E−01
f_5	5.43E−02 ± 1.74E−03	**4.16E−02 ± 1.55E−03**	2.10E−01 ± 2.84E−01	3.39E−01 ± 2.95E−01
	MOEA/D(D)	MOAE/D(P)	CA(D)	CA(P)
f_1	2.12E−02 ± 1.27E−04	**2.12E−02 ± 1.15E−04**	4.14E−01 ± 9.56E−02	5.07E−01 ± 1.25E−01
f_2	8.53E−01 ± 9.22E−12	8.54E−01 ± 3.74E−12	**1.20E+00 ± 3.35E−01**	1.21E+00 ± 5.17E−01
f_3	5.35E−01 ± 1.80E−14	3.69E−01 ± 7.32E−12	**8.10E−01 ± 1.57E−01**	9.91E−01 ± 2.84E−01
f_4	9.00E−01 ± 1.12E−02	**8.92E−01 ± 1.10E−02**	8.18E−01 ± 3.09E−01	**5.39E−01 ± 2.60E−01**
f_5	5.16E−02 ± 7.39E−06	4.07E−02 ± 1.27E−03	**1.97E+00 ± 1.98E+00**	6.01E+00 ± 1.04E+01

A similar trend was observed for MOEA/D. For the CA, the modified Pareto-dominance approach outperformed the penalty-based approach on 3 of the 5 functions. However, both approaches performed equally well for PAES. A similar trend was observed for ϵ_m.

5   Conclusion

This study investigated the effect of two constraint-dealing approaches on the performance of the steady-state non-dominated sorting genetic algorithm II (SNSGA-II), the Pareto-archived evolution strategy (PAES), the multi-objective evolutionary algorithm based on decomposition (MOEA/D) and a cultural algorithm (CA). The two constraint-dealing approaches are: adding a penalty term to each objective function; and using a modified Pareto-dominance approach that incorporates constraint violations. Each multi-objective evolutionary algorithm (MOEA) had two configurations, where each configuration incorporated one of these approaches to deal with constraints.

The results indicated that there was no statistical significant difference between the two constraint-dealing approaches. However, the penalty function approach did slightly outperform the modified Pareto-dominance approach based on the wins and losses. Furthermore, for the CA the modified Pareto-dominance approach performed better on all performance measures. In contrast, the penalty-based MOEA/D outperformed the modified Pareto-dominance version on all performance measures. For PAES, both approaches performed equally well, except on the inverted generational distance (IGD) measure where the modified Pareto-dominance approach slightly outperformed the penalty function approach. The penalty function SNSGA-II slightly outperformed the modified Pareto-dominance SNSGA-II on the hypervolume (HV) and ϵ_m. However, both approaches performed equally well on IGD.

A penalty function modifies the objective space and can lead to changes in the optima. The modified Pareto-dominance approach does not change the objective functions. Since there was not a huge statistical significant difference in performance between these two approaches, it is simpler to rather use the modified Pareto-dominance approach than to add a penalty function where the ideal penalty parameter values, that are problem dependent, should be found.

Future work will include extending the study to incorporate additional multi-objective optimization constrained problems [2], as well as adaptive penalty approaches [10].

References

1. Binh, K.: MOBES: A multi-objective evolution strategy for constrained optimization problems. In: Proceedings of the International Conference on Genetic Algorithm, Brno, Czech Republic, pp. 176–182 (1997)
2. Cheng, P.: A tunable constrained test problems generator for multi-objective optimization. In: Proceedings of the International Conference on Genetic and Evolutionary Computing, Washington, DC, USA, pp. 96–100, September 2008
3. Coello Coello, C.A.: Evolutionary multi-objective optimization: a historical view of the field. IEEE Comput. Intell. Mag. 1(1), 28–36 (2006)
4. Coello Coello, C.A., Becerra, R.L.: Evolutionary multiobjective optimization using a cultural algorithm. In: Proceedings of the IEEE Swarm Intelligence Symposium, pp. 6–13 (2003)
5. Deb, K.: Multi-Objective Optimization Using Evolutionary Algorithms, vol. 16. Wiley, New York (2001)
6. Deb, K., Beyer, H.G.: Self-adaptive genetic algorithms with simulated binary crossover. Technical report CI-61/99, University of Dortmund (1999)
7. Deb, K., Pratap, A., Agarwal, S., Meyarivan, T.: A fast and elitist multiobjective genetic algorithm: NSGA-II. IEEE Trans. Evol. Comput. 6(2), 182–197 (2002)
8. Edgeworth, F.: Mathematical Physics. P Keagan, London (1881)
9. Eiben, A.E., Hinterding, R., Michalewicz, Z.: Parameter control in evolutionary algorithms. IEEE Trans. Evol. Comput. 3(2), 124–141 (1999)
10. Harada, K., Sakuma, J., Ono, I., Kobayashi, S.: Constraint-handling method for multi-objective function optimization: pareto descent repair operator. In: Obayashi, S., Deb, K., Poloni, C., Hiroyasu, T., Murata, T. (eds.) EMO 2007. LNCS, vol. 4403, pp. 156–170. Springer, Heidelberg (2007). doi:10.1007/978-3-540-70928-2_15

11. Jiménez, F., JM Cadenas, J.: Evolutionary techniques for constrained multiobjective optimization problems. In: Proceedings of the Workshop on Multi-Criterion Optimization Using Evolutionary Methods: Genetic and Evolutionary Computation Conference, pp. 115–116 (1999)

12. Kirsch, U.: Optimal Structural Design. McGraw-Hill, New York (1981)

13. Knowles, J., Corne, D.: The pareto archived evolution strategy: a new baseline algorithm for pareto multiobjective optimisation. In: Proceedings of the IEEE Congress on Evolutionary Computation, vol. 1 (1999)

14. Li, H., Zhang, Q.: Multiobjective optimization problems with complicated pareto sets, MOEA/D and NSGA-II. IEEE Trans. Evol. Comput. **13**(2), 284–302 (2009)

15. Osyczka, A., Kundu, S.: A new method to solve generalized multicriteria optimization problems using the simple genetic algorithm. Struct. Optim. **10**, 94–99 (1995)

16. Pareto, V.: Cours D'Economie Politique. F. Rouge, Lausanne (1896)

17. Ragsdell, K., Phullips, D.: Optimal design of a class of welded structures using geometric programming. J. Eng. Ind. Ser. B **98**, 1021–1025 (1975)

18. Ray, T.: An evolutionary algorithm for multiobjective optimization. Eng. Optim. **33**(3), 399–424 (2001)

19. Reynolds, R.: An introduction to cultural algorithms. In: Proceedings of the Annual Conference on Evolutionary Programming, pp. 131–139 (1994)

20. Sierra, M., Coello Coello, C.: Improving PSO-Based multi-objective optimization using crowding, mutation and ϵ-dominance. In: Coello Coello, C.A., Hernández Aguirre, A., Zitzler, E. (eds.) EMO 2005. LNCS, vol. 3410, pp. 505–519. Springer, Heidelberg (2005)

21. Srinivas, N., Deb, K.: Multi-objective function optimization using non-dominated sorting genetic algorithms. Evol. Comput. J. **2**(3), 221–248 (1994)

22. Zhang, Q., Li, H.: MOEA/D: a multiobjective evolutionary algorithm based on decomposition. IEEE Trans. Evol. Comput. **11**(6), 712–731 (2007)

23. Zitzler, E., Thiele, L.: Multiobjective evolutionary algorithms: a comparative case study and the strength pareto approach. IEEE Trans. Evol. Comput. **3**(4), 257–271 (1999)

24. Zitzler, E., Thiele, L.: Multiobjective optimization using evolutionary algorithms — a comparative case study. In: Eiben, A.E., Bäck, T., Schoenauer, M., Schwefel, H.-P. (eds.) PPSN 1998. LNCS, vol. 1498, pp. 292–301. Springer, Heidelberg (1998). doi:10.1007/BFb0056872

25. Zitzler, E., Thiele, L., Laumanns, M., Fonseca, C.M., da Fonseca, V.G.: Performance assessment of multiobjective optimizers: an analysis and review. IEEE Trans. Evol. Comput. **7**(2), 117–132 (2003)

Portfolio Optimization

Multi-objective Comprehensive Learning Bacterial Foraging Optimization for Portfolio Problem

Ben Niu[1,2(✉)], Wenjie Yi[1], Lijing Tan[3(✉)], Jia Liu[1], Ya Li[4], and Hong Wang[1,5(✉)]

[1] College of Management, Shenzhen University,
Shenzhen 518060, China
drniuben@gmail.com, ms.hongwang@gmail.com
[2] School of Computing, Information, and Decision Systems Engineering,
Arizona State University, Tempe, USA
[3] Department of Business Management,
Shenzhen Institute of Information Technology, Shenzhen 518172, China
Mstlj@163.com
[4] School of Computer and Information Science,
Southwest University, Chongqing 400715, China
[5] Department of Mechanical Engineering,
The Hong Kong Polytechnic University, Kowloon, Hong Kong

Abstract. Multi-objective portfolio optimization (PO) problem is always converted into a single objective problem by using the weighted method, which is sensitive to the pareto optimal front and requires that decision makers must have previous experience about the preference for weights. Based on multi-objective comprehensive learning bacterial foraging optimization (MOCLBFO), this paper proposes an algorithm which is specially designed for multi-objective PO problem. The corresponding coding strategy which considers each particle as a feasible solution is also given. In order to test the validity of the algorithm, multi-objective comprehensive learning particle swarm optimization (MOCLPSO) is chosen as the competing algorithm. Comparative experimental tests on ten assets PO problem demonstrate that MOCLBFO is able to find a more well-distributed Pareto set.

Keywords: Multi-objective problem · Comprehensive learning strategy · Bacterial foraging optimization · Portfolio optimization · Pareto solutions

1 Introduction

Portfolio optimization (PO) is the process of arranging the proportion of investment in different assets with the minimum risk to obtain the maximum profit. Many scholars have modified the original portfolio model proposed by Markowiz [1] by adding more constraints, such as no short sales [2, 3], the transaction costs [3, 4]. As the number of constraints increase, the difficulty to deal with such a multi-objective PO model is also increasing. As a result, traditional mathematical methods cannot solve it well. Therefore,

© Springer International Publishing AG 2017
Y. Tan et al. (Eds.): ICSI 2017, Part II, LNCS 10386, pp. 69–76, 2017.
DOI: 10.1007/978-3-319-61833-3_8

scholars begin to use swarm intelligent algorithms, including particle swarm optimization (PSO) [5, 6], bacterial foraging optimization (BFO) and its variants [3], and others.

Our proposed multi-objective comprehensive learning bacterial foraging optimization (MOCLBFO) [7] has been successfully applied to environmental/economic dispatch problem (EED) [7]. Based on MOCLBFO, this paper proposes a specific method for solving multi-objective PO model with two conflicting objectives and two constraints [8].

The rest of the article is organized as follows: Sect. 2 provides a brief description of BFO and MOCLBFO. Section 3 presents the multi-objective PO model and the related computational steps. The results and analyses of the experiment are shown and discussed in Sect. 4. Finally, the concluding remarks are provided in Sect. 5.

2 Multi-objective Comprehensive Learning Bacterial Foraging Optimization

2.1 Bacterial Foraging Optimization

Original BFO tackles problem by using three operators, including chemotaxis, reproduction, and elimination/dispersal. For more detailed information, please refer to [9].

2.2 Multi-objective Comprehensive Learning Bacterial Foraging Optimization

Although BFO has been applied to many single objective optimization problems successfully, it cannot directly tackle multi-objective optimization problems (MOPs) with non-dominated solutions set instead of an absolute global best solution. Inspired by the comprehensive learning strategy used in MOCLPSO [10], it also was incorporated into MOBFO [11] and thus MOCLBFO [7] is put forward.

3 MOCLBFO for Portfolio Optimization Problem

3.1 Portfolio Optimization Model

Markowiz [1] proposed mean-variance model with many assumptions. These strict assumptions made this PO model can't get a good application in practical PO problems. In order to solve these problems, Li L.et al. [8] proposed an improved model, which has considered the transaction cost and no short selling and other factors. The expression of the new PO model is presented in Eq. (1–4) and the related parameters and definitions are shown in Table 1.

$$Return \ f(x) = \sum_{i=1}^{n} r_i x_i - \sum_{i=1}^{n} \left[\mu * k_i^b * \left(x_i - x_i^0 \right) + (1 - \mu) * k_i^s * \left(x_i^0 - x_i \right) \right]$$

$$and \ \mu = \begin{cases} 1, \ldots, x_i \geq x_i^0 \\ 0, \ldots, x_i \leq x_i^0 \end{cases} \tag{1}$$

$$Risk \ g(x) = \sum_{i=1}^{n} \sum_{j=1}^{n} \sigma_{ij} x_i x_j \tag{2}$$

Subject to:

$$0 \leq x_i \leq 1, \forall i = 1, 2, \ldots 10 \tag{3}$$

$$\sum_{i=1}^{n} x_i = 1 \tag{4}$$

Table 1. Parameters and definitions of PO model

Variables	Definitions
$f(x)$	The first objective function which represents profit and pursuits the maximum value
$g(x)$	The second objective function which represents risk and pursuits the minimum value
n	The number of assets; $n = 10$
r_i	The expected yields of asset i; $i = 1, 2, 3 \ldots 10$
x_i	The proportion of investment of asset i; $i = 1, 2, 3 \ldots 10$
x_i^0	The initial holding proportion of investment of asset i; $x_i^0 = 0, \forall i = 1, 2, 3 \ldots 10$
μ	$\mu=1$: buy assets from market $\mu=0$:sell assets to market
k_i^b	The transation cost of buying asset i from market; $k_i^b = 0.00065, \forall i = 1, 2, 3 \ldots 10$
k_i^s	The transation cost of selling asset i to market; $k_i^s = 0.00075, \forall i = 1, 2, 3 \ldots 10$
σ_{ij}	The covariance of r_i and r_j; $i, j = 1, 2, 3 \ldots 10$

3.2 MOCLBFO for Portfolio Optimization Model

3.2.1 Encoding

When MOCLBFO is used to seek solutions to the PO model, each particle is regarded as a potential feasible solution. Three kinds of information are carried by each bacterium, including the proportion of investment, the value of corresponding profit and risk [3].

$$\theta = [x_1, x_2, x_3 \ldots x_n, f(x), g(x)] \tag{5}$$

$$s = \sum_{i-1}^{n} x_i$$

$$\theta' = \left[\frac{1}{s} (x_1, x_2, x_3 \ldots x_n), \quad f(x)', g(x)' \right] \tag{6}$$

Equation (5) shows the coding of the bacteria. As is shown in Eq. (6), we sum up proportions of all asset firstly and then divide every proportion of asset by the s so that the sum of all asset proportions is equal to 1. And the penalty function method is used to guarantee that each proportion is positive number.

3.2.2 Four Key Mechanisms of MOCLBFO

Before describing computational steps of solving PO model, brief introduction about four key mechanisms of MOCLBFO [7] is given as follows.

3.2.2.1 Health Evaluation

Every bacterium's capacity to look for nutrients is different. The health index is calculated according to Eq. (7). The greater health index means that the bacterium is healthier, so greater probability would be given for this individual to reproduce. And the unhealthy bacteria would be given less probability to generate new offspring.

$$J_t^i(health) = 1 / \sum_{j=1}^{Nc} J_t(i,j,k,l) \tag{7}$$

3.2.2.2 Non-dominance Choice

In MOCLBFO, the pareto solutions which are obtained in the optimization process would be stored to external archive, but the size of the external archive is limited. Therefore, the maintenance and management of the external archive is very important. The process of non-dominance choice is the same as the mechanism in [7].

3.2.2.3 Comprehensive Learning Mechanism

In this algorithm, every individual has capacity to learn from other bacterium or the bacterial group by dimensions and the pareto solutions are stored in the external archive. As Eq. (8) shows, the computational formula of the moving direction of ith bacterium at d^{th} dimension consists of four parts, including the original direction, random direction, the best previous position, and the position of random bacterium in the external archive.

$$\theta_d^i(j+1,k,l) = \theta_d^i(j,k,l) + c(i)\frac{\Delta(i)}{\sqrt{\Delta^T(i)\Delta(i)}} + \lambda * r_1 * (pbest_{id} - \theta_d^i(j,k,l))$$
$$+ (1-\lambda) * r_2 * (rep_{id} - \theta_d^i(j,k,l)) \tag{8}$$

$$pbest_{id} = \theta * pbest_{compet} + (1-\theta) * pbest_{id} \tag{9}$$

$$pbest_{compet} = b_i * pbest_n + (1-b_i) * pbest_m \quad n,m \in \{1,2...S\} \tag{10}$$

$$\lambda = ceil(rand - 1 + p_c)c \tag{11}$$

Among the equations above, θ, $b_i, \lambda \in \{0, 1\}$ $rand \in [0, 1]$, and r_1, r_2 are both known numbers. S denotes the size of the bacteria and n, m mean the random individuals in the group. P_c denotes the probability of learning.

3.2.2.4 Constrained Boundary Control

Boundary control affects the validity of the border solutions and the diversity of bacteria species. Bacteria change direction randomly in the process of chemotaxis, which may let them exceed the prescribed area. If bacteria are allowed to leave the given area, the border solutions may not been obtained. Otherwise, the border solutions may be infeasible. The boundary control rules are shown in Eq. (12), D_{min} and D_{max} are the lower and upper boundaries. Besides, P_r is a fixed value given beforehand.

$$\theta^i(j+1, k, l) = \begin{cases} D_{min} + rand * (D_{max} - D_{min}), \text{ if } P > P_r \\ \theta^i(j, k, l) + c(i) \dfrac{\Delta(i)}{\sqrt{\Delta^T(i)\Delta(i)}}, \text{ if } P \le P_r \end{cases} \tag{12}$$

3.2.3 Computational Steps of MOCLBFO Algorithm for PO Model

Based on the mechanisms mentioned above, an algorithm which is designed for solving PO model is proposed and the experiment is performed in MATLAB environment. MOCLPSO is chosen as the competing algorithm. The two algorithms have the same experimental settings, e.g., the population sizes are both set to 200. The pseudo-code of MOCLBFO for PO model is shown in Table 2.

Table 2. The pseudo-code of MOCLBFO

Begin
Initialize parameters and the location of bacteria ($p = 10; S = 200; N_c = 200; \ N_s = 4;$
$N_{re} = 4; N_{ed} = 2; P_{ed} = 0.25; P_r = 0.4; P_c = 0.1; C(i); \theta^i(j, k, l); rep\_size = 100)$
For ($l = 1: Ned$):
For ($k = 1: Nre$):
For ($j = 1: Nc$):
Do chemotaxis steps using Eq.(8);
Update the external archives using non-dominance mechanism in [7];
Update position and fitness;
Do boundary control according to the rules in Eq.(12);
End
Do reproduction steps based on health evaluation mechanism (Section 3.2.2.1);
End
Do elimination/dispersal steps;
End
Output: the pareto solutions in external archives

3.3 Experimental Data

The data of 10 assets which are collected from the financial.sina.com.cn are chosen as the example. The detailed data are from the beginning of 2013 to the third quarter of 2016. After collecting the related data, we use MATLAB to calculate the mean value and the covariance matrix of return on net assets (RONA) of the related companies. The detailed data are presented as follows:

$$r = (0.0942, 0.0992, 0.0181, 0.0216, 0.0558, 0.1089, 0.1254, 0.2528, 0.0541, 0.0336)$$

$$
\begin{aligned}
\sigma_{ij} = [&0.0018, 0.0019, 0.0004, 0.0003, 0.0010, 0.0007, 0.0006, 0.0043, 0.0015, 0.0004; \\
&0.0019, 0.0023, 0.0002, 0.0004, 0.0011, 0.0024, 0.0008, 0.0048, 0.0016, 0.0005; \\
&0.0004, 0.0002, 0.0007, -0.0002, 0.0004, -0.0029, -0.0000, 0.0012, 0.0004, 0.0001; \\
&0.0003, 0.0004, -0.0002, 0.0002, 0.0001, 0.0016, 0.0002, 0.0005, 0.0002, 0.0001; \\
&0.0010, 0.0011, 0.0004, 0.0001, 0.0014, -0.0005, 0.0005, 0.0044, 0.0010, 0.0003; \\
&0.0007, 0.0024, -0.0029, 0.0016, -0.0005, 0.0191, 0.0013, -0.0013, -0.0000, 0.0004; \\
&0.0006, 0.0008, -0.0000, 0.0002, 0.0005, 0.0013, 0.0005, 0.0020, 0.0005, 0.0001; \\
&0.0043, 0.0048, 0.0012, 0.0005, 0.0044, -0.0013, 0.0020, 0.0219, 0.0040, 0.0007; \\
&0.0015, 0.0016, 0.0004, 0.0002, 0.0010, -0.0000, 0.0005, 0.0040, 0.0014, 0.0003; \\
&0.0004, 0.0005, 0.0001, 0.0001, 0.0003, 0.0004, 0.0001, 0.0007, 0.0003, 0.0002;]
\end{aligned}
$$

$$k_i^b = 0.00065, k_i^s = 0.00075$$

4 Experimental Results and Analyses

Table 3 shows the data obtained by MOCLBFO and MOCLPSO when the investment profit is the largest, including the investment proportion of the ten assets and the corresponding value of profit and risk. Similarly, Table 4 presents the corresponding results when the risk is the lowest. The best value is in italics. Figure 1 presents the pareto optimal front. Two conclusions can be drawn.

- MOCLBFO and MOCLPSO can both obtain Pareto set that make objective function satisfied. But much better performance of MOCLBFO can be observed. For example, MOCLBFO's Pareto set has better diversity.
- The convergence curve with increasing trend indicates that the greater the investment profit, the greater the investment risk. It gives investors a revelation that they should allocate the proportion of assets based on their ability to bear the risk, rather than blindly pursuit high returns.

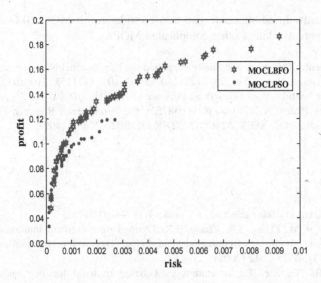

Fig. 1. Convergence curve of MOCLBFO and MOCLPSO

Table 3. The experimental results of the maximum profit

	MOCLBFO	MOCLPSO
x_1	0.0304	0.0425
x_2	0.0579	0.1752
x_3	0.0053	0.0228
x_4	0.0113	0.0241
x_5	0.0286	0.0382
x_6	0.0322	0.1785
x_7	0.1412	0.1636
x_8	0.5945	0.1870
x_9	0.0753	0.1440
x_{10}	0.0233	0.0353
Profit	*0.1862*	0.1198
Risk	0.0092	*0.0027*

Table 4. The experimental results of the minimum risk

	MOCLBFO	MOCLPSO
x_1	0.0698	0.0244
x_2	0.0047	0.0425
x_3	0.2613	0.1003
x_4	0.0698	0.0177
x_5	0.0009	0.0086
x_6	0.0392	0.0002
x_7	0.2188	0.2655
x_8	0.0013	0.0021
x_9	0.0447	0.0082
x_{10}	0.2895	0.1770
Profit	*0.0569*	0.0489
Risk	0.0002	*0.0001*

5   Conclusions

In this paper, based on MOCLBFO, we propose a specific algorithm for solving a multi-objective PO model with two conflicting objectives and two constraints. In order to test the validity of MOCLBFO, MOCLPSO is chosen as the competing algorithm. Experimental tests on ten assets PO problem demonstrate that MOCLBFO is outstanding in addressing MOPs. In the future, attention should be placed on building a new PO model that considers more realistic constraints and studying more powerful

MOBFO variants. In addition, we can also consider using MOCLBFO and another MOBFO variants to address other complicated MOPs.

Acknowledgment. This work is partially supported by The National Natural Science Foundation of China (Grants Nos. 71571120, 71271140, 61603310, 71471158, 71001072, 61472257), Natural Science Foundation of Guangdong Province (2016A030310074) and Shenzhen Science and Technology Plan (CXZZ20140418182638764), the Fundamental Research Funds for the Central Universities Nos. XDJK2014C082, XDJK2013B029, SWU114091.

References

1. Markowitz, H.: Portfolio selection. J. Financ. **7**(1), 77–91 (1952)
2. Shi, N.Z., Lai, M., Zheng, S.R., Zhang, B.X.: Optimal algorithms and intuitive explanations for markowitz's portfolio selection model and sharpe's ratio with no short-selling. Sci. Chin. Math. **51**(11), 2033–2042 (2008)
3. Niu, B., Bi, Y., Xie, T.: Structure-redesign-based bacterial foraging optimization for portfolio selection. In: Han, K., Gromiha, M., Huang, D.-S. (eds.) ICIC 2014. LNCS, vol. 8590, pp. 424–430. Springer, Heidelberg (2014)
4. Davis, M.H.A., Norman, A.R.: Portfolio selection with transaction costs. Math. Oper. Res. **15**(4), 676–713 (1988)
5. Chen, W., Zhang, R.T., Yang, L.: Portfolio selection model based on the improved particle swarm optimization. Comput. sci. **36**(1), 146–147 (2009)
6. Ayodele, A.A., Charles, K.A.: Improved constrained portfolio selection model using particle swarm optimization. Indian J. Sci. Technol. **8**(31), 1–8 (2015)
7. Tan, L.J., Wang, H., Yang, C., Niu, B.: A multi-objective optimization method based on discrete bacterial algorithm for environmental/economic power dispatch. Nat. Comput., 1–7, 22 April 2017. doi:10.1007/s11047-017-9623-4
8. Li, L., Xue, B., Tan, L.J., Niu, B.: Improved particle swarm optimizers with application on constrained portfolio selection. In: Huang, D.-S., Zhao, Z., Bevilacqua, V., Figueroa, J.C. (eds.) ICIC 2010. LNCS, vol. 6215, pp. 579–586. Springer, Heidelberg (2010)
9. Passino, K.M.: Biomimicry of bacterial foraging for distributed optimization and control. IEEE Control Syst. Mag. **22**(3), 52–67 (2002)
10. Huang, V.L., Suganthan, P.N., Liang, J.J.: Comprehensive learning particle swarm optimizer for solving multiobjective optimization problems. Int. J. Intell. Syst. **21**(2), 209–226 (2006)
11. Niu, B., Wang, H., Wang, J., Tan, L.J.: Multi-objective bacterial foraging optimization. Neurocomputing **116**(116), 336–345 (2013)

Metaheuristics for Portfolio Optimization

Sarah El-Bizri and Nashat Mansour[(⌧)]

Department of Computer Science and Mathematics,
Lebanese American University, Beirut, Lebanon
{sara.elbizri,nmansour}@lau.edu

Abstract. Portfolio optimization refers to allocating an amount of investors' wealth to different assets in order to satisfy the investors' preferences for return and risk. We address the portfolio optimization problem with real-world constraints, where traditional optimization methods fail to efficiently find an optimal or near-optional solution. Hence, we design a modified cuckoo search (MCS) meta-heuristic for finding good sub-optimal portfolios. Cuckoo search was inspired by the brood parasitism of cuckoo species by laying their eggs in the nests of other host birds. Our implementation explores the search space using Levy flights and allocates the good sub-optimal distribution of investment weights for a chosen set of assets. The MCS results show a clear improvement in comparison with previously published results, based on Markowitz and Sharpe models.

Keywords: Cuckoo search · Markowitz model · Metaheuristics · Portfolio optimization · Sharpe model

1 Introduction

Billions of dollars are invested in financial markets and investors seek to select a portfolio that generates profitable return. A portfolio is defined as a group of assets. Portfolio optimization (PO) is the process by which parts of different assets are chosen to be put in a portfolio with certain fixed total amount. Investment decisions are made by evaluating the trade-offs between risks and expected returns. Markowitz coined the Modern Portfolio theory (MPT) decades ago [1]. Later, Sharpe also presented a model [2]. Regardless of the model adopted, the key purpose is to find the optimal set of weights in allocating the assets of the portfolio. Investors prefer to invest in a specific number of assets with different allocations and a fixed total value. By adding this constraint, "the problem becomes a mixed integer programming problem with a quadratic objective function" that has no known efficient way in locating a fast solution [3]. The time needed to resolve this issue grows proportionally fast and such a problem is NP-hard [3].

A number of metaheuristics have been proposed for the PO problem. Chang et al. [4] applied a genetic algorithm, which mimics the process of natural selection, to cardinality constrained portfolio optimization (CCPO). The algorithm was tested on 5 datasets that included 31–225 assets. The authors also considered simulated annealing, which is based on a physical phenomenon, and Tabu search for comparing with the GA. Fu et al. [5] examined two implementations of the GA for CCPO, the traditional and hierarchical, in order to improve the performance. Aranha and Iba [6] introduced a

© Springer International Publishing AG 2017
Y. Tan et al. (Eds.): ICSI 2017, Part II, LNCS 10386, pp. 77–84, 2017.
DOI: 10.1007/978-3-319-61833-3_9

tree-based representation of the GA. Babaei [7] adapted a simulated annealing meta-heuristic for extended portfolio selection. Cura [8] presented a particle swarm optimization (PSO), which is inspired by bird swarm behavior, for a cardinality constrained mean–variance PO model. The PSO results were compared with those of tabu search, genetic algorithm, and simulated annealing. Golmakani and Fazel [9] presented a metaheuristic that is a mixture of both improved and binary PSO for answering a stretched Markowitz mean–variance portfolio selection model that is limited by four bounds. Tuba and Bacanin [10] proposed an artificial bee colony (ABC) algorithm enhanced with hybridization by the firefly algorithm in order to improve the process of exploitation. Chen [11] suggested a modified ABC in which the author also introduced alternative measures of risk, including real-world constraints, for constrained PO.

This paper addresses the portfolio optimization problem by employing both the Markowitz model and the Sharpe model with practical realistic constraints. We propose adapting a modified cuckoo search metaheuristic (MCS) [12, 13] for finding suboptimal solutions with reasonable time. This algorithm was inspired by the natural behavior of cuckoo birds where it simulates their behavior for finding new nests that are portfolio solutions in our problem. Recently, cuckoo search was adapted for improving the performance of cyber-physical systems [14]. To the best of our knowledge, Modified cuckoo search is applied, for the first time, for the portfolio optimization problem. The suggested approach is compared with other meta-heuristic approaches that stem from similar practical and existent objectives and constraints. The pilot outcomes undoubtedly convey that MCS delivers sound resolutions.

This paper is organized as follows. Section 2 describes Markowitz's and Sharpe's models. In Sect. 3, we present the MCS metaheuristic and its design for solving the portfolio optimization problem. The implementation results and analysis are presented in Sect. 4. Section 5 concludes the paper.

2 Problem Description

The objective of portfolio theory is to allocate investments to various assets. Allocation is affected by two attributes: the mean of the expected return on the assets which is amounted as the weighted average of the likely profits of the assets in the portfolio; the second is the variance which measures the variability of risk from the assets. The Modern Portfolio Theory (MPT) is a mathematical model of the notion of investment diversification. An early model, developed by Markowitz [1], focuses on the covariance between stock returns in order to improve the risk-return trade-off.

The Markowitz mean-variance optimization model [1] allows the allocation of investments among several assets taking into consideration the trade-off between risk and return. In this model the risk is minimized for a mandatory rate of return R_p. Thus, the goal is to minimize the portfolio's risk $\sigma_p^2 = \sum_{i=1}^{n} \sum_{j=1}^{n} \sigma_{ij} w_i w_j$, Subject to the portfolio's expected return $R_p = \sum_{i=1}^{n} R_i w_i$, where $\sum_{i=1}^{n} w_i = 1$, where

  n: the number of assets in the investable universe
  R_i: the expected return of asset i (i = 1; ...; n) above the risk-free rate
  R_p: the expected return of the portfolio above the risk-free rate

w_i: the weight in the portfolio of asset i (i = 1; ...; n), where $0 \leq w_i \leq 1$
s_i, s_j: the standard deviation of the returns
ρ_{ij}: the correlation between i and j
$\sigma_{ij} = \rho_{ij} \, s_i s_j$ is the covariance between assets i and j (i = 1; ...; n, j = 1; ...; n)

Sharpe [2] introduced an approximation model of Markowitz's. Sharpe's model captured the covariance between securities through one variable, β, which summarizes the prospects regarding stock market index changes in accordance to price fluctuations. For both, Sharpe and Markowitz models, the aim was to identify the efficient frontier by applying different mechanics. Sharpe's model takes into account the constraints of cardinality, floor, and ceiling. If the market includes n assets, the investor will be willing to invest in k out of n assets, which leads to a large number of combinations. Hence, there is a need to add a binary variable z_i used to flag the insertion or omission of asset i in the portfolio. Also, floor constraints are practically applied to evade unnecessary organization costs for negligible capitals that won't affect the portfolio's behavior remarkably. Ceiling constraints are applied based on policies to limit excessive exposure. Adding these constraints, the Sharpe function becomes: Minimize

$$\gamma \left[\left(\sum\nolimits_{i=1}^{n} w_i \beta i \right)^2 \sigma_m^2 + \sum\nolimits_{j=1}^{n} w_i^2 \sigma_{ei}^2 \right] - (1 - \gamma) \left[\sum\nolimits_{i=1}^{n} R_i w_i \right]$$

Subject to $0 \leq w_i \leq 1$, $\sum_{i=1}^{n} w_i = 1$, $z_i \in \{0, 1\}$, $\sum_{i=1}^{n} z_i = k$, $0 \leq f_i . z_i < w_i < c_i . z_i < 1$
where:

n: the number of assets available
f_i: the minimum weight an asset can hold
c_i: the maximum weight an asset can hold
k: the maximum number of assets in a portfolio
w_i: the weight of asset i in the portfolio
z_i: asset flag set to 1 if the asset is held and set to 0 otherwise.

3 Cuckoo Search Metaheuristic

Cuckoo search is not an old metaheuristic, developed by Yang and Deb in 2009 [12]. It is inspired by brood parasitism of some cuckoo species. Cuckoo birds lay their off-spring in shared nests so that the host feed the new born cuckoo. Sometimes, the host bird could recognize that the egg in the nest does not belong to it and end up removing it. Since cuckoo's eggs hatch earlier than their hosts' eggs, this gives the cuckoo an opportunity to hatch and force out the host's egg. Cuckoo birds are identified for laying eggs according to the Levy flight. A Levy flight is a random walk (steps are defined in terms of the step-lengths and have random directions) and has a certain probability distribution. Walton et al. [13] introduced a modification to the original algorithm in order to increase the convergence rate. One modification was adjusting the step length by a factor, α, that drops with the increase in the number of generations; the aim is to stimulate localized searches while the eggs move towards the best nest. The other modification was adding information trade-off between the eggs in order to identify top eggs group as well as the lowest (to be dropped).

Our proposed approach is comprised of a hybrid technique that is based on Modified Cuckoo Search, integrated with a quadratic programming solver – JOptimizer [15]. The master solver will select the best set of assets (k out of n assets) and the slave solver will assign the optimal weights for the selected k assets. The solution for both sub-problems should meet the purpose of selecting the best portfolio that produces maximum return with least risk possible.

Every portfolio/nest has the following components: Index of every asset included in this particular portfolio, as referenced in the assets dataset; Weight of every asset; Portfolio return; Portfolio risk; Portfolio Sharpe ratio. The Modified Cuckoo Search (MCS) steps are described as follows:

Step 1:   Initializing a number (50) of feasible portfolio solutions. The selection of K assets in every candidate portfolio sets the first component of the portfolio. The indices of the K assets in the portfolio are randomly selected from the available dataset

Step 2:   Every possible portfolio is evaluated in order to assign the optimal weights to the K assets in the candidate portfolios. The weight assignment is realized using JOptimizer, which is an open source library implemented in Java programming language that addresses the solution of a minimization problem with equality and inequality constraints [15]

Step 3:   Following the weight assignment for every portfolio of the 50 feasible solutions, every candidate portfolio is measured by fitness. The Sharpe ratio [2] serves as the fitness value for the nest (portfolio). All portfolios will be sorted according to the Sharpe ratio and replaced if a new portfolio with a higher Sharpe ratio is found. The Sharpe ratio is the "average return earned in excess of the risk-free rate per unit of volatility or total risk". As the value of the Sharpe ratio grows, the risk-adjusted return becomes more appealing

Step 4:   For all portfolios with the lowest fitness, that are portfolios numbered 1 to 35 after the sorting, we mutate by Levy flight to generate a new enhanced portfolio to substitute the candidate portfolio. The solutions obtained are mutated by performing Levy flights according to the transformation: $x_i^{(t+1)} = x_i^{(t)} + \alpha \oplus \text{Levy}(\lambda)$. If the portfolio that is subject to Levy flight mutation is among the portfolios to be abandoned then the step size is set to $50/\sqrt{G}$ (G is the current generation number), else if the portfolio subject to Levy flight mutation is among the top nests then the step size is set to 50/G2

Step: 5:   After mutating all the nests to be abandoned, we mutate all (remaining) top nests. Select portfolio j randomly from the pool of portfolios and label it portfolio Pj. If Pj is the same as the current portfolio of the top nests to be mutated, then step 6 will follow. Else, if Pj is different, then step 7 will follow. Steps 5–7 are applied on every portfolio in the set of best nests

Step 6:   For the selected portfolio of the top nests, mutate by Levy flight to generate a new portfolio Pk. In this mutation the step size selected is 50/ G2. We then choose a random portfolio Pl from all nests and compare the fitness of Pl to the fitness of the mutated portfolio Pk. If the fitness of Pk is greater than that of Pl, then Pl will be substituted with Pk

Step 7:   To obtain portfolio Pk, distance dx needs to be shifted from the weakest nest to the optimal one. The distance is calculated as follows: $d_x = |x_i - x_j|/\varphi$ where $\varphi = (1+\sqrt{5})/2$ [15]. The mutated portfolio Pk is generated based on the move distance dx. Then, choose a random portfolio Pl from all nests and compare the fitness of Pl to that of Pk. If the fitness of Pk is greater than that of Pl, then Pl will be substituted with Pk

The process of mutating portfolios in order to create dominating portfolios (steps 4–7) is repeated 100 times to produce dominating portfolios. However, if the fitness of portfolios is not enhanced after 10 consecutive iterations, then the iterations are terminated.

4 Experimental Results and Discussion

4.1 Experimental Procedure and Datasets

We applied the proposed algorithm to find the unconstrained efficient frontier (UEF), which can be calculated using different solvers (e.g., LINGO v16.0.7). Hence, the heuristic outcomes may be compared to ideal solutions. It is safe to consider that our heuristic will not be able to find the cardinality-constrained efficient frontier (CCEF) unless it finds the UEF at an accurate level. In this case, our MCS approach will be applied for k = n. To be able to compare the UEF, generated by LINGO, and UEF generated by our algorithm, we need to compute the ratio deviation of every portfolio from the (linearly interpolated) precise UEF. The distance/space from the benchmark frontier to the computed frontier is measured by the median percentage error and the mean % error. So, the mean percentage error (MPE) is the calculated average of error percentages, where by the predictions of a model vary from the real values of the amount that is predicted. That is, MEP $= \frac{100\%}{n}\sum_{i=1}^{n}\frac{at-ft}{at}$ where a_t is the real value of the amount that is predicted, f_t is the prediction, and n is the number of times the variable is predicted.

The comparison of LINGO-UEF to MCS-UEF yields results of mean return and standard deviation of return errors. The same approach will be applied for the CCPO problem based on both Markowitz model and Sharpe Model. It is also important to show the distribution of the efficient points along the frontier. When thinking about buying portfolio assets, a smart buyer would carefully study the efficient frontier with the better distribution of points across it.

To test our MCS algorithm based on Markowitz model, we refer to five test data sets in OR-Library reflecting the global stocks associated with the various capital market files. The data mentioned above equate to the 7 days prices figures from Mar. 1992 to Sep. 1997 for: Hang Seng 31, DAX 100, FTSE 100, S&P 100, and Nikkei 225. The stocks amounts involved in each set ranges from 31 to 225. However, to test the MCS algorithm based on Sharpe model, we refer to a test dataset that comprises the "top 100 stocks by market capitalization on the Johannesburg Stock Exchange (JSE)" [16]. The input parameters to the model are [16]: the risk-free rate employed is the present 3-months Treasury bill (TB) rate rf = 0.130; and $\sigma_m = 0.276$ having betas and the variance of regression errors calculated by monthly information on the span of the 3 previous years.

4.2   Results Based on Markowitz and Sharpe Models

Our proposed MCS algorithm is applied on the OR library datasets with cardinality constraint K = 10, using the Markowitz model. We compare the results to the GA, TS, and SA algorithms with the same constraints by comparing the standard efficient frontiers of the five real-world standardized series with the heuristic efficient frontier of the exact data series. Table 1 shows the results of the median and mean percentage error of our algorithm for each of the 5 data sets. The results demonstrate that the MCS algorithm scored first for 4 out of 5 data sets. The improvement percentage for every dataset is also indicated in the right-hand column. The results reveal that the MCS outperformed other heuristics by 10% on average. Consequently, MCS is the closest to the UEF.

Table 1. Comparative results of metaheuristics based on Markowitz model.

Index	No. of assets		GA	TS	SA	MCS	Improvement %
Hang Seng	31	Median % error	1.2181	1.2181	1.2181	**1.1946**	1.93
		Mean % error	1.0974	1.1217	1.0957	**1.0462**	5.32
DAX	85	Median % error	2.5466	2.6380	2.5661	**2.4501**	5.17
		Mean % error	2.5424	3.3049	2.9297	**2.4402**	16.56
FTSE	89	Median % error	1.0841	1.0841	1.0841	**1.0781**	0.55
		Mean % error	**1.1076**	1.6080	1.4623	1.4550	−4.48
S&P	98	Median % error	**1.2244**	1.2882	1.1823	**1.2244**	0.59
		Mean % error	1.9328	3.3092	3.0696	**1.6974**	38.73
Nikkei	225	Median % error	**0.6133**	0.6093	0.6066	0.6635	8.82
		Mean % error	0.7961	0.8975	0.6732	**0.6635**	15.90

With regard to finding the Sharpe-model-based CCEF, the MCS and its competitor, GA, are applied on the JSE dataset. We set the cardinality k to 40 in order to have 40 assets in every portfolio. The results of MCS are compared to the results of a genetic algorithm [16]. After 20 runs, we extract the best values for the portfolio risk and portfolio return to plot the efficient frontier (EF). The results of the best deviation values of MCS, GA and UEF are plotted in Fig. 1, which shows that UEF dominates MCS efficient frontier and the latter dominates the GA efficient frontier. Clearly, MCS attained a lower mean and median percentage errors in comparison with GA for this application and problem size. That is, MCS has generated acceptable approximate solutions to the UEF.

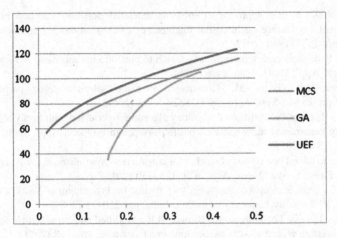

Fig. 1. Cardinality constrained efficient frontier

5 Conclusion

A Modified Cuckoo Search approach was proposed for identifying the efficient frontier for the portfolio optimization problem. Both Markowitz and Sharpe models were generalized to include cardinality and bounding constraints. The suggested MCS approach was tested on various datasets. The test results showed that the MCS algorithm, for the datasets employed for this experiment, increased exploration efficiency and produced portfolios that more effective than those of other metaheuristics (GA, SA, and TS) by an average of 10%. That is, the results of MCS for CCPO can be more valuable for investors.

References

1. Markowitz, H.: Portfolio selection. J. Finan. **7**(1), 77–91 (1952)
2. Sharpe, W.F.: A linear programming approximation for the general portfolio analysis problem. J. Financ. Quant. Anal. **6**(5), 1263 (1971)
3. Moral-Escudero, R., Ruiz-Torrubiano, R., Suarez, A.: Selection of optimal investment portfolios with cardinality constraints. In: Proceeding of the IEEE International Conference on Evolutionary Computation, pp. 2382–2388 (2006)
4. Chang, T.J., Meade, N., Beasley, J.E., Sharaiha, Y.M.: Heuristics for Cardinality constrained portfolio optimisation. Comput. Oper. Res. **27**(13), 1271–1302 (2000)
5. Fu, T., Chung, C., Chung, F.: Adopting genetic algorithms for technical analysis and portfolio management. Comput. Math Appl. **66**(10), 1743–1757 (2013)
6. Aranha, C.C., Iba, H.: A tree-based GA representation for the portfolio optimization problem. In: Proceeding of Genetic and Evolutionary Computation Conference, Atlanta, GA, USA, pp. 873–880 (2008)

7. Babaei, M.H.: A new approach to solve an extended portfolio selection problem. In: International Conference on Industrial Engineering and Operations Management, Istanbul, Turkey, pp. 1954–1960 (2012)
8. Cura, T.: Particle swarm optimization approach to portfolio optimization. Nonlinear Anal. : Real World Appl. **10**(4), 2396–2406 (2009)
9. Golmakani, H.R., Fazel, M.: Constrained portfolio selection using particle swarm optimization. Expert Syst. Appl. **38**(7), 8327–8335 (2011)
10. Tuba, M., Bacanin, N.: Artificial bee colony algorithm hybridized with firefly algorithm for cardinality constrained mean-variance portfolio selection problem. Appl. Math. Inf. Sci. **8** (6), 2831–2844 (2014)
11. Chen, W.: Artificial bee colony algorithm for constrained possibilistic portfolio optimization problem. Phys. A: Stat. Mech. Appl. **429**, 125–139 (2015)
12. Yang, X.S., Deb, S.: Cuckoo search via levy flights. In: Proceeding of World Congress on Nature & Biologically Inspired Computing, India, pp. 210–21 (2009)
13. Walton, S., Hassan, O., Morgan, K., Brown, M.R.: Modified cuckoo search: a new gradient free optimisation algorithm. Chaos, Solitons Fractals **44**(9), 710–718 (2011)
14. Cui, Z.Z., Sun, B., Wang, G.G., Xue, Y., Chen, J.J.: A novel oriented cuckoo search algorithm to improve DV-hop performance for cyber-physical systems. J. Parallel Distrib. Comput. **103**, 42–52 (2017)
15. Trivellato, A.: JOptimizer - java convex optimizer. http://www.joptimizer.com/ (2016)
16. Busetti, F.R.: Metaheuristic Approaches to Realistic Portfolio Optimization. M.S. thesis, Department of Operations Research, University of South Africa (June 2000)

Community Detection

Community Detection Under Exponential Random Graph Model: A Metaheuristic Approach

Tai-Chi Wang[1] and Frederick Kin Hing Phoa[2(⊠)]

[1] National Center for High-Performance Computing, Hsinchu, Taiwan
[2] Academia Sinica, Taipei 115, Taiwan
fredphoa@stat.sinica.edu.tw

Abstract. Community is one of the most important features in social networks. Although many algorithms are developed for detecting communities, most of them hit the computational ceiling when a large-scale network is dealt. The nature-inspired metaheuristic algorithm is a potential solver for optimizing a target function. Its parallel-computing architecture improves the computing efficiency for analyzing the large-scale network. This study proposes a novel swarm intelligence algorithm to detect communities. A new criterion, derived from the exponential random graph model with two communities, is proposed to detect the most significant community within the whole network. The feasibility and the detection accuracy of our algorithm are verified by the simulation studies.

1 Introduction

The researches in social network analysis are approaching towards an elevated plateau recently due to the popularity of social media and the advance in data mining [15, 24, 25]. Community detection is generally categorized into four types according to the community's properties of interests [21]: node-centric, group-centric, network-centric and hierarchy-centric. Unfortunately, few traditional methods provide statistical inferences. On the contrary, the likelihood-based methods, like the exponential random graph model (ERGM) [19], provide statistical significance on the detected communities or make the inference on the goodness of detection [7]. In this study, we use the criterion derived from ERGM with communities, and show that our derived criterion is equivalent to the one provided by [16].

The search of the most important community under a given criterion is highly interested in social network analysis, but it is an NP-hard (non-deterministic polynomial time) problem due to the complexity of a large-scale network [3]. The k-means method is popular for finding communities. It determines the eigenvectors of a specific measurement matrix for communities. Unfortunately, not all criteria can be expressed into matrix forms, and the search of all eigenvectors in a large-scale network is never simple. Thus, some metaheuristic approaches are developed to deal with the community detection problem, like the genetic

© Springer International Publishing AG 2017
Y. Tan et al. (Eds.): ICSI 2017, Part II, LNCS 10386, pp. 87–98, 2017.
DOI: 10.1007/978-3-319-61833-3_10

algorithm (GA) [6], the simulated annealing (SA) and its enhanced approach
[13], the extremal optimization (EO) [2], the Tabu algorithm [4], D-Net [5], the
label propagation algorithm (LPA) [1] and many others.

On the other hand, [22] provided the scan statistics with a systematic procedure, which is originated from the spatial cluster detection and is developed
for determining both the attribute and structure clusters by scanning windows
in social networks. This approach can be executed via parallel computing by
distributing and calculating the test statistics of independent scanning windows.
However, it is restricted to the detection of community without irregular shape
from the scanning windows. To overcome this constraint, we develop a swarm
intelligence based (SIB) algorithm for the community detection problem. The
SIB method was first proposed in [17,18] for optimizing the factor level settings
in various types of experimental plans. It belongs to a class of nature-inspired
metaheuristic algorithm including the particle swarm optimization (PSO) [9]
and many others. In this study, we propose a novel SIB algorithm that can be
used to detect statistically significant communities.

We assume to have undirected and unweighted networks for simple illustration on the operation of the SIB algorithm. The rest of this paper is arranged
as follows. We first review the concepts of ERGM and derive a criterion from
ERGM in Sect. 2. Then, the SIB algorithm is introduced and the hierarchical
structure of communities is discussed in Sect. 3. Section 4 contains simulation
studies on the performance of our method. The last section concludes the whole
paper and provides some future directions extended from this work.

2   Exponential Random Graph Model

2.1   Basic Properties

An ERGM is a statistical model for the ties in a network G. In general, It
has the form $P(G) = \frac{e^{H(G)}}{Z}$, where $H(\cdot)$ is a graph Hamiltonian and Z is a
normalized constant expressed as $Z = \sum_{G=\mathcal{G}} e^{H(G)}$. A graph Hamiltonian is
given by $H(G) = \sum_{i=1}^{r} \theta_i X_i(G)$, where $\{X_i(G)\}$ are the values of the observables
$\{x_i\}$ for G and $\theta = \{\theta_1, \ldots, \theta_r\}$ are the respective parameters. Since the ERGM
follows an exponential-family distribution, the statistics $\{X_i(G)\}$ are complete
and sufficient for θ. The expectation of X_j is $E(X_j) = \frac{\partial \ln Z}{\partial \theta_j}$ and the variance of
X_j is $V(X_j) = \frac{\partial^2 \ln Z}{\partial \theta_j^2}$.

2.2   Erdös-Rényi Model

Erdös-Rényi (ER) model is one of the most classical models for an undirected
graph. Its Hamiltonian is only related to the number of edges and the probability
of G is $P(Y = y) = \frac{\exp(\theta \sum y_{ij})}{Z}$. Suppose there are n nodes and $N = n(n-1)/2$
edges. Let $s = \sum y_{ij}$ be the number of observed edges. The binomial theorem and
reexpression of y suggest that the normalized constant Z is $Z = (1 + \exp(\theta))^N$.

Furthermore, the expectation of S is $E(S) = N\frac{\exp(\theta)}{1+\exp(\theta)}$ and the variance is $V(S) = N\frac{\exp(\theta)}{(1+\exp(\theta))^2}$.

Let $p = \frac{\exp(\theta)}{1+\exp(\theta)}$ be the connection probability. By considering S being a binomial distribution with the parameter p, we express the expectation and variance of S as $E(S) = Np$, and $V(S) = Np(1-p)$ respectively. The large sample theory suggests $\sqrt{N}(\hat{p}-p) \xrightarrow{D} N(0, p(1-p))$ as $N \to \infty$, where $\hat{p} = S/N$. If the original θ is taken into account, the θ is equivalent to the logarithmic odd-ratio, $\theta = logit(p) = \log\frac{p}{1-p}$. By the delta method, the statistical inference of θ is $\sqrt{N}(\hat{\theta} - \theta) \xrightarrow{D} N(0, \frac{1}{p(1-p)})$ as $N \to \infty$.

2.3   ERGM with Independent Communities

Suppose a graph G can be divided into k disjointed and independent parts, that is, $G = \{G_0, G_1, \ldots, G_k\}$, where G_0 is the part that does not belong to any other parts. Assuming that each part follows an Erdös-Rényi model with a different θ_i, we can construct the likelihood of the partitions based on the similar way in the Erdös-Rényi model, $P(G) = \frac{\Pi_{i=0}^{k} \exp(\theta_i s_i)}{Z}$. Suppose the total number of edges is divided into $N = \sum_{i=0}^{k} N_i$, where $N_i = n_i(n_i - 1)/2 \forall i \neq 0$ in which n_i is the number of nodes in G_i, and $N_0 = N - \sum_{i=1}^{k} N_i$. The normalized constant Z is evaluated as $Z = \prod_{i=0}^{k}(1+\exp(\theta_i))^{N_i}$. According to the independent assumption of each group, we can separately make the inference of the different θs. Similar to the approach in the ER model, let $p_i = \frac{\exp(\theta_i)}{1+\exp(\theta_i)}$, then $E(S_i) = N_i p_i$, and $V(S) = N_i p_i(1 - p_i)$. Furthermore, $\sqrt{N_i}(\hat{p}_i - p_i) \xrightarrow{D} N(0, p_i(1-p_i))$ as $N_i \to \infty$, where $\hat{p}_i = S_i/N_i$, and $\sqrt{N_i}(\hat{\theta}_i - \theta_i) \xrightarrow{D} N(0, \frac{1}{p_i(1-p_i)})$ as $N_i \to \infty$, where $\theta_i = logit(p_i) = \log\frac{p_i}{1-p_i}$.

3   The Methodology

The aim of this study is to propose a method that can identify the most influential community in a network. There are two major concerns: (1) how to define the most influential community, and (2) how to find the elements of the most influential community. Based on the ERGM with communities, we can determine the importance of the subgraphs. Furthermore, we adopt the general framework of the swarm intelligence based (SIB) method and construct a new SIB algorithm to fit the structure of network communities.

3.1   Objective Function

Since we consider the best partitions for a given graph, we adopt the model described in Sect. 2.3. Suppose the elements of partitions are given as $\{G_0, G_1, \ldots, G_k\}$. We can directly determine the respective statistics of the likelihood of the partitioned ERGM. The full likelihood is $L(\boldsymbol{\theta}|\{G_0, G_1, \ldots, G_k\}) =$

$\prod_{i=0}^{k} p_i^{s_i}(1-p_i)^{N_i-s_i}$, where $p_i = \frac{\exp(\theta_i)}{1+\exp(\theta_i)}$. When the partitions are given, we can obtain the maximum likelihood by replacing the p_is with their maximum likelihood estimators (MLEs), $L(\hat{\boldsymbol{\theta}}|\{G_0, G_1, \ldots, G_k\}) = \prod_{i=0}^{k} \hat{p}_i^{s_i}(1-\hat{p}_i)^{N_i-s_i}$, where $\hat{p}_i = \frac{S_i}{N_i}$. This measure is exactly the same as the measure provided in [16]. When we consider the model with only one community, the likelihood is $L(\boldsymbol{p}|\boldsymbol{s}, \boldsymbol{z}) = p_c^{s_c}(1-p_c)^{N_c-s_c} p_u^{s_u}(1-p_u)^{N_u-s_u} p_b^{s_b}(1-p_b)^{N_b-s_b}$, where $\boldsymbol{z}$ is an $n \times 1$ group membership vector with elements taking on values $\{0, 1\}$, where $z_i = 1$ when the node i is selected and $z_i = 0$ otherwise, $\boldsymbol{s}|\boldsymbol{z} = [s_c, s_u, s_b]$ is the observed numbers of edges among the selected and unselected nodes, and the edges between the selected and the unselected nodes, which is called "betweenness" here, and $\boldsymbol{N} = [N_c, N_u, N_b]$ are all possible numbers of edges with the similar definition as $\boldsymbol{s}$. The target function can be recast as $\boldsymbol{z} = \hat{p}_c^{s_c}(1-\hat{p}_c)^{N_c-s_c} \hat{p}_u^{s_u}(1-\hat{p}_u)^{N_u-s_u} \hat{p}_b^{s_b}(1-\hat{p}_b)^{N_b-s_b}$, where $\hat{p}_c = s_c/N_c$, $\hat{p}_u = s_u/N_u$, and $\hat{p}_b = s_b/N_b$.

3.2   Swam Intelligence Based Method

The particle swarm optimization (PSO) [9] is a well-established algorithm for finding solutions in high dimensions. The PSO is initialized with a group of random particles (solutions) and then it searches for optimal values by updating iterative computations. In every iteration, each particle is updated by the two "best" values: the **local best** (LB) and **global best** (GB) according to some specified velocities. Suppose there are N particles $X(0) = \{\boldsymbol{x}_1(0), \ldots, \boldsymbol{x}_N(0)\}$ generated in the initial step and the initial velocity vector is defined as $V = \{v_1(0), \ldots, v_N(0)\}$. In the updated procedure, each particle is updated according to the following formulation: $\boldsymbol{x}_i(t+1) = \boldsymbol{x}_i(t) + \boldsymbol{v}_i(t+1)$, in which the velocity vector $\boldsymbol{v}_i$ is determined by $\boldsymbol{v}_i(t+1) = w\boldsymbol{v}_i(t) + \eta_1 \text{rand}()(P_{LB} - \boldsymbol{x}_i(t)) + \eta_2 \text{rand}()(P_{GB} - \boldsymbol{x}_i(t))$, where $rand()$ is a random number between $(0, 1)$, and w, η_1, and η_2 are learning factors. Suppose the target function is J. After all particles are fitted in J, the best value among all $J(x)$ is the global best and denoted as P_{GB}, and the best value of each particle is a local best and is denoted as P_{LB}.

3.3   Modifications for Community Detection

Since the original PSO is designed for the optimization with "continuous" solutions, this approach has some limitations when dealing with "discrete" solutions. Recently, a modified approach has been introduced for discrete cases, and it is called the swarm intelligence based (SIB) method. This method provides many good optimization results in statistics, especially in the design of experiments. For example, [18] constructed a SIB method for optimizing supersaturated designs under the $E(s^2)$ criterion. The difference between SIB and PSO is that the SIB modifies the updating procedures via the MIX and MOVE operations according to the values of LB and GB. Such modifications allow the SIB to search in both discrete and continuous domains, while PSO works only in continuous domain. In general, the stepwise procedure of the SIB is given as follows.

0: Randomly generate a set of initial particles. Evaluate objective function value of each particle. Initialize the local best (LB) for each particle and the global best (GB).

1: For each particle, perform the MIX operation.

2: For each particle, perform the MOVE operation.

3: Evaluate the objective function value of each particle.

4: Update the LB for each particle and the GB among all particles.

5: If the updated value does not converge or the algorithm does not reach the setting iterations, repeat Step 1 to Step 4.

Although this method is feasible to find some good results in experimental designs, it is not ready to be applied for detecting network communities. Since one of the most important features of community is that the communities must be **connected**, the algorithm of the SIB method has to be modified.

In Step 0, the initial particles, which are subgraphs in a network, have to be connected. To make the randomly selected particles connected, an initial particle is generated by selecting a node as center and a length of the shortest path. Suppose the selected node is v_i and the selected length is r. Then the nodes that are distanced with v_i within r constructed a vector of initial particles. Such generator has been verified as a good selection of possible community [22].

The "MIX" operation basically contains randomly "ADD" and "DELETE" procedures. Since the possible community must be connected, we have to carefully decide the "ADD" and "DELETE" procedures to make subgraphs connected. The ADD procedure considers to add q new elements, which are connected to but do not belong to the initial seed, from the LB or GB subgraphs. The DELETE procedure considers to delete q elements of the connected graph that is obtained after the ADD procedure so that the subgraph is still connected. The parameter q is called **exchangeable parameter** here. These steps allow different sizes of subgraphs to be mixed and the mixed subgraph is still connected. In addition, our algorithm contains two additional procedures, "contraction" and "expansion", which are used to polish the subgraph when the current seed contains the mixed subgraph or is contained by the mixed subgraph respectively. Besides, when the current seed is exactly the same as the mixed subgraph, we execute "contraction" and "expansion" separately, and record which procedure can produce the better target value. After executing these procedures, the "MOVE" procedure is adopted by checking the change of target function. Then, the reminder steps are the same as the original framework.

4   Simulation Study

4.1   One Community with Poisson Assumption

According to the results reported in [22], the community size and the connection probability of community play important roles in community detection. Thus, we construct the simulation cases that are constituted by K community nodes and $100 - K$ usual nodes in a studied network, where the values of K are 5, 10, 15, and

20 in this one community case. The connection probability of the usual nodes is 1/20, and that of the community node is set to be 1/4, 1/2, 3/4, and 1. For each combination of community size and connection probability, 100 simulations are conducted. When conducting our SIB algorithm for community detection, we set the number of seeds to be 300, the number of iterations to be 20, and the mixing fraction to be $|V(G)|/10$. Based on the settings of the SIB method, most of these parameters are known to be insensitive to the eventual outcomes. They may slightly affect the iteration steps or times towards convergence.

In this section, we check the accuracy of our SIB method under the likelihood statistic (SIB-LS), and compare the detection results with the traditional modularity method (MOD). The modularity is defined as $Q = \frac{1}{2|E(G)|} \sum(e_{ij} - \frac{k_i k_j}{2|E(G)|}) \delta(v_i, v_j)$, where k_i and k_j are the degrees of v_i and v_j, e_{ij} is 1 if v_i and v_j are adjacent, and 0 otherwise; and $\delta(v_i, v_j)$ is 1 if v_i and v_j are labeled as the same community, and 0 otherwise. When only one community is taken into account, the community is obtained by the signs of the maximum eigenvector.

Since the accuracy of community members is considered, some clearly-defined measurements other than the testing power are also used to evaluate the accuracy of detection results. We define the true positive, false positive, true negative, and false negative as follows. True positive (TP) cells represent the *true community nodes* that are correctly detected as a community; false positive (FP) cells represent the *usual nodes* that are incorrectly detected as a community; true negative (TN) cells represent the *usual nodes* the are **not** identified as a community; false negative (FN) cells represent the *true community nodes* that are **not** identified as a community. Furthermore, since only one ground-truth community is set in the simulation study, we use some common criteria to evaluate our detection results.

To check if the method can identify community nodes, we define TP/(TP+FN) as the recall r and use it to measure the proportion of identified community nodes among all true community nodes. For the measurement of the proportion of true community nodes among the identified community nodes, we use TP/(TP+FP), which is defined as the precision (p). F_1-score and Jaccard similarity are two commonly used criteria that evaluate community detection and are defined respectively as $F_1 = 2\frac{pr}{p+r}$ and $J(C^*, C) = \frac{|C^* \cap C|}{|C^* \cup C|}$, where C^* is the detected community and C is the ground-truth community. In addition, we adopt the modularity proposed by [14] to measure the similarity among the detection groups.

The comparison results are summarized in Fig. 1. All measures show similar but slightly different patterns despite the size of community. In general, when the community size is large (15 or 20), the SIB-LS method provides the best detection accuracy and the gap between SIB-LS and other methods gets larger as the connection probability gets higher. However, when the community size is 10, the connection probability decides the differences between these two methods. The modularity method is slightly better than the SIB-LS when a small community size meets a low connection probability (0.25 or 0.5). On the contrary, our method has significant improvement when the connection probability is high.

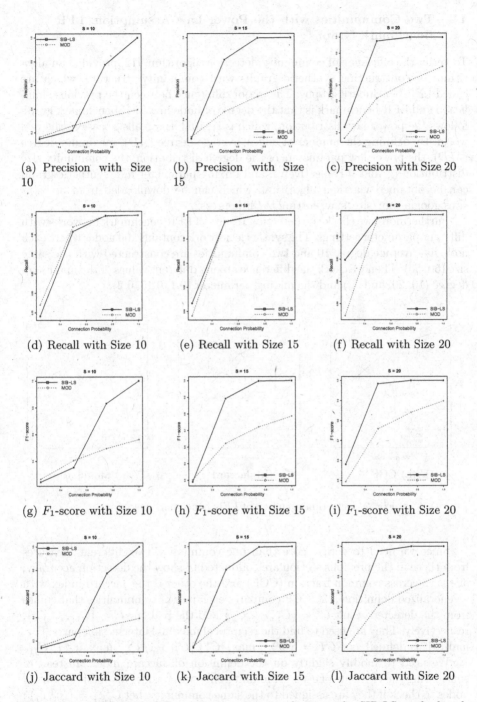

Fig. 1. Comparison with different community sizes between the SIB-LS method and the modularity.

4.2 Two Communities with the Power-Law Assumption: LFR Benchmark Graphs

To verify the efficiency of community detection algorithm, [12] provided an algorithm for constructing synthetic graphs with community structure, which are called LFR benchmark graphs. The major difference between the previous simulation and LFR benchmark is that the degree of node in a LFR benchmark graph follows the power-law distribution, which is treated as a realistic case in the real world. By defining the number of nodes, average degree (AD), maximum degree (MD), the power law parameters of the degree distribution, the community size distribution (γ and β respectively), and mixing parameter μ, a synthetic graph can be obtained via their algorithm, which can be downloaded from https://sites.google.com/site/santofortunato/inthepress2.

Furthermore, we check the detection results of LFR benchmark networks with different parameter settings. The synthetic network contains 100 nodes where each node has average degree 10 and two communities are constructed with the same size (50, 50). Then we check the difference among different values of the maximum degree (10, 20, and 30) and the mixing parameter (0.1, 0.15, 0.2).

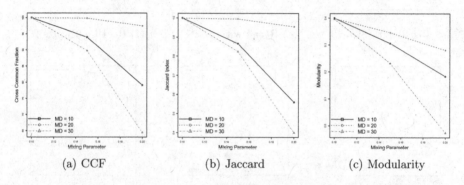

(a) CCF   (b) Jaccard   (c) Modularity

Fig. 2. Detection accuracy of LFR benchmark networks.

Since we need to verify more than one community, two different measures from those in the previous section are conducted to show the detection accuracy; one is the cross common fraction (CCF) and the other is the Jaccard index with a generalized definition [23]. CCF compares each pair of communities that comes from the detected result $C^* = \{C_1^*, \dots, C_s^*\}$ and the real one $C = \{C_1, \dots, C_t\}$, respectively. They are used to find the reciprocal maximal intersection parts. Formally, it is defined as $CCF = 1/2 \sum_i^s \max_j |C_i^* \bigcap C_j| + 1/2 \sum_j^t \max_i |C_i^* \bigcap C_j|$.

We need to modify slightly on the definition of Jaccard index. Instead of considering the elements of each community only, we consider all possible pairs of nodes to check if they are assigned to the same community. Let $C_{pair}^* = \{(v_i, v_j) : \delta(v_i, v_j) = 1\}$ and $C_{pair} = \{(v_i, v_j) : \xi(v_i, v_j) = 1\}$ be the sets of pairs that are assigned to the same communities with respect to the detected communities and

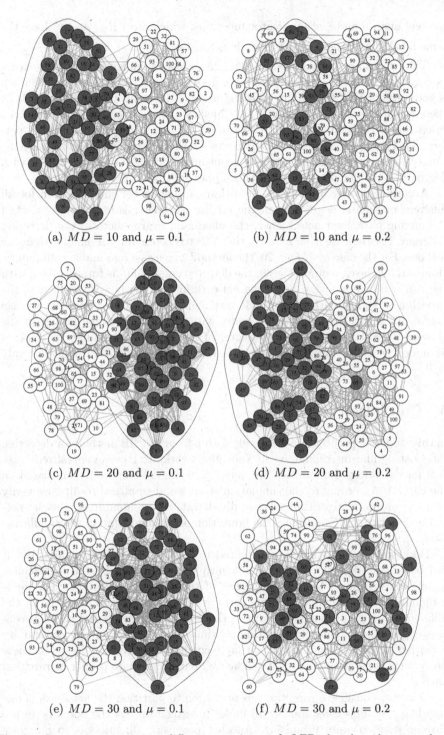

(a) $MD = 10$ and $\mu = 0.1$

(b) $MD = 10$ and $\mu = 0.2$

(c) $MD = 20$ and $\mu = 0.1$

(d) $MD = 20$ and $\mu = 0.2$

(e) $MD = 30$ and $\mu = 0.1$

(f) $MD = 30$ and $\mu = 0.2$

Fig. 3. Comparisons among different settings of LFR benchmark networks. (Color figure online)

the real ones. δ and ξ are indicator functions, which are 1 if v_i and v_j have the same label. Thus, the Jaccard index is defined as $J(C^*, C) = \frac{|C^*_{pair} \cap C_{pair}|}{|C^*_{pair} \cup C_{pair}|}$.

The results are summarized in Fig. 2. An obvious tendency is that the accuracy gets worse as the mixing parameter gets higher. Nevertheless, the accuracy is good except that the mixing parameter is 0.2. In terms of the different MD, there is a surprising result that $MD = 20$ shows great detection accuracy. Even when the mixing parameter increases to 0.15, the accuracy is still almost perfect. This means that some central nodes (with higher degrees) can make the connection stronger than the homogeneous nodes do ($MD = 10$, each node with degree 10). We demonstrate this special phenomenon in Fig. 3.

According to Fig. 3, the SIB algorithm can detect communities well for all different maximum degrees when the mixing parameter is low. However, when the mixing parameter gets higher, the maximum degree dictates the detection accuracy. For the case of $MD = 10$, the detected community is smaller than the real one. For the case of $MD = 20$, the method suggests a reasonable community. However, for the case of $MD = 30$, the detected community is totally mixed with the two real communities. Since the network follows the power-law degree, the variation of degrees with $MD = 30$ is very high in order to maintain the average degree 10, and this results in the nodes with higher degrees gathering. In the above $MD = 30$ case, the average degree of the detected community nodes (red 31 nodes) is 18.03, but that of the unselected nodes (white 69 nodes) is only 6.80.

5   Conclusion

In this study, we propose a modified algorithm from the SIB method for detecting the most significant community in a social network. To have a generalized criterion for detecting communities, we derive a likelihood-based criterion based on the ERGM. According to the simulation studies and empirical results, we verify that the proposed algorithm successfully detects the communities in social networks. Furthermore, we discuss the limitation of our algorithm in the simulation study.

Besides, this algorithm is not limited to detect a single community, but it can be enhanced to the detection of multiple communities by the hierarchical regime due to the theoretical properties of the provided criterion. Each single community can be divided into two parts by our algorithm and be tested by the difference of edge proportions between the selected subgraph and the betweenness edges. The community partition continues until each resulting community is neither statistically significant nor large enough to be divided. Based on these procedures, a network of interest can be well partitioned into proper hierarchical parts.

One may consider a simultaneous operation to partition the networks in one step like [13]. This direction is still under investigation but the outcome in the current stage is quite promising. Another promising direction is to integrate the current optimization procedure with the greedy algorithm or dynamic programming, rather than a simple and arithmetic update in the MIX and MOVE operations.

There are some studies about the convergence properties of PSO methods [8,20]. The SIB method contains the random jumps that lead to complications in deriving the theoretical convergence rate. Unless one is willing to make the assumptions on the random part, the success on the derivation is questionable.

Acknowledgement. The authors would like to thank three reviewers for their constructive suggestions to improve the quality of this paper. This work is supported by Career Development Award of Academia Sinica (Taiwan) grant number 103-CDA-M04 and National Science Council (Taiwan) grant numbers 104-2118-M-001-016-MY2 and 105-2118-M-001-007-MY2.

References

1. Barber, M.J., Clark, J.W.: Detecting network communities by propagating labels under constraints. Phys. Rev. E **80**(2), 026129 (2009)
2. Boettcher, S., Percus, A.G.: Optimization with extremal dynamics. Phys. Rev. Lett. **86**(23), 5211–5214 (2001)
3. Fortunato, S.: Community detection in graphs. Phys. Rep. **486**(3), 75–174 (2010)
4. Glover, F.: Future paths for integer programming and links to artificial intelligence. Comput. Oper. Res. **13**(5), 533–549 (1986)
5. Gong, M.G., Ma, L.J., Zhang, Q.F., Jian, L.C.: Community detection in networks by using meltiobjective evolutionary algorithm with decomposition. Phys. A **391**(15), 4050–4060 (2012)
6. Holland, J.H.: Adaptation in natural and artificial system. MIT Press, Cambridge (1975)
7. Hunter, D.R., Goodreau, S.M., Handcock, M.S.: Goodnewss of fit of social network models. J. Am. Stat. Assoc. **103**(481), 248–258 (2008)
8. Jiang, M., Luo, Y., Yang, S.: Stochastic convergence analysis and parameter selection of the standard particle swarm optimization algorithm. Inf. Process. Lett. **102**(1), 8–16 (2007)
9. Kennedy, J.: Particle swarm optimization. In: Encyclopedia of Machine Learning, pp. 760–766, Springer (2010)
10. Lusseau, D.: The emergent properties of a dolphin social network. Proc. R. Soc. Lond. B Biol. Sci. **270**(2), S186–S188 (2003)
11. Lusseau, D., Schneider, K., Boisseau, O.J., Haase, P., Slooten, E., Dawson, S.M.: The bottlenose dolphin community of doubtful sound features a large proportion of long-lasting associations. Beahavioral Ecol. Sociobiol. **54**(4), 396–405 (2003)
12. Lancichinetti, A., Fortunato, S., Radicchi, F.: Benchmark graphs for testing community detection algorithm. Phys. Rev. E **78**(4), 046110 (2008)
13. Liu, J., Liu, T.: Detecting community structure in complex networks using simulated annealing with k-means algorithms. Physica A **389**(11), 2300–2309 (2010)
14. Newman, M.E.: Fast algorithm for detecting community structure in networks. Phys. Rev. E **69**(6), 066133 (2004)
15. Newman, M.: Networks: An Introduction. Oxford University Press, New York (2010)

16. Perry, M.B., Michaelson, G.V., Ballard, M.A.: On the statistical detection of clusters in undirected networks. Comput. Stat. Data Anal. **68**, 170–189 (2013)
17. Phoa, F.K.H. A swarm intelligence based (SIB) method for optimization in designs of experiments. Natural Computing (Accepted 2016)
18. Phoa, F.K.H., Chen, R.B., Wang, W.C., Wong, W.K.: Optimizing two-level supersaturated designs via swarm intelligence techniques. Technometrics **58**, 43–49 (2016)
19. Robins, G., Pattison, P., Kalish, Y., Lusher, D.: An introduction to exponential random graph (p^*) models for social networks. Soc. Netw. **29**(2), 173–191 (2007)
20. Trelea, I.C.: The particle swarm optimization algorithm: convergence analysis and parameter selection. Inf. Process. Lett. **85**(6), 317–325 (2003)
21. Tang, L., Liu, H.: Toward predicting collective behavior via social dimension extraction. IEEE Intell. Syst. **25**(4), 19–25 (2010)
22. Wang, T.C., Phoa, F.K.H.: A scanning method for detecting clustering pattern of both attribute and structure in social networks. Physica A **445**, 295–309 (2016)
23. Wang, M., Wang, C., Yu, J.X., Zhang, J.: Community detection in social networks: an in-depth benchmarking study with a procedure-oriented framework. Proc. VLDB Endowment **8**(10), 998–1009 (2015)
24. Wasserman, L.: All of Statistics: A Concise Corse in Statistical Inference. Springer, New York (2004)
25. Wasserman, S., Fraser, A.: Social Network Analysis: Methods and Applications, vol. 8. Cambridge University Press, New York (1994)
26. Zachary, W.W.: An information flow model for conflict and fission in small groups. J. Anthropol. Res. **33**(4), 452–473 (1977)

An Enhanced Particle Swarm Optimization Based on *Physarum* Model for Community Detection

Zhengpeng Chen[1], Fanzhen Liu[1], Chao Gao[1,2,3(✉)], Xianghua Li[1], and Zili Zhang[1,4]

[1] School of Computer and Information Science, Southwest University,
Chongqing 400715, China
cgao@swu.edu.cn
[2] Potsdam Institute for Climate Impact Research (PIK), 14473 Potsdam, Germany
[3] Institute of Physics, Humboldt-University zu Berlin, Berlin 12489, Germany
[4] School of Information Technology, Deakin University,
Geelong, VIC 3220, Australia

Abstract. Community detection, an effective tool to analyze and understand network data, has been paid more and more attention in recent years. One of the most popular methods of detecting community structure is to find the division with the maximal modularity. However, the modularity maximization is an NP-complete problem. In the field of swarm intelligence algorithm, particle swarm optimization (PSO) has been widely used to solve such NP-complete problem. Nevertheless, premature convergence and lower accuracy limit its performance in community detection. In order to overcome these shortcomings, this paper proposes a novel PSO called P-PSO for community detection through combining the computational ability of *Physarum*, a kind of slime. The proposed algorithm improves the efficiency of PSO by recognizing inter-community edges based on *Physarum*-inspired network model (PNM). Experiments in eight networks show that the proposed algorithm is effective and promising for community detection, compared with other algorithms.

Keywords: Community detection · PSO · *Physarum* network model

1 Introduction

Complex networks have numerous characteristics, among which the community structure is an important one. Community detection, a powerful tool to discover community structures, has a wide application prospect, like predicting protein functions [1] and analyzing the information dissemination [2].

In the past few decades, a large number of algorithms have been proposed for community detection. They can be classified into optimization algorithm and

© Springer International Publishing AG 2017
Y. Tan et al. (Eds.): ICSI 2017, Part II, LNCS 10386, pp. 99–108, 2017.
DOI: 10.1007/978-3-319-61833-3_11

heuristic algorithm. Meanwhile, a modularity measure Q [3] is proposed to evaluate the quality of community divisions, which has been widely used. It has been proved that swarm intelligence optimization algorithms including particle swarm optimization algorithm (PSO) [4] show their superiority in local learning and global search. Recently, Cai et al. have successfully used greedy discrete particle swarm optimization algorithm (GDPOS) [5] to detect the community structures in a network. However, failing to make full use of prior knowledge of network and generate high-quality initial population, this algorithm does not lead to the good enough performance of global search and relatively high accuracy.

According to the latest reports, a large number of biological experiments have demonstrated that a slime named *Physarum* has an intelligence of solving mazes and constructing efficient and robust networks [6,9]. Meanwhile, the *Physarum*-inspired Mathematical Model (PM) has been proposed by Tero et al. [7], which has been used for optimizing the heuristic algorithms [8]. Thus, a *Physarum*-inspired network model (PNM) is proposed for initializing the PSO based on the PM model, which is utilized to distinguish inter-community edges from intra-community edges. Furthermore, we attempt to optimize the phase of PSO's initialization for higher quality in community detection.

The remaining of this paper is organized as follows: Sect. 2 illustrates the related background and introduces the particle swarm optimization algorithm for community detection. Section 3 proposes the *Physarum*-inspired particle swarm optimization algorithm. Section 4 reports the experiments in eight real-world networks and the comparisons with state of the art algorithms. Section 5 concludes this paper.

2   Related Work

2.1   Community Detection

A network can be composed of nodes and edges, in which nodes usually stand for members and edges represent relationships between members. Let $G = (V, E)$ denote a network, where V and E are the aggregations of nodes and edges, respectively. Aiming at dividing the nodes in a network into different communities, community detection results in that nodes across communities are sparsely connected, while nodes within a community are relatively densely connected. Under the premise that a community is a subset of V and n_c is defined as the number of communities, a community division is a set of communities, $C_i \subset G, C = \{C_1, C_2, \ldots, C_{n_c}\}$, where $C_i \neq \varnothing, \bigcap_{i=1}^{n_c} C_i = \varnothing, \bigcup_{i=1}^{n_c} C_i = G$.

In this paper, the proposed fitness function is the widely used modularity (normally denoted as Q) [3]. The Q function can be written as Eq. (1), where $|V|$ and $|E|$ are the number of nodes and edges of a network, respectively; A is the adjacency matrix of a network and $A_{ij} = 1$ if there exists an edge between node i and j; k_i is the degree of node i, and $\delta(i, j) = 1$ if the nodes i and j are in the same group, otherwise $\delta(i, j) = 0$. Without the loss of generality, we assume that the better division corresponds to the higher Q value. Therefore, the community detection can be transformed into an optimization problem formulated as Eq. (2).

$$fit(\cdot) = Q = \frac{1}{2|E|}\sum_{i,j}^{|V|}(A_{ij} - \frac{k_i \cdot k_j}{2|E|})\delta(i,j) \tag{1}$$

$$C^* = \arg\max_C Q(C, G) \tag{2}$$

2.2 PSO for Community Detection

Derived from the social behavior seen in some animal populations, like fish school and birds flock, PSO is a type of swarm intelligence algorithm proposed by Eberhart and Kennedy in 1995 [4]. The concise framework, simple principle and fast convergence make PSO a popular algorithm for solving continuous optimization problems. Each particle has a position and velocity vector. The position vector usually stimulates a candidate solution to the optimized problem, and the velocity vector denotes the tendency of position updating. A particle updates its status iteratively according to its own and the other particles' experiences to search for the optimal solution. Here, we take a typical PSO for network clustering, termed GDPSO, as an example to introduce the basic parts of PSO for community detection.

Particle representation: Considering that the community detection is a discrete optimized problem, we have to redefine the particle positions. One position vector represents a network division and the position vector of the particle i is defined as $X_i = \{x_i^1, x_i^2, \ldots, x_i^n\}$, where $x_i^j \in [1, n]$ is an integer.

In such definition, x_i^j is called a label identifier standing for the community the node j belongs to. If $x_i^j = x_i^k$, then node j and k belong to the same community. Not only is this coding scheme easy to decode, but also it can determine the number of the communities after division directly. As a result, the computational complexity will be reduced. The coding scheme of the particle is shown in Fig. 1.

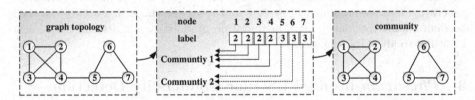

Fig. 1. The coding scheme of the particle in GDPSO. Each particle is coded as a string of integers, which represents the label identifier of the corresponding node.

Particle-status-updating rules: The operation of updating status must be redefined under the discrete background in order to make GDPSO practicable for community detection. The updating rules are put forward as follows:

$$V_i = \omega V_i \oplus (c_1 r_1 (Pbest_i \ominus X_i) + c_2 r_2 (Gbest_i \ominus X_i)) \tag{3}$$

$$X_i = X_i \otimes V_i \tag{4}$$

In the above equations, the $Pbest_i = \{pbest_i^1, pbest_i^2 \ldots, pbest_i^D\}$ and $Gbest_i = \{gbest_i, gbest_i \ldots, gbest_i\}$ are the ith particle's best personal position and the best global position of the swarm, respectively; the inertia weight ω, the learning factors c_1 and c_2 are set typical values of 0.7298, 1.4961 and 1.4961; the r_1 and r_1 are random numbers ranging from 0 to 1.

In Eq. (3), $\ominus$ is defined as an XOR operator. Provided with two velocity vectors $V_1 = \{v_1^1, v_1^2, \ldots, V_1^n\}$ and $V_2 = \{v_2^1, v_2^2, \ldots, v_2^n\}$, $V_1 \oplus V_2 = V_3 = \{v_3^1, v_3^2, \ldots, v_3^n\}$ is a velocity vector with a detailed operation shown as follows:

$$\begin{cases} v_3^i = 0, & rand(0,1) \geqslant \dfrac{1}{1 + e^{-(v_1^i + v_2^i)}} \\ v_3^i = 1, & rand(0,1) < \dfrac{1}{1 + e^{-(v_1^i + v_2^i)}} \end{cases} \tag{5}$$

In Eq. (4), given an old position $X_{old} = \{x_{old}^1, x_{old}^2, \ldots, x_{old}^n\}$ and a velocity $V = \{v_1, v_2, \ldots, v_n\}$, $X_{old} \otimes V = X_{new} = \{x_{new}^1, x_{new}^2, \ldots, x_{new}^n\}$ is a position vector whose element is defined as follows:

$$\begin{cases} x_{new}^i = x_{old}^i, & if \quad v_i = 0 \\ x_{new}^i = \arg\max_j \Delta Q(x_{old}^i, j | j \in L_i), & if \quad v_i = 1 \end{cases} \tag{6}$$

where $L_i = \{l_1, l_2, \ldots, l_k\}$ is the set of label identifiers of node $i's$ neighbors. The ΔQ is calcuated using the following equation:

$$\Delta Q(x_{old}^i, j | j \in L_i) = fit(X_{old} | x_{old}^i \leftarrow j) - fit(X_{old}) \tag{7}$$

In general, each node chooses the community identifier which contributes to the largest increase or the smallest decrease of Q value based on its neighbors.

Mutation: GDPSO implements the mutation operation so as to preserve diversity and avoid falling into local optima. The procedure can be depicted as follows: generating a random number between 0 and 1; for each node in a network, if the random number is smaller than the mutation probability pm, assigning its label identifier to all of its neighbors.

3   *Physarum*-inspired PSO for Community Detection

3.1   The *Physarum*-based network mathematical model

In this paper, PM model is modified into *Physarum*-based network model (PNM) which could be used to recognize the intra-community edges in a network. The key mechanism of PM model is the feedback system between the fluxes and conductivities of tubes based on the Posieuille flow.

First, let $Q_{i,j}^t$, $D_{i,j}^t$, $L_{i,j}$ and p_i^t stand for the flux, the conductivity, the length of $e_{i,j}$ and the pressure of v_i at time step t, respectively. The relationship among these parameters can be represented as Eq. (8). Second, according to the Kirchhoff's law formulated in Eq. (9), the pressure and fluxes can be obtained by solving such equations at each iteration step. Third, $Q_{i,j}^t$ feeds back to $D_{i,j}^t$ based on Eq. (10), and as iteration step t is completed, the iteration step $t+1$ repeats the above procedures on the basis of the data iteration step t returns. Finally, as such positive feedback continues, a highly efficient network is generated [7].

$$Q_{i,j}^t = \frac{Q_{i,j}^t}{L_{i,j}} |p_i^t - p_j^t| \tag{8}$$

$$\sum_i Q_{i,j}^{t-1} = \begin{cases} I_0, & \text{if } v_j \text{ is an inlet} \\ -I_0, & \text{if } v_j \text{ is an outlet} \\ 0, & \text{others} \end{cases} \tag{9}$$

$$D_{i,j}^t = \frac{(Q_{i,j}^t + D_{i,j}^{t-1})}{k} \tag{10}$$

PNM is based on the *Physarum*-inspired Mathematical Model (PM), whose major modification is the scheme of choosing inlets/outlets in each iteration. In such model, a vertice is chosen as an inlet, while the others are chosen as outlets. Namely, Eq. (9) is modified as Eq. (11), where D and L are known. Given a certain inlet and outlet, a set of equations based on Eq. (11) can be obtained. By solving such equations, we get p_i of node i, where i ranges from 1 to $|V|$. Besides, every vertice is chosen as the inlet once in each iteration step of PNM. When v_i is chosen as the inlet, a local conductivity matrix denoted as $D^t(i)$ is calculated based on the feedback system. Eventually, after all local conductivity matrices are obtained, the global conductivity matrix is updated by the average of $D^t(i)$ based on Eq. (12).

$$\sum_i \frac{Q^{t-1}(i)_{i,j}}{L_{i,j}} |p_i^t - p_j^t| = \begin{cases} -I_0, & \text{if } v_j \text{ is an inlet} \\ \dfrac{-I_0}{|V|-1}, & \text{others} \end{cases} \tag{11}$$

$$D^t = \frac{1}{|V|} \sum_i^{|V|} D^t(i) \tag{12}$$

3.2 *Physarum*-Inspired Network Model for Community Detection

Taking advantage of PNM, we roughly distinguish the inter-community edges from intra-community through conductivities. Then, we adopt PNM optimize initialization generating a high-quality initial solution and accelerating convergence.

We can obtain a matrix D through PNM, and suppose that node i has a neighbor set $L(i) = \{l_1, l_2, \ldots, l_k\}$ and let $label(i)$ be the community label which node i belongs to. First, for each node i, we initialize $label(i)$ as i. In addition, we assume that $\Omega_i = \{label(j)|j \in L(i) \text{ and } D_{i,j} < (1 - R\%) * D_{max}\}$ includes the community labels of neighbors of node i. Namely, the top $R\%$ conductivities $D_{i,j}$ denote that the edges between node i and j are inter-community edges. Then, each node randomly selects an element from Ω_i as its new label.

For the next step, the label propagation is utilized to optimize preliminary initial solution further. Each node determines its community label based on the labels of its neighbors. We assume that each node in the network chooses to join the community with the largest number of its neighbors, which can be represented as Eq. (13), where $\delta(i,j)$ is 1, if node i and j belong to the same community, otherwise $\delta(i,j)$ is 0. This step is executed *iters* times where *iters* is the number of propagation iteration. For a clear expression, with a prefix (i.e., P−) added to the original GDPSO algorithm for distinction, the novel algorithm is denoted as P-PSO. The detailed process of P-PSO is shown in Algorithm 1.

$$label(i) = \arg\max_r \sum_{j\in L(i)} \delta(label(j), r) \tag{13}$$

Algorithm 1. The framework of P-PSO

Input: An adjacent matrix A and the label propagation iterations: *iters*
Output: The community division of a network
1. Calculating the conductivity matrix D;
2. Initializing the population that each node has unique label in each particle;
3. **for** each *particle* $\in$ *population* **do**
4.   **for** $i = 1 : nodes$ **do**
5. $label(i) \leftarrow$ *choose a label randomly from* Ω_i;
6.   **for** $j = 1 : iters$ **do**
7.   **for** $j = 1 : nodes$ **do**
8. $label(i) \leftarrow$ *formula* (13);
9. Evaluating the fitness of population and initializing the *Gbest* particle;
10. **while** not satisfy the terminal condition **do**
11.   **for** each *particle* **do**
12.   Updating particle status, see Sect. 2.2 for more information;
13.   Operateing mutation on particle, see Sect. 2.2 for more information;
14.   Evaluating the fitness of particle and updating the *Pbest* particle;
15.   Evaluating the fitness of swarm and updating the *Gbest* particle;

---

4   Experiments and Results

All experiments are executed in the same environment to enable fair comparisons between our algorithm and other algorithms including GDPSO [5], IACO-Net [12] and PNGACD [13]. All results are averaged over 30 repeated runnings in order to eliminate fluctuation. There are two popular metrics for evaluating the performance of community detection: the modularity Q and normalized mutual information (NMI) [10].

4.1 Results on Benchmark Networks

Some experiments are carried out in the GN benchmark network proposed by Lancichinetti et al. [11]. α denotes the mixing parameter which controlls the proportion of links within and out of a community. We test all algorithms in eleven computer-generated networks with the value of α ranging from 0 to 0.5.

As shown in Fig. 2, when the mixing parameter is no larger than 0.1, all algorithms except PNGACD can discover the correct communities ($NMI = 1$). With the mixing parameter increasing, the IACO-Net fails to detect the true partitions. For $\alpha = 0.4$, P-PSO and GDPSO still obtain $NMI = 1$. When α is larger than 0.4, the NMI of GDPSO decreases more quickly than the proposed P-PSO. The experiments in the GN benchmark networks prove that P-PSO is feasible for community detection.

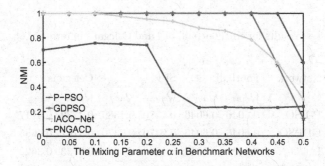

Fig. 2. The experimental results from the GN benchmark networks.

4.2 Results on Real-World Networks

Table 1 shows the structural characteristics of eight real-world networks used in our experiments for evaluating the performance of our proposed method.

Figure 3 shows the results that some experiments are implemented to verify the robustness of P-PSO in the four networks. It can be concluded that P-PSO has a better stability than that of GDPSO. Table 2 reports the maximal and

Table 1. Networks used in this paper. *Clusters* stands for the number of communities in standard divisions, in which "–" means that the standard division is non-existent.

Network	Nodes	Edges	Clusters	Network	Nodes	Edges	Clusters
Karate	34	78	4	Dolphins	62	159	2
Polbooks	105	441	3	Football	115	613	12
Lesmis	77	254	–	Adjnoun	112	425	–
SFI	118	200	–	Celegans	297	1540	–

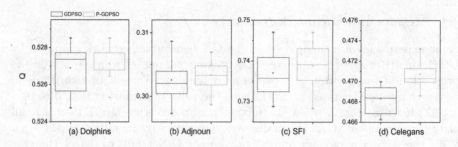

Fig. 3. The average Q of the final iteration in four real-world networks. The upper and lower ends of whiskers represent the maximum and minimum of Q, and the vertical height of the box ranges from the first and the third quartiles. Besides, the small square and band inside the box denote the average and median of Q, respectively. These box charts demonstrate that P-PSO is inclined to a better robustness in community detection.

Table 2. The test results for the Football, SFI and Celegans in terms of Q_{max} and Q_{avg}

Network	Football		SFI		Celegans	
	Q_{max}	Q_{avg}	Q_{max}	Q_{avg}	Q_{max}	Q_{avg}
P-PSO	0.6046	0.6046	0.7470	0.7389	0.4732	0.4717
GDPSO	0.6046	0.6046	0.7470	0.7370	0.4707	0.4685
IACO-Net	0.6032	0.5817	0.1940	0.1969	0.3733	0.3622
PNGACD	0.5973	0.5856	0.7457	0.7400	0.2914	0.2903

mean values of Q in other real-world networks. Results show that P-PSO is substantially better than the compared algorithms.

Figure 4 reports the dynamic average modularity with the increment of iteration. The optimized algorithm P-PSO has a higher growth rate than the original GDPSO at the initial phase. The difference between them becomes smaller with the increment of iteration, and yet P-PSO converges faster than GDPSO. Above all, P-PSO shows a superiority in Q value during the whole iteration process.

Figure 5 shows the community divisions in Polbooks and Football. In Fig. 5(a), the geometric figures denote the real communities and the colors denote communities detected by P-PSO. Due to the context of books, some books are connected more closely and form smaller communities, which disorganizes the original divisions in the real world. In terms of the Football network, the positions are denoted as the real division and the colors mean five communities in the division of P-PSO. Each node represents a football team in the real world, and an edge stands for a game they have together. The marked circle emphasizes the main difference between the detected communities by P-PSO and the real communities.

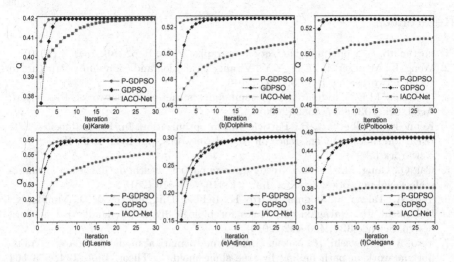

Fig. 4. The dynamic Q with the increment of iteration. The results show that the proposed algorithm can accelerate the convergence, compared with GDPSO and IACO-Net.

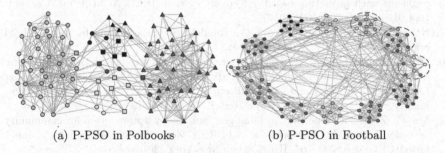

(a) P-PSO in Polbooks   (b) P-PSO in Football

Fig. 5. The visualizations of community divisions in two networks

5   Conclusion

The research about community detection is helpful for us to analyze the basic characteristics of networks. Taking advantage of the *Physarum* network model (PNM) and greedy discrete particle swarm optimization algorithm (GDPSO), we propose a particle swarm optimization algorithm (P-PSO). The experimental results in eight real-world networks demonstrate that P-PSO shows a better ability in optimizing the initial solution and can obtain effective and promising results than other state of the art algorithms.

Acknowledgments. Zhengpeng Chen and Fanzhen Liu contributed equally to this work and should be considered as co-first authors. This work is supported by the National Natural Science Foundation of China (Nos. 61402379, 61403315), Fundamental Research Funds for the Central Universities (No. XDJK2016A008, XDJK2016B029, XDJK2016E074), CQ CSTC (cstc2015gjhz40002).

References

1. Fortunato, S.: Community detection in graphs. Phys. Rep. **486**, 75–174 (2010)
2. Weng, L., Menczer, F., Ahn, Y.Y.: Virality prediction and community structure in social networks. Sci. Rep. **3**, 2522 (2013)
3. Newman, M.E.: Modularity and community structure in networks. Proc. Natl. Acad. Sci. **103**, 8577–8582 (2006)
4. Kennedy, J., Eberhart, R.: Particle swarm optimization. In: Proceedings of 1995 IEEE International Conference on Neural Networks, pp. 1942–1948. IEEE Press, New York (1995)
5. Cai, Q., Gong, M., Ma, L.: Greedy discrete particle swarm optimization for large-scale social network clustering. Inf. Sci. **316**, 503–516 (2015)
6. Tero, A., Takagi, S., Saigusa, T., Ito, K., Bebber, D.P., Fricker, M.D., Yumiki, K., Kobayashi, R., Nakagaki, T.: Rules for biologically inspired adaptive network design. Science **327**, 439–442 (2010)
7. Tero, A., Kobayashi, R., Nakagaki, T.: A mathematical model for adaptive transport network in path finding by true slime mold. J. Theor. Biol. **224**, 553–564 (2007)
8. Liu, Y., Gao, C., Zhang, Z., Lu, Y., Chen, S., Liang, M., Tao, L.: Solving np-hard problems with physarum-based ant colony system. IEEE/ACM Trans. Comput. Biol. Bioinf. **14**, 108–120 (2017)
9. Nakagaki, T., Yamada, H., Tóth, Á.: Intelligence: maze-solving by an amoeboid organism. Nature **407**, 470–470 (2000)
10. Danon, L., Daz-Guilera, A., Duch, J., Arenas, A.: Comparing community structure identification. J. Stat. Mech. Theory Exp. **2005**, P09008 (2005)
11. Lancichinetti, A., Fortunato, S., Radicchi, F.: Benchmark graphs for testing community detection algorithms. Phys. Rev. E **78**, 046110 (2008)
12. Mu, C., Zhang, J., Jiao, L.: An intelligent ant colony optimization for community detection in complex networks. In: 2014 IEEE Congress on Evolutionary Computation(CEC), pp. 700–706. IEEE Press, New York (2014)
13. Gao, C., Liang, M., Li, X., Zhang, Z., Wang, Z.: Network community detection based on the Physarum-inspired computational framework. IEEE/ACM Trans. Comput. Biol. Bioinf. (2016). doi:10.1109/TCBB.2016.2638824

The Design and Development of the Virtual Learning Community for Teaching Resources Personalized Recommendation

Bo Song[1(⊠)], Haihui Wu[1], Xiaomei Li[2], Liyan Guo[3], and Chang Liu[3]

[1] College of Software, Shenyang Normal University,
Shenyang 110034, Liaoning, China
songbo63@aliyun.com, wuhaihui0305@126.com
[2] Basic Education in Liaoning Province Inquest into the Center,
Shenyang 110034, China
lxm63@Yahoo.com.vn
[3] Institute for Basic Education, Shenyang Normal University,
Shenyang 110034, China

Abstract. With the integration of advanced concepts and technologies, such as Web2.0, Web3.0 and cloud computing, the teaching resources of the virtual learning community are constantly enriched and expanded to bring diversification and autonomy to the users. It also to different backgrounds and different preferences of the user's teaching resources required for the inconvenience in a large number of teaching resources can not accurately choose their own teaching resources. In order to more effectively recommend the products of interest to users, this paper uses Java EE 5 technology to achieve a user interface with good interactivity. At the same time, an ALS recommendation algorithm based on Spark is proposed, and the performance and effect evaluation of the system are made in the virtual learning community.

Keywords: Virtual learning community · Java EE 5 · Spark · ALS

1 Introduction

Virtual Learning Community, also known as Online Learning Community, Electronic Learning Community and so on, It is based on the constructivist learning theory, based on computer information processing technology, computer network resource sharing technology and multimedia information display technology of new distance education network teaching support platform, while the virtual learning community is also a new type of learning organization, so it not only has the sociological properties, but also has the basic properties of human-machine system characteristics [1]. The existing types of virtual communities are structured and procedural static groups based on curriculum, research topics or learning tasks under the auspices of teachers. The composition of the group is the random grouping of learners or the designated grouping of teachers. Whether these group members have similar cognitive characteristics and learning needs. However, this kind of learning system is too much emphasis on the teacher's dominance and the completion of the curriculum tasks, members of the personalized

© Springer International Publishing AG 2017
Y. Tan et al. (Eds.): ICSI 2017, Part II, LNCS 10386, pp. 109–118, 2017.
DOI: 10.1007/978-3-319-61833-3_12

learning needs and depth of the socialization of knowledge construction process is ignored, did not well reflect the learners in the collective wisdom spread and social cognitive construction side of the great advantages [2].

This article will be personalized recommendation ideas into the virtual community, used to recommend personalized teaching resources for users. On the one hand, virtual learning community environment, a variety of teaching resources, a large number of users in the process of resource selection decision-making is very easy to choose the phenomenon; the other hand, the virtual learning community environment users are mostly different professional background, Different learning preferences and different learning needs of users, their demand for teaching resources are very different, users need their own types of resources and resource access path, the community design and platform for the consideration of the user's personality and preferences, The ability to take the initiative to push personalized learning support services to help reduce their time and effort investment. For these two reasons, it is very necessary to study the personalized recommendation of teaching resources in the virtual learning community, and it is also very useful for reference [3].

2 Spark

Spark is UC Berkeley University AMP Laboratory open source similar MapReduce computing framework, it is a memory-based cluster computing system, the initial goal is to solve the cost of MapReduce disk read and write. Its core logic organizational structure RDD (Resilient Distributed Datasets),Spark Streaming (Real-time large-scale streaming computing framework), Spark SQL (Support for structured data SQL query engine), MLlib (Distributed machine learning library) and other modules. Compared with the programming model provided by Mapreduce, Spark framework is based on memory calculation, so the calculation speed is faster, speed can be increased by 10 to 100 times [4]. Fault-tolerant mechanism is also more complete, the calculation process, if a problem occurs, the cost of event recovery and recurrence is far less than Mapreduce. It also outperforms Mapreduce in implementing policy and task scheduling overhead. Spark will be unified into various types of data structures are abstracted as RDD structure, MapReduce, Spark Streaming, Spark SQL, Machine Learning on Spark, GraphX and other models into a computing framework, the public also use a unified API, so we You can use Spark one-stop solution to a variety of large data areas of the problem, not only shortened the learning time of the developer, but also greatly improve its processing efficiency [5].

3 Implementation of Personalized Recommendation System of Teaching Resources in Virtual Learning Community

3.1 System Structure

In the design of the system architecture, the hierarchical system is adopted to improve the logic and maintainability of the system. The virtual learning community teaching

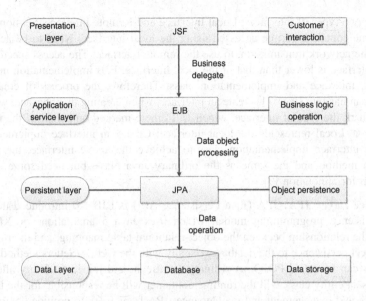

Fig. 1. System architecture diagram

resources recommendation system is divided into four layers: presentation layer, application service layer, persistence layer and database layer. The architecture of JSF2.0, EJB3.0 and JPA 2.0 is chosen as the technical framework of the 4-layer architecture, which is based on Java EE5 platform and B/S mode, according to the idea of security, stability and layering in the technical architecture design. Based on JSF2.0, EJB3.0 and JPA 2.0 integration framework of the system architecture shown in Fig. 1 [6, 7].

3.2 Key Technology Design

Presentation Layer JSF. JSF full name of Java Server Faces, which is the Java EE MVC specification, to promote the way the page components to hide the traditional Web application development HTTP details, allowing developers to traditional desktop programming approach to the development of Web applications. JSF by the managed bean properties, methods directly bound to the page component value attribute or action attribute, you can very easily achieve the system MVC control [8]. In JSF 1.x, Facelets is only used as a replacement for JSP-based views. Since JSF 2.0, Facelets has replaced JSP, JSF's default view technology. Each Web application has its own unique style, a Web site, all the pages have the same layout and style. Facelets can be a common layout and style into a template, so you can change the template to update the face of the entire site, without having to individually update the page.

Business Kogic Layer EJB. In this article, only the session bean is used, and the interface of the session bean specifies that the session bean provides service content.

There are two types of interfaces: Local Interface and Remote Interface. Session beans in the same container can be accessed using the local interface. Need to be accessed through the network transmission, to use the remote interface. The access speed of the remote interface is lower than that of the local interface. EJB implementation includes two parts: interface and implementation class. Therefore, the process of creating a session bean has two steps: (1) create an interface, when creating the interface, with the mark to mark the type of interface, which @ Remote marked on behalf of the remote interface, @ Local represents the local interface; Create an interface implementation class, the interface implementation class to achieve the above interface, the implementation method and the same as the ordinary Java class, but need some special annotation for annotation [9].

Persistence Layer JPA. JPA (Java Persistence API) is EJB 3.0 introduced the new data persistence programming model, which uses Java 5 annotations or XML to describe the relationship between the object-relational table mapping, and the run-time entity object persistence to the database. Usually with the Getters/Setters method of the Java class can be called POJO, as ordinary JavaBean, and joined @ Entity after it is defined as an entity class, edit the runtime container will be responsible for the type of persistence and management and maintenance. JPA framework to provide flexible and rich API, used to operate the entity object, the implementation of, delete, change, check and other operations, all operations are completed by the framework of the background, so developers from cumbersome JDBC and SQL code freed. The main advantages of the JPA framework are its standardization, support for container-level features, ease of integration, efficient query capabilities, support for object-oriented [10].

4   Recommendation Algorithm Design

4.1   ALS Algorithm

ALS (Alternating Least Squares), is an optimization method to solve the matrix decomposition problem. This method is often used in the recommendation system based on matrix decomposition. For example: the user (user) on the product (item) score matrix is divided into two matrix: one for the goods contained in the implied feature matrix, a user for the product implied characteristics of the preference matrix. In the process of matrix decomposition, the score missing item is filled, so the user can be based on the fill of the score for the product recommended [11].

4.2   ALS Algorithm Problems

Collaborative filtering algorithm based on ALS is a highly efficient matrix decomposition algorithm, can recommend a good application and the actual system, but the ALS algorithm has some characteristics, its every iterative calculation to input a lot of score data, also need to transport the characteristic matrix, if run on a single node will meet the bottleneck of computing time and computing resources. Especially when the user behavior data is very large, the recommended calculation on a single node will take a lot of time, so the recommendation system can not get the results in time. So we need to implement ALS algorithm on the distributed platform. In addition, the ALS algorithm has some information (such as the similarity between users and the similarity between learning resources) of the lost user or learning resource. Therefore, this paper improves the ALS algorithm, The similarity between learning resources, and the ALS algorithm in parallel to improve the efficiency and recommended accuracy.

4.3   Improved Design of ALS Algorithm

In this paper, the ALS algorithm is implemented on the Spark platform. The process of the algorithm is shown in Fig. 2:

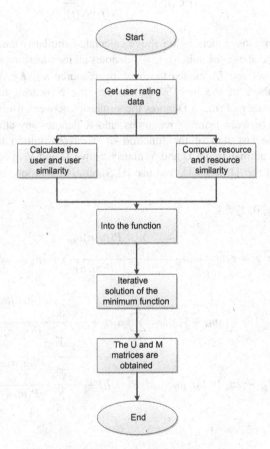

Fig. 2. Flow chart for optimizing ALS algorithm

(1) Step 1: First, the user's score matrix is obtained by user rating history data, and the similarity between users and learning resources is calculated. In this paper, we use spatial similarity method (VSS), also called vector cosine method, where $N(m)$ is the learning resource set owned by user m, that is, user m has positive feedback to learning resource, N(n), The similarity formula between user m and n is:

$$PC(m,n) = \frac{|N(m) \cap N(n)|}{\sqrt{|N(m)||N(n)|}} \tag{1}$$

Similarly, the similarity formula between learning resources i and j is:

$$PC(i,j) = \frac{|N(i) \cap N(j)|}{\sqrt{|N(i)||N(j)|}} \tag{2}$$

(2) Step 2, design a loss function, the above calculated similarity data fusion into the function. In the above formula, $(i,j) \in I$ denotes all user-learning resource pairs, I_j denotes the user set, I_i denotes the learning resource set, $K(u_i)$ denotes the N nearest neighbors of the user, $K(g_j)$ denotes the N nearest neighbors of the learning resource g_j, P(m, n) Denotes the similarity between users, P(i, j) denotes the similarity between learning resources, and K denotes any attribute.

(3) Step 3, in the last step of the function of iterative solution to minimize the function, and ultimately get U and V matrix. Solve U and V in concrete, solve U according to fixed U, V fix U, and use ALS iterative solution, as follows:

$$\frac{1}{2} - \frac{\partial f}{\partial u_{ki}} = 0, \quad \forall i,k$$

$$\Rightarrow \sum_{j \in I_i} (u_i^T g_j - r_{ij}) g_{kj} + (u_{ki} - \frac{\sum_{u_p \in K(u_i)} P(m,n) u_{kp}}{\sum_{u_p \in K(u_i)} P(m,n)}) + \lambda n_{u_i} u_{ki} = 0, \quad \forall i,k$$

$$\Rightarrow \sum_{j \in I_i} g_{kj} g_j^T u_i + (\lambda n_{ui} + 1) u_{ki} = \sum_{j \in I_i} g_{kj} r_{ij} + \frac{\sum_{u_p \in K(u_i)} P(m,n) u_{kp}}{\sum_{u_p \in K(u_i)} P(m,n)}, \quad \forall i,k \tag{3}$$

$$\Rightarrow (M_{I_i} M_{I_i}^T + (\lambda n_{u_i} + 1)E) u_i = M_{I_i} R^T(i, I_i) + \frac{\sum_{u_p \in K(u_i)} P(m,n) u_p}{\sum_{u_p \in K(u_i)} P(m,n)}, \quad \forall i$$

$$\Rightarrow u_i = A_i'^{-1} V_i', \quad \forall i$$

The above is to find each iteration u_j, the same token, for g_j,

$$g_j = (U_{I_i}U_{I_i}^T + (\lambda n_{g_j} + 1)E)^{-1} \left[V_I, R^T(j, I_J) + \frac{\sum\limits_{g_q \in K(g_j)} P(i,j)g_q}{\sum\limits_{g_q \in K(g_j)} P(i,j)} \right], \quad \forall j \qquad (4)$$

4.4 ALS Algorithm Evaluation Index

After designing and implementing the personalized recommendation system for teaching resources, it is necessary to adjust and evaluate the recommended model, so we need to use the evaluation algorithm. This experiment adopts the off-line evaluation algorithm. The evaluation algorithm plays a very important role as a measure of the recommended effect of the recommendation system. So the choice of the evaluation algorithm and the characteristics of the algorithm of different evaluation need to be taken into account so as to compare the recommendation effect of a recommendation system comprehensively and compare it with other recommendation system and get a comparison result of the recommendation system Comprehensive understanding, which for the recommendation system tuning and the next step quickly, provides an important reference.

In this paper, MAE and NDCG evaluation algorithms are selected as the evaluation algorithm of the proposed system. Where MAE is the average absolute error. By calculating the error between the predicted user score and the actual user score, the specific definitions are as follows:

$$MAE = \frac{\sum_g test \subset G^{test} |g^{test}(prdeiction) - g^{test}(authentic)|}{|G^{test}|} \qquad (5)$$

Where $g^{test}(authentic)$ is the true score, $g^{test}(prediction)$ is the predicted score, and G^{test} is the entire subscriber score to be predicted. NDCG evaluation algorithm is mainly used in information retrieval, is in the search engine search results in the accuracy of the commonly used evaluation algorithm. The main basic principle is that the importance of different relevance documents in the information retrieval is not the same, so the different importance of the document to give a different score. From the user's point of view, the user will not read all the documents, usually just look at the first few documents. After getting a sorted document sequence, the NDCG value at the rth bit is calculated as:

$$NCDG@r = N_r \cdot \sum\nolimits_j^r = 1 \frac{2^{r(j)-1}}{\log(1+j)} \qquad (6)$$

Where 1 is the scoring of the jth document and 2 is the normalized parameter to ensure that the result of the optimal NDCG is 1. If the number n of documents in the result sequence is less than r, the formula returns the value of NDCG @ n [12].

4.5   Experiment and Result Analysis

The result of MAE is shown in Fig. 3:

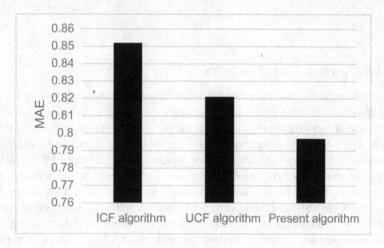

Fig. 3. MAE algorithm comparison chart

It can be seen from the above figure that the proposed algorithm proposed in this paper is more accurate than the other two algorithms. NDCG @ r evaluation algorithm comparison results shown in Fig. 4, the r value of this article take 4.

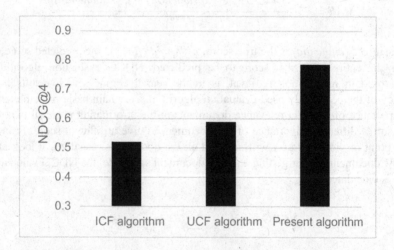

Fig. 4. NCDG @ 4 evaluation algorithm comparison chart

From the above figure, it can be shown that the recommended algorithm in this paper predicts the user's preference for teaching resources, and it is far more than the collaborative filtering algorithm based on the project and the collaborative filtering algorithm based on the user. According to the analysis, it can be seen that the MAE evaluation algorithm can describe the accuracy of the recommended prediction, but the effect of the order of the recommended results on the recommended effect can not be realized. The NDCG evaluation algorithm has the advantage of this recommendation. Algorithm in the recommendation of the results of the order of the advantages reflected, so in the NDCG evaluation algorithm on the proposed algorithm is better than the MAE evaluation algorithm contrast effect. In summary, the proposed algorithm proposed in this paper can meet the requirements of teaching resources recommended, can be applied to the system.

5 Conclusion

Based on Java EE technology, this paper constructs a virtual learning community teaching resource recommendation system. System according to the specific circumstances of the student user is the user recommended teaching resources, to achieve a personalized education for students with a good learning environment, according to each student's specific circumstances, recommend learning content to meet the learning needs of students. Teaching methods to overcome the single-sided disadvantage. Its use can effectively shorten the time for each student learning, improve teaching quality and efficiency, optimize the teaching objectives. Figure 5 shows the average response time based on the Java EE framework and SSH systems, test scenario is the client information query program.

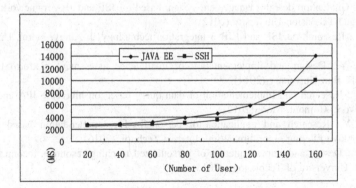

Fig. 5. AART curve.

In analysis, curve in Fig. 5 we can find that, this paper puts forward the solution of the system response time according to the curve of slow increase, until you reach 100 users, growth curve. My solution compared with synthetic solution, the solution of SSH framework of system response time increases according to the linear way, until you reach about 100 users, while the Java EE framework of solutions of the system response time is growing faster than SSH, is relatively stable. Through the analysis of the changing trend of curve, in performance test, maximum limit the number of users is 100, so meet the needs of the application service.

The Java EE framework and SSH framework and there is no essential difference, the difference between them before just the differences of implementation technology, and the Java EE framework suitable for application in large enterprises, while the SSH framework suitable for application in small and medium enterprises.

Acknowledgments. Project supported by the Education Department of Liaoning province science and Technology Research Fund Project (L2013417).

References

1. Wang, L.: Virtual Learning Community Principles and Applications. Higher Education Press, Beijing (2006)
2. Yang, L., Meng, Z., Yan, Z.: Dynamic construction of virtual community learning community based on personalized recommendation. Mod. Educ. Technol. (2012)
3. Sun, L.: Personalized recommendation technology and its application in distance education. Science Weekly (2015)
4. Zaharia, M., Chowdhury, M., Franklin, M.J., et al.: Spark: cluster computing with working sets. In: USENIX Conference on Hot Topics in Cloud Computing. USENIX Association (2010)
5. Wendong L.: The research and implementation of mining large data based on spark. Shandong University (2015)
6. Yuan, M., Wang, H.: Java EE detailed explanation of enterprise programming development example. Tsinghua University Press (2013)
7. He, L.: Graduation design management system based on JSF and JPA framework. Jiangsu Radio and Television University (2012)
8. Qin, L.: Research on JSF and EJB 3 integration technology. J. Xingyi Norm. Univ. Natl. (2011)
9. Zhang, Y.: Design and implementation of intelligence assessment platform based on Java EE. Jilin University (2014)
10. Tan, Y.: Research and implementation of data query based on hibernate JPA and JQuery Framework. Comput. Mod. (2012)
11. Dong, Y.: Research and application of personalized recommendation based on ALS collaborative filtering algorithm. Wirel. Internet Technol. (2016)
12. Niu, Y.: Design and implementation of personalized teaching resources recommendation system. University of Technology (2013)

Effects of Event Sentiment on Product Recommendations in a Microblog Platform

Ping-Yu Hsu[1], Ming-Chia Hsu[1(✉)], Tien-Hao Wei[1], Yao-Chung Lo[1], Chin-Chun Lo[1], Ming Shien Cheng[2], and Hong Tsuen Lei[1]

[1] Department of Business Administration, National Central University, No. 300, Jhongda Road, Jhongli City 32001, Taoyuan County, Taiwan (R.O.C.)
mingjia.hsu@gmail.com
[2] Department of Industrial Engineering and Management, Ming Chi University of Technology, No. 84, Gongzhuan Road, Taishan District, New Taipei City 24301, Taiwan (R.O.C.)
mscheng@mail.mcut.edu.tw

Abstract. The occurrence of a planned event often closely correlates with human emotions. Since words describing an emotion are important vehicles for capturing an event, to understand how emotions affect subsequent product recommendations, in this study, we use the microblogging platform "Plurk" in conjunction with a Chinese emotional lexicon developed previously. More specifically, we identify emotional keywords as a basis for distinguishing user sentiment in event-based posts; in the meanwhile, we use these emotional keywords as a gauge for product recommendation. Further, we utilize T-test, one-way ANOVA, and two-way ANOVA approaches to explore the effects different emotional variables have on product recommendations. Our experimental results show that under the influence of positive events, negative emotional posts of a microblog user can enhance the product recommendation effect better than the positive posts themselves. In product recommendations with various emotional words, positive emotional recommendations are able to achieve better recommendations than that of either negative or neutral recommendations. The results of our study can effectively help service and product suppliers carry out social marketing during a given event period and utilize emotional variables to effectively attract user interest in clicking on advertisements on social network sites, thereby achieving the goal of enhancing the product recommendation effect.

Keywords: Event · Emotion · Microblog · Plurk · Product recommendation

1 Introduction

In recent years, social media's rapid development has become a major part of many people's lives worldwide, not only because of the time and space free advantages while sending messages but also because of the significant social media features, such as social contagiousness, real-time interactions, and social integration. Among social networking sites (SNSs), the "microblog," which differs from general SNSs, consists of content that is markedly low in terms of richness, but it is very easy to post and is

© Springer International Publishing AG 2017
Y. Tan et al. (Eds.): ICSI 2017, Part II, LNCS 10386, pp. 119–131, 2017.
DOI: 10.1007/978-3-319-61833-3_13

highly confidential (i.e., anonymous). Given this, the frequency of posts and interactions between microblog users provides, at the moment a message is transmitted, a true presentation of the users' moods, thoughts, opinions, and intentions.

Regardless of whether we take the perspective of sociology, psychology, or business, the occurrence of events and human emotions are closely related. Emotional words are therefore important vehicles for describing these events. Banerjee et al. [2] noted that compared with other SNSs, in microblogs, users are able to convey their current mood, opinions, ideas, and intentions in a more representative manner through their posts. Meanwhile, they can stimulate and enhance the interaction level within their circle of friends. Huffaker [10] noted that in a network group, compared with non-emotional messages, emotional messages and language can trigger more responses.

In related studies of service and product recommendations, results show that recommendations based on "sentiment" and "reputation" are the most suitable for meeting user demands. Dong et al. [6] developed a product recommendation methodology based on sentiment that significantly improved the success of product recommendations based on similarity.

Summarizing, we note that emotion plays a key role in the occurrence of a special event, in a user's message, and in a corresponding product recommendation advertisement. Utilizing microblogs could efficiently convey users' emotional characteristics, thereby providing service and product suppliers excellent channels and opportunities for text mining technologies to perform emotional analysis. Further, combined with psychological theory, we can couple emotional events, message sentiment moods, and product recommendation moods with each other, thereby enhancing social marketing impact on users.

2   Literature Review

2.1   Special Events Versus Emotional Effects

In natural language research, special events are defined as "actions under occurring or the imminent action". Similarly, Popescu et al. define a particular event as "an activity or an action that occurs within a definite and finite duration of time" and involves a specific "entity" or "object" as the main focus at the time of the event. In this study, we define events as "activities that occur or are about to take place within a limited time period with specific entities and/or objects as the primary subject."

Naveed et al. [19] noted in their study on the microblog "Twitter" that compared with personal topics between users, economic issues, Christmas, social media, and public-related events are more likely to trigger motivations to forward such messages. In the process of emotional analysis, Choudhury et al. [7] found that the conversion and intensity of positive and negative emotions between a given network group can be affected by real-world event impact, thus producing significant differences. Kramer noted that the emotional status of users within the social network function would change based on the occurrence of the given event or events.

Therefore, emotional terms are a crucial medium that people use to characterize and effectively describe events, as the occurrence of these events are closely related to human emotions, whether from the point of view of sociology, psychology, or business [14]. In our study, we use "Christmas" and "Political election" as two special events as examples. Here Christmas can generate positive emotions in the microblogging platform, regardless of whether such posts occur in Eastern or Western countries. On the other hand, during the recent US political election, the negative emotion was twice that of the positive emotion on Twitter, and up to 58% of the articles contained negative emotions toward some people, events, and things [18].

Based on the above, we suggest that the occurrence of special events leads to an emotional effect on the social platform and also causes a positive or negative emotional status for most users. Therefore, the crucial issue in this study is how to improve the product recommendation effect under the influence of event emotion.

2.2 Social Networking Sites Versus Social Marketing

SNSs provide an "Internet platform" that allows users to interact with one another; this platform is similar to social interactions in the "real world" in which participants achieve functions of spreading information and sending messages to one another. Over the past few years, SNSs have attracted an enormous number of users worldwide and become one of the most visited sites on the Internet [8].

The diversified community generated by these SNSs has impelled academics and marketers to study social interactions, social relationships, and social influence among users. They expect to leverage social attribution studies to traditional promotions and product advertisements applicable on SNSs; through this kind of "social marketing," service and product suppliers could successfully ramp up user interest and achieve increased marketing and sales.

In a related study of emotional referrals, Liao noted that emotional terms in microblogs can improve the advertisement's effectiveness. Further, the simpler and stronger the emotion itself, the stronger the effect of the referral; however, a number of hypotheses in the results of the study were interpreted at a 90% confidence level. From this research, we realize that user sentiment has a crucial influence on product recommendation. But in view of Liao's 90% confidence level apparently counteracting its reliability, our study advocates adding event mood and recommendation emotion as two variables to improve the products recommended effect on microblogs.

2.3 Emotional and Psychological Theory

According to Scherer's "appraisal theory," our study considers that the generation of emotion is caused by individual cognitive differences stemming from events, things, or situations. More specifically, when facing the same event (i.e., stimulation), exactly what kind of sentiment (i.e., emotional state) that will arise depends on the individual's "cognitive appraisal" of the event. This implies that even if one encounters the same situation, the state and intensity of the resulting emotion of each person varies. Meanwhile, the occurrence of events are considered the main cause of the sentiment (i.e., emotional state), and it can immediately trigger the generation of or change in emotion [21].

2.4   Sentiment Analysis Versus Emotional Lexicon

Strictly considering the psychological and emotional analysis theses, we divide user sentiment into positive and negative emotions. Further, we divided the advertisements into positive, negative, and neutral emotions as our approach to handle recommendations. We also focus on how the emotional effects and emotional variables impact recommendation results. Following on the heels of the majority, we utilize the Chinese emotional lexicon of Zhuo et al. [24] as the basis of our emotional analysis. This lexicon specifies both the "emotional descriptor" and the "emotional inductor." Among these, the emotional descriptor describes terms that can directly describe and express the feeling of the given emotion; each term contains six orientation attribute scores, i.e., valence, arousal, continuance, dominance, frequency, and typicality. The emotional inductors are terms that can indirectly trigger an emotional state; each term contains four orientation attribute scores, i.e., valence, continuance, dominance, and frequency. Finally, according to the standard of the valance in the lexicon to calculate and to proceed the estimation and classification of positive and negative sentiments (i.e., emotions).

3   Research Methodology

3.1   Plurk Versus Data Collection

Plurk is a favorite microblog in Taiwan. Plurk users are called plurkers. There are various functional robots on the Plurk platform, and these robots detect keywords issued by plurkers in their messages, then produce corresponding responses. This study utilizes the phrasing robot "Han Kyrgyzstan Qian" to produce product recommendations. More specifically, when plurkers in the friends list of "Han Kyrgyzstan Qian" issue a post, the phrasing robot automatically recommends products in the discussion thread; there is also a bonus offering in which the robot invites active users (i.e., "those with high Karma value plurkers") to make friends with our phrasing robot and increase the product recommendation frequency and recommended reach.

3.2   Product Recommendation Method

In this study, we utilize the aforementioned phrasing robot together with the Chinese emotional lexicon of Zhuo et al. [24] to recommend website products on the books. com.tw site to Plurk users. Here, there are two kinds of product recommendations, i.e., sentimental recommendations and non-sentimental recommendations. For sentimental recommendations, the system randomly selects keywords from the emotional lexicon at once, and these keywords are transferred to books.com.tw via Google's search engine API. For non-sentimental recommendations, the system randomly recommends website products on books.com.tw without any emotional terms. We collected 410 data points during the time period covering Christmas and the US election, i.e., from November 9, 2015 through January 20, 2016. The collected data included the post content, emotional keywords, time of release, product description, the number of clicks, and other such data obtained via either sentiment or non-sentimental recommendations.

3.3 Hypothesis

Based on the influence of commercialism, "shopping" is a major conversation topic during Christmastime. We utilized the Chinese emotional lexicon of Zhuo et al. [24] as the basis in which both "shopping" and "window-shopping" are positive emotional terms. Therefore, in our study, we suggest that Christmas is an event with positive emotions. Hence, we proposed our first control variable as follows:

Control Variable 1 (CV1): Christmas is a positive emotional event.

Joyce and Kraut [11] found that messages with negative sentiment could motivate members to participate in the given discussion. Similarly, Thelwall et al. [22] used Twitter to show that messages with negative sentiment could efficiently attract users. Schumaker et al. also noted that negative emotions are more influential than positive or neutral emotions. Liao's research on the microblog Plurk also showed that users posting with negative emotions could significantly boost product click-through rates. In summary, we deduce that during the occurrence period of positive events, posts with negative emotions could better resonate with user feelings, thereby provoking the interests of users clicking on advertisements. Hence, our first hypothesis is as follows:

Hypothesis 1 (H1): During the occurrence of a positive event, posts by plurkers with negative sentiments could better boost the click-through rate of product recommendations.

Schwarz and Clore's research on aversion theory (1996) argued that people may adopt different behaviors and modes of thinking that differ from their emotional status. Similarly, Lang [16] noted that when facing positive sentimental stimulus, people tend to approach it. Based on these studies, we deduce that during the occurrence of a given positive event, microblog users tend to remain in a positive emotional mood; therefore, product recommendations with positive sentiments could attract user attention. Hence, our second hypothesis is as follows:

Hypothesis 2 (H2): During the occurrence of a positive event, product recommendations with positive sentiments could better boost click-through rates of product recommendations versus those with negative or neutral sentiments.

According to aversion theory, people may adopt different behaviors and modes of thinking that differ from their emotional status; these emotional states could trigger the process of an automatic reaction, and the negative sentiments could evoke avoidance actions. According to the aversion theory, our study holds that to break out of the negative emotional state, people would more likely click on product recommendations with positive sentiments. Hence, our third hypothesis is as follows:

Hypothesis 3 (H3): During the occurrence of positive events, promoting product recommendations with positive sentiments under negative sentimental posts of plurkers could better boost click-through rates of product recommendations rather than forming the variable combinations of other posts or promoting sentiments.

Mohammad et al. [18] analyzed the sentiment of Twitter users during the US election period; results showed that negative emotions were twice that of positive emotions during the campaign; further, up to 58% of user posts contained negative sentiments toward certain people, events, and surroundings. In analyzing the sentiment of Taiwan's political parties, Lin and Cong-ji noted that Taiwan has an anti-party sentiment that is similar to that of the present Western democracies. Based on the above, we suggest that political elections are negative emotional events. Hence, we propose our second control variable as follows:

Control Variable 2 (CV2): Elections are negative emotional events.

Psychological research has long found that human beings have "negative bias" in which the concern of positive events is not only beyond compare of negative messages, while the impact from the negative messages is more significant. Therefore, in our study, during the occurrence of negative events, negative sentimental user postings could better arouse audience interests. Hence, we propose our fourth hypothesis as follows:

Hypothesis 4 (H4): During the occurrence of negative events, plurker posts with negative sentiments could better boost click-through rates of product recommendations versus that of posts with positive sentiments.

Many studies indicate that manifestations of the entertainment include both positive and negative emotions. Bartsch et al. also noted that negative emotions can lead to pleasure, i.e., negative emotions can also contribute to the emergence of a positive emotional state. Similarly, Oliver found that grievers will have positive emotions, such as pleasure, after watching a sad movie. Conversely, from the point of view of psychological entertainment, if the political participants could be satisfied in access to negative messages, then a positive emotional state is generated [23]. This implies that audiences tend to use negative messages as a trigger of positive sentiments while participating in political issues. Based on the above, we deduce that during the occurrence of negative events, such as political elections, microblog users with negative sentiments serve the recommendations those with negative sentiments as the entertainment medium trigger off the positive sentiments, thereby increasing user interest in clicking. Hence, we proposed our fifth hypothesis as follows:

Hypothesis 5 (H5): During the occurrence of negative events, product recommendations with negative sentiments could better boost click-through rates of product recommendations versus those with positive or neutral sentiments.

Asch [1] conformity experiments indicate that when facing an obvious wrong answer, people will continue to believe the wrong answer only because most people are giving the same answer. Malatesta also showed that even on the anonymous Internet, people still bet on the wrong horse because of the impact of peer pressure. Based on the above, we conclude that during the occurrence of negative events, users with negative sentiments impact groups in the discussion thread, and emotion "emotional contagion" other audiences, thereby force users break out of the negative sentimental state "Aversion theory" and, thereby creating negative speech to declare their dissatisfaction. Under the pressure of negative sentiments, we suggest utilizing a recommendation that can fully cope with the conformity to better boost user interest in clicking. Hence, we propose our sixth hypothesis as follows:

Hypothesis 6 (H6): During the occurrence of negative events, promoting product recommendations with negative sentiments under negative sentimental posts of plurkers could better boost click-through rates of product recommendations versus that of forming variable combinations of other posts or promoting sentiments.

3.4   Variables and Analytical Methods

3.4.1   Explanation of Variables

Variables in our study are summarized as follows:

- Clicks: The higher the number of clicks on the product advertisement links, the better the performance of the corresponding product recommendation.
- Sole emotion: This post only contains either positive emotional terms or negative emotional terms, not both.
- Multiple emotion: This post contains both positive and negative emotional terms.
- Positive emotion: This product recommendation contains any of the positive emotional terms in the emotion lexicon.
- Negative emotion: This product recommendation contains any of the negative emotional terms in the emotion lexicon.
- Neutral emotion: This product recommendation contains none of the emotional terms in the emotion lexicon.

Therefore, in this study, we utilize the valence provided by Zhuo et al. [24] and the "weighted average" method of Liao for our calculations. The valence here represents the positive and negative degree of each emotional word. The higher the score, the stronger the positive intensity of the emotion word, where the level five represents neutral emotion. The following formula calculates weighted average:

C_i: The number of the occurrence of the ith emotional term
E_i: The valence of the ith emotional term
n: The number of each emotional term

$$\overline{V} = \frac{\sum_{i=1}^{n} C_i E_i}{\sum_{i=1}^{n} C_i}$$

The value calculated by the weighted average method could be used to determine the emotional valence of the post, and we use five grades as the sole criterion to judge the positive emotion as well as the negative emotion.

3.4.2 Analytical Method

We used three analytical methods to evaluate our approach. First, we utilized an independent sample T-test to compare whether the average difference between two groups reached a significant level. As the independent variables of H1 and H4 are all binary data, we utilized an independent sample T-test to check whether if the two types of variables in the independent variables would create significant difference in the average number of click-throughs.

Second, we utilized one-way ANOVA to compare whether the average of the dependent variables between groups has a significant difference. As the independent variables in H2 and H5 are all ternary data, we utilized one-way ANOVA to analyze whether the influences on the number of click-throughs are the same between groups within the three types. In the meanwhile, we utilized posteriori tests to check which two groups had significant differences.

Third, we utilized two-way ANOVA to test the interaction effects, which represent the effects of all independent variables on the dependent variables. H3 and H6 were used to test the effects of the two independent variables on the click-through rate. Hence, we utilized two-way ANOVA to test whether these two independent variables would create interaction effects. In the meanwhile, we utilized a simple main effect to determine under which circumstances the independent variables occurred, the click-through rates would have significant differences.

4   Research and Implementation

4.1   Data Preprocessing

In this study, we primarily focus on the emotional effects (i.e., control variables) of a positive or negative event during a particular event. Therefore, over the course of the experiment, posts (i.e., articles) without emotional words were not excluded from the dataset of this study. In addition, to avoid the influence of outliers in the dataset, we excluded data with exceptional hits. After applying the above data-processing procedure, we used a total of 285 emotional data.

To select the data interval for the special events, the data were segmented according to the time interval defined in the relevant literature. The data interval of the positive event Christmas was based on the research of Hu [9], with a total of 147 data points collected from December 1, 2015 through January 1, 2016. The data interval of the negative political election event was based on the research of Bermingham and Smeaton which covered the election period for 16 days, with a total of 138 data points collected from January 2, 2016 through January 16, 2016.

4.2   Experimental Results

4.2.1   Control Variable: Positive Event (Christmas)

First, to verify H1, we utilized data collected during the occurrence of positive events and conducted an independent sample T-test with independent variable "sentiment class" and dependent variable "hits"; further, we aimed to observe whether the posts (i.e., articles) with negative emotions and positive emotions had significant differences toward hits. We observed that given a 95% confidence level, negative sentimental posts had a significant impact on increasing the number of hits versus that of positive sentimental posts during the occurrence of the positive events.

To verify H2, we utilized data collected during the occurrence of positive events and conducted a one-way ANOVA test using independent variable "recommended emotion categories" and dependent variable "hits"; further, we aimed to observe whether product recommendations within these three emotional categories had significant differences toward hits. When a significant difference occurred, we utilized a posteriori test to compare each of the two categories to distinguish which two categories made this significant difference occur. We observe that given a 95% confidence level, positive sentimental product recommendations had a significant impact on increasing the number of hits during the occurrence of positive events.

To verify H3, we utilized data collected during the occurrence of positive events and conducted a two-way ANOVA test to observe whether the posting and recommended emotions created any interaction effect toward the number of hits. In the meanwhile, if the interaction effect was significant, we utilized a simple main effect analysis to further differentiate under what combination of posting and recommended emotions could produce a significant difference.

Interpreting the entire results of the two-way ANOVA given a 95% confidence level, we found that promoting positive sentimental product recommendations toward posts with negative emotions during the occurrence of positive events would create a significant influence on increasing the number of hits. Further, by observing the section profile of Fig. 1, we realized that the hit performance was worst at promoting negative sentimental recommendations toward negative postings, which is consistent with the aversion theory applied to this study.

4.2.2   Control Variable: Negative Event (Political Election)

First, to verify H4, we used data collected during negative events to conduct an independent sample T-test with independent variable "Posting Emotion Category" and dependent variable "hits"; we attempted to observe whether posts with negative and positive emotions had significant differences in the number of hits. We observed that given a 95% confidence level, negative sentimental posts had a significant impact on increasing the number of hits versus that of positive sentimental posts.

To verify H5, we used data collected during negative events to conduct a one-way ANOVA test with independent variable "Recommendation Emotion Category" and dependent variable "hits"; we attempted to observe whether there was a significant difference in terms of the number of hits between the three categories of product recommendation emotions. Further, when the significant difference occurred, by utilizing posteriori tests to compare each pair of categories, we distinguished which two

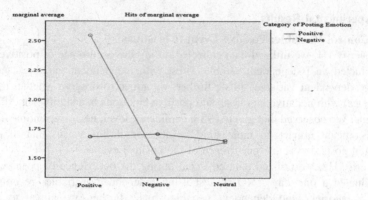

Fig. 1. Section profile of the interaction effect

categories made this significant difference occur. At a 95% confidence level, positive sentimental product recommendations had a significant impact on the number of hits during the occurrence of the positive event.

To verify H6, we used data collected during the occurrence of negative events to conduct a two-way ANOVA test to observe whether there was an interaction effect between the post and recommendation emotions on the number of hits given the occurrence of negative events. In the case of significant interaction effects, we utilizing a simple main effect to further determine under what combination of post and recommendation emotion categories we attain significant differences. We observed that when the interaction was not significant in H6, we were unable to verify that negative sentimental recommendations would have a significant impact on the number of hits during the occurrence of negative events.

Table 1 summarizes our results; from our results, we realize that regardless of whether we have positive or negative events, user posts with negative emotions indeed could arouse user click-through interest more efficiently than positive sentimental posts (i.e., H1 and H4). Under the influence of positive event control variables, the positive emotional recommendations could better enhance the hits (i.e., H2) versus that of the negative or neutral emotional recommendations. Further, there was an interaction effect between the post emotion and the recommendation emotion; under the negative sentimental postings propose with positive emotional product recommendation could create enormous influence on hits (i.e., H3). Under control of the negative event variables, negative emotional recommendations could better attract users versus either the positive or neutral emotional recommendations (i.e., H5); however, promoting negative emotional recommendations under negative postings had no significant effect on increasing the number of hits (i.e., H6).

Table 1. Hypothesis tests and corresponding results

Hypotheses	Analytical method	Results
Control Variable 1 (CV1): Positive events		
H1: Negative sentimental postings > Positive ones	T Test	Supported
H2: Positive emotional recommendations > Negative & Neutral ones	One-way ANOVA	Supported
H3: Negative sentimental postings together with positive emotional recommendations > Other combinations	Two-way ANOVA	Supported
Control Variable 1 (CV1): Negative events		
H4: Negative sentimental postings > Positive ones	T Test	Supported
H5: Negative emotional recommendations > Positive & Neutral ones	One-way ANOVA	Supported
H6: Negative sentimental postings together with negative emotional recommendations > Other combinations	Two-way ANOVA	False

5 Conclusions

Based on the theories of psychology and social science, our study sets out to verify whether event sentiments, post sentiments, and recommendation sentiments have significant influence on the enhancement of product recommendations. Results of H1, H2, and H3 indicate that during positive events, microblog users tend to focus on negative sentimental posts, thus producing an emotional contagion that leads to the generation of negative emotional states. According to aversion theory, user moods will turn to a negative emotional state; hence, product recommendation with positive sentiments could better increase the interest of users. Conversely, based on the results of H4, H5, and H6, our study suggests that negative events are likely to lead to a herd mentality. When users focus on negative sentimental posts, users form network groups based on the force of public opinion; hence, microblogs turn up to the ways for the audiences releasing their emotions by words. Thus, our study is in line with the herd mentality recommendation to prove that negative sentimental product recommendations could better increase user attention under negative posts, thereby becoming an emotional release for audiences that leads to an overall positive mood.

The primary contribution of our study is to join event sentiment and recommendation emotion in our analysis. Using a number of variables to test the impact of emotions on the number of advertisement clicks, our results showed that at a 95% confidence level, those variables can significantly enhance the effectiveness of product recommendations. In view of the effectiveness of emotional variables on product recommendation, our study suggests that a follow-up study of product recommendations in social networking platforms should use emotional variables as theoretical considerations to identify a more accurate recommendation method for product promotion.

In the future, our present study suggests that the classification of products can be carried out to further observe which specific categories of products could effectively lead to user concerns during the occurrence of special events. If "hot" goods exist for a specific period of time that could be identified with the right emotional recommendations, we think our approach should be able to increase product visibility as well as sales promotions.

References

1. Asch, S.E.: Studies of independence and conformity: I. A minority of one against a unanimous majority. Psychol. Monogr. Gen. Appl. **70**(9), 1 (1956)
2. Banerjee, N., et al.: User interests in social media sites: an exploration with micro-blogs. In: Proceedings of the 18th ACM Conference on Information and Knowledge Management. ACM (2009)
3. Bavelas, J.B., Black, A., Lemery, C.R., Mullett, J.: 14 motor mimicry as primitive empathy. In: Eisenberg, N., Strayer, J. (eds.) Empathy and Its Development, p. 317. Cambridge University Press, New York (1990)
4. Bermingham, A., Smeaton, A.F.: On using Twitter to monitor political sentiment and predict election results (2011)
5. Davis, M.R.: Perceptual and affective reverberation components. In: Goldstein, A.B., Michaels, G.Y. (eds.) Empathy: Development, Training, and Consequences, pp. 62–108. Erlbaum, Hillsdale (1985)
6. Dong, R., Schaal, M., O'Mahony, Michael P., McCarthy, K., Smyth, B.: Opinionated product recommendation. In: Delany, S.J., Ontañón, S. (eds.) ICCBR 2013. LNCS, vol. 7969, pp. 44–58. Springer, Heidelberg (2013). doi:10.1007/978-3-642-39056-2_4
7. De Choudhury, M., et al.: Multi-scale characterization of social network dynamics in the blogosphere. In: Proceedings of the 17th ACM Conference on Information and Knowledge Management. ACM (2008)
8. Gillin, P.: Secrets of Social Media Marketing: How to Use Online Conversations and Customer Communities to Turbo-Charge Your Business! Linden Publishing, Chicago (2008)
9. Hu, W.: Real-time Twitter sentiment toward thanksgiving and Christmas holidays. Soc. Netw. **2**, 77–86 (2013)
10. Huffaker, D.: Dimensions of leadership and social influence in online communities. Hum. Commun. Res. **36**(4), 593–617 (2010)
11. Joyce, E., Kraut, R.E.: Predicting continued participation in newsgroups. J. Comput. Mediat. Commun. **11**(3), 723–747 (2006)
12. Kim, Y., Kim, M., Kim, K.: Factors influencing the adoption of social media in the perspective of information needs (2010)
13. Kouloumpis, E., Wilson, T., Moore, J.D.: Twitter sentiment analysis: the good the bad and the OMG! In: ICWSM, vol. 11, pp. 538–541 (2011)
14. Kolya, A.K., et al.: Identifying event: sentiment association using lexical equivalence and co-reference approaches. In: Proceedings of the ACL 2011 Workshop on Relational Models of Semantics. Association for Computational Linguistics (2011)
15. Li, Y.M., Shiu, Y.L.: A diffusion mechanism for social advertising over microblogs. Decis. Support Syst. **54**(1), 9–22 (2012)

16. Lang, A.: Using the limited capacity model of motivated mediated message processing to design effective cancer communication messages. J. Commun. **56**(s1), S57–S80 (2006)
17. Malatesta, G.: Conformity rules in cyberspace. The Australian, p. 40 (2001)
18. Mohammad, S.M., Zhu, X., Kiritchenko, S., Martin, J.: Sentiment, emotion, purpose, and style in electoral tweets. Inf. Process Manag. **51**(4), 480–499 (2015)
19. Naveed, N., et al.: Bad news travel fast: a content-based analysis of interestingness on Twitter. In: Proceedings of the 3rd International Web Science Conference. ACM (2011)
20. Schumaker, R.P., Zhang, Y., Huang, C.N., Chen, H.: Evaluating sentiment in financial news articles. Decis. Support Syst. **53**(3), 458–464 (2012)
21. Talmy, L.: Toward a Cognitive Semantics: Concept Structuring Systems. Language, Speech, and Communication, vol. 1. MIT Press, Cambridge (2003)
22. Thelwall, M., Buckley, K., Paltoglou, G.: Sentiment in Twitter events. J. Am. Soc. Inf. Sci. Technol. **62**(2), 406–418 (2011)
23. Vorderer, P.: It's all entertainment—sure. But what exactly is entertainment? Communication research, media psychology, and the explanation of entertainment experiences. Poetics **29**(4), 247–261 (2001)
24. Zhuo, S., et al.: Chin. J. Psychol. **55**(4), 493–523 (2013)

Multi-agent Systems and Swarm Robotics

Solar Irradiance Forecasting Based on the Multi-agent Adaptive Fuzzy Neuronet

Ekaterina A. Engel[1]([✉]) and Igor V. Kovalev[2]

[1] Katanov State University of Khakassia,
Shetinkina. 61, 655017 Abakan, Russia
ekaterina.en@gmail.com
[2] Siberian State Aerospace University,
Krasnoyarsky Rabochy Av. 31, 660014 Krasnoyarsk, Russia
kovalev.fsu@mail.ru

Abstract. This paper presents a multi-agent adaptive fuzzy neuronet for hourly solar irradiance forecasting under random perturbations. The training algorithm of the multi-agent adaptive fuzzy neuronet must find the optimal network configuration within an architecture space. In a multidimensional search space where the optimum dimension is unknown, the training algorithm must seek both positional and dimensional optima. The simulation results show that multi-dimensional Particle Swarm Optimization outperforms Genetic algorithm in training the effective multi-agent adaptive fuzzy neuronet for hourly solar irradiance forecasting.

Keywords: Particle swarm optimization · Evolutionary intelligent agents · Solar irradiance forecasting

1 Introduction

The fast growth of solar photovoltaic generation can interfere with network stability. The two days ahead forecasting of hourly irradiance is the first and most important step for scheduling of power plants and planning transactions in the electricity market in order to balance the supply and demand of energy and to assure reliable grid operation. For real-life photovoltaic applications the solar irradiation fluctuations under random perturbations of cloudiness have complex dynamics. The neuronet-based solutions have been proposed to overcome these difficulties and show good performance [1]. But there is a growing demand for more accurate forecasting models of solar irradiance. The most important approaches are those that provide effective data processing based on intelligent methods. The aforementioned approaches include combining evolutionary intelligent agents, neural networks and fuzzy logic.

This paper presents a multi-agent adaptive fuzzy neuronet for two days ahead hourly irradiance forecasts. The agents of the multi-agent adaptive fuzzy neuronet are fulfilled based on recurrent networks. The training algorithm of the multi-agent adaptive fuzzy neuronet must find the optimal network configuration within an architecture space. In a multidimensional search space where the optimum dimension is unknown, the training algorithm must find both positional and dimensional optima.

© Springer International Publishing AG 2017
Y. Tan et al. (Eds.): ICSI 2017, Part II, LNCS 10386, pp. 135–140, 2017.
DOI: 10.1007/978-3-319-61833-3_14

A Comparison of the multi-dimensional Particle Swarm Optimization (PSO) and Genetic Algorithm (GA) for training the effective agents of the multi-agent adaptive fuzzy neuronet for hourly solar irradiance forecasting is performed.

2   Clear-Sky Irradiance Modeling

Solar irradiance is the amount of electromagnetic energy incident on a surface per unit time and per unit area. Solar irradiance is calculated as the amount of electromagnetic energy incident on a surface per unit time and per unit area. The total rate of radiation GC striking a PV system on a clear day can be calculated as follows:

$$Gc = Ae^{-km} \left[\cos\beta\cos(\varphi_s - \varphi_C)\sin\Sigma + \sin\beta\cos\Sigma + C\left(\frac{1+\cos\Sigma}{2}\right) + p(\sin\beta + C)\left(\frac{1-\cos\Sigma}{2}\right) \right] \quad (1)$$

where m is the air mass, β is the altitude angle, φ_S is the solar azimuth angle, φ_C is the PV module azimuth angle, Σ is the PV module tilt angle, p is the reflection factor, C is the sky diffuse factor, and A and k are parameters dependent on the Julian day number.

3   The Multi-agent Adaptive Fuzzy Neuronet for Hourly Solar Irradiance Forecasting

The agents of the multi-agent adaptive fuzzy neuronet are fulfilled as two-layered recurrent networks. The two-layered recurrent networks architecture's parameters (delays, weights and biases) have been coded into particles a. In order to train the effective agents of the multi-agent adaptive fuzzy neuronet for hourly solar irradiance forecasting the GA (Fig. 1) and the multi-dimensional PSO (Fig. 2) have been elaborated. The dimension range of the multi-dimensional PSO is $(D_{min} = 34,\ D_{max} = 144)$. The multi-agent adaptive fuzzy neuronet $Fes_{jhq}(x_h^t)$ (Fig. 3 and 4) is fulfilled based on the data

$$s_h^t = (x_h^t = (G0_h^t, Gd_h^{t-2}, C_h^{t-2}, Cl_h^t, T_h^t, P_h^t, W_h^t, Wd_h^t), G_h^t), \quad (2)$$

where $G0_h^t$ is extraterrestrial irradiance, Gd_h^t is the historical data of solar irradiance difference, G_h^t is the target solar irradiance, C_{h-m}^{t-2} is the historical data of clear-sky index, Cl_h^t is cloudiness (%), P_h^t is pressure, W_h^t and Wd_h^t are wind speed and direction, respectively, T_h^t is ambient temperature, m is embedding dimension, $h = \overline{5..23}$, $t = \overline{1..153}$. The number of samples is 2907 ($h*t = 19*153 = 2907$). This database was collected at the site of Abakan (91.4° of longitude East, 53.7° of latitude North and 246 m of altitude) from April through August 2016. The fitness function is

$$f(Fes_{jhq}(x_h^t)) = 1/hm^*tm \sum_{h=1,t=1}^{hm,tm} \left| Fes_{jhq}(x_h^t) - G_h^t \right|. \quad (3)$$

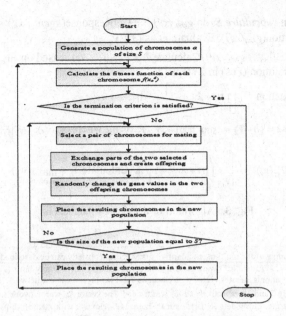

Fig. 1. A genetic algorithm.

MD PSO (termination criteria: $\{IterNo, \varepsilon_C, ...\}$) For $\forall a \in \{1, S\}$ do : Randomize $xd_a(0)$: Dimension component of particle a,

$vd_a(0)$: Velocity component of dimension of particle a, Initialize $\tilde{x d_a}(0) = xd_a(0)$, $x\tilde{d}_a$: Personal best dimension component of particle a,

For $\forall d \in \{D_{min}, D_{max}\}$ do : Randomize $xx_a^d(0)$, $xv_a^d(0)$,   Initialize $xy_a^d(0) = xx_a^d(0)$. End For. End For.

For $\forall t \in \{1, IterNo\}$ do :   For $\forall a \in \{1, S\}$ do : If $(f(xx_a^{xd_a(t)}(t)) < min(f(xy_a^{xd_a(t)}(t-1)), min_{p \in s-\{s\}}(f(xx_p^{xd_a(t)}(t)))))$ then do $xy_a^{xd_a(t)}(t) = xx_a^{xd_a(t)}(t)$

If $f(xx_a^{xd_a(t)}(t)) < f(xy_{gbest(xd_a(t))}^{xd_a(t)}(t-1))$ then $gbest(xd_a(t)) = a$, $gbest(d)$: Global best particle index in dimension d,

If $f(xx_a^{xd_a(t)}(t)) < f(xy_a^{\tilde{d}_a(t-1)}(t-1))$ then $xd_a(t) = xd_a(t)$, If $f(xx_a^{xd_a(t)}(t)) < f(x\hat{y}_{dbest}(t-1))$ then $dbest = xd_a(t)$. End If. End For.

If the termination criteria are met, then Stop. $xy_{aj}^{xd_a(t)}(t-1)$: jth component of the personal best (pbest) position of particle a, in dimension $xd_a(t)$

For $\forall a \in \{1, S\}$ do :   For $\forall j \in \{1, xd_a(t)\}$ do :

Compute $uxx_{a,j}^{xd_a(t)}(t+1) = w(t)uxx_{a,j}^{xd_a(t)}(t) + c_1 r_{1,j}(t)(xy_{a,j}^{xd_a(t)}(t) - xx_{a,j}^{xd_a(t)}(t)) + c_2 r_{2,j}(t)(x\hat{y}_j^{xd_a(t)}(t) - xx_{a,j}^{xd_a(t)}(t))$,

$xx_{a,j}^{xd_a(t)}(t+1) = \begin{cases} xx_{a,j}^{xd_a(t)}(t+1) + uxx_{a,j}^{xd_a(t)}(t+1) & \text{if } X_{min} \leq uxx_{a,j}^{xd_a(t)}(t+1) \leq X_{max} \\ U(X_{min}, X_{max}) + xx_{a,j}^{xd_a(t)}(t+1) & \text{else} \end{cases}$,

$xx_{a,j}^{xd_a(t)}(t+1) \leftarrow \begin{cases} xx_{a,j}^{xd_a(t)}(t+1) & \text{if } X_{min} \leq xx_{a,j}^{xd_a(t)}(t+1) \leq X_{max} \\ U(X_{min}, X_{max}) & \text{else} \end{cases}$. End For.

Compute $vd_a(t+1) = \left[vd_a(t) + c_1 r_1(t)(xd_a - xd_a(t)) + c_2 r_2(t)(dbest - xd_a(t)) \right]$,

$xd_a(t+1) = \begin{cases} xd_a(t) + vd_a(t+1) & \text{if } VD_{min} \leq vd_a(t+1) \leq VD_{max} \\ xd_a(t) + VD_{min} & \text{if } vd_a(t+1) < VD_{min} \\ xd_a(t) + VD_{max} & \text{if } vd_a(t+1) > VD_{max} \end{cases}$, $xd_a(t+1) \leftarrow \begin{cases} xd_a(t) & \text{if } P_d(t+1) \geq max(15, xd_a(t+1)) \\ xd_a(t) & \text{if } xd_a(t+1) < D_{min} \\ xd_a(t) & \text{if } xd_a(t+1) > D_{max} \\ xd_a(t+1) & \text{else} \end{cases}$. End For. End For

Fig. 2. A multi-dimensional PSO.

Notice that Cl_h^t, P_h^t, W_h^t, Wd_h^t, T_h^t are daily average parameters of the weather forecast. The algorithm of the agent's interaction (Fig. 3) uses a fuzzy-possibilistic method [2].

for each $agent_q$ in $subculture$ S_k do g_{khq} $(x_h{}^t)$ $\leftarrow$ GetResponse($agent_q$; $x_h{}^t$);

$\quad v_q \leftarrow$ TakeAction($g_{khq}(x_h{}^t)$): Evaluate e_q as (3); $v_q = 1 - e_q$.

end for. $w = [g_{kh1}(x_h{}^t), ..., g_{khq}(x_h{}^t)]$ Calculate $I_h = Fes_{jh}(g_{jhq}(s))$ based on $(w, [v_1, ..., v_q])$ as fuzzy expected solution (Fes) in 2 steps [2]

Step 1: Solve equation $\quad \left[\prod_{i=1}^{q}(1 + \lambda w_i) - 1 \right] / \lambda = 1, \quad -1 < \lambda < \infty.$

Step 2: Calculate $s = \int h \circ W_\lambda = \sup_{\alpha \in [0,1]} \min\{\alpha, W_\lambda(F_\alpha(v_j))\}$, where $F_\alpha(v_j) = \{F_i | F_i, v_j \geq \alpha\}, v_j \in V,$

$$W_\lambda(F_\alpha(v_j)) = \left[\prod_{F_i \in F_\alpha(v_j)}^{k}(1 + \lambda w_i) - 1 \right] / \lambda. \text{ Calculate } I_h = \max_{v_j \in V} s(w_j)$$

Fig. 3. Algorithm of the agent's interaction.

I unit: Training of the multi-agent adaptive fuzzy neuronet briefly can be described as follows

All samples ($N = h*t = 19*153 = 2907$) were classified into two groups: A_1 – sunny hour ($T_h{}^t = 1$), A_2 – cloudy ($T_h{}^t = -1$). This classification generates vector with elements $T_h{}^t$. Two-layer recurrent network (number of hidden neurons and delays are 7 and 2, respectively): $F(S)$ was trained. The vector $S_h{}^t$ was network's input. The vector $T_h{}^t$ was network's target. Fuzzy sets A_j, (A_1 – sunny hour, A_2 – cloudy) with membership function $\mu_j(s)$ are formed base on aforementioned two-layer recurrent network $F(s_h{}^t)$, $j = \overline{1..2}$.

We train (based on multi-dimensional PSO or GA) three two-layered recurrent neural networks: $g_{jhq}(x_h{}^t)$, $h = \overline{1..19}$, $j = \overline{1..2}$, $q = \overline{1..3}$, based on the data (2). This step provides recurrent neural networks which create the forecasting value of solar irradiance $g_{jhq}(x_h{}^t)$. Two agent's subcultures S_j are formed base on aforementioned two-layer recurrent networks.

If-then rules are defined as: $\qquad \Pi_j$: IF X is A_j THEN $I_h = Fes_{jh}(g_{jhq}(x_h{}^t))$. (4)

II unit: Simulation of the trained multi-agent adaptive fuzzy neuronet $\forall c \in \{122..153\}$ for $h = 5:23$

Aggregation antecedents of the rules (4) maps input data $x_h{}^c$ into their membership functions and matches data with conditions of rules. These mappings are then activates the k rule, which indicates the k hour's state $k = \overline{1..2}$ and k agent's subcultures – S_k.

According the k hour's state the multiagentny adaptive fuzzy neuronet (trained base on the data $s_h{}^d$, where $d = \overline{1..c} - 1$) creates the forecasting value of solar irradiance $I_h = Fes(f_{jhq}(s_h{}^c))$ as a result of multi-agent interaction (Fig.3) of subculture S_k.

Fig. 4. The multi-agent adaptive fuzzy neuronet.

Figure 4 briefly describes fulfillment of the multi-agent adaptive fuzzy neuronet.

Figure 4 shows the units of the proposed multi-agent adaptive fuzzy neuronet. The fuzzy-possibilistic method allows for the forecasting of the value of solar irradiance in a flexible manner, so as to take into account the responses of all agents based on fuzzy measures and the fuzzy integral. In this research the task's solution is the predicted value, two days ahead, of the hourly irradiance base on the trained multi-agent adaptive fuzzy neuronet. Based on the cloudiness hour's state, the trained multi-agent adaptive fuzzy neuronet effectively forecasts a solar irradiance under random perturbations.

4   Results

To illustrate the benefits of the multi-agent adaptive fuzzy neuronet in two days ahead hourly solar irradiance forecasting, the numerical examples from the previous Sect. 3 are revisited using software "The multi-agent adaptive fuzzy neuronet" [3]. There the two multi-agent adaptive fuzzy neuronet with three agents were fulfilled based on the data (2) $t = \overline{1..121}$ (Fig. 4). First, the multi-agent adaptive fuzzy neuronet (MAFN1) was trained using multi-dimensional PSO. Table 1 shows that only one set of MAFN1 architecture with $dbest = 124$ can achieve the root mean square error under 8.72 over test data (1), $t = \overline{122..153}$. The second multi-agent adaptive fuzzy neuronet (MAFN2) was trained by GA. Table 2 shows that the multi-dimensional PSO requires much more computational time for training MAFN than GA. But the multi-dimensional PSO provided within the architecture space the best multi-agent adaptive fuzzy neuronet configuration (MAFN1). The MAFN1 have three agents of each subculture S_k. The above mentioned agents are the two-layered recurrent neural network with two delays. The first agent's number of hidden neurons are 5. The second agent's number of hidden neurons are 4. The third agent's number of hidden neurons are 3. MAFN2 has same configuration.

Table 1. A two days ahead forecasting of hourly irradiance - results of multi-dimensional PSO

The MAFN's *dbest* dimension	34	44	54	64	74	84	94	114	124	134	144
The root mean square error – RMSE (W/m^2)	12.8	13.2	12.7	11.64	10.71	12.3	9.8	9.3	8.57	12.4	17.4

Figure 5 shows that MAFNN1 is definitely more accurate than MAFNN2 in two days ahead hourly solar irradiance forecasting. Figure 5 shows that the forecasted values of the MAFN1 are close to the measured irradiance. In Fig. 5, the ineffectiveness of the MAFN2 can be seen.

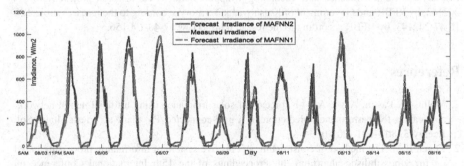

Fig. 5. Plot of the measured irradiance in comparison to forecast irradiance of the MAFNN1 and the MAFNN2 at the site of Abakan.

Table 2. A two days ahead forecasting of hourly irradiance - comparison of results

Model	MAFN2 solution		MAFN1 solution	
The computational time needed for training MAFN	8.1 (sec)		23.4 (sec)	
The root mean square error – RMSE (W/m^2)	Sunny	Cloudy	Sunny	Cloudy
	30.89	71.24	5.81	11.34

Table 2 shows two days ahead forecasting of hourly irradiance in August 2016, comparing the results between the MAFNN1 and the MAFNN2. Comparisons between irradiation models show that the MAFN1 is definitely more accurate.

Table 2 shows that BIAS, RMSE MABE of the two multi-agent adaptive fuzzy neuronets in sunny hours are quite small. The performance of the MAFN1 is changing in sunny and cloudy hours (Table 2). Nevertheless, the forecast values of the MAFN1 effectively track the complex dynamics of real measured data effectively in cloudy hours. Table 2 indicates that the MAFN1 outperforms the MAFN2, especially in the cloudy hours. The performance of the proposed multi-agent adaptive fuzzy neuronet trained by multi-dimensional PSO is superior to the same one trained by GA, especially during fast fluctuations of cloudiness.

5 Conclusions

This paper presents a multi-agent adaptive fuzzy neuronet for solar irradiance forecasting. Simulation comparison results for a two days ahead forecasting of hourly irradiance demonstrates the effectiveness of the multi-agent adaptive fuzzy neuronet trained by multi-dimensional PSO as compared with the same one trained by GA. The analysis of the evolving errors shows the potential of the multi-agent adaptive fuzzy neuronet in solar irradiation forecasts.

Acknowledgements. The authors wish to thank Daniel Foty and the reviewer for valuable comments. The reported study was funded by RFBR, Government of Krasnoyarsk Territory, Krasnoyarsk Region Science and Technology Support Fund to the research project №16-47-242143, by RFBR and Government of Krasnoyarsk Territory according to the research project № 16-47-242143, by RFBR according to the research project № 17-48-190156.

References

1. Mellit, A., Pavan, A.M.: A 24-h forecast of solar irradiance using artificial neural network: application for performance prediction of a grid-connected PV plant at Trieste, Italy. Sol. Energy **5**, 807–821 (2010)
2. Engel, E.A.: The method of constructing an effective system of information processing based on fuzzy-possibilistic algorithm. In: Proceedings of the 15th International Conference on Artificial Neural Networks (Neuroinformatics-2013), Part 3, pp. 107–117 (2013)
3. Engel, E. A.: The multi-agent adaptive fuzzy neuronet. M.: Federal Service for Intellectual Property (Rospatent). Certificate about State registration of computer programs № 2016662951

Passive Field Dynamics Method: An Advanced Physics-Based Approach for Formation Control of Robot Swarm

Zhu Weixu and Yuan Zhiyong[✉]

School of Computer, Wuhan University, Wuhan, China
zhuweixu_harry@126.com , zhiyongyuan@whu.edu.cn

Abstract. We have proposed the Passive Field Dynamics method, an upgraded physics-based approach for formation control of robot swarm. Firstly, in this method, the force field generated by each individual in a swarm is not exerting forces on the neighboring individuals but is being moved by the existence of the neighbors. Secondly, each individual is no longer considered as a particle but as a rigid-body with the ability to rotate with its force field. Thirdly, the force field consists of not only attractive or repulsive forces but also non-radical forces to generate individuals' rotation, and the force field is optimized by genetic algorithm. Experiments show that the Passive Field Dynamics method can achieve complicated swarm behavior pattern which can not be achieved using standard physics-based approaches.

Keywords: Passive Field Dynamics method · Robot swarm control · Virtual physics-based approach · Lattice configuration

1 Introduction

Compared with single robot or multi-robot systems with centralized control, decentralized robot swarm systems have many advantages, such as adaptability, scalability and redundancy. A swarm has the ability to perform excellent responses to environment disturbances, as well as to perform more complicated tasks by absorbing new individuals, and to maintain robust status even if some individuals break down. These merits of swarm are able to benefit many applications. The key problem for swarm robot system is how to find the relation between the low level rules for each individual and the high level collective behavior of the whole swarm.

Among all the swarm control methods, virtual physics-based approach draws inspiration from physics. Each individual in a swarm is considered as a physical particle and generate a virtual force field in the surrounding area to exert forces on neighboring individuals. Virtual physics-based approach has its unique advantages. Instead of multiple behavior rules, a single mathematical equation is sufficient to guide individuals' movement, and certain physical tools can be used to analyze the properties of a robot swarm.

© Springer International Publishing AG 2017
Y. Tan et al. (Eds.): ICSI 2017, Part II, LNCS 10386, pp. 141–148, 2017.
DOI: 10.1007/978-3-319-61833-3_15

However, in the standard virtual physics-based approaches, individuals of a swarm are only considered as particles without the ability of rotation, and only radical forces (attraction and repulsion) are exerted between individuals. This makes the swarm can only form itself into simple patterns of hexagonal lattice. Researchers have to make extra efforts, such as using heterogeneous swarms, to achieve more complicated behavior control.

We propose Passive Field Dynamics method to make the following development: (i). In our method, the force field around each individual is no longer used to exert forces on neighboring individuals, but to exert forces on its owner. (ii). Each individual is no longer considered as a particle, but as a rigid-body with the ability to rotate along with its force field. (iii). The force field consists of not only attractive or repulsive forces but also non-radical forces to generate individuals' rotation, and genetic algorithm is exploited to optimize the force field. With this upgraded method, we can achieve more complicated swarm behavior patterns.

2   Related Work

One of the first researches on swarm simulation is in 1987, Reynold et al. [7] created a flock by computer simulation. He showed that flocking can be achieved by individuals following three rules: collision avoidance, velocity matching, and flock centering.

Spears et al. [3, 10–13] proposed *Physicomimetics*, a standard virtual physics-based design method for robot swarm control. They use virtual physical forces to form a hexagonal lattice swarm. Their work showed that in the framework of *Physicomimetics*, geometric lattice configurations can be achieved by following natural physics laws, and robots can organize themselves into collision-free and obstacle-avoidant swarm.

Turgut et al. [1] implemented flocking on a real robot system using two rules called proximal control and alignment control. Proximal control is the combination of separation and cohesion, and alignment control is realized through a sensing system which allows each robot to sense the heading direction of the other robots.

Olfati et al. [5, 6] proposed a framework for formation stabilization of multiple autonomous vehicles in a distributed fashion. They also presented researches about consensus problem and provided a theoretical framework for analysis of consensus algorithms for multi-agent networked systems. He saw formation reconfiguration and flocking as switching networks which possess dynamic topology structures. Olfati emphasized the importance of consensus problem in switching networks.

Other researches on swarm formation control include: Flocchini et al. [2] focused on a theoretical analysis of pattern formation. They proved that not all possible patterns can be achieved by a group of robots who is asynchronous. Yasuhiro Nishimura et al. [4] presented an adaptive triangle generation algorithm to allow individual robots to form different equilateral triangular configurations.

Rubenstein el al. [8,9] presented a method to enable a robot swarm to form a pre-defined shape. Vasarhelyi et al. [14] present the first decentralized multi-copter flock that performs stable autonomous outdoor flight with up to 10 flying agents.

These researches are all considering individual robots as particles and only control the distance between individuals. We, in the other hand, proposed Passive Field Dynamics Method that considers individual robots as rigid-bodies and control both distance and bearing.

3   Passive Field Dynamics Method

3.1   Standard Physics-Based Approach

In the standard physics-based approach, each individual in a swarm is considered as a particle with a location x, a velocity v, an acceleration a, a mass m and a surrounding force field $f(d)$, where d represents the distances between the focal individual and the neighbors, and the direction of vector f is radical. In more usual cases, the force field $f(d)$ is substituted by a potential energy field $P(d)$ whose gradient is $f(d)$. The movement of the individual within each time step Δt is calculated by the following equations, which are basically derived from traditional physics law: $\Delta x = v\Delta t$, $\Delta v = a\Delta t$ and $a = \frac{F}{m}$, where F is the sum of all the forces exerted by neighboring individuals' force field.

Therefore, the key to the high level behavior of the swarm is the design of the force fields (or potential energy force fields). In the standard physics-based approaches, Lennard-Jones potential energy field is frequently adopted (Eq. 1).

$$P(d) = \varepsilon[(\frac{\sigma}{d})^{12} - (\frac{\sigma}{d})^{6}].$$ (1)

3.2   The Development of Passive Field Dynamics Method

The assumption of the physics-based approach is that each individual is able to detect the bearing and distance of each neighboring individual so that the magnitude and direction of $f(d)$ can be calculated.

However, since the relative location in the nearby 2D space of each individual is accessible, there is no need for the force field (or potential energy force field) to be radical like Lennard-Jones potential energy field. We are able to design the force field (or potential energy force field) arbitrarily in 2D space, with all kinds of forces including non-radical force field. Figure 1(a) shows the gradient of Lennard-Jones potential energy field, in which radical forces only produce attraction and repulsion. Figure 1(b) shows an example of arbitrarily set force field, in which non-radical forces exist.

Therefore, by designing non-radical force fields, we are able to make more complicated lattice patterns that can not be achieved by standard physics-based approach, as shown in Fig. 2.

However, because of the arbitrariness of the non-radical forces, the coordination of different individuals' force fields should not be in the same direction, like

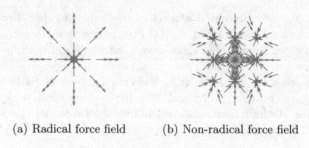

(a) Radical force field   (b) Non-radical force field

Fig. 1. Radical and Non-radical force field

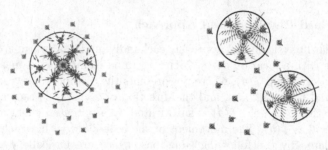

(a) Square lattice configuration   (b) Honeycomb lattice configuration

Fig. 2. The swarm lattice configurations and the corresponding force fields of individuals. The circles show the range of the fields, the red and green lines represent the x-axis and y-axis of the coordination systems of force fields, respectively. (Colr figure online)

the honeycomb case shown in Fig. 2(b). Thus we have to make the force field rotatable. Therefore, the standard approach, in which each particle's movement comes from its neighbors' force fields, is no longer suitable, so we propose the Passive Field Dynamics, making the following development.

In Passive Field Dynamics method, other than x, v, a and m, each individual also has a moment of inertia I, an orientation l_c, an angular velocity ω and an angular acceleration α, which follow $\Delta l_c = \omega \Delta t$, $\Delta \omega = \alpha \Delta t$ and $\alpha = M/I$, where M is the torque. (l_c, ω, α and M are vectors. In 2D cases the direction of these vectors are vertical to the 2D plain.)

In order to generate a rotation torque, we modify the calculation of F and M. Instead of exerted by neighboring individuals' force field, in Passive Field Dynamics method, F and M are calculated by the focal individual's own force field.

The basic idea of Passive Field Dynamics method can be described using an intuitive illustration: If we consider the field of an individual as a mountainous terrain, whose altitude at a certain point represents the value of the potential energy, a neighboring individual in this region will tend to slide down along the gradient of the terrain to low altitude area. In the meantime, let's consider this mountainous terrain is floating on a water surface, then the sliding down of a neighboring individual will generate a reverse force to the mountainous terrain, and the mountain will move. If the gradient is non-radical, the mountain will rotate. We call this "passive" because the force field is not pushing the neighboring individuals like the normal way but is being moved by the existence of the neighboring individual.

In Passive Field Dynamics method, we model the non-radical force field of individual i by $G_i(r)$, where r is the relative location of neighboring individual, i.e.

$$r_{ij} = C(l_{ci})(x_j - x_i). \tag{2}$$

where $C(l_{ci})$ is the transformation matrix between global coordination system and individual i's coordination system. Define the set of individual i's neighboring individuals as N_i, then the F and M of individual i are calculated by

$$M_i = \sum_{j \in N_i} C(-l_{ci})(r_{ij} \times G_i(r_{ij})). \tag{3}$$

$$F_i = \sum_{j \in N_i} C(-l_{ci})(G_i(r_{ij}) \cdot \frac{r_{ij}}{\|r_{ij}\|}). \tag{4}$$

3.3   Optimization of the Force Field

In addition to the upgraded dynamics, we have also used genetic algorithm to optimize the force field to improve the performance of the swarm. We first represent the force field discretely by generating a mesh on the force field and storing only the values on the nodes of the mesh into a matrix. We consider this matrix as a representation of the force field. Then we use standard genetic algorithm to optimize this matrix.

In this genetic algorithm case, the population number is 300, and we use weighted sum as the crossover method. During each mutation, every elements in the matrix will be multiplied by a random double float from 0.7 to 1.3. To test the performance of the optimized force field, we generated 1000 random initial swarm states, and run each for 1000 time steps. After each time step, the states are evaluated by calculating the sum of the errors between the desired positions and real positions of all the individuals.

4   Simulations and Analysis

We first show some lattice configurations that can only be achieved with heterogeneous swarm using standard physics-based approaches but can be achieved with homogeneous swarm using our method. Then we show some lattice configurations that can only be achieved using our method.

Figure 3 shows the process of the emerging of square lattice and honeycomb lattice. These two lattice configurations can be achieved using standard method by divide the swarm into two groups with different attraction/repulsion threshold [10], while it is sufficient in our method to use homogeneous swarm with uniform force field. Further more, if dividing the swarm into groups, the Passive Field Dynamics method can achieve more complicated lattice configurations that cannot be achieved before, such as hollow-square, as shown in Fig. 4. The ability comparison between standard physics-based approaches and the Passive Field Dynamics method is shown in Table 1.

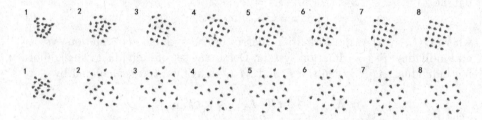

Fig. 3. Emerging process of square lattice and honeycomb lattice, the tip of each individual shows the x-axis direction of the individual's force field.

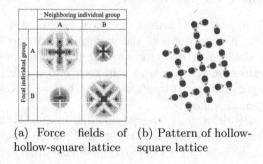

(a) Force fields of hollow-square lattice   (b) Pattern of hollow-square lattice

Fig. 4. Force fields and pattern of hollow-square lattice. The tip of each individual shows the x-axis direction of the individual's force field. Individuals with blue tips are in Group A, Individuals with red tips are in Group B.

We also tested our force field optimization method. As shown in Fig. 5(a), the red line is the standard function of L-J potential energy field, and blue line is the potential energy field optimized by our genetic algorithm method. Figure 5(b) shows one of the final patterns after 1000 time steps, and Fig. 5(c) shows the performance comparison of standard function of L-J potential energy field and optimized function on forming hexagonal lattice.

Table 1. Comparison between standard physics-based approaches and the Passive Field Dynamics method

	Standard method		The Passive Field Dynamics method	
	Homogeneous	Heterogeneous	Homogeneous	Heterogeneous
Hexagonal lattice	Yes	Yes	Yes	Yes
Square lattice	No	Yes	Yes	Yes
Honeycomb lattice	No	Yes	Yes	Yes
Hollow-square lattice	No	No	No	Yes

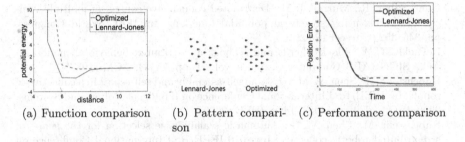

(a) Function comparison   (b) Pattern compari-   (c) Performance comparison
son

Fig. 5. Comparison between before and after optimization of genetic algorithm.

5   Conclusion and Discussion

We proposed the Passive Field Dynamics method, an upgraded dynamics to enable complicated swarm lattice pattern. Experiments showed that this new method can achieve new complicated lattice configurations. In addition to lattice configuration control, the Passive Field Dynamics method can also produce typical swarm behaviours such as obstacle avoidance, pursuit and collision avoidance.

However, there are still certain problems that require further research. The behaviour of the swarm are still crucially depending on the design of the force field. A good design will make the swarm form the desired lattice configuration successfully, but bad designs will lead to undesirable states. Although the genetic algorithm worked well in this paper, further research on methods for designing the fields is needed.

Acknowledgements. This work is supported by the Science and Technology Program of Wuhan, China under Grant No. 2016010101010022; National Natural Science Foundation of China under Grant No. 61373107.

References

1. Ferrante, E., Turgut, A.E., Huepe, C., Stranieri, A., Pinciroli, C., Dorigo, M.: Self-organized flocking with a mobile robot swarm: a novel motion control method. Adaptive Behavior, p. 1059712312462248 (2012)

2. Flocchini, P., Prencipe, G., Santoro, N., Widmayer, P.: Arbitrary pattern formation by asynchronous, anonymous, oblivious robots. Theor. Comput. Sci. **407**(1), 412–447 (2008)
3. Hettiarachchi, S., Spears, W.M., Hettiarachchi, S., Spears, W.M.: Distributed adaptive swarm for obstacle avoidance. Int. J. Intell. Comput. Cybern. **2**(4), 644–671 (2009)
4. Nishimura, Y., Lee, G., Chong, N.Y.: Adaptive lattice deployment of robot swarms based on local triangular interactions. In: 2012 9th International Conference on Ubiquitous Robots and Ambient Intelligence (URAI), pp. 279–284. IEEE (2012)
5. Olfati-Saber, R., Fax, A., Murray, R.M.: Consensus and cooperation in networked multi-agent systems. Proc. IEEE **95**(1), 215–233 (2007)
6. Olfati-Saber, R., Murray, R.M.: Distributed cooperative control of multiple vehicle formations using structural potential functions. In: IFAC World Congress, pp. 346–352. Citeseer (2002)
7. Reynolds, C.W.: Flocks, herds and schools: a distributed behavioral model. In: ACM SIGGRAPH Computer Graphics, vol. 21, pp. 25–34. ACM (1987)
8. Rubenstein, M., Shen, W.M.: Scalable self-assembly and self-repair in a collective of robots. In: IEEE/RSJ International Conference on Intelligent Robots and Systems, 2009. IROS 2009, pp. 1484–1489. IEEE (2009)
9. Rubenstein, M., Shen, W.M.: Automatic scalable size selection for the shape of a distributed robotic collective. In: 2010 IEEE/RSJ International Conference on Intelligent Robots and Systems (IROS), pp. 508–513. IEEE (2010)
10. Spears, W.M., Spears, D.F.: Physicomimetics: Physics-Based Swarm Intelligence. Springer Science & Business Media, Heidelberg (2012)
11. Spears, W.M., Spears, D.F., Hamann, J.C., Heil, R.: Distributed, physics-based control of swarms of vehicles. Auton. Robots **17**(2–3), 137–162 (2004)
12. Spears, W.M., Spears, D.F., Heil, R.: A formal analysis of potential energy in a multi-agent system. In: Hinchey, M.G., Rash, J.L., Truszkowski, W.F., Rouff, C.A. (eds.) FAABS 2004. LNCS, vol. 3228, pp. 131–145. Springer, Heidelberg (2004). doi:10.1007/978-3-540-30960-4_9
13. Spears, W.M., Spears, D.F., Heil, R., Kerr, W., Hettiarachchi, S.: An overview of physicomimetics. In: Şahin, E., Spears, W.M. (eds.) SR 2004. LNCS, vol. 3342, pp. 84–97. Springer, Heidelberg (2005). doi:10.1007/978-3-540-30552-1_8
14. Vásárhelyi, G., Virágh, C., Somorjai, G., Tarcai, N., Szorenyi, T., Nepusz, T., Vicsek, T.: Outdoor flocking and formation flight with autonomous aerial robots. In: 2014 IEEE/RSJ International Conference on Intelligent Robots and Systems (IROS 2014), pp. 3866–3873. IEEE (2014)

Adaptive Potential Fields Model for Solving Distributed Area Coverage Problem in Swarm Robotics

Xiangyu Liu and Ying Tan[✉]

Key Laboratory of Machine Perception (MOE),
and Department of Machine Intelligence
School of Electronics Engineering and Computer Science,
Peking University, Beijing 100871, China
{xiangyu.liu,ytan}@pku.edu.cn

Abstract. Complete coverage of a given region has become a fundamental problem addressed in the field of swarm robots. Currently available approaches to the coverage problem are typically of computational complexity, and are manually specified with different map settings, which are not scalable and flexible. To address these shortcomings, this paper describes an efficient distributed approach based on potential fields method and self-adaptive control. It makes no assumptions about prior knowledge on global map, and need few manual intervention during execution. Although the motion policy of each robot is very simple, efficient coverage behavior is achieved at team level. We evaluate the approach against a traditional rule-based method and pheromone method under different target area scenarios. It shows state-of-the-art performance, both in the percentage of coverage and the degree of connectivity.

Keywords: Swarm robotics · Distributed area coverage problem · Potential fields · Adaptive control

1 Introduction

In recent years, there has been a rapid growth of progress in Swarm Robotics. Past works have demonstrated that using a team of less complex robots to solve tasks in a distributed manner is more efficient than using a sophisticated, well-designed individual agent [3,12,16]. Many applications of swarm robots require them to disperse and cover throughout their environments, such as exploration [8], surveillance [14], patrolling [2], and multiple target searching [7,17]. The coverage task is usually used as a sub-task of more complex activities. Triggered by these interests, area coverage problem today has became an attractive topic in swarm robotics research, which is considered to be highly relevant in practical applications.

In the absence of any centralized control, it is often challenging to monitor the system's global behavior when using swarm-based approaches. In this paper,

© Springer International Publishing AG 2017
Y. Tan et al. (Eds.): ICSI 2017, Part II, LNCS 10386, pp. 149–157, 2017.
DOI: 10.1007/978-3-319-61833-3_16

we concentrate on the problem of distributed coverage of an unknown environment using a swarm of mobile robots. What we are interested in, is how the robots can disperse in a distributed, self-organised way. We propose a motion policy based on adaptive potential fields method, which doesn't need manual intervention (e.g. parameter tuning) when executing in a real scenario. The policy also maximizes the use of potential information from nearby robots within a local communication.

This paper is organized as follows. We start by discussing previous work in Sect. 2. In Sect. 3, we introduce the distributed area coverage problem and several definitions. We describe the details of the adaptive potential fields model in Sect. 4, and show the experimental results and discussions in Sect. 5. Finally, we conclude in Sect. 6.

2   Related Work

Conventional approaches for distributed coverage are realized with large robots that have considerable computation and memory capabilities on-board. A common feature underlying almost all these approaches is that they do not assume any limitations of the robots while executing the coverage task. A large number of these algorithms also assume that robots have a priori information about the environment [4,5,13]. Some researchers have obtained theoretical results for the coverage time and redundancy for multi-robot coverage problems [1]. However, no information about these parameters is provided in these papers.

Besides, several researchers have used swarming techniques to achieve distributed area coverage with multiple robots. A common feature of these swarm-based coverage algorithms is that they require localization capabilities on the robots to enable them to record or remember locations that are already covered. [15] describes ant-inspired heuristics for distributed area coverage. The environment is decomposed into cells using a grid and robots deposit virtual pheromone when visiting a cell. [9] defines four basic motion behaviours (random walk, wall following, avoiding all obstacles, and avoiding other robots), and designs a mechanism to switch between the four behaviours to maximize the area coverage.

3   Problem Formulation

In this section, we'll introduce the distributed area coverage problem and several definitions. In the area coverage task, swarm members must position themselves away from one another, with the objective of maximizing the area covered globally by the swarm. Also, the degree of swarm connectivity [6] should be minimized. We evaluate the effect of algorithms in these two metrics. This section defines and clarifies some key terms which are relevant to this intention and idea, and will be used throughout this article.

- **Environment**: A screenshot of the area coverage problem at the beginning of a simulation is shown in Fig. 1. $M \subset \mathbb{R}^2$ is an allowable environment area. We assume the robots have no priori information about the environment in advance, which requires exploration. We also assume the coverage area is an enclosed space.
- **Robot**: We use the foot-bot model defined in [10]. The foot-bot is a ground-based robot that moves with a combination of wheels and tracks. It is also equipped with numerous sensors and actuators.
- **Sensor**: The foot-bots can communicate with each other through a *range-and-bearing communication device* [10,11], allowing robots in line-of-sight to exchange messages within a limited range. Also, the robots are equipped with proximity sensors on board, which can detect objects around the robots.
- **Motion Policy**: The motion policy tells a robot what to do at each iteration. Therefore, when a robot detects the objects (obstacles or wall) and gathers information from other robots, it will decide what to do next based on motion policy.
- **Coverage**: We consider an environment to be covered as a condition that every place can be detected by at least one robot. Therefore, the motion policy should guide the robots in a way that their territory intersections decrease as time passes, but keep it in an ideal distance in order to communicate with each other. Different from [2], where each robot patrols by moving on the territory border when the full coverage is achieved, here we consider a static final state.

To conclude, we give a full definition of approximate distributed swarm robotics distributed coverage problem based on all the terms above.

Definition 1. *Approximate Swarm Robotics Distributed Area Coverage Problem:*

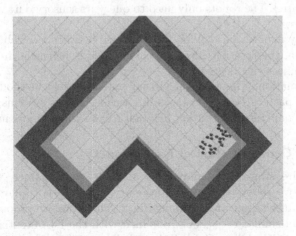

Fig. 1. A screenshot of the problem at the beginning of a simulation.

Given a set of R robots, with communication radius c_r and local communication graph g_r in an initially unknown environment, find a set of actions $a_r^1, a_r^2, ..., a_r^T$ to be performed by each robot $r \in R$ based on the motion policy, such that the maximum complete coverage criterion is satisfied: $\max \bigcup_{r \in R} \bigcup_{t=1...T} c_r^t$ and the degree of connectivity is minimized: $\min \bigcap_{r \in R} \bigcap_{t=1...T} g_r^t$.

4   Adaptive Potential Fields Model

In this section, the main method of this paper is presented. Particularly, we first introduce a mathematically simple model used in molecules mechanics, which we find convenient to model the interactive virtual force between robots. Next, we design an adaptive control policy upon the potential parameter, which does not need any manual intervention while the coverage task is executing.

4.1   Lennard-Jones Potential Fields

The Lennard-Jones potential (also termed the L-J Potential) is a mathematically simple model that approximates the interaction between a pair of neutral atoms or molecules. The most common expression of the L-J potential is:

$$V_{LJ}(\rho) = \varepsilon[(\frac{\delta}{\rho})^{12} - 2(\frac{\delta}{\rho})^6]. \tag{1}$$

from which we can derive the force:

$$F_{LJ}(\rho) = -\nabla V_{LJ}(\rho) = -\frac{12\varepsilon}{\rho}[(\frac{\delta}{\rho})^{12} - (\frac{\delta}{\rho})^6]. \tag{2}$$

where ε is the depth of the potential well, δ is the distance at which the potential reaches its minimum, and ρ is the distance between particles. The reasons why we choose the LJ-Potential Fields to model the interactive force are mainly in three aspects:

- Easy calculated. The robots only need to query its sensory data and calculate the joint force.
- Connectivity maintenance. When the distance between two robots exceeds a certain value, the traction effect appears, which is beneficial to the maintenance of connectivity degree.
- Pattern formation. The artificial virtual potential field methods are widely used in the pattern formation tasks in swarm robotics systems, and it contributes to cooperative execution between robots when in an emergency.

4.2   Guided Growth Potential Field Model

When the entire swarm system reaches a stable state, namely the distance between each two robots is kept in stationary, they are not guaranteed to be fully covered in the target area, such as Fig. 2. Therefore, we propose a "guided growth" method similar to [8], but combined with the LJ-Potential model and adaptive control. We call it Guided Growth Potential Field method (GGPF). The key issues are as follows:

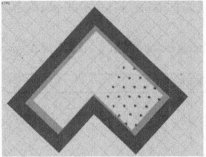

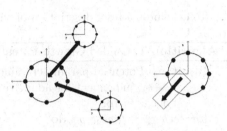

Fig. 2. Partially covered example   **Fig. 3.** A local vector graph

- **Local vector graph.** For each robot, when its local potential force tends to become zero, it'll construct a local vector graph as Fig. 3 shows. The local graph contains all the entities detected via the proximity sensors on-board. If the population is not enough to cover the entire area, the local vector graph may not be balanced from the geometric configuration view as Fig. 3.
- **Guided balance force.** We define the balance force as the supplementary of the joint LJ-Potential force in the local vector graph, which makes robot's velocity vector zero. The angle of the balance force can be calculated as:

$$\theta = atan\frac{-F_y}{-F_x},$$ (3)

in which the F_x and F_y are the joint force decomposed into x-axis and y-axis respectively. The robots on edge of the swarm system start to move away to explore uncovered area and doing so "pulls" the entire swarm because the other robots will follow the exploring robots in order to keep the potential field defined by formula 1. Thus the connectivity is kept.
- **Adaptive control of potential parameter.** The parameter δ in formula 2 determines where the potential reaches its minimum. It surely has an impact on the position where the interactive force becomes zero as well. When the balance force is applied to a robot, which means a "pull" effect is applied to the entire swarm, a constant will be added to δ. This will make the potential fields expanding until all robots extend to the entire target area. When the balance force is near zero, it will be reduced in turn. This is the core of the adaptive mechanism in the guided growth model.
- **Information sharing via sensing.** For each iteration when executing, robot shares its potential information (δ) through range and bearing sensor, and collects all the potential parameter δ_i from the neighborhood robots. It will adjust its δ with the average. This also embodies the essence of cooperation between robots.
- **Energy Decay.** In experiments, we have found an oscillation effect emerged in swarm robotics system when all robots fully cover the target area, due to the fact that potential energy converted to kinetic energy. We just add an

energy decay to the velocity of robots to make it stable, and it really performs well in simulation.

To conclude, a brief description of the proposed algorithm is shown in Algorithm 1.

Algorithm 1. Guided Growth Potential Field Model

Input: δ_i^0: Potential parameter value for each robot r_i;
$\quad\quad\quad$ ϵ_1, ϵ_2: judgement for adaptive control;
$\quad\quad\quad$ decay factor

1 **for** *each r_i in timestep t* **do**
2 $\quad$ **Loop**
3 $\quad\quad$ Gather Sensor information $(d_0^t, \theta_0^t, \delta_0^t), ..., (d_N^t, \theta_N^t, \delta_N^t)$;
4 $\quad\quad$ Get new potential parameter δ_i^{t+1} : $average_{j=1...N}(\delta_j^t)$;
5 $\quad\quad$ Update joint LJ-Potential force $\overrightarrow{F_{i\_LJ}^t}$ with formula 2;
6 $\quad\quad$ Construct local vector graph G_i^t and guided balance force $\overrightarrow{F_{i\_BL}^t}$;
7 $\quad\quad$ Calculate $\Delta x_i^t = (\overrightarrow{F_{i\_LJ}^t} + \overrightarrow{F_{i\_BL}^t})_x$; $\Delta y_i^t = (\overrightarrow{F_{i\_LJ}^t} + \overrightarrow{F_{i\_BL}^t})_y$;
8 $\quad\quad$ **if** $\|\Delta x_i^t\|^2 + \|\Delta y_i^t\|^2 < \epsilon_1$ **then**
9 $\quad\quad\quad$ $\delta_i^{t+1} - = constant$
10 $\quad\quad$ **else**
11 $\quad\quad\quad$ $\delta_i^{t+1} + = constant$
12 $\quad\quad$ **if** $\overrightarrow{F_{i\_BL}^t} < \epsilon_2$ **then**
13 $\quad\quad\quad$ $\Delta x_i^t* = decay$; $\Delta y_i^t* = decay$

---

5 Simulation Results and Discussions

In this section, we demonstrate the GGPF algorithm on simulation experiments. We have used the ARGoS [10] robot simulation platform for our simulations. ARGoS is a multi-physics robot simulator. It can simulate large-scale swarms of robots of any kind efficiently. We use the foot-bot model to perform the experiments and verify the algorithm. All the experimental scenarios are random generated with identical members, and the swarm robots know nothing about global information. In our experiment scenario, robots are initialized in a corner of an obstacle-free field and disperse according to the motion policy defined above. Each test is repeated for 20 times with 20 random closed maps with random obstacles, and the default number of robots is 25. Moreover, we pay careful attention to the percentage covered and the degree of swarm connectivity, with the variation of map size.

5.1 Algorithms for Comparison

Two algorithms are chosen for comparison, which are Rule-based Random Walk (RBRW) [9], and Ant-Robot Node Counting (ARNC) [15]. The RBRW algorithm defines four basic motion behaviours (random walk, wall following, avoiding all obstacles, and avoiding other robots), and designs a mechanism to switch between the four behaviours to maximize the area coverage. The ARNC algorithm devises the Ant-Colony Optimization (ACO) algorithm, where robots drop evaporating pheromone along their path, and when choosing their walking path give precedence to areas with the lowest pheromone level.

5.2 Simulation Results and Discussion

We first verify the scalability and self-adaption of the GGPF algorithm. As shown in Fig. 4, this set of simulation records one robot's potential parameter δ as the iteration increases. The map size is set to 100, which is relatively large compared with the swarm population. Figure 4 confirms our prediction in Sect. 4.2, that as the swarm expands, the potential parameter δ will decrease after robots reaching the edge of the target area. It is intuitively clear the entire swarm system has the dynamic perception to the target area, and it doesn't need any manual intervention of parameter tuning during the policy execution. When the local vector map of a robot is balanced, the information will spread to the inside, and finally the entire swarm remains stable persistently.

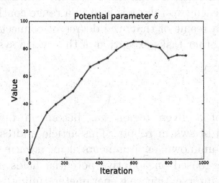

Fig. 4. The evolution of potential parameter δ.

The performance of percentage covered with the variation of map size is shown in Fig. 5a. The RBRW algorithm performs very poorly as the map size increases, and only achieves good performance when the target area is crowded. This is reasonable because of the total absence of coordination among the robots. The connectivity performance is also bad because robots will leave each other when they are get closer (Fig. 5b). The pheromone-based algorithm performs well in percentage of coverage, indicating that the pheromone information is useful for coverage task. But it also performs poorly with swarm connectivity. This is because each robot selects its path only depending on the pheromone value of the local position. A robot tends to choose the direction with rare pheromone, and thus breaking the integrity of swarm connectivity.

Our method, combining the potential fields method with guided balance force outperforms the other two algorithms. Robots will detect the edge of the target

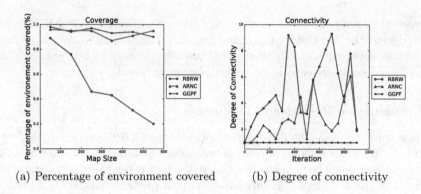

(a) Percentage of environment covered   (b) Degree of connectivity

Fig. 5. Simulation results

area, and expand the coverage through implicit communication. Each robot in close proximity of other robots is repelled by nearby entities until its local communication graph is balanced. The percentage of environment covered is always high, as the map size increases. Meanwhile, with the limitation of potential field, the distance between each two robots is kept in a range controlled by the potential parameter δ. As a result of that, the degree of connectivity is always one, which is beneficial to other tasks execution of the swarm system.

6   Conclusion

Complete coverage of a given region has become a fundamental problem addressed in the field of swarm robots. This article addressed the distributed coverage problem in unknown environments using swarm robotics, and proposed a motion policy based on adaptive potential fields method. The policy doesn't need manual intervention (e.g. parameter tuning) and maximizes the use of potential information gathering from nearby robots within a local communication. Experimental results showed that the adaptive potential field based algorithm is efficient and is superior to a rule-based random walk approach and pheromone method under different scenarios.

Acknowledgement. This work was supported by the Natural Science Foundation of China (NSFC) under grant no. 61375119 and the Beijing Natural Science Foundation under grant no. 4162029, and partially supported by the Natural Science Foundation of China (NSFC) under grant no. 61673025, and National Key Basic Research Development Plan (973 Plan) Project of China under grant no. 2015CB352302.

References

1. Agmon, N., Hazon, N., Kaminka, G.A.: Constructing spanning trees for efficient multi-robot coverage. In: Proceedings 2006 IEEE International Conference on Robotics and Automation. ICRA 2006, pp. 1698–1703. IEEE (2006)

2. Agmon, N., Kaminka, G.A., Kraus, S.: Multi-robot adversarial patrolling: facing a full-knowledge opponent. J. Artif. Intell. Res. **42**, 887–916 (2011)
3. Bayındır, L.: A review of swarm robotics tasks. Neurocomputing **172**, 292–321 (2016)
4. Hert, S., Tiwari, S., Lumelsky, V.: A terrain-covering algorithm for an AUV. Auton. Robots **3**(2), 91–119 (1996)
5. Kantaros, Y., Thanou, M., Tzes, A.: Distributed coverage control for concave areas by a heterogeneous robot-swarm with visibility sensing constraints. Automatica **53**, 195–207 (2015)
6. Kernbach, S.: Structural Self-Organization in Multi-agents and Multi-robotic Systems. Logos Verlag Berlin GmbH, Berlin (2008)
7. Li, J., Tan, Y.: The multi-target search problem with environmental restrictions in swarm robotics. In: 2014 IEEE International Conference on Robotics and Biomimetics (ROBIO), pp. 2685–2690. IEEE (2014)
8. McLurkin, J., Smith, J.: Distributed algorithms for dispersion in indoor environments using a swarm of autonomous mobile robots. In: 7th International Symposium on Distributed Autonomous Robotic Systems (DARS). Citeseer (2004)
9. Morlok, R., Gini, M.: Dispersing robots in an unknown environment. Distrib. Auton. Robot. Syst. **6**, 253–262 (2007)
10. Pinciroli, C., Trianni, V., O'Grady, R., Pini, G., Brutschy, A., Brambilla, M., Mathews, N., Ferrante, E., Di Caro, G., Ducatelle, F., Birattari, M., Gambardella, L.M., Dorigo, M.: ARGoS: a modular, parallel, multi-engine simulator for multi-robot systems. Swarm Intell. **6**(4), 271–295 (2012)
11. Roberts, J.F., Stirling, T.S., Zufferey, J.C., Floreano, D.: 2.5 d infrared range and bearing system for collective robotics. In: IEEE/RSJ International Conference on Intelligent Robots and Systems. IROS 2009, pp. 3659–3664. IEEE (2009)
12. Şahin, E.: Swarm robotics: from sources of inspiration to domains of application. In: Şahin, E., Spears, W.M. (eds.) SR 2004. LNCS, vol. 3342, pp. 10–20. Springer, Heidelberg (2005). doi:10.1007/978-3-540-30552-1_2
13. Schwager, M., Rus, D., Slotine, J.J.: Decentralized, adaptive coverage control for networked robots. Int. J. Robot. Res. **28**(3), 357–375 (2009)
14. Spears, W.M., Spears, D.F., Hamann, J.C., Heil, R.: Distributed, physics-based control of swarms of vehicles. Auton. Robots **17**(2), 137–162 (2004)
15. Svennebring, J., Koenig, S.: Building terrain-covering ant robots: a feasibility study. Auton. Robots **16**(3), 313–332 (2004)
16. Tan, Y., Zheng, Z.Y.: Research advance in swarm robotics. Defence Technol. **9**(1), 18–39 (2013)
17. Zheng, Z., Tan, Y.: Group explosion strategy for searching multiple targets using swarm robotic. In: 2013 IEEE Congress on Evolutionary Computation (CEC), pp. 821–828. IEEE (2013)

Swarm-Based Spreading Points

Xiangyang Huang[1](✉), LiGuo Huang[2](✉), Shudong Zhang[1],
and Lijuan Zhou[1]

[1] College of Information Engineering,
Capital Normal University, Beijing 100048, China
hxy@cnu.edu.cn
[2] Department of Computer Science and Engineering,
Southern Methodist University, Dallas 750122, TX, USA
lghuang@lyle.smu.edu

Abstract. In this paper we propose a Swarm-based Spreading Points algorithm (SSP) for improving the solutions for packing problems. The SSP repositions the initial set of points and evolves it to improve the minimum distance between points. During the evolving process, for each point, a feasible direction of movement is computed according to its nearest neighbors so that the shortest pairwise distance between the point and other points can be increased along this direction (if any). Our experiments showed that the SSP algorithm can improve certain best-known solutions for some problems previously reported in the literature.

Keywords: Spreading points · Packing · Particle swarm optimization

1 Introduction

Spreading points (or packing) problems are unsolved problems in geometry [1]. They are very challenging optimization problems with a wide variety of applications including facility dispersion, communication networks, dashboard layout and etc. [2, 3]. These problems can be modelled as spreading n points into a predefined shape region so that the minimal distance d between any pair of points is greatest [2, 4]. The spreading problem has been formulated as a DC programming problem [5] or as an all-quadratic optimization problem (sometimes called Nonlinear Programming problem, NLP problem) [6], with the hope of achieving high quality approximate solutions by using available NLP solvers [2, 7].

However, the NLP problem has uncountable stationary points that are not necessarily a local minimum [8], and it is so complex that even solving local optimization is a hard task. Direct and simple gradient methods can stop their searches for various reasons such as errors in evaluating gradient values due to inaccurate derivatives, meeting a zero gradient, violation of regularity conditions, etc. [7, 8]. In this paper, we propose a Swarm-based Spreading Points algorithm (SSP) as an alternative to the gradient methods. The SSP is a swarm-based optimization technique [9]. It computes feasible directions of movement for every point according to its nearest neighbors. Along these directions of movement (if any), the point can increase its shortest pairwise

© Springer International Publishing AG 2017
Y. Tan et al. (Eds.): ICSI 2017, Part II, LNCS 10386, pp. 158–166, 2017.
DOI: 10.1007/978-3-319-61833-3_17

distance. Although many movements of points do not change the minimal distance d and only make space for movements of other points, the movements of the latter might increase the minimal distance d at next iteration.

Several particle systems have been proposed to find approximate packings, such as the billiard simulation [10] and the repulsion simulation [11]. In the billiard simulation, the points are regarded as billiard balls which grow gradually in a uniform manner and collide with each other. In the repulsion simulation, the points are regarded as electrical charges (all positive or all negative) which repulse each other. These particle systems use stochastic search. Compared with them, the SSP uses the local exact search. It can converge to a local optimum within a number of iterations, even without descent direction. Our experiments showed that the SSP algorithm can improve certain optimal solutions to some problems previously published by other researchers (http://www.packomania.com/).

The rest of paper is organized as follows: Sect. 2 describes the proposed Swarm-based Spreading Points algorithm (SSP), including how to solve a feasible direction of movement. Section 3 provides the computation results produced by the SSP algorithm, which are provided to outperform existing solutions to certain optimization problems. Section 4 concludes and envisages the future work.

2   Proposed Approach

2.1   Optimization Models

The spreading points problem can be stated as scattering n points within a container in two or three-dimensional geometrical shape (e.g., a unit square, cubic, circle or sphere) such that the minimum distance d between any two points is maximized [2]. Assuming that the container is in the shape of unit square, this problem can be formulated as follows:

$$maximize\ d$$
$$subject\ to\ 0 \le x_i, y_i \le 1, i \in I = \{1, 2, \ldots, n\},$$
$$d^2 \le d_{ij}^2, i, j \in I, i < j.$$

where (x_i, y_i) denotes the coordinates of point i, and $d_{ij}^2 = \left(x_i - x_j\right)^2 + \left(y_i - y_j\right)^2$ denotes the square of the Euclidean distance between points i and j.

This spreading points problem is equivalent to the problem of packing n identical and non-overlapping circles within a unit square with the objective of maximizing the radius r of the n circles [2, 3]. This packing problem is formulated as follows:

maximize r

subject to $r \leq x_i, y_i \leq 1 - r, i \in I = \{1, 2, \ldots, n\},$

$$4r^2 \leq d_{ij}^2, i, j \in I, i < j.$$

where (x_i, y_i) denotes the center of circle i, and $d_{ij}^2 = (x_i - x_j)^2 + (y_i - y_j)^2$ denotes the square of the Euclidean distance between the centers of circles i and j.

The spreading points problem and the packing problem are closely related. The optimal solution of the packing problem can be obtained from that of the spreading points problem as follows: Let the optimal values of the spreading problem and the packing problem respectively be d^* and r^*. Let the optimal coordinates of the n points in the spreading problem be (x_i, y_i), and the optimal coordinates of centers of the n circles in the packing problem be (x_i', y_i'). Then, the optimal solutions to the parking problem and the spreading-points problem can be related as $r^* = d^*/2(1 + d^*)$, $x_i' = (2x_i + d^*)/2(1 + d^*)$ and $y_i' = (2y_i + d^*)/2(1 + d^*)$ for the unit square container.

The spreading points problem (or the packing problem) is a continuous nonlinear global optimization problem with the large number of nonlinear constraints ($n(n-1)/2$) when it has been formulated as a nonconvex optimization problem, and most available NLP solvers fail to identify global optima [2] because it has a large number of local optima and an uncountable stationary points that satisfy the KKT (Karush, Kuhn, Tucker) conditions [8]. The recent trend to solve this problem is to combine local searches and global heuristic procedures. We focus on methods of local exact search in this paper.

2.2   Swarm-Based Spreading Points Algorithm

We first present some notions and definitions before introduce the SSP algorithm.

Let d_i be the shortest pairwise distance between any point p_i and other points, i.e., the distance between the point p_i and its nearest neighbors (denoted as n_i), d_i^+ be the second shortest pairwise distance between the point p_i and other points, i.e., the distance between the point p_i and its second nearest neighbors (denoted as n_i^+). If there doesn't exist any second nearest neighbors for the point p_i, let $d_i^+ = d_i + \varepsilon$ where ε is a predefined positive number.

A feasible direction of movement (r_i) for point p_i can be in any direction, along which the d_i of the point p_i can increase for any small positive step.

Let $d_i(t), d_i^+(t), n_i(t), n_i^+(t)$ and $r_i(t)$ be the values of d_i, d_i^+, n_i, n_i^+ and r_i, respectively, at iteration t.

We will describe the SSP in more detail by the following pseudo-code. Unlike other meta-heuristics, the SSP is simple and requires few parameters.

Algorithm 1. Swarm-based Spreading Points (SSP)
1. Initialize the number n of points, the number m of dimensions that a point has, and the lower and upper bound of each dimension l_k and h_k, $k \in \{1, 2, ..., m\}$
2. Read n points (coordinates) from other algorithms (if any), or randomly generate n points (coordinates) in the container domain
3. Set the step-size parameter $0<a<1$ (the default is 0.5) and a flag variable *isContinue = true*
4. Repeatedly execute the following steps 5-10 while the variable *isContinue* is *true*
5. For each point p_i (in some order without repetition), execute the following steps 6-9
6. Compute the nearest neighbors n_i and the shortest pairwise distance d_i
7. Compute the second nearest neighbors n_i^+ and the second shortest pairwise distance d_i^+
8. Compute a feasible direction of movement r_i according to the nearest neighbors n_i
9. Move this point by a step $a(d_i^+ - d_i)$ along the direction r_i (if the point reaches any boundary, it will stop moving), if the direction r_i is not equal to *zero*
10. Set the flag variable *isContinue = false* if no points can move

In the SSP, every point moves away from other points as far as possible and make space for the movement of other points. Thus, maybe SSP finds a better solution than simple gradient methods. The SSP algorithm is greedy and it is convergent (see Appendix for proof).

SSP may be parallelized. It is also an anytime algorithm, which can return a (an improved) solution even though it is interrupted at any time. It is expected to find a better solution if it can take more time.

Note that we use an $n(n-1)/2$-dimensional vector to save the distances between points (the square of distance). When a point can be moved, the distances between the point and other $n-1$ points are updated after the movement occurs. This leads to a significant reduction of computational costs. Computing a feasible direction of movement is a somewhat difficult task and we will introduce it in the next section.

2.3   Computing a Feasible Direction

A feasible direction of movement for a point is computed based on its nearest neighbors. Moving the point along the direction can increase the shortest pairwise distance of the point no matter how small the step is. We first illustrate the process and introduce some definitions with a simple example (see Fig. 1).

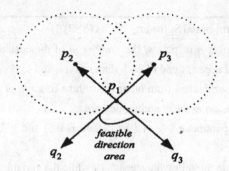

Fig. 1. A simple example for illustrating feasible directions

In Fig. 1, points p_2 and p_3 are the nearest neighbors of point p_1. Vectors $\overline{p_1 p_2}$ and $\overline{p_1 p_3}$ are defined from the point p_1 to points p_2 and p_3 respectively. Hence the equalities, $d_1 = |\overline{p_1 p_2}| = |\overline{p_1 p_3}|$, are satisfied, where d_1 is the shortest pairwise distance of the point p_1. Let vectors $\overline{p_1 p_2}$ and $\overline{p_1 q_2}$ be two orthogonal vectors, as well as vectors $\overline{p_1 p_3}$ and $\overline{p_1 q_3}$, i.e., $\overline{p_1 p_2} \cdot \overline{p_1 q_2} = 0$ and $\overline{p_1 p_3} \cdot \overline{p_1 q_3} = 0$. Any vector r that starts from the point p_1 and satisfies $r = \lambda_1 \cdot \overline{p_1 q_2} + \lambda_2 \cdot \overline{p_1 q_3}$, where $\lambda_1, \lambda_2 \geq 0$ and $\lambda_1 + \lambda_2 > 0$, is a feasible direction vector, i.e., feasible directions are in between the vectors $\overline{p_1 q_2}$ and $\overline{p_1 q_3}$ (counter-clockwise in Fig. 1) and include them (do not include *zero* vector). Moving along a non-feasible direction (not a *zero* vector) with a small step, the point p_1 will enter any circle with center point p_2 or p_3 and, as a result, the shortest pairwise distance of the point p_1 decreases. We call the vectors $\overline{p_1 q_2}$ and $\overline{p_1 q_3}$ basic feasible direction vectors, which construct a feasible direction pyramid (or cone).

Let P be a matrix, each row representing an m-dimensional vector from the point p to one of its nearest neighbors. Any feasible direction vector for the point p is defined as an m-dimensional column vector r. It satisfies the following inequality according to the above description:

$$P\,r \leq 0. \tag{1}$$

Note that the right hand side of (1) means that each component of the resulting vector is less than or equal to *zero*. If the point p is within the boundary, we have the following constraint for r because the point is not allowed to be out of the container (any convex container).

$$s \bullet r \geq 0. \tag{2}$$

where s is an inward-pointing normal vector (pointing towards the interior of the container) to the boundary at the point p. If container is a square or a cubic, (2) is equivalent to (3):

$$r[i] \geq 0 \text{ if } p[i] = l_i \text{ or } r[i] \leq 0 \text{ if } p[i] = h_i. \tag{3}$$

where $r[i]$ denotes the i-dimensional component of the vector r and $p[i]$ denotes the i-dimensional coordinate of the point p.

Additionally, we only care for the direction of the vector r rather than its magnitude in the SSP algorithm. Hence we can add additional linear constraints for the magnitude of the vector r so that classic linear programming methods, for example the simplex method [12], can be utilized. An m-dimensional unit box is defined as (4)

$$|r_i| \leq 1. \tag{4}$$

We can solve a special solution r using the simplex method, combining (1), (2), (3) and (4) and defining one or more objective functions. If r is not a *zero* vector, then it is a feasible direction vector. We can also solve all basic feasible direction vectors, and then assign weights (i.e., λ) to the basic feasible direction vectors and sum them up.

Intuitively speaking, when the points within the boundaries move along the boundaries, it is likely to be useful in improving the effectiveness of the SSP algorithm. In fact, if a point p within a boundary has a feasible direction of movement r (r_x, r_y), then its projection direction r', along the tangent to the boundary at the point p, is also a feasible direction for the point. For the circle container with the radius equal to R, if a point p within a boundary has a feasible direction of movement r (the direction may be along the tangent to the boundary according to (2)) and the shortest pairwise distance d_p of the point is less than R, then it can move along the boundary to improve its shortest pairwise distance.

3 Computation Results

In this section the SSP algorithm is applied to packings of equal circles in a square (CSPs). We examine some best-known packings (the problem size $n < 1000$) found in website (http://www.packomania.com/), in which all coordinates of all packings, as well as all the values of radius etc., can be obtained (the precision of all values, including the coordinates, is 10^{-30}). We found improvements for some packings[1].

SSP accepts as the input the coordinates of any packing, which are downloaded from the website mentioned above, and tries to evolve them to improve the minimum distance among points. The SSP algorithm was implemented in C ++ and the precision is 10^{-15}. Hence, when these coordinates (the precision is 10^{-30}) are fed into SSP, it is equivalent to generate an initial solution from a neighborhood of the original solution (all coordinates). Then the SSP algorithm performs a local search starting from the initial solution with 2000 iterations (a predefined upper bound of iterations in our experiment). If the SSP algorithm improves the best-known radius, the best-known radius is updated.

Table 1 show some improved results obtained by SSP within 2000 iterations. Column 1 gives the problem size n. Column 2 gives the best-known radii reported in the literature with the precision 10^{-15} (in fact these radii are intentionally increased

---

[1] The check program is available for download at http://huangxiangyang.ie.cnu.edu.cn/papers/checkpackings.rar.

Table 1. Comparing the SSP's result to the best known radii

n	Best- known radii	New radii	Difference
254	0.0326400137267399	0.0326400137547596	2.80196968739094e−011
682	0.0201539663593881	0.0201539663646636	5.27548064130304e−012
686	0.0201124351080246	0.0201124351162787	8.25410840241073e−012
791	0.0187496687267892	0.0187496687301134	3.32426020607737e−012
816	0.0185262655749775	0.0185262742555956	8.68061805343054e−009
818	0.0185197925355186	0.0185197926027333	6.72147433878606e−011
821	0.0185185206063483	0.0185185206372511	3.09027864743011e−011
992	0.0158730158730169	0.0162128400921169	3.398242190999900e−04
999	0.0165135791514728	0.0165135791605391	9.06632675054478e−012

from their original values). Column 3 reports the results obtained by the SSP algorithm. Column 4 indicates the difference between the results obtained by our algorithm and the best-known results in column 2 (our result improves the previously published results listed in this table).

4   Conclusions and Future Work

Spreading points (or packing) problems have usually been formulated as nonconvex optimization problems such that high quality approximate solutions can be identified by any existing Nonlinear Programming (NLP) solver. However, the NLP solvers leveraging gradients for the search will stop in a stationary point that is not necessarily a local minimum. In this paper we propose a Swarm-based (Locally) Spreading Points algorithm (SSP) as an alternative to the gradient methods. The SSP algorithm computes a feasible direction of movement for every point according to its nearest neighbors, evolves the set of points, and tries to output an improved set of points (we call the approach Locally Spreading, LS in short). It showed that the SSP algorithm can improve best-known solutions for certain packings previously reported by other researchers (http://www.packomania.com/).

Future work includes (1) applying the SSP algorithm on other packing problems such as packings of equal circles in a circle, and (2) investigating the application of the SSP algorithm within some global heuristic procedures such as Monotonic Basin Hopping (MBH) [13], Variable Neighborhood Search (VNS) [8], or Greedy Vacancy Search [14].

Acknowledgments. Part of this work was developed while Xiangyang Huang worked in Professor LiGuo Huang's research group as a visiting scholar at Southern Methodist University. This work was supported in part by National Natural Science Foundation of China under Grants 61371194, 61672361 and 61402033, and Beijing Natural Science Foundation under Grant 4152012.

Appendix

Given a predefined calculation precision, SSP is convergent. A proof of the convergence is given below.

Let c be an n-dimensional vector and the arrangement of its elements, from its first element denoted by $c\,[1]$ to the last $c[n]$, corresponds to the arrangement of d_i for all n points ordered by ascend, i.e., $c[i] \leq c[i+1]$ for any $1 \leq i \leq n\text{-}1$. Given a specified calculation precision and the number of points n, the set of c, denoted by C, is bounded (finite). The inequality $c_1 > c_2$ holds if and only if there exists an integer i ($1 \leq i \leq n$) such that $c_1[j] \geq c_2[j]\ \forall j < i$ and $c_1[i] > c_2[i]$. where $c_1, c_2 \in C$.

Let c_t and $c_{t+1} \in C$ be arrangements of shortest pairwise distances for n points at iterations t and $t+1$ respectively. The elements in c_t and c_{t+1} are ordered by ascend.

Convergence. Let t be the current iteration and $t+1$ be the iteration after movement of a point p_i (suppose the position index of the point p_i is also i in c_t). When the point p_i moves along the direction $r_i(t)$, for any point p_k in the nearest neighbors $n_i(t)$ of p_i, the d_k does not decrease (increases if the point p_i also belongs to $n_k(t)$); for other points p_j (p_j is not in $n_i(t)$), let d_{ij} be the distance from the point p_i to the point p_j, and then $d_{ij} \geq d_i^+$ according to the definition of d_i^+. After the point p_i moves a step $a(d_i^+ - d_i)$ from iteration t to $t+1$, the distance $d_{ij}(t+1) \geq d_i^+(t) - a(d_i^+(t) - d_i(t)) > d_i(t)$ since a is less than 1. So for p_j, if $d_j(t) \leq d_i(t)$ at iteration t, $d_j(t+1)$ does not change at iteration $t+1$ since $d_{ij}(t+1) > d_i(t)$ and the point p_i does not become a neighbor of p_j at iteration $t+1$. If $d_j(t) > d_i(t)$ at iteration t, $d_j(t+1)$ is still larger than $d_i(t)$ at iteration $t+1$ even though the movement of the point p_i affects d_j (i.e., the point p_i becomes a nearest neighbor of the point p_j at iteration $t+1$). In summary, (1) for the point p_u, the position of which in c_t is before that of the point p_i ($d_u(t) \leq d_i(t)$), the $d_u(t+1)$ does not decrease at iteration $t+1$; (2) for the point p_v, the position of which in c_t is after that of the point p_i ($d_v(t) > d_i(t)$), the $d_v(t+1)$ is still larger than $d_i(t)$; (3) for the point p_i, its $d_i(t+1)$ increases. Hence c_{t+1} increases according to the above definition when there exists any movable point. Because C is bounded (finite), c_t will converge with $t \to \infty$.

References

1. Croft, H.T., Falconer, K.J., Guy, R.K.: Unsolved Problems in Geometry: Unsolved Problems in Intuitive Mathematics. Springer Science & Business Media, New York (2012)
2. Hifi, M., M'hallah, R.: A literature review on circle and sphere packing problems: models and methodologies. Adv. Oper. Res. **2009**(4), 22 (2009)
3. Castillo, I., Kampas, F.J., Pintér, J.D.: Solving circle packing problems by global optimization: numerical results and industrial applications. Eur. J. Oper. Res. **191**(3), 786–802 (2008)
4. Cabello, S.: Approximation algorithms for spreading points. J. Algorithms **62**(2), 49–73 (2007)
5. Horst, R., Thoai, N.V.: DC Programming overview. J. of Optim. Theory Appl. **103**, 1–43 (1999)

6. Raber, U.: Nonconvex All-quadratic global optimization problems: solution methods, applications, and related topics. Ph.D. thesis, University of Trier, Germany (1999)
7. Mladenović, N., Plastria, F., Urošević, D.: Reformulation descent applied to circle packing problems. Comput. Oper. Res. **32**(9), 2419–2434 (2005)
8. M'Hallah, R., Alkandari, A., Mladenovic, N.: Packing unit spheres into the smallest sphere using VNS and NLP. Comput. Oper. Res. **40**(2), 603–615 (2013)
9. Tan, Y., Shi, Y., Niu, B. (eds.): ICSI 2016. LNCS, vol. 9712. Springer, Heidelberg (2016)
10. Boll, D.W., Donovan, J., Graham, R.L., et al.: Improving dense packings of equal disks in a square[J]. Electron. J. Com. **7**(1), R46 (2000)
11. Szabó, P.G., Markót, M.C., Csendes, T.: Global Optimization in Geometry—Circle Packing into the Square: Essays and Surveys in Global Optimization, pp. 233–265. Springer, US (2005)
12. Dantzig, G.B., Orden, A., Wolfe, P.: The generalized simplex method for minimizing a linear form under linear inequality restraints. Pac. J. Math. **5**(2), 183–195 (1955)
13. Addis, B., Locatelli, M., Schoen, F.: Disk packing in a square: a new global optimization approach. INFORMS J. Comput. **20**(4), 516–524 (2008)
14. Huang, W., Ye, T.: Greedy vacancy search algorithm for packing equal circles in a square. Oper. Res. Lett. **38**(5), 378–382 (2010)

A Survivability Enhanced Swarm Robotic Searching System Using Multi-objective Particle Swarm Optimization

Cheuk Ho Yuen[✉] and Kam Tim Woo

Department of Electronic and Computer Engineering,
The Hong Kong University of Science and Technology, Hong Kong, China
{chyuenaa,eetim}@ust.hk

Abstract. This paper aims at outlining an algorithm for groups of swarm robots solely powered by light energy to survive and complete target searching tasks in unknown fields where light energy charging points and targets are scattered. To sustain the searching operation and solve energy consumption conflicts between surviving and searching, this paper introduces a multi-robot algorithm based on Multi-Objective Particle Swarm Optimization (MOPSO) and energy-saving decision rules. A novel mechanism of selecting the best performing particle in PSO is introduced. Several sets of simulation experiments were conducted and results show that a 15-robot swarm system running this algorithm is able to search a single target and stabilize the energy level for the long-term simultaneously. It demonstrates the feasibility of applying this energy-optimized MOPSO as a design framework for a long-term searching swarm robot system.

Keywords: Particle Swarm Optimization · Swarm robotics · Survivability · Multi-robot searching · Foraging

1 Introduction

In current long-term outdoor robotic applications like Unmanned Aerial Vehicles (UAVs) or self-driving cars, a common solution for a long-term power source is to carry a large capacity battery or fuel tank. However, energy constraints drastically limit the real-world applications of system with scare payload like robotic-WSN and remote area operation. Energy harvesting during operations is considered a good methodology to prolong or sustain the running time of the system.

A swarm robotics system which can adapt its action policy to survive under different environments and perform user-defined tasks (e.g. maintain a spatial formation for routing, or exploring the fields) simultaneously and efficiently provides a strong framework for searching and foraging robotics operations. This paper proposes a system that can achieve this goal using multi-objective Particle Swarm Optimization (MOPSO). The potential applications of this system could be long-term searching operations in remote areas, for example foraging in the agricultural industry, or search-and-rescue operations under debris. In our assumption, each robot can sense light intensity and obstacles and harvest the light energy within a finite range. They can

© Springer International Publishing AG 2017
Y. Tan et al. (Eds.): ICSI 2017, Part II, LNCS 10386, pp. 167–175, 2017.
DOI: 10.1007/978-3-319-61833-3_18

form a mesh-network with a limited communication range. Complete drain of battery power is considered to be 'death'. The core task is to search and locate the target by staying adjacent to it. In the meantime, keeping a certain level of energy reserve for long-term operation is required. Therefore, two PSOs, one for searching the energy station and one for the target, run concurrently. The final output is a combination of two PSOs with different weight factors. The parameter set will change per different internal energy state, e.g. the user-task-PSO should be dominant in high-energy state, the light-searching-PSO should be dominant in low-energy state, etc.

This paper focuses on the design of the survival mechanism and the solution of conflicts between surviving and performing user-defined tasks simultaneously and effectively. The first part is justification of that using our modified MOPSO can help the whole swarm to sustain without greatly sacrificing the ability to search. The second part, studies whether the collective behavior of whole swarm can adapt to a variety of energy availability condition and be efficient in terms of user tasks.

2 Related Work

Liu et al. [1, 2] presented an adaption mechanism for a group of autonomous robots and discussed how the combination of simple cues affect the efficiency of foraging.

Kernbach et al. [3, 4] investigated cooperating a swarm of robots with local communication ability to adjust their action plan to maintain energy homeostasis within the swarm. They summed up several rules on the basis of a simplified swarm foraging kinetic model built and compared the results with a bio-inspired optimal foraging algorithm. Results from their work justified that collective behaviors that emerge from a swarm of robots help sustain a certain level of collective energy.

The major challenge of surviving and performing tasks concurrently is that optimization based on these two objectives may conflict. Performing tasks inevitably means the consumption of energy, which makes the explicitly specified objective function of energy not an appropriate approach. Bredeche et al. [5] assumed that the environmental part can be represented by an implicit fitness function hidden. To solve the stated conflicts, Haasdijk et al. [6, 7] proposed the MONEE framework in which task-centric fitness functions are added to mEDEA. Results from simulation experiments show that the proposed algorithm is able to solve the conflicts.

In contrast to a genetic algorithm (GA), the particle swarm optimization (PSO), is less computationally-demanding and can be applied to real-value problems without encoding need. [8] Pugh and Martinoli et al. [9, 10] proposed an adapted version of PSO to distributed unsupervised robotic learning problems with only local communication. The results of those experiments also show that systems with a limited communication range can perform as well as those with an unlimited range. Couceiro et al. [11] modified and implemented it into a real world multi-robot exploration problem. Different types of extended multi-robot PSO searching algorithms which target various aspects were proposed. [12, 13] Since the idea of multi-objective PSOs (MOPSOs) was introduced by Moore et al. [14], there were many examples of applying and mathematically analyzing in MOPSO [15, 16].

3 Problem, Approach, Algorithm

3.1 System Settings

Target and energy stations are considered virtual beings which emit signals to the robot sensor, for example, odor, light, heat source or RF source; while obstacles except robots themselves are not existing. Two distances, D_{ST} and D_{RT}, as illustrated in Fig. 1., are used as the performance metrics of this searching system. Every robot is homogenous in hardware design and carries charging equipment that harvests energy from energy stations and sensors of signals from the energy station and target. The energy source is a recharging battery unit. Based on the percentage of energy reserve, there are 6 energy states: {FULL, GOOD, FINE, HUNGRY, STARVE, DEATH}.

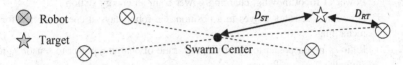

Fig. 1. Illustrations of individual minimum distance (D_{RT}) & swarm-to-target distance (D_{ST})

3.2 Particle Swarm Optimization in Multi-robot Searching

Particle Swarm Optimization is an optimization method that tries to find the solution of a candidate function subjected to a fitness function. The main idea is to generate a swarm of pointless particles and let them move in the search space. In every iteration, the performance of the solution with the current position as input is calculated. The best performed particle among the whole swarm will share its own position to all particles as g^{best} and the best position a particle has ever reached is its own l^{best}. $W_t, C^l, r_t^l, C^g, r_t^g$ are inertial coefficient, acceleration coefficients and random factors of local part and global part respectively. Finally, a new velocity will be calculated as shown in Eq. (1) [8].

Using PSO in robotic searching tasks assumes that each robot is a particle and the velocity calculated will be the new velocity of the real robot. [9, 10] The fitness function is usually virtual signal emitted by the target. Unlike ordinary PSO, g^{best} is the best valued position among robots within the communication range.

$$V_t = W_t V_{t-1} + C^l r^l (X_t - P_t^l) + C^g r^g (X_t - P_t^g) \tag{1}$$

3.3 Energy-Optimized MOPSO

In our system, we implement two PSOs in one single robot. L-PSO is used to search for a light energy station and T-PSO is used to search for the target. The fitness functions in L-PSO and T-PSO are light intensity and radio signal strength. Unlike usual MOPSO settings, they are running independently. The final velocity update is calculated by the

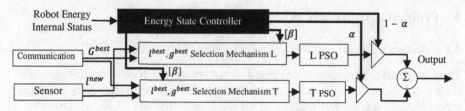

Fig. 2. Diagram of energy-optimized MOPSO algorithm

Table 1. Parameters in robot hardware specification

SYMBOL	DEFINITION
P_m, P_L	Power of robot moving, charging power from an energy station
t_m, t_L	How long the robot moves to a position, the robot stays at energy station for recharging
K_m, K_L	Ratio of motor power to robot velocity, ratio of charging power to sensor signal strength
S	Distance travelled in every movement
$F(X_t)$	Sensor signal strength at position X_t

weighted sum of updates from L-PSO and T-PSO. As shown in Fig. 2., robots will adjust its tendency of approaching light energy station or target under different α values. Our goal is to maximize the α values under the survival constraint (Table 1).

$$P_m t_m < P_L t_L \tag{2}$$

$$K_m S < K_L F(X_t) t_L \tag{3}$$

To maintain a stable intake of energy, each robot should estimate whether it can harvest more energy than the amount such a searching operation consumes. In this paper, we will focus on consumption of energy, for moving, which is usually motor in robotic systems. In terms of L-PSO, the fitness value, light intensity is directly proportional to the light energy charging rate. From Eqs. (2) and (3), the determinant of this inequality is the reciprocal of t_L. Intuitively, t_L also means how long does it take before reaching a break-even point, i.e. the point at which it gains back the energy it consumed. A robot with scarce energy reserves should have a smaller t_L.

Finally, we define a β-value, "Worth value", for every energy state. This value works as a selection mechanism to filter out g^{best} or l^{best} which may not favor the surviving of the robot from the list. The selection procedure can be seen in Fig. 3.

There are two novel components in our modified algorithm, g^{best} & l^{best} selection mechanism and weight factors for the final output from two PSOs. Parameter sets α, β and the maximum velocity are adjusted by an energy state controller.

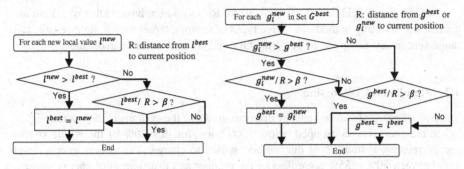

Fig. 3. l^{best} & g^{best} Selection mechanism

4 Experiments and Results

4.1 Setup

To validate the survivability and robustness of our proposed energy-optimized MOPSO, simulation experiments were conducted under varied energy availability and distribution. There are 6 cases which can be seen in Table 2. Robots in swarm in Case 1 and Case 6 were running the original PSO searching algorithm proposed by Pugh et al. [9, 10] and the algorithm proposed by Kernbach et al. [3] respectively. In Cases 2–5, they were running our energy-optimized MOPSO in different environments as illustrated in Fig. 4. A multi-robot simulation environment, Stage is used [17].

Table 2. Experiments Conditions

Label	Case 1	Case 2	Case 3	Case 4	Case 5	Case 6
Environment	A	A	B	C	D	B
Algorithm	Pugh et al. [9, 10]	Ours	Ours	Ours	Ours	Kernbach et al. [3]
No. of energy station	0	1	3	5	8	3
Distance from closet station to target	N/A	19.7 m	23.06 m	6.02 m	9.52 m	23.06 m

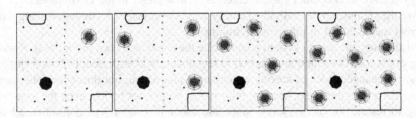

Fig. 4. Distribution of energy stations and target in environment A, B, C, D (from left)

The simulated field size is 30 m × 30 m. Robots which have radius of 20 cm are distributed randomly in the field. Three types of sensors: target sensor, light sensor, and range sensor, are equipped to perform PSO updates and obstacle avoidance.

4.2   Results and Discussion

The overall energy level and energy dissipation rate of the system can be seen in Fig. 5. When the energy level dropped below 5000 units, roughly 30% of the whole swarm energy reserve, system using our method started to charge. The system energy level kept between 30% – 55% regardless of the number and distribution of energy stations in the field. Considering the Case-1 and Case-6, the energy dropped gradually to 0 after the 310th s and converged to a stable value respectively. It is consistent that only system using our method can charge and work in balanced time as shown in Fig. 5(b). Once an individual energy dropped below FINE or HUNGRY states, the robot adjusted the PSO velocity update weights of T-PSO and L-PSO accordingly.

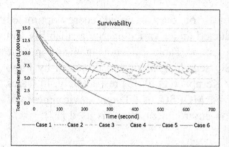 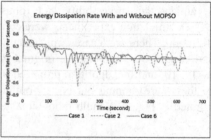

Fig. 5. System energy level (a) & Energy dissipation rate (b) versus time

By assigning different α-value to each energy level, our system can keep the whole swarm continuously working long-term. These values are essentially the efficiency of the searching part. Depending on the hardware design, α values cannot be set higher than a threshold since an individual robot may not have enough energy reserve to go back from the current position to the energy station. Besides this, high α values in low-energy states may induce oscillations in the system energy and therefore, less energy reserve to encounter sudden changes or new target in the environment.

We then compared the individual and system performance of simulation using three different methods and in different environments.

In all cases, the swarm system could locate the target as shown in Fig. 6(a). System using original PSO method approached to target shortly as expected. Robots using our method exchanged roles of staying at the target and recharging throughout the simulation. Therefore, the target location information would not be lost even the swarm is trying to restore its energy level. Since the robot in Case-1 had consumed all the reserve, they stopped operating after the 310th s.

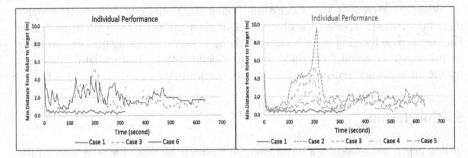

Fig. 6. Minimum distance of different methods (a) & different environments (b)

The system performance over time is presented in Fig. 7. The system in Case-6 could not locate the target collaboratively and the swarm-to-target distance did not decrease since it is using random search and has no exchanges of target information.

The system using original PSO searching algorithm performed better the system using our method. Since the swarm system using our method recharges itself and does tasks simultaneously, there is a tradeoff in searching performance. In different conditions, our system shares a similar trend. It could search and stay at the target successfully in the beginning of the simulation and return to the energy station once the energy level dropped. It achieved a relatively stable point after the 300th s. Considering these bounded values, we found that they are smaller than the corresponding mid-point distance between the closest energy station and target as stated in Table 2. In simulation settings, each station shares the same sensor output. The introduction of β-value imposes selection pressure on distant energy stations. To validate such a hypothesis, an extra set of experiments was conducted with all β values reduced to zero. The results show that appropriate β values can help to improve the performance in the energy-optimized MOPSO. Furthermore, we observed that several robots ran out of power in this extra experiment when they tried to move to distant energy stations.

An adaptive version of this system that copes with dynamic changing conditions or multi-target could be investigated in future work. In addition, this work shows the possibility of using this algorithm as a framework for along-term operating swarm system. Further research on these parts could also be done.

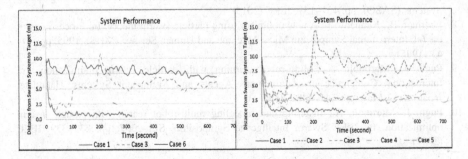

Fig. 7. Swarm-to-target distance of different methods (a) & different environments (b)

When the swarm system successfully found target and energy stations, a stable route from these two points was observed. This emergent behavior could be applied to long-term crops foraging work in the agricultural industry. Adopting the same principle in this paper, we could further add more user-defined tasks, like spatial formation in remote areas for self-powered mesh-network connectivity, etc.

5   Conclusion

In this paper, we proposed a novel energy-optimized MOPSO for long term swarm robotic searching tasks. Several sets of simulations of 15 robots were performed. The experimental results justify the survivability and robustness of this algorithm. The trend of energy level against runtime remains at a relatively stable value regardless of the availability of energy stations. For most of the cases in the simulation, the whole swarm could search and locate the target successfully in the beginning and return to energy stations at low-energy states.

References

1. Liu, W., Winfield, A.F., Sa, J., Chen, J., Dou, L.: Towards energy optimization: emergent task allocation in a swarm of foraging robots. Adapt. Behav. 15(3), 289–305 (2007)
2. Liu, W., Winfield, A.F.: Modeling and optimization of adaptive foraging in swarm robotic systems. Int. J. Robot. Res. 29(14), 1743–1760 (2010)
3. Kernbach, S., Nepomnyashchikh, V.A., Kancheva, T., Kernbach, O.: Specialization and generalization of robot behaviour in swarm energy foraging. Math. Comput. Model. Dyn. Syst. 18(1), 131–152 (2012)
4. Kernbach, S., Kernbach, O.: Collective energy homeostasis in a large-scale microrobotic swarm. Robot. Auton. Syst. 59(12), 1090–1101 (2011)
5. Bredeche, N., Montanier, J.M., Liu, W., Winfield, A.F.: Environment-driven distributed evolutionary adaptation in a population of autonomous robotic agents. Math. Comput. Model. Dyn. Syst. 18(1), 101–129 (2012)
6. Haasdijk, E., Bredeche, N., Eiben, A.E.: Combining environment-driven adaptation and task-driven optimisation in evolutionary robotics. PLoS ONE 9(6), e98466 (2014)
7. Haasdijk, E.: Combining conflicting environmental and task requirements in evolutionary robotics. In: 2015 IEEE 9th International Conference on Self-Adaptive and Self-Organizing Systems (SASO), pp. 131–137. IEEE (2015)
8. Eberhart, R., Kennedy, J.: A new optimizer using particle swarm theory. In: Proceedings of the 6th International Symposium Micro Machine and Human Science (MHS), 1995, pp. 39–43 (1995)
9. Pugh, J., Martinoli, A.: Multi-robot learning with particle swarm optimization. In: International Conference on Autonomous Agents and Multiagent Systems, Hakodate, Japan, 8–12 May, pp. 441–448 (2006)
10. Pugh, J., Martinoli, A.: Inspiring and modeling multi-robot search with particle swarm optimization. In: IEEE Swarm Intelligence Symposium, SIS 2007, pp. 332–339. IEEE (2007)

11. Couceiro, M.S., Rocha, R.P., Ferreira, N.M.: A novel multi-robot exploration approach based on particle swarm optimization algorithms. In: 2011 IEEE International Symposium on Safety, Security, and Rescue Robotics (SSRR), pp. 327–332. IEEE (2011)
12. Couceiro, M.S., Vargas, P.A., Rocha, R.P., Ferreira, N.M.: Benchmark of swarm robotics distributed techniques in a search task. Robot. Auton. Syst. **62**(2), 200–213 (2014)
13. Derr, K., Manic, M.: Multi-robot, multi-target particle swarm optimization search in noisy wireless environments. In: Proceedings of the 2nd Conference on Human System Interactions, Catania, Italy, pp. 78–83 (2009)
14. Moore, J., Chapman, R.: Application of particle swarm to multiobjective optimization. Department of Computer Science and Software Engineering, Auburn University, vol. 32 (1999)
15. Reyes-Sierra, M., Coello, C.C.: Multi-objective particle swarm optimizers: a survey of the state-of-the-art. Int. J. Comput. Intell. Res. **2**(3), 287–308 (2006)
16. Gong, D.W., Zhang, J.H., Zhang, Y.: Multi-objective particle swarm optimization for robot path planning in environment with danger sources. J. Comput. **6**(8), 1554–1561 (2011)
17. Vaughan, R.T.: Massively multi-robot simulation in stage. Swarm Intell. **2**(2–4), 189–208 (2008)

Autonomous Coordinated Navigation of Virtual Swarm Bots in Dynamic Indoor Environments by Bat Algorithm

Patricia Suárez[1], Akemi Gálvez[1,2], and Andrés Iglesias[1,2(✉)]

[1] Department of Applied Mathematics and Computational Sciences,
University of Cantabria, Avenida de los Castros s/n, 39005 Santander, Spain
iglesias@unican.es

[2] Department of Information Science, Faculty of Sciences, Toho University,
Narashino Campus, 2-2-1 Miyama, Funabashi 274-8510, Japan

Abstract. Autonomous navigation (i.e., without human intervention) in indoor spaces such as houses and office buildings has many important applications; for instance, in areas affected by building collapse due to natural or artificial disasters. However, it is also a difficult task because any prescribed trajectory can be suddenly interrupted by unexpected obstacles. Arguably, a group of simple autonomous drones driven by swarm intelligence might be more efficient than a sophisticated robot for navigation within such environments. Based on this idea, this work presents a method that applies a powerful swarm intelligence technique called bat algorithm to the autonomous coordinated navigation of a swarm of virtual bots in dynamic indoor environments. Some computational experiments are conducted to test the performance of this approach.

Keywords: Swarm intelligence · Bat algorithm · Autonomous coordinated navigation · Dynamic indoor scenes · Virtual swarm bots

1   Introduction

One of the most surprising findings in artificial intelligence during the last few decades is the fact that a swarm of simple agents with extremely limited intelligence but collaborating together might lead to the emergence of highly sophisticated behaviors [1,8]. This exciting research field, called *swarm intelligence* (SI), is based on the principle that there is no a centralized intelligence controlling the swarm, taking decisions, and defining the behavioral patterns of the swarm units. Instead, the limited intelligence of such swarm units is amplified through interaction and information sharing with each other and the environment.

Nowadays, swarm intelligence is attracting increasing attention due to its ability to achieve complex tasks [2,9]. An interesting example is the navigation in dynamic indoor spaces such as houses and office buildings. In this problem, a swarm of simple interconnected mobile robots deployed throughout the

© Springer International Publishing AG 2017
Y. Tan et al. (Eds.): ICSI 2017, Part II, LNCS 10386, pp. 176–184, 2017.
DOI: 10.1007/978-3-319-61833-3_19

search space possesses greater exploratory capacity than a single but sophisticated robot. This makes the swarm much more effective in navigation tasks. This
feature is very valuable in areas affected by building collapse due to natural or
artificial disasters such as earthquakes or terrorist attacks, where the primary
concern is to find and rescue victims as efficiently and safely as possible without
risking the lives of survivors and human rescue workers. In such critical situations, the swarm of simple robots has many potential advantages over a single
sophisticated robot, such as greater flexibility, adaptability, and robustness.

Two major requirements of swarm members in navigation tasks are *autonomy*
and *coordination*. The former is particularly important for unknown or dynamic
environments, where any prescribed trajectory can be suddenly interrupted by
obstacles. Think, for instance, about home spaces affected by earthquakes, where
we could find many unexpected objects (e.g., door frames, jambs, or pillars) lying
on the ground, requiring frequent on-the-fly detours. Coordination is also very
important, not only to avoid collisions with other robots but also to circum-
vent situations leading to bottlenecks, such as a swarm of robots trying to pass
through a very narrow place (such as a corridor or a wall hole).

Although generally regarded as conflicting features, autonomy and coordination can still be made compatible. In fact, they are key features of all SI methods,
where they are typically balanced via the dynamic evolution equations. Moreover, SI methods are relatively simple to implement and understand, and are
quite affordable in terms of computing resources. Owing to these reasons, in this
paper we apply a powerful SI approach called *bat algorithm* to efficient navigation in dynamic indoor environments. The algorithm has shown to be very
effective to address difficult multimodal continuous optimization problems (such
as that described in this work) involving a large number of variables [3–6].

The structure of this paper is as follows: in Sect. 2 we provide a gentle overview about the bat algorithm. Our bat algorithm-based approach
for autonomous coordinated navigation in dynamic indoor environments is
explained in Sect. 3. The method is illustrated through its application to a swarm
of virtual bots deployed throughout a large and fully furnished house as reported
in Sect. 4. The paper closes with the conclusions and future work in the field.

2   The Bat Algorithm

The *bat algorithm* is a bio-inspired swarm intelligence algorithm originally proposed by Xin-She Yang to solve optimization problems [7,10,11]. The algorithm
is based on the echolocation behavior of bats. The author focused on micro-
bats, as they use a type of sonar called *echolocation*, with varying pulse rates of
emission and loudness, to detect prey, avoid obstacles, and locate their roosting
crevices in the dark. The interested reader is referred to [12] for a comprehensive,
updated review of the bat algorithm, its variants and applications.

This paper considers the standard bat algorithm in [10], whose pseudo-code
is shown in Algorithm 1. Basically, the algorithm considers an initial population of $\mathcal{P}$ individuals (bats) representing potential solutions of the optimization

Require: (Initial Parameters)
  Population size: $\mathcal{P}$; Maximum number of generations: $\mathcal{G}_{max}$
  Loudness: $\mathcal{A}$; Pulse rate: r ; Maximum frequency: f_{max}
  Dimension of the problem: d ; Random number: $\theta \in U(0,1)$; $g = 0$
  Objective function: $\phi(\mathbf{x})$, with $\mathbf{x} = (x_1, \ldots, x_d)^T$
 1: Initialize the bat population $\mathbf{x}_i$ and $\mathbf{v}_i$, $(i = 1, \ldots, n)$
 2: Define pulse frequency f_i at $\mathbf{x}_i$
 3: Initialize pulse rates r_i and loudness $\mathcal{A}_i$
 4: **while** $g < \mathcal{G}_{max}$ **do**
 5: **for** $i = 1$ **to** $\mathcal{P}$ **do**
 6: Generate new solutions by adjusting frequency,
 7: and updating velocities and locations //eqns. (1)-(3)
 8: **if** $\theta > r_i$ **then**
 9:   $\mathbf{s}^{best} \leftarrow \mathbf{s}^g$ //select the best current solution
 10:   $\mathbf{ls}^{best} \leftarrow \mathbf{ls}^g$ //generate a local solution around $\mathbf{s}^{best}$
 11: **end if**
 12: Generate a new solution by local random walk
 13: **if** $\theta < \mathcal{A}_i$ *and* $\phi(\mathbf{x_i}) < \phi(\mathbf{x}^*)$ **then**
 14:   Accept new solutions
 15:   Increase r_i and decrease $\mathcal{A}_i$
 16: **end if**
 17: **end for**
 18: $g \leftarrow g + 1$
 19: **end while**
 20: Rank the bats and find current best $\mathbf{x}^*$
 21: **return** $\mathbf{x}^*$

Algorithm 1. Bat algorithm pseudocode

problem. Each bat has a location $\mathbf{x}_i$ and velocity $\mathbf{v}_i$, initialized with random values. Also a pulse frequency, pulse rate, and loudness, which are computed for each individual bat (lines 2–3). Then, the swarm evolves over generations (line 4) until the maximum number of generations, $\mathcal{G}_{max}$, is reached (line 18). For each generation g and each bat (line 5), new frequency, location and velocity are computed (lines 6–7) as:

$$f_i^g = f_{min}^g + \beta(f_{max}^g - f_{min}^g) \tag{1}$$
$$\mathbf{v}_i^g = \mathbf{v}_i^{g-1} + [\mathbf{x}_i^{g-1} - \mathbf{x}^*] f_i^g \tag{2}$$
$$\mathbf{x}_i^g = \mathbf{x}_i^{g-1} + \mathbf{v}_i^g \tag{3}$$

where $\beta \in [0,1]$ follows the random uniform distribution, and $\mathbf{x}^*$ represents the current global best location (solution). The superscript $(.)^g$ is used to denote the current generation g. Then, a local solution around $\mathbf{x}^*$ is selected (lines 7–10). The search is intensified by a local random walk (line 11): once a solution is selected among the current best solutions, it is perturbed locally through a random walk of the form: $\mathbf{x}_{new} = \mathbf{x}_{old} + \epsilon \mathcal{A}^g$ where ϵ is a uniform random number on the interval $[-1, 1]$ and $\mathcal{A}^g$ is the average loudness of all the bats at generation g. If the new solution achieved is better than the previous best one, it is probabilistically accepted depending on the value of the loudness. In that case, the algorithm increases the pulse rate and decreases the loudness (lines 13–16). The evolution rules for loudness and pulse rate are as: $\mathcal{A}_i^{g+1} = \alpha \mathcal{A}_i^g$ and $r_i^{g+1} = r_i^0 [1 - exp(-\gamma g)]$ respectively, where α and γ are constants. We can take an initial loudness $\mathcal{A}_i^0 \in (0, 2)$ while the initial emission rate r_i^0 can be any value in the interval $[0, 1]$. For simplicity, we take $\mathcal{A}_0 = 1$ and $\mathcal{A}_{min} = 0$. Loudness and emission rates will be updated only if the new solutions are improved, an indication that the bats are moving towards the optimal solution.

3   Bat Algorithm Method for Dynamic Indoor Navigation

The goal of *indoor navigation* is to determine a trajectory from an initial source point to a given target point within an indoor environment. In this work, we assume that the environment is *dynamic*, so that we allow different spatial configurations of the different interior areas (such as rooms or offices) along with different locations for the objects within such areas (e.g., home/office furniture or personal items of the people living/working inside). In this problem, we assume that the bots know the value of the fitness function at any specific location of the physical space. However, *they do not have any information about the indoor environment* itself, such as its geometry or the spatial configuration of the objects within. Although the environment is a 3D space, the dynamic navigation is performed at a 2D level, to replicate the most common case of walking robots.

In our method, a swarm of bots is distributed randomly throughout the whole indoor space. Each bot moves autonomously, according to the current values of its fitness function and its absolute position. To this aim, each virtual bot b_i is mathematically described at each time instance j by a vector $\mathbf{B}_{i,j} = \{f_{i,j}, \mathbf{X}_{i,j}, \mathbf{V}_{i,j}\}$, where $f_{i,j}$, $\mathbf{P}_{i,j} = (x_{i,j}, y_{i,j})$ and $\mathbf{V}_{i,j} = (v_{i,j}^x, v_{i,j}^y)$ represent the fitness value, position, and velocity of bot b_i at time j, respectively. Vectors $\mathbf{P}$ and $\mathbf{V}$ are initialized with uniform random values within the indoor environment. The dynamic indoor navigation can then be seen as a optimization problem, that of minimizing the distance from the current location to the target point. Note that the bots do not know the geometry of the environment, so they can know the distance but not the path leading to the target point. This minimization problem is solved by applying the bat algorithm described in Sect. 2.

Our method has been implemented in *Unity 5*, a popular multi-platform game engine supporting personal computers, video consoles, mobile devices and websites. It provides a nice graphical editor (see Fig. 1) with different windows and workspaces for the project, scene view, game view, hierarchy, toolbars, and many other features. It also supports the programming languages *C#* and *JavaScript*. All programming code in this paper, including our method, has been created in *JavaScript* using the *Visual Studio* programming framework. The parameter tuning is performed empirically. We tested a population size ranging from 10 to 50 bots and finally set this value to 20 bots. The initial and minimum loudness,

Fig. 1. Screenshot of the graphical editor of game engine *Unity 5*.

parameter α, initial pulse rate, and parameter γ are set to 0.5, 0, 0.6, 0.5 and 0.4, respectively. All executions are performed until all bots reach the target point.

4   Experimental Results

The proposed method has been tested with several experiments. Only one is described here for limitations of space (see Fig. 2). It consists of a fully furnished house with eleven rooms and an attached garage. Rooms (connected by open doors to allow free movement) are quite different from each other in terms of length, width, geometry, or spatial distribution of objects within. Some are small (i.e., the toilet), others are large (i.e., the dining room), some are empty (i.e., the main corridor), others are usually cluttered (i.e., the kids' room), and others have complex walkable geometry (i.e., the kitchen or the large bathroom). The environment is highly dynamic, with many changes for different executions, making it particularly well suited to check the performance of our approach.

The goal in the experiment is to reach a target point (the exterior garage door in Fig. 2) without any knowledge about its location or possible paths leading to it. A swarm of bots is distributed randomly throughout the house. Each bot moves autonomously (but in coordination with the other bots) at a speed initially set by the user or automatically computed by the game engine. In both cases the effective speed actually used is re-computed and updated during runtime according to several factors, such as obstacles and the like.

To ensure a fair assessment of our approach, 100 computer executions with different random initial locations for the bots have been carried out.

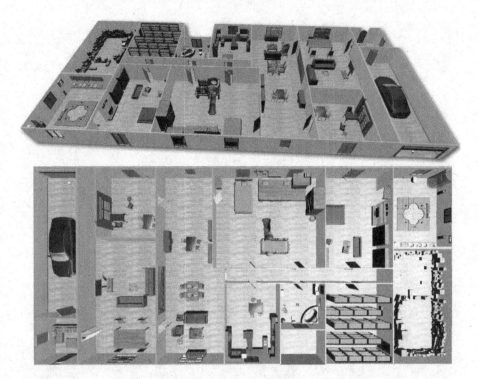

Fig. 2. Our virtual indoor environment: perspective view (above) and top view (below).

All them were recorded to generate 100 movies, lasting from about 3 to 25 min at a resolution of 1920×1000 and frame rate of 24 fps. An illustrative example is displayed in Fig. 3. It shows the convergence diagram of the simulation consisting of the value of the distance to the target point for one of the bots of the swarm (chosen randomly) against the number of iterations. Although we have computed a similar diagram for all bots of the swarm, only one is depicted here for better visualization. The figure also includes four significant screenshots, used to show the evolution of the swarm over the iterations. Since the target point is in the leftmo st part of the house, we distribute the bots (displayed as small yellow squares) randomly on the right half of the house at the beginning (shot 1) to make the problem more interesting and challenging. The bots explore the environment autonomously, and move around the different rooms trying to improve their fitness. Some bots move alone in rooms with a difficult configuration (i.e., the kitchen or the library) but most of them tend to cluster in areas around the local bests. Once some bots move towards the left half of the house, their fitness improves, making them strengthen this strategy even further. Because these better positions are communicated to the other members of the swarm, they begin to follow a similar motion pattern (shot 2). Once the swarm flocks around the most promising areas, they explore those rooms and (after a transient

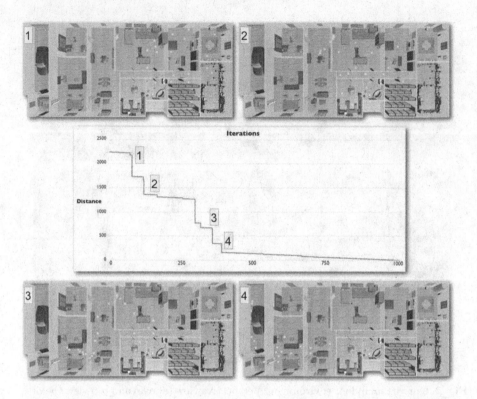

Fig. 3. Convergence diagram and four screenshots of an illustrative simulation.

time) they find the best path to further approach the final target point. As a result, the fitness of best bots improves; such improvement is propagated over the swarm owing to the coordination feature of the algorithm (shot 3). As time increases, the best bots are able to find the right path and reach the garage area, avoiding the parked car and eventually landing at the target point (shot 4). The rest of the swarm replicate this successful path with some delay. This is the case for the chosen bot in Fig. 3, which is not the best but follows it, reaching the target point in less than 1000 iterations (actually, 984 iterations were needed). This simple but illustrative example shows that the bat algorithm is very well suited to solve this indoor navigation problem.

5   Conclusions and Future Work

Our experimental results have been amazingly good, actually much better than expected. Even although we modified the configuration of the indoor environment (for instance, moving some objects and obstacles) and the initial locations of the bots for different executions, the swarm of bots could actually find the target point in all cases. Furthermore, we did not find any configuration unsolvable for the bots so far, provided that at least one walkable solution exists.

We also found surprisingly intelligent behaviors where some bots leave the swarm to explore other alternative ways. Sometimes, a bots moves away a promising area without any apparent reason for doing so; however, it becomes clear later on that this unexpected action allowed it to find a better path than the other bots, even although it initially implied to leave the swarm for no benefit at short-term.

Although not shown in the paper because of limitations of space, in addition to the numerical tracking we also attached a front camera to each bot during the simulation so we can watch what is ahead of them on our screen as if the bots themselves could actually see. We also set some fixed cameras at different locations (i.e., at the target point, or the top camera used to generate the screenshots) for monitoring purposes. All these nice features along with the numerical data we collected provide an invaluable source of data and information that has still to be analyzed at full extent. This is part of our future work in the field.

Acknowledgements. This work has been supported by the Computer Science National Program of the Spanish Ministry of Economy and Competitiveness, Project Ref. #TIN2012-30768, and Toho University.

References

1. Engelbrecht, A.P.: Fundamentals of Computational Swarm Intelligence. Wiley, Chichester (2005)
2. Iglesias, A., Gálvez, A.: Nature-inspired swarm intelligence for data fitting in reverse engineering: recent advances and future trends. In: Yang, X.-S. (ed.) Nature-Inspired Computation in Engineering. SCI, vol. 637, pp. 151–175. Springer, Cham (2016). doi:10.1007/978-3-319-30235-5_8
3. Iglesias, A., Gálvez, A., Collantes, M.: Bat algorithm for curve parameterization in data fitting with polynomial Bézier curves. In: Proceedings of Cyberworlds 2015, Visby, Sweden, CA, pp. 107–114. IEEE Computer Society Press, Los Alamitos (2015)
4. Iglesias, A., Gálvez, A., Collantes, M.: Global-support rational curve method for data approximation with bat algorithm. In: Chbeir, R., Manolopoulos, Y., Maglogiannis, I., Alhajj, R. (eds.) AIAI 2015. IAICT, vol. 458, pp. 191–205. Springer, Cham (2015). doi:10.1007/978-3-319-23868-5_14
5. Iglesias, A., Gálvez, A., Collantes, M.: A Bat algorithm for polynomial Bézier surface parameterization from clouds of irregularly sampled data points. In: Proceedings of ICNC 2015, CA, pp. 1034–1039. IEEE Computer Society Press, Los Alamitos (2015)
6. Iglesias, A., Gálvez, A., Collantes, M.: Four adaptive memetic bat algorithm schemes for Bézier curve parameterization. In: Gavrilova, M.L., Tan, C.J.K., Sourin, A. (eds.) Trans. on Comput. Sci. XXVIII. LNCS, vol. 9590, pp. 127–145. Springer, Heidelberg (2016). doi:10.1007/978-3-662-53090-0_7
7. Iglesias, A., Gálvez, A., Collantes, M.: Iterative sequential bat algorithm for free-form rational Bézier surface reconstruction. Int. J. Bio-Inspired Comput. (in press)
8. Kennedy, J., Eberhart, R.C., Shi, Y.: Swarm Intelligence. Morgan Kaufmann Publishers, San Francisco (2001)
9. Yang, X.-S.: Nature-Inspired Metaheuristic Algorithms, 2nd edn. Luniver Press, Frome (2010)

10. Yang, X.S.: A new metaheuristic bat-inspired algorithm. In: González, J.R., Pelta, D.A., Cruz, C., Terrazas, G., Krasnogor, N. (eds.) NICSO 2010. SCI, vol. 284, pp. 65–74. Springer, Heidelberg (2010)
11. Yang, X.S., Gandomi, A.H.: Bat algorithm: a novel approach for global engineering optimization. Eng. Comput. **29**(5), 464–483 (2012)
12. Yang, X.S.: Bat algorithm: literature review and applications. Int. J. Bio-Inspired Comput. **5**(3), 141–149 (2013)

Building Fractals with a Robot Swarm

Yu Zhou[✉] and Ron Goldman

Computer Science Department, Rice University, Houston, TX 77005, USA
{yuzhou,rng}@rice.edu

Abstract. Fractals are common in nature, and can be used as well for both art and engineering. We classify those fractals that can be represented by line segments into several types: tree-based fractals, curve-based fractals, and space filling fractals. We develop a set of methods to generate fractals with a swarm of robots by using robots as vertices, and line segments between selected robots as edges. We then generalize our algorithms so that new fractals can be built with only a few parameters, and we expand our methods to generate some shape-based fractals.

Keywords: Fractal formation · Distributed algorithm · Multi-agent path planning · Swarm intelligence

1 Motivation and Related Work

Fractals are self-similar shapes [12]. For example, a minor branch of a tree is similar to a major branch at a smaller scale, and a snowflake has complicated details that have small structures similar to the structure of the snowflake as a whole. Fractals are common in nature, and include such diverse objects as plants, crystals, mountain surfaces, lightning bolts, and tracheobronchial trees.

We want robots to form fractals because fractals are useful in engineering – certain antennas have fractal shapes [7,8,18,25,28] (Fig. 1); civil utility structures [24] and the Internet [33] demonstrate self-similar behaviors. Fractal formations can be useful as well for generating aesthetic shapes [4,16,32], such as artificial trees, or beautiful patterns of flowers in a botanical garden.

Two traditional methods for generating fractals are Recursive Turtle Programs [1] and Iterated Function Systems [15]. Recursive Turtle Programs require a pre-set depth, and a single turtle draws the fractal in a single thread, like a Depth-First Search. Iterated Function Systems require keeping track of a detailed description of a large number of objects so that at each iteration each object can be replaced with a self-similar structure.

In the real world people manufacture fractals with several different techniques. One approach is to draw the fractal with a pen or to stitch the fractal with a sewing machine, following the trajectory generated by a recursive turtle program. The Valiant Turtle [31] was introduced in 1983 as a robot toy to draw fractals on paper. Another approach is to build the fractal as a whole by printing or casting; this method is typically used to manufacture fractal antennas.

© Springer International Publishing AG 2017
Y. Tan et al. (Eds.): ICSI 2017, Part II, LNCS 10386, pp. 185–198, 2017.
DOI: 10.1007/978-3-319-61833-3_20

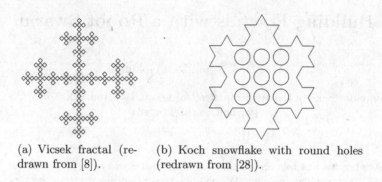

(a) Vicsek fractal (redrawn from [8]).   (b) Koch snowflake with round holes (redrawn from [28]).

Fig. 1. Fractal antennas with different designs.

Fractals can also be assembled with pre-made building blocks, such as cellular base stations disguised as trees. The assembly process is usually single-threaded either by a human or by a robot arm.

We are interested in generating fractals with multi-robot systems, because robot swarms are a cheap, effective, and reconfigurable tool for generating complex shapes. With the help of multiple robots, we can generate fractals in parallel, and the detailed description can be distributed to different robots so that each robot needs to remember only a small number of states.

There are several articles related to robots forming randomized and deterministic fractals. Rold [26] discovers that multiple robots can exhibit fractal dimensions by programming the robots to simulate attractor behaviors. Sugawara and Watanabe [30] gather robots towards the center of the environment, and the clustering process builds a fractal tree. Aznar, Pujol, and Rizo [5] describe a multi-robot self-assembly procedure with L-Systems.

We are also inspired by algorithms that place robots into other formations. Lee, Fekete, and McLurkin [20] develop an algorithm to build a triangular mesh with a robot swarm. This algorithm can be used to explore unknown spaces and guide other robots for security patrols. Alonso-Mora et al. [3] move multiple robots into solid geometric shapes in order to generate artistic patterns. Guo, Meng, and Jin [13] deploy swarm robots on a NURBS curve to approximate the boundary of an arbitrary shape.

2   Model and Assumptions

Fractals have complicated structures, but these structures can be generated from simple growth rules. Based on how to apply these rules with multi-robot systems, we classify fractals that can be approximated by line segments into four major types:

- Tree-based fractals: fractals with branching growth rules and a tree skeleton.
- Curve-based fractals: fractals formed by polygonal chains, or represented by subsets of polygonal chains.

- Space-filling fractals: similar to curve-based fractals, these fractals are continuous and dense in the unit square.
- Shape-based fractals: fractals that are often drawn with colored fills.

There are many fractals outside these categories, especially those fractals that cannot be effectively approximated by line segments or polygons, like the Mandelbrot set [10] and the Lorenz attractors [23]. This paper focuses on 2-dimensional fractals for which robots can be used to outline their skeletons.

Fractals are defined as a limit of an infinite number of iterations, but fractals can only be printed or manufactured with some finite number of vertices and edges. Therefore, we are more interested in approximating a fractal with a limited number of iterations. We define the *level* of a fractal as the depth of the iterations or recursions used to approximate the fractal. We also define the *level* of a vertex as the lowest level at which the vertex appears in the fractal. A lower level is a level with fewer iterations; a higher level is a level with more iterations.

3 Algorithms

3.1 Tree-Based Fractals

Fractal trees are fractals with a branching growth rule. The growth rule should have two or more branches in non-trivial directions with a scale factor < 1 so that the fractal converges to a tree-like shape. Fractal trees differ from each other by a set of parameters (Fig. 2): the number of branches, the angle between branches, the scale factor, symmetric or asymmetric rules, and the activeness of each node. We need to define a set of rules so that the robots generate the desired fractal.

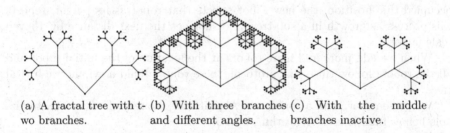

(a) A fractal tree with t- (b) With three branches (c) With   the   middle
wo branches.   and different angles.   branches inactive.

Fig. 2. Fractal trees with different parameters (shown at level 5).

For iterated function systems the base case is usually represented by a line segment, and this line segment is replaced by a set of line segments described in the growth rule. To implement these growth rules using robots, we denote the base case by a single vertex and an expansion direction. We manually place a stationary initial robot into the workspace, heading towards a fixed direction. We define a polar coordinate system whose pole O lies on the center of the initial robot and whose polar axis L points in the direction of the heading of the initial robot. We also set up an initial distance d_0 either manually or by placing another auxiliary robot whose distance to the initial robot is d_0.

For the simplest fractal tree, a k-branch symmetric fractal tree, we describe the growth rule with k vectors extending from the pole O, embodied as k robot children forming branches from their parent robot. Each vector v_i can be described by its angle θ_i from the polar axis L, in addition to its scale factor s_i. For a symmetric fractal tree, the angles $\{\theta_i\}$ form an arithmetic sequence symmetric about zero, and all the scale factors $\{s_i\}$ are equal (Fig. 3(a)).

To simplify the description of our algorithm, we shall initially assume that the robots have a sensing range larger than the length of any edge in the fractal tree. In addition, we will assume that a robot can always move accurately a given distance in a given direction without colliding with other robots. We will handle limited sensing range, errors in accuracy, and collisions later in this section.

When a new robot joins the building process, the robot starts near the initial robot with the same heading as the initial robot. The new robot treats the initial robot as its parent. Now there are k directions in which to grow the fractal tree. One strategy is to pick a branch randomly from $1, ..., k$; another strategy is to ask the parent robot to select a branch, and have the parent robot reply sequentially with $1, ..., k$ to each request. These different strategies determine whether the fractal tree grows probabilistically or deterministically. Once the new robot knows in which branch i to grow, the robot rotates by θ_i to face towards that branch, where $\theta_i = \frac{k+1-2i}{2}\alpha$ if the angle between two branches is given by α, or $\theta_i = \frac{k+1-2i}{2k-2}\beta$ if the angle between the leftmost and the rightmost branches is given by β.

The robot moves a distance d in that direction. When the robot arrives at its target, the robot becomes a vertex of the fractal tree and remains stationary during the remainder of this algorithm. However, if another robot has already occupied this position, the new robot selects that robot as its parent, repeats this process as growth in a sub-tree, and shortens the next distance by a given scale factor s.

When we add more and more robots at the location of the initial robot and all the robots follow this same protocol, these robots build a symmetric fractal tree.

We present our main algorithm in Algorithm 1, along with several customizable helper functions in Algorithms 2 to 5.

Next we enhance our algorithm to deal with more complicated fractal trees:

Asymmetric angles. Instead of calculating θ_i from the adjacent branch angle α or the total branch angle β and the number of branches k, the user can specify an array of angles to describe the directions of the branches.

Asymmetric scale factors. Instead of all the robot children moving the same distance, the user can specify an array of scale factors. Individual scale factors apply to each branch, and the size difference of sub-trees amplifies in higher levels. All of the scale factors should be less than 1 to ensure convergence. We can combine the angles and the scale factors into vectors (Fig. 3(b)).

Algorithm 1. FRACTALTREE(u)

1: INITROBOT(u)
2: **while** $u.state \neq STOP$ **do**
3:   **if** $u.state = EXPAND$ **then**
4:     **if** $\exists v$ such that $v.parent = u.parent$ **and** $v.branch = u.branch$ **and** $v.level = u.level$ **and** $v.state = STOP$ **then**
5:       $u.state \leftarrow FOLLOW$
6:       **if** $u.branch$ is active branch **then**
7:         $u.parent \leftarrow v$
8:         $u.level \leftarrow v.level + 1$
9:         u moves towards v
10:       **else**
11:         u backs to parent and selects another branch
12:       **end if**
13:     **else if** u reaches expansion destination **then**
14:       $u.state \leftarrow STOP$
15:     **else**
16:       u continues current expansion motion
17:     **end if**
18:   **else if** $u.state = FOLLOW$ **then**
19:     **if** u is close enough to v **then**
20:       $u.branch \leftarrow$ GETBRANCH(u)
21:       **if** $u.branch$ is active or inactive branch **then**
22:         $u.state \leftarrow EXPAND$
23:         $u.length \leftarrow u.length \times$ GETSCALE(u)
24:         u rotates GETANGLE($u.branch$)
25:         u moves forward for a distance up to $u.length$
26:       **else**
27:         $u.length \leftarrow u.length \times$ GETSCALE(u)
28:         $u.level \leftarrow u.level + 1$
29:       **end if**
30:     **else**
31:       u continues current following motion
32:     **end if**
33:   **end if**
34: **end while**

Algorithm 2. INITROBOT(u)

1: **if** $\exists v$ such that $v.level = 0$ **then**
2:   $u.parent \leftarrow v$
3:   $u.level \leftarrow 1$
4:   $u.state \leftarrow FOLLOW$
5: **else**
6:   $u.parent \leftarrow \varnothing$
7:   $u.level \leftarrow 0$
8:   $u.state \leftarrow STOP$
9: **end if**
10: $u.length \leftarrow$ an initial length

Algorithm 3. GETSCALE(u)

1: *scale* ← a constant value less than 1
2: **return** *scale*

Algorithm 4. GETBRANCH(u)

1: **if** random branch mode **then**
2:    *branch* ← a random branch among all active, inactive, and head branches
3: **else**
4:    *branch* ← ask *u.parent* about the next branch at *u.level*
5: **end if**
6: **return** *branch*

Algorithm 5. GETANGLE(u)

1: *angle* ← $(k/2 + 1/2 - u.branch)\alpha$ where k is the number of active and inactive branches
2: **return** *angle*

Inactive branches. Some branches can be inactive, i.e. no fractal sub-trees grow from inactive branches. The user can specify an array of Boolean values to indicate whether each branch is active or inactive (Fig. 3(c)). If a robot selects a branch randomly and a robot is already present in this inactive branch, the robot must return to its parent and reselect another branch. If the parent assigns branches for robot children, an inactive branch can be assigned only once.

Vertices with multiple degrees. Fractal trees usually grow only at the vertices on the highest level, but some fractals such as Vicsek fractals grow at all the vertices. These fractals can be treated as if they have additional branches above the plane extending into three-dimensional space, and the original fractal is the parallel projection onto the plane of a three-dimensional fractal. Thus a parent robot can have children robots at various levels. When a moving robot approaches a stationary robot and selects this stationary robot as its parent, the moving robot can either branch immediately or fall into the higher level by applying

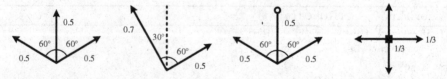

(a) Growth rule for 3 symmetric branches.
(b) Growth rule with arbitrary vectors.
(c) No iteration at inactive branch.
(d) Growth rule for Vicsek fractal with an additional iteration at the root.

Fig. 3. Fractal tree growth rules.

the scale factor to its future motion and choosing a branch again at the same parent (Fig. 3(d)). Since any stationary robot is considered as multiple instances in the different levels, if a child robot asks the parent robot to assign a branch, the parent robot needs to maintain multiple states for different levels in order to answer at the child's level.

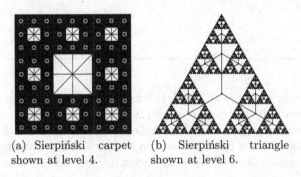

(a) Sierpiński carpet shown at level 4.   (b) Sierpiński triangle shown at level 6.

Fig. 4. Shape-based fractals approximated with fractal trees.

Complicated growth rules. Some shape-based fractals, such as the Sierpiński carpet and the Sierpiński triangle, can be approximated with fractal trees (Fig. 4). For the Sierpiński carpet, we use one robot to represent each square (and the robot may have the ability to paint the square). Each node has 8 branches, but these branches have different scale factors from the scale factor in the parent node. Furthermore, scale factors alternate in every level and all the squares must be oriented along the same directions. We also allow the user to override some helper functions and attach an event handler to state change in order to build more complicated fractals.

For a Sierpiński carpet, we define even branches as the branches in the forward direction, and these branches increment at right angles (0°, 90°, 180°, 270°). Odd branches are those branches with an angle of 45 deg to even branches (45°, 135°, 225°, 315°). We set the number of branches to 8 and override GETANGLE to return these angles. Each robot u needs to count how many times u enters an odd branch in $u.counter$, and this counter increments whenever $u.state$ changes to $EXPAND$ and $u.branch$ is odd. Finally, we override GETSCALE as in Algorithm 6.

Limited sensing range and motion errors. Even when the robots have a limited sensing range and even when the robots have motion errors, the robots can still form the correct fractal topology if the maximum length of edge d satisfies the following condition (in any Cartesian coordinate system):

$$\forall a, b \in \{(x, y) | (d - \Delta d)^2 < x^2 + y^2 < (d + \Delta d)^2,$$
$$\tan(\theta - \Delta\theta) < \frac{y}{x} < \tan(\theta + \Delta\theta)\} \Rightarrow \|a, b\| < r. \quad (1)$$

Algorithm 6. SIERPINSKICARPET::GETSCALE(u)

1: $scale \leftarrow 1/3$
2: **if** $u.branch$ is odd **then**
3:    **if** $u.counter$ is odd **then**
4:       $scale \leftarrow scale \times \sqrt{2}$
5:    **else**
6:       $scale \leftarrow scale/\sqrt{2}$
7:    **end if**
8: **end if**
9: **return** $scale$

This condition means that if two robots try to reach the same destination (d, θ) but separate due to motion error $(\Delta d, \Delta \theta)$, these robots will fall within each other's sensing range r when both believe that they have reached their destination. In this case, if a robot moves into a branch that is already established, the robot can always discover the next vertex robot and become its child.

Collision avoidance. To implement this algorithm on real robots, we need to keep the robots at a safe distance from each other in order to avoid collisions. We pack the robot swarm into a dense formation near the initial robot, and we run the algorithm in [27] so that the robots enter the workspace one after another without becoming disconnected.

Once a robot moves along edges in the fractal tree, we define a set of routes around the edges so that robots only move sequentially on these routes (Fig. 5). There are four types of routes with descending priority:

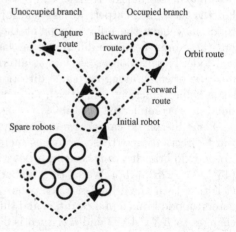

Fig. 5. Routes for collision avoidance.

- Capture route (robot becomes a new vertex),
- Backward route (robot returns to its parent due to an inactive branch or for other reasons),
- Orbit route (circular connection between other routes),
- Forward route (to reach a higher level).

Robots in a route with a lower priority must yield to robots with a higher priority, as well as to any robot in front of the robot. If a robot encounters an oncoming robot or is about to collide with a vertex robot due to motion errors, the robot must yield to the right to avoid a collision.

If the fractal to be built has enough clearance between branches at the desired level, the robots will be collision-free and can organize themselves automatically into the fractal.

3.2   Curve-Based Fractals

Curve-based fractals are those fractals formed by a polygonal chain, such as the Koch Curve [2]; in particular, those fractals that are continuous and do not have branches. This category can be expanded to fractals that self-intersect, such as the Lévy C curve [21], but can still be represented as an Eulerian path. This category can be further expanded to fractals that are not continuous, such as the Cantor set [6], but are subsets of a continuous curve. Space-filling fractals will be discussed in a later section because the iteration rule for space-filling fractals requires interfaces between components.

(a) Growth rule for distorted Koch curve.
(b) Skipping a middle section (shown as a dashed line).
(c) Mirroring the middle section (shown as a dotted line).

(d) Reversing the middle section (shown as an arrow).
(e) Flipping (mirroring and reversing) the middle section.
(f) Growth rule for the Sierpiński arrowhead curve.

Fig. 6. Growth rules for curve-based fractals.

We describe the base case by a line segment with two vertices, and for the growth rule we replace the line segment with a polygonal chain (Fig. 6(a)). We denote the number of vertices in the growth rule (including both endpoints) by $t + 1$, and the level of iterations by k. For a perfect fractal with n vertices, we should have $n = t^k + 1$. The base shape with two vertices and one line segment has $k = 0$ and $n = 2$.

Suppose we already have n robots placed along a line, and each robot knows both of its adjacent neighbors in opposite directions (except for the two robots at each end which have only one adjacent neighbor). The linear ordering can be achieved with chain-formation algorithms [9,11,14,19], or with physical sorting algorithms [17,22,34,35]. From some of these algorithms each robot also knows its topological distances to one end h and to the other end h'; thus all the robots know the total number of robots $n = h + h' + 1$.

To generate curve-based fractals with a robot swarm, we first describe the iterating shape as a set of vectors. The iterating shape maps to $t + 1$ vectors, where we denote the first vector by the zero vector $(0, 0)$ and the last vector by the unit vector $(1, 0)$. Other vectors have their own scales of the unit vector; for

---

Algorithm 7. GETSEQUENCE($h, begin, end, k$)

1: **if** $end - begin \geq k$ **then**
2:    $step \leftarrow \lfloor (end - begin)/k \rfloor$
3:    **if** $h - begin \equiv 0 \pmod{step}$ **then**
4:        **return** $(h - begin)/step$
5:    **else**
6:        **return** GETSEQUENCE($h, \lfloor (h - begin)/step \rfloor, \lceil (h - begin)/step \rceil, k$)
7:    **end if**
8: **else**
9:    Not enough robots to form a new level. The robot simply moves to the midpoint of the closest two neighbors.
10:    **return** $\varnothing$
11: **end if**

---

example, the three middle vectors describing the Koch curve are $(\frac{1}{3}, 0)$, $(\frac{1}{2}, \frac{\sqrt{3}}{2})$, and $(\frac{2}{3}, 0)$. Then each robot is assigned its level and sequence in the fractal. Since each robot knows the total number of robots n and its relative position h along the line, the robots can run an iterated function (Algorithm 7) locally to determine its sequence as well as the closest two robots on the previous level.

If each robot has a sensing range large enough to cover at any time the two closest robots on the previous level, each robot can move to the relative position defined by its corresponding vector. When all the robots move in the same manner (except the two robots on each end that remain stationary), the line of robots transforms progressively into a curve-based fractal.

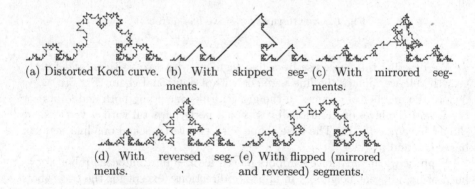

(a) Distorted Koch curve. (b) With skipped segments. (c) With mirrored segments.

(d) With reversed segments. (e) With flipped (mirrored and reversed) segments.

Fig. 7. Curve-based fractals (shown at level 4).

In order to build more complicated curve-based fractals, we introduce some properties attached to each line segment in the growth rule. *Skip* determines whether a line segment is skipped instead of replaced (Fig. 6(b)), and can be used for erasing the line segments in a Cantor set. A customized GETSEQUENCE may be provided to avoid assigning robots into such void zones. *Mirror* determines whether the replacement rule is mirrored over the original line segment

(Fig. 6(c)). *Reverse* determines whether the replacement rule is rotated 180° (Fig. 6(d)). Mirror and reverse can be combined into a *flip* for the same line segment (Fig. 6(e)). The corresponding fractals are shown in Fig. 7.

Some shape-based fractals can be approximated with curve-based fractals. For example, the Sierpiński triangle can be constructed using a Sierpiński arrowhead curve. The vectors for constructing a Sierpiński arrowhead curve are:

$$v_0 = (0,0), \ v_1 = (\tfrac{1}{4}, \tfrac{\sqrt{3}}{4}), \ v_2 = (\tfrac{3}{4}, \tfrac{\sqrt{3}}{4}), \ v_3 = (1,0).$$

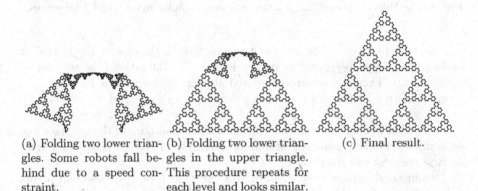

(a) Folding two lower triangles. Some robots fall behind due to a speed constraint.

(b) Folding two lower triangles in the upper triangle. This procedure repeats for each level and looks similar.

(c) Final result.

Fig. 8. Procedure for building the Sierpiński arrowhead curve (shown at level 6).

The line segments (v_0 to v_1) and (v_2 to v_3) are mirrored (Fig. 6(f)). When the program starts, the lower two triangles fold inward until their destination, while the other robots for the upper triangle move above. Then the upper triangle gets constructed with a similar procedure. The process for building this fractal (Fig. 8) is a fractal!

This algorithm requires some robots to have a longer sensing range than other robots. This assumption is still practical since the swarm can have different robots with different costs, and still keep the total cost low. Removing the requirement of long sensing range is possible, but may introduce accumulated errors.

3.3 Space-Filling Fractals

Space-filling fractals are more difficult to generate than curve-based fractals because of the existence of interfaces (Fig. 9). *Interfaces* are the line segments that connect adjacent sub-fractals. At a given level these interfaces usually have the same size as the shortest line segments, which depend on the number of levels and the number of robots used to build the space-filling fractals. Therefore, the number of levels must be determined in advance, and the ratio of interfaces to sub-fractals must be predetermined.

For the Hilbert curve, we use an algorithm similar to the algorithm for a curve-based fractal, but we define the vectors based on the highest level $l_{\max}$

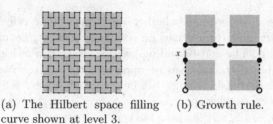

(a) The Hilbert space filling curve shown at level 3.   (b) Growth rule.

Fig. 9. The Hilbert space-filling fractals contains 4 similar squares and 3 interfaces.

and the current level l. The ratio of the interface to the size of the fractal at level l is $x = \frac{1}{2^{l_{\max}-l+2}-1}$, and we denote by $y = \frac{1-x}{2}$ the ratio of the sub-fractal to the fractal. Then vectors are calculated at each level by setting:

$v_1 = (0, y)$, $v_2 = (0, y+x)$, $v_3 = (y, y+x)$, $v_4 = (y+x, y+x)$, $v_5 = (1, y+x)$, $v_6 = (1, y)$.

To avoid assigning robots inside the interfaces, we also mark the interface line segments as skipped, using a customized assignment function. Then we mirror the first and last sub-fractals.

This method applies to space-filling fractals that have start and end points distinct from each other. An alternative method is to pack the robots into a square first [29], and then assign the virtual edges between robots.

4   Simulation Results

All the algorithms discussed in this paper are implemented with our Multi-Robot Simulator written in C#. Running on a Windows laptop (1.8 GHz CPU) and using 25 MB memory, this simulator can generate the Sierpiński arrowhead curve up to level 8 (with 6,562 robots) in a two-minute animation with smooth motion. All the fractal figures in this paper are exported directly from this simulator in SVG format.

5   Conclusion

We presented a distributed method that allows swarm robots to self-assemble into fractals. Starting from a dense swarm, robots move into fractal trees and fractal curves, and our algorithm adapts to different fractals by setting a small set of parameters. We validated our algorithm with simulations using thousands of robots.

References

1. Abelson, H., DiSessa, A.: Turtle Geometry. MIT Press, Cambridge (1981)
2. Addison, P.: Fractals and Chaos: An Illustrated Course. Taylor & Francis, Gloucester (1997)

3. Alonso-Mora, J., Breitenmoser, A., Rufli, M., Siegwart, R., Beardsley, P.: Multi-robot system for artistic pattern formation. In: 2011 IEEE International Conference on Robotics and Automation, pp. 4512–4517, May 2011

4. Ashlock, D., Tsang, J.: Evolving fractal art with a directed acyclic graph genetic programming representation. In: 2015 IEEE Congress on Evolutionary Computation (CEC), pp. 2137–2144, May 2015

5. Aznar, F., Pujol, M., Rizo, R.: L-system-driven self-assembly for swarm robotics. In: Lozano, J.A., Gámez, J.A., Moreno, J.A. (eds.) CAEPIA 2011. LNCS, vol. 7023, pp. 303–312. Springer, Heidelberg (2011). doi:10.1007/978-3-642-25274-7_31

6. Cantor, G.: Ueber unendliche, lineare punktmannichfaltigkeiten. Math. Ann. **15**, 1–7 (1879)

7. Choudhary, R., Yadav, S., Jain, P., Sharma, M.M.: Full composite fractal antenna with dual band used for wireless applications. In: 2014 International Conference on Advances in Computing, Communications and Informatics (ICACCI), pp. 2517–2520, September 2014

8. Cohen, N.: Fractal antennas and fractal resonators. US Patent 6,452,553, 17 September 2002

9. Degener, B., Kempkes, B., Kling, P., auf der Heide, M.F.: A continuous, local strategy for constructing a short chain of mobile robots. In: Patt-Shamir, B., Ekim, T. (eds.) SIROCCO 2010. LNCS, vol. 6058, pp. 168–182. Springer, Heidelberg (2010). doi:10.1007/978-3-642-13284-1_14

10. Douady, A., Hubbard, J.: Étude dynamique des polynômes complexes. No. v. 1–2 in Publications mathématiques d'Orsay, Département de Mathématique, Université de Paris-Sud (1984)

11. Dynia, M., Kutylowski, J., auf der Heide, M.F., Schrieb, J.: Local strategies for maintaining a chain of relay stations between an explorer and a base station. In: Proceedings of the Nineteenth Annual ACM Symposium on Parallel Algorithms and Architectures, SPAA 2007, pp. 260–269. ACM, New York (2007)

12. Goldman, R.: An Integrated Introduction to Computer Graphics and Geometric Modeling, 1st edn. CRC Press Inc., Boca Raton (2009)

13. Guo, H., Meng, Y., Jin, Y.: Swarm robot pattern formation using a morphogenetic multi-cellular based self-organizing algorithm. In: 2011 IEEE International Conference on Robotics and Automation, pp. 3205–3210, May 2011

14. auf der Heide, F.M., Schneider, B.: Local strategies for connecting stations by small robotic networks. In: Hinchey, M., Pagnoni, A., Rammig, F.J., Schmeck, H. (eds.) Biologically-Inspired Collaborative Computing. IFIP – The International Federation for Information Processing, vol. 268. Springer, Boston (2008)

15. Ju, T., Schaefer, S., Goldman, R.: Recursive turtle programs and iterated affine transformations. Comput. Graph. **28**(6), 991–1004 (2004)

16. Kharbanda, M., Bajaj, N.: An exploration of fractal art in fashion design. In: 2013 International Conference on Communication and Signal Processing, pp. 226–230, April 2013

17. Krupke, D., Hemmer, M., McLurkin, J., Zhou, Y., Fekete, S.: A parallel distributed strategy for arraying a scattered robot swarm. In: 2015 IEEE/RSJ International Conference on Intelligent Robots and Systems (IROS), pp. 2795–2802, September 2015

18. Kumar, D., Manmohan, Ahmed, S.: Modified ring shaped Sierpinski triangle fractal antenna for c-band and x-band applications. In: 2014 International Conference on Computational Intelligence and Communication Networks, pp. 78–82, November 2014

19. Kutyłowski, J., auf der Heide, F.M.: Optimal strategies for maintaining a chain of relays between an explorer and a base camp. Theor. Comput. Sci. **410**(36), 3391–3405 (2009)

20. Lee, S.K., Fekete, S.P., McLurkin, J.: Geodesic topological voronoi tessellations in triangulated environments with multi-robot systems. In: 2014 IEEE/RSJ International Conference on Intelligent Robots and Systems, pp. 3858–3865, September 2014

21. Lévy, P.: Les courbes planes ou gauches et les surfaces composées de parties semblables au tout. J. de l'Ecole Polytechnique Série III(7-8), 227–291 (1938). Reprinted in "Oeuvres de Paul Lévy", vol. II, (Dugué, D., Deheuvels, P., Ibéro, M. (eds.)), pp. 331–394, Gauthier Villars, Paris, Bruxelles and Montréal (1973)

22. Litus, Y., Vaughan, R.: Fall in! sorting a group of robots with a continuous controller. In: 2010 Canadian Conference on Computer and Robot Vision (CRV), pp. 269–276, May 2010

23. Lorenz, E.N.: Deterministic nonperiodic flow. J. Atmos. Sci. **20**(2), 130–141 (1963)

24. Peiravian, F., Kermanshah, A., Derrible, S.: Spatial data analysis of complex urban systems. In: 2014 IEEE International Conference on Big Data (Big Data), pp. 1–6, October 2014

25. Puente-Baliarda, C., Romeu, J., Pous, R., Cardama, A.: On the behavior of the Sierpinski multiband fractal antenna. IEEE Trans. Antennas Propag. **46**(4), 517–524 (1998)

26. Rold, F.D.: Deterministic chaos in mobile robots. In: 2015 International Joint Conference on Neural Networks (IJCNN), pp. 1–7, July 2015

27. Rubenstein, M., Cornejo, A., Nagpal, R.: Programmable self-assembly in a thousand-robot swarm. Science **345**(6198), 795–799 (2014)

28. Sankaranarayanan, D., Venkatakiran, D., Mukherjee, B.: Koch snowflake dielectric resonator antenna with periodic circular slots for high gain and wideband applications. In: 2016 URSI Asia-Pacific Radio Science Conference (URSI AP-RASC), pp. 1418–1421, August 2016

29. Spears, W.M., Spears, D.F., Hamann, J.C., Heil, R.: Distributed, physics-based control of swarms of vehicles. Auton. Robots **17**(2), 137–162 (2004)

30. Sugawara, K., Watanabe, T.: A study on foraging behavior of simple multi-robot system. In: IEEE 2002 28th Annual Conference of the Industrial Electronics Society, IECON 2002, vol. 4, pp. 3085–3090, November 2002

31. Valiant Designs Limited: The valiant turtle (1983)

32. Xu, S., Yang, J., Wang, Y., Liu, H., Gao, J.: Application of fractal art for the package decoration design. In: 2009 IEEE 10th International Conference on Computer-Aided Industrial Design Conceptual Design, pp. 705–709, November 2009

33. Zhang, J., Zhao, H., Luo, G., Zhou, Y.: The study on fractals of internet router-level topology. In: 2008 The 9th International Conference for Young Computer Scientists, pp. 2743–2747, November 2008

34. Zhou, Y., Goldman, R., McLurkin, J.: An asymmetric distributed method for sorting a robot swarm. IEEE Robot. Autom. Lett. **2**(1), 261–268 (2017)

35. Zhou, Y.: Swarm Robotics: Measurement and Sorting. Master's thesis, Rice University, Houston, TX, USA (2005)

A Stigmergy Based Search Method for Swarm Robots

Qirong Tang[1]([✉]), Fangchao Yu[1], Yuan Zhang[1], Lu Ding[1], and Peter Eberhard[2]

[1] Laboratory of Robotics and Multibody System, School of Mechanical Engineering, Tongji University, Shanghai 201804, People's Republic of China
qirong.tang@outlook.com
[2] Institute of Engineering and Computational Mechanics, University of Stuttgart, Pfaffenwaldring 9, 70569 Stuttgart, Germany

Abstract. This paper presents a stigmergetic search method for swarm robots, in which an indirect information interaction mechanism, namely stigmergy, is used for the coordination of swarm robots via the environment. RFID tags are arranged in the environment as a carrier of the pheromone. Robots move in the environment while reading and writing pheromone that contain the search experience of robots. The reading and writing algorithms are established. The pheromone map can be used to guide robots' motion without any localization system, and it can be built by robots according to their search experience. The method is verified by ample numerical experiments. Results show the applicability of the stigmergy mechanism for swarm robots target search.

Keywords: Swarm robots · Stigmergy · Target search · Pheromone · RFID

1 Introduction

The problem of target searching in unknown environments is one of the fundamental problems in robotics. Finding the target is the precondition to perform tasks there, e.g., to rescue the survivors after a disaster or to perform planetary exploration [1]. Autonomous mobile robots are often used to execute these tasks, since they can save much time and resources. Compared to a single but high performance robot, simple swarm robots may have higher efficiency in the implementation of these tasks due to swarm effects. Swarm robots make up for the weak point of a single robot and expand the scope of their abilities by sharing their information and knowledge. In recent years, the research on swarm robotic systems mainly focus on how to organize the system structure, and how to make good use of collaboration, e.g., a Bacterial Foraging Optimization-based search method in [2]. Effective communication in collaboration can greatly improve system performance [3]. The communication mechanism of swarm robots usually includes two types, the explicit and the implicit ones. Robots using explicit communication can communicate with each other quickly and effectively, and can

© Springer International Publishing AG 2017
Y. Tan et al. (Eds.): ICSI 2017, Part II, LNCS 10386, pp. 199–209, 2017.
DOI: 10.1007/978-3-319-61833-3_21

achieve some advanced coordination and cooperation strategies. However, when applied to actual robots, this method faces many difficulties. For example, the limitation of bandwidth in explicit communication makes the robot prone to bottlenecks when exchanging information, especially when the scale of the robot swarm is relatively large [4].

The indirect information exchange between individuals in a natural environment can show superior swarm performance. The stigmergy mechanism which is similar to the "actionless governance" thought of Laozi, is the most typical and widespread coordination mechanism [5]. The key of using this mechanism is to correspond bi-directional information interaction between biota and environment (medium) to actual physical driving of the robot. It allows robots in a swarm to learn and influence each other through environment, and the "information explosion" problem caused by the expansion of the swarm size may be avoided. Then, it yields the coordination of swarm robots.

This paper proposes a stigmergetic search method for swarm robots. The rest of the paper is organized as follows. Section 2 describes the RFID system and the stigmergy mechanism. The stigmergetic search method is given in Sect. 3. Numerical experiments based on the proposed method are given in Sect. 4, while Sect. 5 concludes the paper.

2   Overall Principle

2.1   RFID System and Search Area

In recent years, RFID technology has begun to enter the field of mobile robotics. The use of RFID tags in robot navigation is becoming increasingly popular [6,7]. An RFID reader can detect RFID tags within a few meters. In most cases, tags must be initialized by writing some predefined information, e.g., their positions. The proposed approach does not require the storage or predefined information in the tags since these running robots will collect information into tags. In this paper, the robots are searching in a two-dimensional space which is divided into 10×10 grids, see Fig. 1(a). In the center of each grid, there is an RFID tag indicated by a black square in Fig. 1. Each robot carries an RFID reader. It is assumed that a robot can only read and write the RFID tag on the grid in which it is located. Different from the agents/particles in a swarm intelligence algorithm, the robots in our research have physical properties, e.g., volumes. An obstacle avoidance strategy [8] is used to make a robot keep a safe distance to the others, see Fig. 1(b).

2.2   Stigmergy and Pheromone

The stigmergy mechanism, which was first proposed by Grassé in [9], is a mechanism of indirect coordination, through the environment, between agents or actions [10]. The stigmergy principle is that the trace left in the environment by an agent stimulates the actions of the others, then, subsequent actions tend

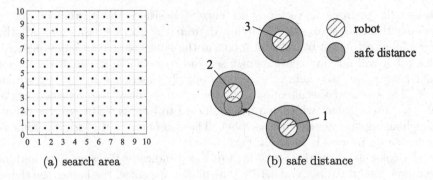

(a) search area (b) safe distance

Fig. 1. Search area and safe distance between robots

to reinforce and build on each other, leading to the spontaneous emergence of a systematic behavior. Obviously, the environment is the medium of information interaction, and the trace is the carrier of information in this mechanism. In nature, the trace may be an ant pheromone, a nest of termites, or even the target. Different studies, due to their perspective, result in different understandings of the stigmergy mechanism and different models. In this paper, pheromones are treated as the trace and simulated by RFID tags. The structure of the stigmergy mechanism is shown in Fig. 2.

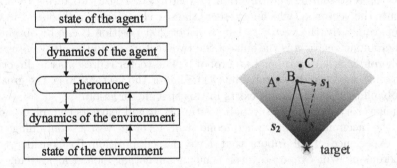

Fig. 2. Structure of the stigmergy **Fig. 3.** Pheromone superposition

Some search algorithms based on ACO (Ant Colony Optimization) also use pheromone. In these algorithms, however, the robot needs to detect the pheromone concentrations at several positions and compare them. After some complex probabilistic operations, the robot can decide which direction to move. It increases the computational cost. In this paper, the pheromone vector is innovatively used. The advantages of pheromone vector are that they have both size and direction to contain more information, and the vector can simplify operation. In addition, RFID tags are very suitable for mimicking vectors. Then, a stigmergy-based search model can be established. In this model, each robot releases the pheromone vector at its current position, the pheromone vector contains the search experience obtained from the robot's last step. All the pheromone vectors left by different robots at the same position are superimposed. Each robot will not only leave its own search experience in the environment for other robots, but also gets other robots' search experience from the environment. The effect of the pheromone on the robot is ultimately reflected in the behavior of the robot, which can be considered to impose a force on the robot, thus changing the velocity of the robot. The demonstration of the pheromone superposition process is shown in Fig. 3.

The color depth in the sector area of Fig. 3 indicates the signal strength of the target that the robot can detect. The darker the color, the higher the signal strength. It is assumed that a robot (robot 1) moves from position A to position B, the pheromone is left at position B which is represented by a pheromone vector s_1. The vector is given by

$$s_1 = \frac{f_B - f_A}{|AB|} \frac{X_B - X_A}{|AB|},$$ (1)

where f_A and f_B represent the signal strength detected by robot 1 at position A and B, respectively, while X_A and X_B represent the coordinates of position A and B, $|AB|$ is the distance from position A to position B. In Eq. (1), the term $(f_B - f_A)/|AB|$ which is the increase of the signal along the robot's moving direction, represents the size of vector s_1, and the term $(X_B - X_A)/|AB|$ is a unit vector representing the direction of vector s_1. These two terms together constitute the vector s_1, which represents the search experience of robot 1 in the last step. Similarly, the vector s_2 left by robot 2 at position B can be obtained. The pheromone vector s is the sum of the two vectors s_1 and s_2. It denotes the total pheromones left by robot 1 and robot 2. The vector represents the direction of a possible target and its probability. The larger the vector size is, the greater the probability that the target exists in this specific direction.

Whenever a robot moves through position B, it will leave its own pheromone there, the accumulation of pheromone will form a vector containing the search experience of all robots that have passed position B. Consider that the pheromone will evaporate over time in an actual biological system, the pheromone is actually superimposed by

$$s_n^{m+1} = (1 - c_d)s_n^m + c_a s_i,$$ (2)

where s_n^m represents the m-th writing operation of the n-th RFID tag, and s_i represents the pheromone generated by the i-th robot in current step. Parameters c_d and c_a represent the pheromone evaporation factor and superposition factor, respectively. The pheromone produced by each robot will be affected and due by the sensors. Therefore, the superposition factor c_a is used to tuning the vector in a reasonable range in case robots are equipped with different sensors.

3   Search Method

3.1   Velocity and Position Update

The emergence of swarm intelligent behavior is due to the fact that the individuals in the swarm get the global information or local small-scale global information, e.g., the global optimal position in PSO (Particle Swarm Optimization) [11] and the path with the greatest concentration of pheromone in ACO. Pheromone vectors in this research released by robots in the environment constitute a form of the global information. Inspired by the PSO algorithm, the velocity and position of a robot are updated in this study by

$$V_i^{k+1} = \omega V_i^k + c_1 r_1 (P_i - X_i^k) + c_2 r_2 s_n, \tag{3}$$

$$X_i^{k+1} = X_i^k + V_i^{k+1} \Delta t, \tag{4}$$

where V_i^k and X_i^k are the velocity and position of robot i at the k-th iteration, respectively. One saves P_i as the individual best position, and s_n as the pheromone vector detected by robot i from the n-th RFID tag. Here ω is an inertia weight, c_1, c_2 are referred to as self learning factor and social learning factor, r_1, r_2 are two independent random parameters, Δt is a time step that usually is 1 s and thus is usually omitted. In addition, for the actual physical system constraints, the velocity of a robot is limited to V_{max}.

3.2   Reading and Writing Algorithm

This study establishes the reading and writing algorithms of pheromone vector in the process of robot motion. A robot reads the pheromone at each position from the RFID tag and updates it with its own search experience obtained in the last steps. Algorithm 1 formalizes the reading process while Algorithm 2 formalizes the writing process, see Table 1 for details. In practice, a robot's reading and writing are often carried out simultaneously.

Table 1. The reading and writing processes of pheromone

Algorithm 1 ReadPheromones()	Algorithm 2 WritePheromones()
1: **while** *Explore* **do**	1: **while** *Explore* **do**
2: $s = ReadTag()$	2: $f = f_{curr} - f_{pre}$
3: $V_{pre} = V_{curr}$	3: $d = X_{curr} - X_{pre}$
4: $V_{curr} = \omega V_{pre} + c_1 r_1 (P_i - X_{pre}) + c_2 r_2 s$	4: $s_{curr} = f d / |d|^2$
5: **if** $|V_{curr}| > V_{max}$ **then**	5: $s_{pre} = ReadTag()$
6: $|V_{curr}| = V_{max}$	6: $s_{curr} = (1 - c_d) s_{pre} + c_a s_{curr}$
7: **end if**	7: WriteTag(s_{curr})
8: $X_{curr} = X_{pre} + V_{curr}$	8: $X_{pre} = X_{curr}$
9: **end while**	9: **end while**

4   Simulation

In order to verify the effectiveness of the proposed method thoroughly, this research starts from simple scenes to integrated searches. Firstly, the pheromone reading algorithm is tested by making robots moving on a predefined pheromone map. Then, some robots move randomly in the environment to build a pheromone map autonomously, which can verify the pheromone writing algorithm. Finally, the integrated searches are performed.

4.1   Robots Move on a Predefined Pheromone Map

In this section, robots move on a predefined pheromone map. In this map, each RFID tag is written with a pheromone vector. This vector can be considered as the gradient of the tested function at the specific RFID tag. The test function is here

$$f(x,y) = -(x - 8)^2 - (y - 8)^2 + 128. \tag{5}$$

The target locates at coordinates of $(8,8)$ and is represented by a pentagram. Vectors are indicated by arrows with different lengths, see Fig. 4.

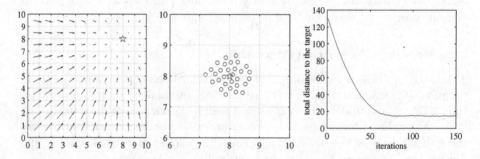

Fig. 4. Predefined map   **Fig. 5.** Robots converge   **Fig. 6.** Total distance

Due to the safe distance between robots, the robots will not converge to the target exactly, see Fig. 5. The total distance from the robots to the target is studied to judge if all the robot reach close enough to the target. Parameters ω and c_2 are set to 0.5, while c_1 is set to 0 to eliminate the effects of robots' self-learning. Results are obtained with a 30 robots swarm and with 200 iterations, see Fig. 6. It can be seen that after 70 iterations the total distance curve tends to be stable which means all the robots are moving near the target.

4.2   Build Pheromone Map

Having assessed the potentiality of a pheromone map stored in the RFID system, how a pheromone map can be built autonomously by a robot swarm is studied. Now the test function is used again. However, it is not aiming for building a pheromone map with its gradient. The value of the test function can be considered as the signal strength detected by the robots. In this experiment, parameters c_1 and c_2 are set to 0, while ω is set to 1 to ensure that robots move at a certain speed. Here 30 random-motion robots are initially set in the search area. They continually write pheromones to the environment. After a period of time, the pheromone map will be generated by the robots quite completely. Moreover, this pheromone map should be similar to the previous map which is built directly by the gradient of the test function. Several maps are obtained under different number of iterations, e.g., 200 iterations, 400 iterations, or 800 iterations. They are shown in Fig. 7(a)–(c).

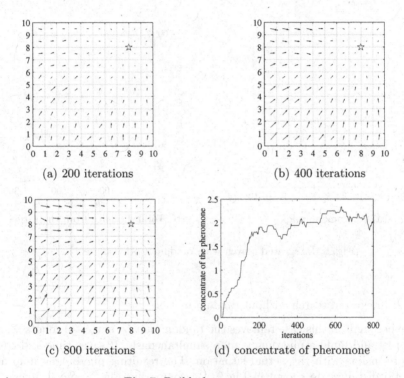

(a) 200 iterations   (b) 400 iterations

(c) 800 iterations   (d) concentrate of pheromone

Fig. 7. Build pheromone maps

Figure 7(d) shows a typical pheromone superposition process for a specific RFID tag. It shows that the concentrate of a pheromone vector increases with the number of iterations. At the beginning, the superposition rate is faster, after 400 iterations, it tends to be stable. Finally, a pheromone map is built by the robots autonomously which is similar to the map in Fig. 4.

4.3   Integrated Search

In this section, the situation that robots are performing the search and building maps simultaneously is studied. It is organized in three cases.

Case 1: Integrated search with random motion

The experiment in this case is divided into two stages. In the initial stage, 30 robots move randomly in the environment to build the pheromone map. 100 iterations later, it comes into the second stage, in which robots search the target by the signal strength and the pheromone map built in the first stage. The pheromone map is continuously built throughout the experiment. The final pheromone map and the total distance are shown in Fig. 8.

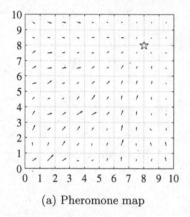 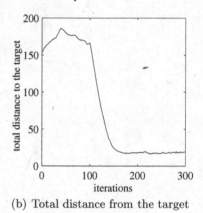

(a) Pheromone map   (b) Total distance from the target

Fig. 8. Integrated search with random motion (case 1)

Case 2: Integrated search without random motion

The experiment in this case removes the random motion stage, 30 robots search the target and build a pheromone map simultaneously. It saves time and could be more in line with the actual situation. The resulting pheromone map and the total distance are shown in Fig. 9. By comparing Figs. 8 and 9, it can be concluded that even if there is no pheromone in some positions, the robots can still reach close to the target.

Case 3: Integrated search with some robots without search capability

We want to tap more potential of this approach, so the 3rd case is performed. In this case, half of the robots have no search capability, they can only move using the pheromone map built by those robots who can detect the signal of the target. The initial random motion stage is still necessary here. Results are shown in Fig. 10. All robots can still reach close to the target, but the convergence process takes longer compared with the Case 1.

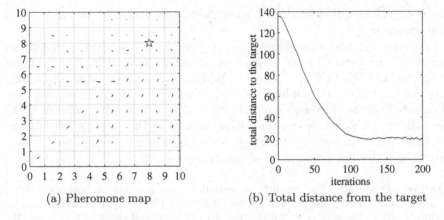

(a) Pheromone map (b) Total distance from the target

Fig. 9. Integrated search without random motion (case 2)

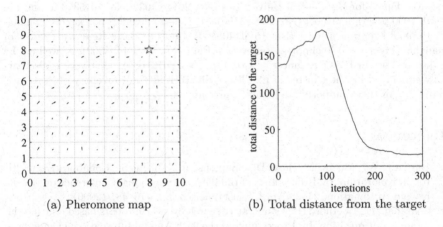

(a) Pheromone map (b) Total distance from the target

Fig. 10. Integrated search with some robots without search capability (case 3)

5   Conclusion

A new method based on stigmergy is proposed in this study for swarm robots to search a target. Pheromones are innovatively presented by vectors, which allows them to contain more information than traditional methods. The RFID tags are used as a carrier of pheromones. Robots are equipped with RFID readers so that they can read and write the information in the RFID tags at the current positions. The method also includes a velocity and position updating algorithm and pheromone reading and writing algorithms. Numerical experiments are performed thoroughly to verify the proposed method. Some conclusions are obtained according to the results. The stigmergy mechanism can be used to guide the robots' motion. A pheromone map can be built based on the search experience of the robots, and the robots can search the target on a pheromone map combining with their own search capability. Even if some robots have no search capabilities, they can still use the pheromone map built by other robots to reach the target.

The most significant advantage of this stigmergetic method is that direct communication between robots is not necessary. Therefore, the scale of the robot swarm will not be restricted and better performance may be obtained. In addition, the robot does not even have to be described in the same coordinate system. Each robot can use its own coordinate system, as long as their axes are in the same directions, which is very easy to implement in the actual swarm robots. As for future work, we plan to apply this method in more complicated search situations and using actual physical swarm robots.

Acknowledgements. This work is supported by the projects of National Science Foundation of China (No. 61603277; No. 51579053), the State Key Laboratory of Robotics and Systems (HIT) (No. SKLRS-2015-ZD-03), and the SAST Project (No. 2016017). Meanwhile, this work is also partially supported by the Fundamental Research Funds for the Central Universities (No. 2014KJ032; No. 20153683), and the Youth 1000 program project (No. 1000231901). It is also partially sponsored by the Shanghai Pujiang Program project (No. 15PJ1408400), the Key Basic Research Project of Shanghai Science and Technology Innovation Plan (No. 15JC1403300), as well as the project of Nuclear Power Engineering Co., Ltd. (No. 20161686). This work is also partially supported by the Cluster of Excellence SimTech at the University of Stuttgart, Germany. All these supports are highly appreciated.

References

1. Burgard, W., Moors, M., Fox, D., Simmons, R., Thrun, S.: Collaborative multi-robot exploration. In: Proceedings of the IEEE International Conference on Robotics and Automation, 24–28 April, San Francisco, pp. 476–481 (2000)
2. Sharafian, E., Eberhard, P.: Cooperative search by swarm robots based on bacterial foraging optimization. In: Proceedings of the 24th Annual International Conference on Mechanical Engineering, 26–28 April, Yazd (2016)
3. Balch, T., Arkin, R.: Communication in reactive multiagent robotic systems. Auton. Robot. **1**, 27–52 (1994)
4. Wang, Y., Xue, S., Zeng, J.: Survey of communication of swarm robotics. Ind. Control Comput. **24**(9), 64–66 (2011)
5. Zhang, G.: Actionless governance: view of Taoist about sport class administration. J. Beijing Sport Univ. **31**(2), 234–236 (2008)
6. Park, S., Saegusa, R., Hashimoto, S.: Autonomous navigation of a mobile robot based on passive RFID. In: Proceedings of the IEEE International Symposium on Robot and Human Interactive Communication, 26–29 August, Jeju, pp. 218–223 (2007)
7. Johansson, R., Saffiotti, A.: Navigating by stigmergy: a realization on an RFID floor for minimalistic robots. In: Proceedings of the IEEE International Conference on Robotics and Automation, 12–17 May, Koba, pp. 245–252 (2009)
8. Tang, Q., Zhang, L., Luo, W., Ding, L., Yu, F., Zhang, J.: A comparative study of biology-inspired algorithms applied to swarm robots target searching. In: Proceedings of the International Conference on Swarm Intelligence, 25–30 June, Bali, pp. 479–490 (2016)

9. Grassé, P.: La reconstruction du nid et les coordinations interindividuelles chez Bellicositermes natalensis, et Cubitermes sp. la théorie de la stigmergie: Essai d'interprétation du comportement des termites constructeurs. Insectes Sociaux 6(1), 41–80 (1959). (in French)
10. Marsh, L., Onof, C.: Stigmergic epistemology, stigmergic cognition. Cognitive Syst. Res. 9(1), 136–149 (2009)
11. Kennedy, J., Eberhart, R.: Particle swarm optimization. In: Proceedings of the IEEE International Conference on Neural Networks, 27 November–1 December, Perth, vol. 4, pp. 1942–1948 (1995)

Cooperative Control of Multi-robot System Using Mobile Agent for Multiple Source Localization

Naoya Ishiwatari[1](✉), Yasunobu Sumikawa[1](✉), Munehiro Takimoto[1](✉),
and Yasushi Kambayashi[2](✉)

[1] Department of Information Sciences, Tokyo University of Science, Tokyo, Japan
{n-ishiwatari,yas,mune}@cs.is.noda.tus.ac.jp
[2] Department of Computer and Information Engineering,
Nippon Institute of Technology, Miyashiro, Japan
yasushi@nit.ac.jp

Abstract. In this paper, we propose a new approach for multiple source localization tasks of multiple robots. Our approach controls robots by mobile agents that behave based on particle swarm optimization. The key process of our algorithm is to make subgroups in the population. In order to make subgroups, we indirectly transfer information using mobile agents instead of through direct communications. We have implemented our approach in a simulator, and conducted experiments to show effectiveness of our approach. Through the experimental results, we have confirmed that our approach is more efficient and less susceptible to initial placement than other approaches.

Keywords: Swarm intelligence · Swarm robotics · Mobile agent

1 Introduction

Swarm-robotic systems are attracting attention in various tasks [3,19,20,22,23] to make a system robust, scalable, and flexible [24]. One of such tasks is the source localization. In this task, robots localize for targets based on only relative intensity of signals emitted from the targets, such as locating sources of hazards like light, sound, heat, and leaks in pressurized systems [11], hazardous plumes/aerosols resulting from nuclear/chemical spills [26], fire-origins in forest fires [5], oil spills [6], and others.

Recently, it is found that particle swarm optimization (PSO) is useful for controlling robot's behaviors in the source localization task hunting a single target [7,8,10]. PSO makes all the robots move toward to the robot that is the closest to a target, and this behavior gives the robots a tendency being located near each other; thus, if the PSO is simply applied to hunt multiple targets, the robots can be regarded as single robot after the first hunting. However, applying PSO might fall into local optimum in complex environments that have multiple targets [1]. For searching in such complex environments, researchers have proposed

© Springer International Publishing AG 2017
Y. Tan et al. (Eds.): ICSI 2017, Part II, LNCS 10386, pp. 210–221, 2017.
DOI: 10.1007/978-3-319-61833-3_22

PSO based approaches [4,7,8]. Especially, Adaptive Robotic PSO (A-RPSO) [10], which is the state-of-the-art of this task, has a mechanism to escape from local optima for searching a target in an environment with obstacles. This approach has a property of diverging immediately before the robots converge. This property is important for escaping local optima.

In this paper, we propose a novel A-RPSO based approach for the multiple sources localization task by using mobile agents as the communication means. In the mobile agent system, each agent can actively migrate from one site to another site. Since a mobile agent can bring the necessary functionalities with it, and perform its own tasks autonomously, it can reduce interaction with the other side of the communication. In the minimal case, a mobile agent requires the connection only when it performs migration [2].

The key properties of our approach are to suppress convergence by mobile agents and divergence by A-RPSO. Each robot indirectly communicates with other robots through migrations of the mobile agents instead of through direct communications. In particular, mobile agents delay information propagation and suppress rapid convergence. This property is important for the multiple sources localization task, because it leads robots to the target without falling into local optima. The PSO-based multiple source localization study [4] solves this problem by using information on multiple targets, whereas our research uses only the information on only source that is closest to the robot. Our approach is more effective than other approaches in two aspects. First, our approach searches more efficiently than other approaches in a sparse environment such as a large field or a field with a small number of robots. Second, it behaves independently of the initial placement of robots. We have confirmed these properties through experiments.

The structure of the balance of this paper is as follows. In Sect. 2, we describe related works. In Sect. 3, we outline PSO and A-RPSO based hunting. We then extend the hunting approaches with mobile agents in Sect. 4. In Sect. 5, we show our experimental results. We conclude our discussions in Sect. 6.

2   Related Works

Cooperative source localization task using multiple robots has recently received attentions in the collective robotics community. Krishnanand et al. have proposed glowworm swarm optimization (GSO) algorithm for multiple sources localization task [16]. Shaukat et al. have proposed bio-inspired model which invokes collective behaviors using passive sensing [25]. It shows better results than other approaches in higher background noise levels without any explicit inter-robot communication.

Many studies have used PSO for the cooperative source localization task. Pugeh et al. and Nighot et al. have shown a system that displays self-organizational, robust, and flexible collective behaviors by using PSO [18,21]. Cai et al. have proposed a potential field-based particle swarm optimization (PPSO) for the multi-robot system to explore unknown areas [4]. Couceiro et al. have proposed two extensions of PSO that are named Robotic PSO (RPSO) and Robotic Darwinian PSO (RDPSO) [7,8]. RPSO is the extensions of PSO by taking into account of obstacle avoidance in the real-world. RDPSO is an approach based on RPSO which is equipped with a reward and punishment technique. This approach outperforms the other approaches for search task in almost all the experiments [9].

2.1   Aggregations of Foraging Swarm (AFS)

Gazi and Passino have presented an approach based on attractant and repellent profiles as aggregations of foraging swarm (AFS) [12,13]. The inter-individual attractant and repellent profiles are used on the interplay between attraction and repulsion forces with the attraction dominating on large distances, and the repulsion dominating on short distances. This approach has high communication complexity; all robots in the population need to communicate with each other. For small population, the performance is almost the same as GSO [9]. To that end, the authors carried out a behavioral analysis followed by several simulation experiments so as to define the most adequate values of the system parameters.

2.2   Glowworm Swarm Optimization (GSO)

Krishnanand et al. propose GSO for multiple sources localization task [16]. This approach makes formation of subgroups in the population of robots to track multiple signal sources. For this, in the approach, each robot moves towards another robot that locates in better position in dynamic communication range. According to [9,17], it is known that GSO tends to fail the task compared with other algorithms.

3   PSO and A-RPSO Based Hunt

For defining our approach, we detail in this section past two studies on source localization, PSO and A-RPSO.

3.1   Particle Swarm Optimization (PSO)

Kennedy et al. have proposed PSO for optimization problems by leading the particles toward the area with the best position in swarm [15]. PSO assumes that all the particles are randomly located initially. Each particle is represented by its position x_i^t, velocity v_i^t, current iteration t, and particle number i. The position and velocity of each particle are updated by the following equations:

$$x_i^{t+1} = x_i^t + v_i^{t+1}. \tag{1}$$

$$v_i^{t+1} = \omega v_i^t + C_1 Rand_1(pBest_i^t - x_i^t) + C_2 Rand_2(gBest_i^t - x_i^t). \tag{2}$$

These values are handled as vectors. x_i^{t+1} is a new position at iteration $t+1$, which is obtained by just updating the current position x_i^t with the new velocity v_i^{t+1}. v_i^{t+1} is calculated by compounding the current velocity v_i^t at iteration t, the vector to the locally best position, $pBest_i^t - x_i^t$, and the vector to the globally best position, $gBest^t - x_i^t$ iteration, $pBest_i^t - x_i^t$ and $gBest^t - x_i^t$, which are adjusted with weight ω, $C_1 Rand_1$ and $C_2 Rand_2$, respectively. C_1 and C_2 are empirically determined values, and $Rand_{1/2}$ is a random number over the range $[0, 1]$. Notice that ω is the inertial constant, which is determined to be more than or equal to 0 and less than 1.

These approaches have two critical disadvantages. First, they may get stuck on local optima; in other words, multiple sources localization task may fail. Second, robots have slow progress because their speed values are determined by ωv_i^t.

3.2 A-RPSO

Masoud Dadga et al. have proposed the A-RPSO for target searching on unknown environments with obstacles [10]. A-RPSO utilizes "evolution speed" and "aggregation degree" to escape from local optima.

In A-RPSO, the value of velocity is updated by the following equations:

$$v_i^{t+1} = w_i^t v_i^t + C_1 Rand_1 (pBest_i^t - x_i^t) + c_2 Rand_2 (gBest^t - x_i^t) + C_3 Rand_3 (p_i^t - x_i^t). \tag{3}$$

$$w_i^t = y_i^t (w_{ini} - \alpha h_i^t + \beta s). \tag{4}$$

$$h_i^t = 1 - \left[\frac{F(pBest_i^t)}{F(pBest_i^{t-1})} \right]. \tag{5}$$

$$s = \left[\frac{F_{tBest}}{F_t} \right]. \tag{6}$$

$$y_i^t = \begin{cases} U - \left[(U - L) \frac{t}{maxItr} \right] & \text{if } i \text{ is the best robot} \\ 1 & \text{otherwise} \end{cases} \tag{7}$$

$$c_2 = min((NR \times \rho + c_1), 2c_1). \tag{8}$$

In Eq. 3, C_3 is the coefficient to keep away from obstacles. However our study sets zero to this parameter because we assume that their is no obstacle on the field. The difference between Eqs. 3 and 2 is that A-RPSO determines the values of inertial weight w_i^t and c_2 by Eqs. 4 and 8. The value of inertial weight depends on evolutionary speed factor h_i^t and aggregation degree s. The inertia weight on the robot's velocity is increased in the two situations. First, the evolutionary speed decreases. That is, the robots slowly progress. Second, the aggregation degree is increased, that is, the robots are close each other. As a result, robots keep diversity in the reminder searching mission. w_{ini} is the initial value for the inertia weight. The values of α and β are constant in the range of $[0, 1]$. y_i^i is the coefficient for the inertia weight of a robot in the best position. The evolutionary speed factor is determined by Eq. 6 where $F(pBest_i^t)$ is the fitness value of $pBest_i^t$. The aggregation degree is determined by Eq. 6 where $\overline{F_t}$ is the mean fitness of all robots in the tth iteration and F_{tBest} is the best fitness in the current iteration. The coefficient for inertia weight is determined by Eq. 7 where U, L are the upper and lower bounds for y_i^t. $maxItr$ is the maximum iteration. Because the robot moves quickly at the start of searching mission and slowly at the end, the chance of the robot to find better position increases. On the other hand, c_2 is determined by Eq. 8 where NR is the number of robots and ρ is a constant. Generally, it is better to slightly increase the value of c1 and slightly decrease the value of c_2 after each iteration.

This approach is particularly faster than other approaches in two distinctive states: large environments and a small number of robots. For the multiple sources localization task, most searching algorithms divide swarm into several subgroups and since this approach has good performance for small population, this approach is expected to perform properly in multiple targets searching.

4   Multi-target Approaches

4.1   Multi Mobile Agent Behaviors

Our approach extends formation of subgroups in the population of robots under the A-RPSO for the multiple sources localization task. In the formation, each robot communicates with other robots through migrations of multiple mobile agents. Mobile agents contribute to delay propagation of information and suppress rapid convergence. We show how robots behave to approach targets in Fig. 1. Let the red-cross sign and the green-cross sign be the targets and robots represented by blue circles. The robots on which mobile agent resides are represented by a double circle. As shown in Fig. 1(a), the red mobile agent (agent A)

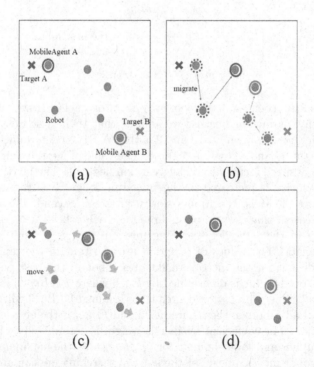

Fig. 1. A searching process in our approach. (Color figure online)

and the green mobile agent (agent B) record the positions of the robot that are closest to the red target (target A) and the green target (target B), respectively. As shown in Fig. 1(b), the agents A and B traverse robots to share recorded information. As shown in Fig. 1(c), the robots that are sharing information with agent A and agent B move toward target A and target B, respectively. As shown in Fig. 1(d), robots make formation of subgroups in the population, and then, the robots move based on mobile agents information. As a result, robots under the PSO or A-RPSO can move toward several targets without converging into only a signal target.

Mobile agents traverse robots and collect three kinds of information: robot's ids, the locally best positions ($pBest$), and the strength of signal at $pBest$. We call those information "propagated information." Robots record the propagated information. At this time, the mobile agents adjust the information shared with robots through three functionalities: sharing propagated information, deleting propagated information of found targets, and selecting the destination to where the mobile agent migrates.

For sharing propagated information, each mobile agent combines the information held on the robot which it migrates with its own information. If the mobile agent has information that the robot does not have, it is shared with the robot. If both of the mobile agent and the robot have information, and values of $pBest$ are different, the information closer to the target is shared. Similarly, when several mobile agents simultaneously migrate to the same robot, they merge the information into one mobile agent.

For deleting propagated information, a mobile agent deletes information about found targets ("deleted information"), so that robots with deleted information stop moving toward the target. Mobile agents and robots share the deleted information. The robot recognizes that the target has been captured when the value of the sensor changes beyond a certain amount. When robots capture a target, they create their own propagated information as deleted information. The deleted information is shared by mobile agents as well as the other propagated information. Once a robot receives the deleted information, the robot deletes corresponding information.

As selecting a mobile agent's migration destination, if several robots are within the view range of the camera, the mobile agent selects only one robot in the forward direction. The migration manner allows the mobile agent to migrate to others robots closer to the target because a lot of robots face around a target based on PSO or A-RPSO. In addition, the migration manner allows the mobile agent to record the locations of robots visited in the process of migrations to prevent it from revisiting the same robots.

4.2 A-RPSO with Multi Mobile Agent Approach (MA-PSO)

We propose an approach using mobile agent for communications between robots based on A-RPSO for the multiple sources localization task. We call this approach MA-$ARPSO$. Robots search for several targets simultaneously by using mobile agents without rapidly converging for a singe target. $pBest_i^t$ and $\overline{F_t}$ are determined by the best value of the information shared by robots and the average of the fitness values of the share information. $maxItr$ is updated using the following equations:

$$maxItr = \frac{NT}{maxT - t}.$$ (9)

where NT is the number of unfound targets, and $maxT$ is the maximum number of iterations.

5   Experiments and Results

We have implemented a simulator to demonstrate the effectiveness of our mobile agent based robot hunting system. The simulator assumes that there is no obstacle on the field as shown in Fig. 2. The red circles represent targets. The blue circles, each of which has the white line showing its forward direction, represent robots. The circle around each robot represents a mobile agent residing on it. In the simulator, robots are initially located around the targets. Each robot can just sense the strongest of signal from targets without knowing the coordinate of the target. Once some robots reach the target position, and find it, the signal is lost. The simulation terminates when satisfying one of the following two termination conditions. First, all the target are captured. Second, the simulation works beyond the maximum number of iterations (we use 1000 as the number of iterations). We have conducted a number of experiments where the number of targets is 10. Each experiment has been repeated 100 times.

We have conducted each experiment with five searching algorithms: AFS, GSO, A-RPSO and two new algorithms MA-ARPSO and MA-PSO extended by mobile agents. AFS sets parameter value determined by Artificial Bee Colony (ABC) [14] for the experiment. MA-ARPSO is the extension of A-RPSO which

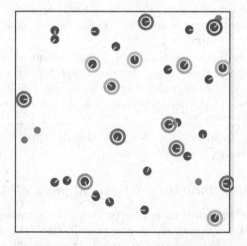

Fig. 2. A snapshot of the simulation. (Color figure online)

was discussed in Sect. 4. MA-PSO is the extension of PSO using mobile agents for communications. We also give the results for MA-PSO to show the impact of A-RPSO against PSO in the extension. To investigate what kinds of factors affect the effectiveness of our algorithm, we present three kinds of experiments. The first one is for the impact of the number of robots in the search. The second one is for the impact of the work field size. The last one is for the impact of the initial positions of the robots.

5.1 Impact of the Number of Robots

Figure 3 shows the results with the difference of the number of robots. The size of the work field is 1000 × 1000 squares. The MA-ARPSO and MA-PSO have searched the targets more quickly than AFS and GSO. In the cases of 50–70 robots, MA-ARPSO has searched about 40% more quickly than the GSO. Especially, in the cases of 10 and 20 robots, AFS and GSO have failed the search; they exceeded 1000 iterations. In the case of 100 robots, the GSO has searched as quickly as the MA-ARPSO and MA-PSO. These results show that MA-ARPSO is effective when searching with a few robots. Also, MA-ARPSO becomes more efficient as the number of robots increases. Furthermore, the effectiveness is improved by increasing the number of robots. In the case of 50–100 robots, the A-RPSO does not change performance because the robots converges during the search. We can conclude that MA-ARPSO is scalable.

In the approaches such as GSO or AFS, each robot moves towards another neighbor robot closer to a target, so that robots are divided into several groups. This feature suppresses the robots falling into a local optima, although convergence becomes slow. On the other hand, in the approaches such as PSO, each robot moves towards the best one in the group, so that robots convergences

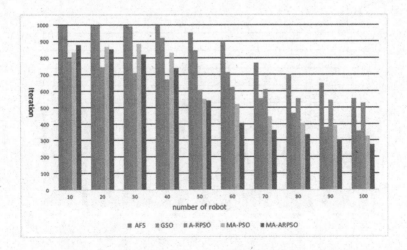

Fig. 3. Performance comparison with the number of robots.

quickly, although it tends to fall into a local optima. In our approach, communications through mobile agents enable each robot to search quickly, because the robot can move around the target based on the target information that mobile agents collect through traversing other robots.

5.2   Impact of the Work Field Size

Figure 4 shows the results with the difference of the sizes of the work field. The number of robots is 50. The MA-ARPSO and MA-PSO have searched more slowly than GSO in the work fields that are less than 400×400. The MA-ARPSO and MA-PSO have searched quickly than GSO, AFS and A-RPSO in the work fields that are more than 800×800. For example, in GSO and AFS, the difference between iterations they took for 300×300 and 1000×1000 is about 450. On the other hand, in MA-ARPSO, the difference between iterations for 300×300 and 1000×1000 is about 100. Thus, the MA-ARPSO is useful in large work fields and has flexibility for the size of work fields.

5.3   Impact of Initial Placement of Robots

In the experiments with the difference of initial positions, we have compared MA-ARPSO and MA-PSO. The number of robots is 50, and the size of the work field is 1000×1000. We have conducted experiments with random initial placement where robots are randomly located and a specific initial arrangement of robots as shown in Fig. 5. Figure 6 shows that, in MA-PSO, the specific initial placement is about 300 iterations more than that random initial positions does, and in MA-ARPSO, the specific initial placement produces only about 100 iteration more than that random initial placements does. The results show that MA-ARPSO does not depend on initial configuration as strongly as MA-PSO does.

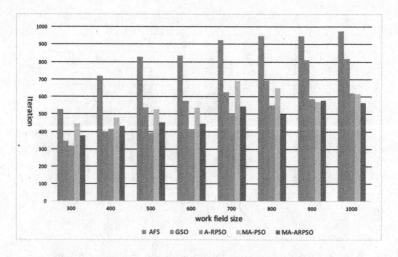

Fig. 4. Performance comparison with the work field sizes.

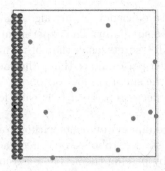

Fig. 5. Screen shot of the specific initial placement.

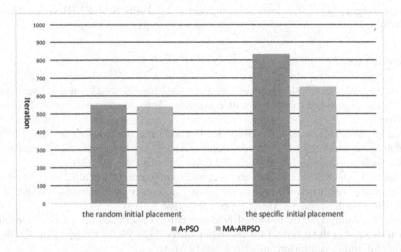

Fig. 6. Performance comparison in the specific initial placement of robots.

MA-ARPSO and MA-PSO in the case of the specific initial placement take longer for searching than them in the case of the random initial positions. In the case of the specific initial placement, robots gather quickly at a certain point despite of suppression of mobile agents for convergence of the robots. On the other hand, the A-RPSO has the feature of dispersing robots which are gathering. Thus, MA-ARPSO searches faster than MA-PSO without strongly depending on initial configuration. Practically, it is difficult to randomly locate robots in the entire search space. Therefore, it is an important for performance independent on initial positions in the real world.

6   Conclusions

In this paper, we have proposed an approach for multiple sources localization task based on A-RPSO with communications through mobile agents. Our approach forms subgroups of robots by delaying propagation of information through agent migrations. To demonstrate efficiency of our algorithm, we have conducted a number of experiments on a simulator. The experimental results show that our searching method capture the targets faster than other methods do. In particular, our method has shown higher performance than other methods for environments with large work fields and the small number of robots. We also have shown that the search speed does not decrease even if it is easy for robots to converge in their initial positions.

In this paper, we focused on environments without obstacles. In order to deal with environments with obstacles, robots would need additional functionalities: sensing the obstacles and by passing them. These extensions are our important future topics for the multiple sources localization task.

Acknowledgments. This work was supported by JSPS KAKENHI Grant Number 26750076.

References

1. Alireza, A.: PSO with adaptive mutation and inertia weight and its application in parameter estimation of dynamic systems. Acta Automatica Sinica **37**(5), 541 (2011)
2. Binder, W., Hulaas, J.G., Villaz, A.: Portable resource control in the J-SEAL2 mobile agent system. In: Proceedings of the Fifth International Conference on Autonomous Agents (AGENTS 2001), pp. 222–223. ACM (2001)
3. Bogert, K., Doshi, P.: Multi-robot inverse reinforcement learning under occlusion with state transition estimation. In: Joint Conference on Autonomous Agents and Multi-agent Systems (AAMAS), pp. 1837–1838 (2015)
4. Cai, Y., Yang, S.X.: A potential field-based PSO approach for cooperative target searching of multi-robots. In: Proceeding of the 11th World Congress on Intelligent Control and Automation, pp. 1029–1034 (2014)
5. Casbeer, D.W., Beard, R.W., McLain, T.W., Li, S.M., Mehra, R.K.: Forest fire monitoring with multiple small UAVs. In: Proceedings of the 2005 American Control Conference, vol. 5, pp. 3530–3535 (2005)
6. Clark, J., Fierro, R.: Cooperative hybrid control of robotic sensors for perimeter detection and tracking. In: Proceedings of the 2005 American Control Conference, vol. 5, pp. 3500–3505 (2005)
7. Couceiro, M.S., Rocha, R.P., Ferreira, N.M.F.: Ensuring ad hoc connectivity in distributed search with robotic darwinian particle swarms. In: 2011 IEEE International Symposium on Safety, Security, and Rescue Robotics, pp. 284–289 (2011)
8. Couceiro, M.S., Rocha, R.P., Ferreira, N.M.F.: A novel multi-robot exploration approach based on particle swarm optimization algorithms. In: IEEE International Symposium on Safety, Security, and Rescue Robotics, pp. 327–332 (2011)
9. Couceiro, M.S., Vargas, P.A., Rocha, R.P., Ferreira, N.M.F.: Benchmark of swarm robotics distributed techniques in a search task. Robot. Auton. Syst. **62**(2), 200–213 (2014)
10. Dadgar, M., Jafari, S., Hamzeh, A.: A PSO-based multi-robot cooperation method for target searching in unknown environments. Neurocomputing **177**, 62–74 (2016)
11. Fronczek, J.W., Prasad, N.R.: Bio-inspired sensor swarms to detect leaks in pressurized systems. In: 2005 IEEE International Conference on Systems, Man and Cybernetics, vol. 2, pp. 1967–1972 (2005)

12. Gazi, V., Passino, K.M.: Stability analysis of social foraging swarms. Trans. Sys. Man Cyber. Part B **34**(1), 539–557 (2004)
13. Gazi, V., Passino, K.M.: Stability analysis of swarms. IEEE Trans. Autom. Control **48**(4), 692–697 (2003)
14. Karaboga, D., Basturk, B.: A powerful and efficient algorithm for numerical function optimization: artificial bee colony (ABC) algorithm. J. Glob. Optim. **39**(3), 459–471 (2007)
15. Kennedy, J., Eberhart, R.: Particle swarm optimization. Proc. IEEE Int. Conf. Neural Netw. **4**, 1942–1948 (1995)
16. Krishnanand, K.N., Ghose, D.: A glowworm swarm optimization based multi-robot system for signal source localization. In: Liu, D., Wang, L., Tan, K.C. (eds.) Design and Control of Intelligent Robotic Systems, vol. 177, pp. 49–68. Springer, Heidelberg (2009)
17. McGill, K., Taylor, S.: Comparing swarm algorithms for large scale multi-source localization. In: 2009 IEEE International Conference on Technologies for Practical Robot Applications, pp. 48–54, November 2009
18. Nighot, M., Patil, V., Mani, G.: Multi-robot hunting based on swarm intelligence. In: 2012 12th International Conference on Hybrid Intelligent Systems (HIS), pp. 203–206 (2012)
19. Petersen, K., Nagpal, R., Werfel, J.: TERMES: an autonomous robotic system for three-dimensional collective construction. In: Robotics: Science and Systems VII, University of Southern California, Los Angeles, 27–30 June 2011
20. Portugal, D., Rocha, R.P.: Cooperative multi-robot patrol with Bayesian learning. Auton. Robot. **40**(5), 929–953 (2016)
21. Pugh, J., Martinoli, A.: Inspiring and modeling multi-robot search with particle swarm optimization. In: 2007 IEEE Swarm Intelligence Symposium (SIS 2007), Honolulu, 1–5 April 2007, pp. 332–339 (2007)
22. Rosenfeld, A., Agmon, N., Maksimov, O., Azaria, A., Kraus, S.: Intelligent agent supporting human-multi-robot team collaboration. In: Proceedings of the Twenty-Fourth International Joint Conference on Artificial Intelligence (IJCAI 2015), Buenos Aires, 25–31 July 2015, pp. 1902–1908 (2015)
23. Rubenstein, M., Cabrera, A., Werfel, J., Habibi, G., McLurkin, J., Nagpal, R.: Collective transport of complex objects by simple robots. In: Proceedings of the 2013 International Conference on Autonomous Agents and Multi-agent Systems, pp. 47–54 (2013)
24. Şahin, E.: Swarm robotics: from sources of inspiration to domains of application. In: Şahin, E., Spears, W.M. (eds.) SR 2004. LNCS, vol. 3342, pp. 10–20. Springer, Heidelberg (2005). doi:10.1007/978-3-540-30552-1_2
25. Shaukat, M., Chitre, M.: Bio-inspired practicalities: collective behaviour using passive neighbourhood sensing. In: Proceedings of the 2015 International Conference on Autonomous Agents and Multiagent Systems (AAMAS 2015), Richland, pp. 267–277 (2015)
26. Zarzhitsky, D., Spears, D.F., Spears, W.M.: Swarms for chemical plume tracing. In: Proceedings of the 2005 IEEE Swarm Intelligence Symposium (SIS 2005), pp. 249–256 (2005)

Hybrid Optimization Algorithms and Applications

Evolutionary Fuzzy Control of Three Robots Cooperatively Carrying an Object for Wall Following Through the Fusion of Continuous ACO and PSO

Min-Ge Lai[1], Chia-Feng Juang[1(✉)], and I-Fang Chung[2]

[1] Department of Electrical Engineering,
National Chung-Hsing University, Taichung, Taiwan
cfjuang@dragon.nchu.edu.tw
[2] Institute of Biomedical Informatics,
National Yang-Ming University, Taipei, Taiwan
ifchung@ym.edu.tw

Abstract. This paper proposes evolutionary fuzzy control of three robots cooperatively carrying an object in executing a convex wall following behavior. The object is not connected to the robots and may fall off for a failed control. Evolutionary fuzzy control of a single robot for wall following is first performed. Then, evolutionary fuzzy control of two robots cooperatively carrying a long strip object along the wall is performed. For the carry of a larger object for wall following, a third robot is included to cooperate with the two successfully controlled robots. Fuzzy controller of the third robot is also learned through a data-driven evolutionary learning approach. To improve learning efficiency of the FC, the swarm intelligence algorithm of adaptive fusion of continuous ant colony optimization and particle swarm optimization (AF-CACPSO) is employed. Simulations show the effectiveness of the evolutionary fuzzy control approach for the three cooperative object-carrying robots.

Keywords: Ant colony optimization · Particle Swarm Optimization · Evolutionary fuzzy systems · Fuzzy Control · Evolutionary robots

1 Introduction

This paper considers the problem of controlling three wheeled robots cooperatively carrying an object to move along a convex wall. This wall-following behavior enables the robots to bypass obstacles (walls) and avoids collision in navigation applications. The control of a single wheeled or hexapod robot to perform the wall-following behavior has been extensively studied [1–7]. One popular control approach is using a fuzzy controller (FC). For the design of the FC, the approaches of manual design [1, 6], neural learning [2], and evolutionary learning [3, 4, 7] have been proposed. In contrast to the former two approaches, the evolutionary learning shows the advantage of avoiding the exhaustive manual tuning of rule parameters and collection of supervised input-output training data. To improve evolutionary learning efficiency, the adaptive

© Springer International Publishing AG 2017
Y. Tan et al. (Eds.): ICSI 2017, Part II, LNCS 10386, pp. 225–232, 2017.
DOI: 10.1007/978-3-319-61833-3_23

fusion of continuous ant colony optimization and particle swarm optimization (AF-CACPSO) has been proposed and its performance has been shown to outperform various particle swarm optimization (PSO) and continuous ant colony optimization (CACO) algorithms [8]. Therefore, the AF-CACPSO algorithm is used in this paper for evolutionary learning of FCs in controlling three cooperative robots.

Wheeled robots can be applied to move objects in home service or factories demanding automation. Different types of object moving methods have been proposed [9–11]. Carry of objects placed on the top of two cooperative robots was proposed in [8, 11], where the object is not connected to the robots and may fall off if there is an improper control activity. The cooperative schemes in [8, 11] were manually designed and learned through evolutionary fuzzy control, respectively. This paper considers wall-following control of three robots cooperatively carrying an object. Cooperative robots with more than two robots involved may be used for hunting [12] and entrapment/escorting and patrolling missions [13]. In the wall-following object-carrying application, the FCs for cooperative wall-following and triangle formation are automatically learned through the AF-CACPSO. This behavior can be applied to bypass obstacles in the navigation of object-carrying robots.

This paper is organized as follows. Section 2 describes the FC and its optimization using the AF-CACPSO. Section 3 describes evolutionary fuzzy wall-following control of a single robot and two cooperative object-carrying robots through data-driven learning. Section 4 describes evolutionary fuzzy wall-following control of three cooperative robots carrying a larger object. Section 5 presents simulation results. Finally, Sect. 6 presents the conclusion.

2 Fuzzy Controller and Adaptive Fusion of Continuous Ant Colony Optimization and Particle Swarm Optimization (AF-CACPSO)

This paper uses zero-order Takagi-Surgeon (TS)-type FCs to control robots for wall following. In the FC, the antecedent and consequent parts use Gaussian fuzzy sets and fuzzy singletons, respectively. The mathematical functions in computing the FC outputs are the same as those in [8]. In the evolutionary fuzzy control approach, the number of rules is assigned in advance. All of the free parameters in the antecedent and consequent parameters are optimized using the AF-CACPSO [8].

The AF-CACPSO is a population-based optimization algorithm. Figure 1 shows the block diagram of the algorithm. Each new individual in a population may be generated either from PSO or CACO. At each iteration, the number of new solutions generated from PSO and CACO decreases and increases, respectively. The setting of the iteration-varying number is to enhance exploration and exploitation ability at the initial and final optimization stages, respectively. The AF-CACPSO has been shown to outperform various population-based optimization algorithms in FC optimization. Therefore, this paper uses the AF-CACPO to optimize all of the FCs in controlling three cooperative robots in object carrying and moving along a wall.

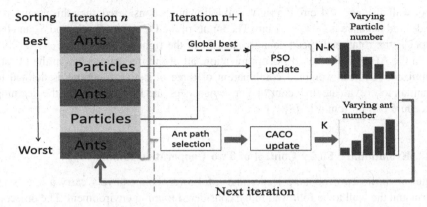

Fig. 1. The block diagram and update of population of solutions in AF-CACPSO.

3 Evolutionary Fuzzy Control of a Single and Two Cooperative Object-Carrying Robots

This paper extends the evolutionary fuzzy control of two cooperative robots carrying a long strip object [8] to the control of three robots carrying a larger object. The controlled robot is the e-puck mobile robot (http://www.e-puck.org/) and simulations are performed using the Webots 5.8 simulator [14]. The e-puck has two wheels and eight infra-red distance sensors $S_0,...,S_7$, as shown in Fig. 2. The sensor range is from 1 cm to 6 cm.

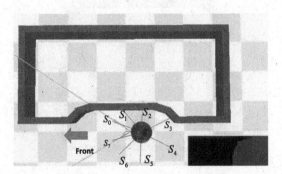

Fig. 2. The training environment for evolutionary fuzzy control of a single robot and the deployment of the infra-red distance sensors in the robot.

3.1 Evolutionary Fuzzy Control of a Single Robot

Figure 2 shows the training environment for evolutionary fuzzy control of a single robot. The robot moves with a constant speed at $v = 10$ rad/s. An FC is designed to control the robot orientation to move along the wall boundary with a constant robot-wall distance of 4 cm. If a right wall following is considered, then this means that the desired distance is $S_2(k) = 4$ cm. The inputs of the FC are S_0, S_1, S_2, and S_3 and the output is the rotational speed change, $\Delta\omega_{right}$, of the right wheel.

In the AF-CACPO-based FC optimization, all of the free parameters in the FC are optimized. A linear weighted combination of three objective functions is defined to quantitatively evaluate the control performance of an FC. Details of the learning procedure can be found in [8].

3.2   Evolutionary Fuzzy Control of Two Cooperative Robots

Figure 3 shows two robots, a leader and a follower, cooperatively carry a long strip object and the wall to be followed in the considered training environment. The object is not connected to the two robots and may fall off when the two robots move. Each robot is controlled by the FC designed for the control of a single robot as described above. For the two robots to successfully carry the object, an auxiliary FC, FC_{AUX}, is included to control the speed and orientation of the follower robot. The FC_{AUX} has six inputs, including the four right sensor measurements, the output $\Delta\omega_{right}$ (leader) of the FC controlling the leader robot, and the change of the distance, Δd, between the two robots. The FC_{AUX} sends two outputs to control the follower robot, including the change in the moving speed Δv and the additional rotational speed change of the right wheel $\Delta\omega_{right}^{AUX}$. The FC_{AUX} is also learned through the AF-CACPSO. Five requirements are imposed in the definition of a successful FC and the detailed descriptions can be found in [8].

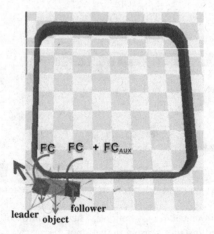

Fig. 3. Two robots cooperatively carry a long strip object and the wall to be followed in the training environment.

4 Evolutionary Fuzzy Control of Three Cooperative Robots

Figure 4 shows the three robots named as the leader, follower, and accompanying robots and the carried object placed on their tops in the considered training environment. In the wall following behavior, the direction of the accompanying robot is opposite to the other two robots. The object is not connected to the robots and, therefore, may fall off if there is an improper control activity. The leader and follower robots are controlled by the FCs and FC$_{AUX}$ introduced in Sect. 3. An accompanying FC, FC$_{ACC}$, is designed to control the accompanying robot so that the distances d_{13} and d_{23} between the accompanying and the leader and the follower robots, respectively, are fixed at a normal distance $\bar{d}$ (set to 15 cm) during moving, as shown in Fig. 4.

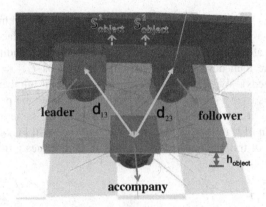

Fig. 4. The sensor measurements used in detecting a control failure in the learning process of the cooperative wall-following behavior of three robots.

The two inputs of the FC$_{ACC}$ that controls the accompanying robot are the distance differences $\Delta d_{13} = d_{13} - \bar{d}$ and $\Delta d_{23} = d_{23} - \bar{d}$. The FC$_{ACC}$ controls the orientation and moving speed of the accompanying robot. A control is deemed successful if none of the following conditions occur at each time step:

Condition 1: The carried object falls off, i.e., the height of the object h_{object}, as Fig. 5 shows, is smaller than 6 cm.
Condition 2: The distance between the leader and the accompanying robots is too short or too long, i.e., $|\Delta d_{13}| > 3$ cm.
Condition 3: The distance between the follower and the accompanying robot is too short or too long, i.e., $|\Delta d_{23}| > 3$ cm.
Condition 4: The carried object is too close to the obstacle, i.e., the measurements S_{object}^1 and S_{object}^2 of the two distance sensors set on the right side of the carried object, as Fig. 4 shows, are both smaller than 1.5 cm.

The FC$_{ACC}$ is learned through the AF-CACPSO. In the AF-CACPSO designed FC$_{ACC}$, an individual represents an FC$_{ACC}$. To evaluate the performance of each new

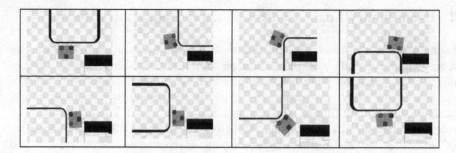

Fig. 5. Screenshots (left-to-right, top-to-down) of the fuzzy control of three robots cooperatively carry a large object and move along a wall.

FC_{ACC}, an FC_{ACC} is applied to control the accompanying robot working with the leader and follower robots in the training environment with an object placed on them. The three robots move along the wall and stop if one of the above conditions occurs (i.e., control fails) or the total number of pre-assigned time steps T_{suc} (=1200) is reached (i.e., control succeeds). When a robot control fails, the total number of control time steps until failure is used as a reinforcement signal to evaluate the performance of the FC_{ACC} (individual). If an FC_{ACC} fails, the robot is reset to the same initial position for the next trial. A run of learning process ends when a successful FC_{ACC} is found or the maximum number of trails is reached, where the latter indicates a failed run.

5 Simulations

In the simulations, the population size in each of the AF-CACPSO-designed FC, FC_{AUX}, and FC_{ACC} was identically set to 50. The iteration number was set to 200 and so the number of control trials was $50 \times 200 = 100000$.

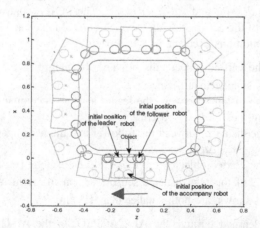

Fig. 6. Trajectories of the three cooperative robots and the carried object in the wall following task.

Based on the designed FC and FC_{AUX}, an FC_{ACC} that controlled the accompanying robot was optimized. Figure 5 shows screenshots of the wall-following result in the typical map. Figure 6 shows the trajectory of the carried object and the three robots cooperatively carrying the object while moving along the wall. All robots successfully move along the wall with the object on all of them without collision with the wall.

6 Conclusion

This paper has proposed a new evolutionary fuzzy control approach for the control of three robots cooperatively carrying a large object in executing the wall-following task. In this approach, the control problem is formulated as an optimization problem and the FC solution is found through the AF-CACPSO. This data-driven evolutionary learning approach avoids the explicit use of the mathematical model of the robot to derive the controllers. Simulations verify the effectiveness of the evolutionary fuzzy control of the accompanying robot to cooperate with the other two robots. In the feature, navigation of the three cooperative object-carrying robots with experimental implementation will be studied.

References

1. Cupertino, F., Giordano, V., Naso, D., Delfine, L.: Fuzzy control of a mobile robot. IEEE Robot. Autom. Mag. **13**(4), 74–81 (2006)
2. Zhu, A., Yang, S.X.: Neurofuzzy-based approach to mobile robot navigation in unknown environments. IEEE Trans. Syst. Man Cybern. C Appl. Rev. **37**(4), 610–621 (2007)
3. Juang, C.F., Hsu, C.H.: Reinforcement ant optimized fuzzy controller for mobile-robot wall-following control. IEEE Trans. Ind. Electron. **56**(10), 3931–3940 (2009)
4. Hsu, C.H., Juang, C.F.: Evolutionary robot wall-following control using type-2 fuzzy controller with species-DE activated continuous ACO. IEEE Trans. Fuzzy Syst. **21**(1), 100–112 (2013)
5. Paredes, A.C., Malfaz, M., Salichs, M.A.: Signage system for the navigation of autonomous robots in indoor environments. IEEE Trans. Ind. Inform. **10**(1), 680–688 (2014)
6. Zaheer, S.A., Choi, S.H., Jung, C.Y., Kim, J.H.: A modular implementation scheme for nonsingleton type-2 fuzzy logic systems with input uncertainties. IEEE/ASME Trans. Mechatron. **20**(6), 3182–3193 (2015)
7. Juang, C.F., Chen, Y.H., Jhan, Y.H.: Wall-following control of a hexapod robot using a data-driven fuzzy controller learned through differential evolution. IEEE Trans. Ind. Electron. **62**(1), 611–619 (2015)
8. Juang, C.F., Lai, M.G., Zeng, W.T.: Evolutionary fuzzy control and navigation for two wheeled robots cooperatively carrying an object in unknown environments. IEEE Trans. Cybern. **45**(9), 1731–1743 (2015)
9. Ghosh, A., Ghosh, A., Konar, A., Janarthanan, R.: Multi-robot cooperative box-pushing problem using multi-objective particle swarm optimization technique. In: Proceedings of 2012 World Congress on Information and Communication Technologies, pp. 272–277 (2012)

10. Liu, Z., Kamogawa, H., Ota, J.: Manipulation of an irregularly shaped object by two mobile robots. In: Proceedings of IEEE/SICE International Symposium on System Integration, Sendai, Japan, pp. 218–223 (2010)
11. Udomkun, M., Tangamchit, P.: Cooperative overhead transportation of a box by decentralized mobile robots. In: Proceedings of IEEE International Conference on Robotics, Automation and Mechatronics, Chengdu, China, pp. 1161–1166, September 2008
12. Yamaguchi, H.: A cooperative hunting behavior by mobile robot troops. In: Proceedings of IEEE International Conference on Robotics and Automation, vol. 4, pp. 3204–3209, May 1998
13. Mas, I., Li, S., Acain, J., Kitts, C.: Entrapment/escorting and patrolling missions in multi-robot cluster space control. In: Proceedings of IEEE/RSJ International Conference on Intelligent Robots and Systems, St. Louis, MO, pp. 5855–5861, October 2009
14. Michel, O.: Webots: professional mobile robot simulation. Int. J. Adv. Robot. Syst. 1(1), 39–42 (2004)

Optimal Operational Planning of Energy Plants by Multi-population Differential Evolutionary Particle Swarm Optimization

Norihiro Nishimura[2], Yoshikazu Fukuyama[1(✉)], and Tetsuro Matsui[2]

[1] Meiji University, Room 1206, Meiji University Nakano Campus, 4-21-1,
Nakano, Nakano-ku, Tokyo, Japan
yfukuyam@meiji.ac.jp
[2] Tokyo, Japan

Abstract. This paper presents optimal operation planning of energy plants by multi-population differential evolutionary particle swarm optimization (DEEPSO). The problem can be formulated as a mixed integer nonlinear optimization problem and various evolutionary computation techniques such as particle swarm optimization (PSO) and differential evolution (DE) have been applied. However, solution quality can be improved. Multi-population is known as one of a way of increasing solution quality. This paper applies multi-population DEEPSO for optimal operational planning of energy plants in order to improve solution quality.

Keywords: Energy Management System · Evolutionary computation · Optimal operational planning of energy plants · Multi-population

1 Introduction

An energy plant, which integrates these facilities, supplies not only electric power, but also steam, and cold and heat energy by various generators, boilers, and refrigerators. An energy plant is built up with an appropriate choice of generators, boilers, and refrigerators considering electric power, steam, and heat energy loads of target facilities.

For example, in Japan, since about 40% of energy consumption is in factories and about 20% of that is in business and commercial buildings [1]. The energy consumption in these facilities leads purchase of huge amount of the primary energy and the cost reduction is one of the big challenges. In order to reduce the huge operational cost, the least amount of primary energy should be purchased and supplied to the secondary energy (energy loads) required in factories, and business and commercial buildings. Namely, optimal operational planning of energy plants is crucially important.

Researches on optimal operational planning of energy plants have been done by energy optimization of FEMS (Factory Energy Management System) and BEMS (Building Energy Management System). The problem has been solved by mathematical programming previously [2–4]. However, the problem should be solved by

© Springer International Publishing AG 2017
Y. Tan et al. (Eds.): ICSI 2017, Part II, LNCS 10386, pp. 233–241, 2017.
DOI: 10.1007/978-3-319-61833-3_24

evolutionary computation techniques because it problem can be formulated as a mixed integer nonlinear optimization problem (MINLP) considering startup/shutdown status (discrete variables) and input-output values of facilities (continuous variables), and nonlinear characteristics of facilities. Various evolutionary computation techniques have been applied to the problems. PSO has been applied firstly [5, 6]. Then, DE has been applied [7]. Recently, the authors have applied DEEPSO to the problem [8]. However, there is still room for improvement on solution quality.

Multi-population is known as one of a way of increasing solution quality and applies to various evolutionary computation techniques. In this method, one swam of agents is divided into several sub-swarms and calculation is performed independently at each sub-swarm. Solution quality can be improved because of maintenance of diversification and intensification.

This paper expands the original DEEPSO to multi-population DEEPSO and presents optimal operation planning of energy plants by the expanded multi-population DEEPSO toward improvement of solution quality.

2 Formulation of the Optimal Operational Planning of Energy Plants

2.1 A Target Energy Plant

A target energy plant is shown in Fig. 1. In this plant, electric loads are supplied by an electric power purchase and electric power outputs of gas turbine generators. Steam loads are supplied by steam outputs of exhaust gas boilers of gas turbine generators and steam outputs of boilers. Heat energy loads are supplied by outputs of turbo and steam refrigerators [9].

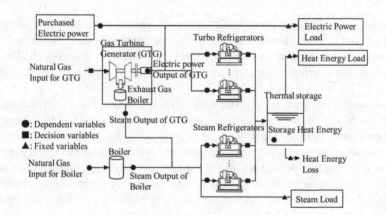

Fig. 1. A target energy plant [9].

2.2 Decision Variables

The decision variables are shown below. Each variable has 24 points a day ($N = 24$).

- Gas turbine generators (GTGs): The amount of natural gas input ($x^i_{gj} : i = 1, \cdots, N, j = 1, \cdots, N_g$ (the number of gas turbine generators)) and startup/shutdown status $\left(y^i_{gj} \in \{0, 1\} : i = 1, \cdots, N, j = 1, \cdots, N_g \right)$.
- Turbo refrigerators (TRs): The amount of heat energy output ($x^i_{tj} : i = 1, \cdots, N, j = 1, \cdots, N_t$ (the number of turbo refrigerators)) and startup/shutdown status $\left(y^i_{tj} \in \{0, 1\} : i = 1, \cdots, N, j = 1, \cdots, N_t \right)$.
- Steam refrigerators (SRs): The amount of heat energy output ($x^i_{sj} : i = 1, \cdots, N, j = 1, \cdots, N_s$ (the number of steam refrigerators)) and startup/shutdown status $\left(y^i_{sj} \in \{0, 1\} : i = 1, \cdots, N, j = 1, \cdots, N_s \right)$.

2.3 Objective Function

The objective function is minimization of total costs (electric power and natural gas purchase costs) a day.

$$min \sum_{i=1}^{N} \left\{ C^i_{Ep} E^i_p + C^i_{Gp} \left(\sum_{j=1}^{N_g} x^i_{gj} + \sum_{j=1}^{N_b} x^i_{bj} \right) \right\} \tag{1}$$

where, C^i_{Ep} is an electric power purchase unit price at time i, E^i_p is the amount of electric power purchase at time i, C^i_{Gp} is a natural gas purchase unit price at time i, N_b is the number of boilers, x^i_{bj} is the amount of natural gas input to the jth boiler at time i.

2.4 Constraints

The following constraints are considered in the formulation:

(1) Supply and demand balance
    Electric power, natural gas, and steam supply values should be equal to their load values, which are set by fixed values.
(2) Facility constraints
    Input-output characteristics and operational limits of facilities are considered.
(3) Operational constraints

    - If a facility is started up, then it should not be shut downed for a certain period (Minimum up time).
    - If a facility is shut downed, then it should not be started up for a certain period (Minimum down time).

3 Multi-population Differential Evolutionary Particle Swarm Optimization

3.1 The Processes and Update Equations Utilized in DEEPSO Pb [10]

DEEPSO is a combined method with Evolutionary PSO (EPSO), which is an improvement method of PSO and DE. The method can generate higher quality solutions than conventional PSOs. In [10], four variations of DEEPSO was proposed. This paper applies DEEPSO Pb. Its effectiveness can be found through pre-simulation.

Various processes utilized by DEEPSO are shown below.

(1) REPLICATION - each particle is replicated (cloned) R times.
(2) MUTATION - all R particles have weights which are mutated by the following equations:

$$A^{k+1} = A^k + \tau N(0,1) \tag{2}$$

$$B^{k+1} = B^k + \tau N(0,1) \tag{3}$$

$$C^{k+1} = C^k + \tau N(0,1) \tag{4}$$

$$b_G^* = b_G + \tau' N(0,1) \tag{5}$$

where, $A^k, B^k, and\ C^k$ are weight coefficients at iteration k, b_G is the best searching point so far found by the agents (*gbest*), $\tau\ and\ \tau'$ are learning parameters, $N(0,1)$ is a random variable with the Gaussian distribution (mean is 0 and variance is 1).

(3) REPRODUCTION - R + 1 offspring (original and clones) are updated by the following equations:

$$x_i^{k+1} = x_i^k + v_i^{k+1} \tag{6}$$

$$v_i^{k+1} = A^{k+1}v_i^k + B^{k+1}\left(b_{r1}^k - x_i^k\right) + P\left[C^{k+1}(b_G^* - x_i^k)\right] \tag{7}$$

where, x_i^k is a current searching point of agent i at iteration k, v_i^k is a current velocity of agent i at iteration k, b_{r1}^k is the best searching point of each agent (*pbest*).

b_{r1}^k is selected among the stored *pbests* of a selected searching point except x_i^k randomly.

(4) EVALUATION - objective function values of all offspring are calculated.
(5) SELECTION - the offspring forming the next generation is selected by stochastic tournament or other selection methods.

In (22), $b_{r1}^k\ and\ x_i^k$ have to satisfy the following condition. Therefore, if the condition is not satisfied, The two variables are swapped.

$$f\left(b_{r1}^{k}\right) < f\left(x_{i}^{k}\right) \tag{8}$$

In (22), P is set to 0 or 1 by probability p as shown below.

$$P = \begin{cases} 1(rand(0,1) \leq p) \\ 0(rand(0,1) > p) \end{cases} \tag{9}$$

where, rand $(0, 1)$ is a uniform random number between 0 and 1.

3.2 Multi-population DEEPSO

This section expands the original DEEPSO to multi-population DEEPSO. In the multi-population model, one swarm of searching agents are divided into several sub-swarms and optimized calculation is performed at each sub-swarm. Searching agents are exchanged and replaced among sub-swarms every certain intervals. This process is called migration [11–14].

When implementing the model, the parameters which should be defined are shown below.

(1) Number of sub-swarms: the number of sub-swarms.
(2) Migration topology: the communication structure among sub-swarms when searching agents are migrated.
(3) Migration interval: how often migration occurs.
(4) Migration policy: which searching agents are migrated and which searching agents are replaced at the receiving side sub-swarm.

In the paper, unidirectional ring topology is applied as migration topology as shown in Fig. 2. The ring topology is unidirectional closed structure [11]. When *gbest* of sub-swarm i is migrated to the neighboring sub-swarm $i+1$, the worst current searching point of the sub-swarm $i+1$ is replaced to *gbest* of sub-swarm i. In (7), b_{r1}^{k} is selected among the stored *pbests* of all searching points.

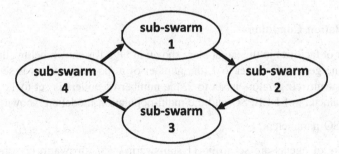

Fig. 2. An example of unidirectional ring topology.

3.3 Optimal Operational Planning of Energy Plants by Multi-population DEEPSO Algorithm

Step 1. Initial searching points of agents are generated by uniform random numbers within facility constraints.

Step 2. The object function value of each agent is calculated with the actual input and output values of the initial searching point. *pbests* and *gbest* are set with the objective function values.

Step 3. REPLICATION: Each agent is replicated (cloned) R times.

Step 4. MUTATION: Weights (A, B, and C) of all R clones and b_G are mutated by (2) to (5).

Step 5. REPRODUCTION: R + 1 offspring (the original and clones) are updated by (6) to (9).

Step 6. EVALUATION: Objective function values of all offspring are calculated with the transformed actual input and output values of facilities.

Step 7. SELECTION: The agent with the best fitness is selected to the next generation among the updated original and cloned agents.

Step 8. If the objective function value of each selected offspring is smaller than a past *pbest*, *pbest* is updated. *gbest* is also updated.

Step 9. When the current iteration number reaches the pre-determined maximum iteration number, the procedure can be stopped and the last *gbest* can be output as a solution. Otherwise, go to Step 4 and repeat the procedures or when the current iteration number reaches defined migration interval, the migration is performed as described in the previous section.

4   Simulations

In this paper, multi-population DEEPSO (the proposed method) is applied to a target plant in Fig. 1 and compared with various numbers of sub-swarms including single sub-swarm (the original DEEPSO).

4.1   Simulation Conditions

The number of facilities in the target plant shown in Fig. 1 is shown below: the number of gas turbine generators is set to 1, the number of turbo refrigerators is set to 1, the number of steam refrigerators is set to 2, the number of boilers is set to 1.

The parameters of DEEPSO and the multi-population model are shown below.

(1) DEEPSO parameters

The numbers of agents are set to 96 (1 sub-swarm), 48/sub-swarm (2 sub-swarms), 24/sub-swarm (4 sub-swarms), 12/sub-swarm (8 sub-swarms), 6/sub-swarm (16 sub-swarms). The maximum iteration number is set to 1000. The learning parameter (τ) is set to 0.2. The learning parameter (τ') is set to 0.006. Probability p is set to 0.8. The initial weight coefficients of each term (A, B, and C) are set to 0.5. The number of clones are set to 1.

(2) Multi-population model parameters

The number of sub-swarms are set to 1, 2, 4, 8, and 16. Migration topology is the unidirectional ring topology. Migration interval is set to 90.

The number of trials is set to 100 and averages and standard deviations are compared in the simulation.

4.2 Simulation Results

Table 1 shows comparison of the average objective function value rates and standard deviation rates by various number of sub-swarms when those values of the original DEEPSO based method (the number of sub-swarms is 1) are set to 100. As a result, when the number of sub-swarm is 8, the best average objective function values are obtained. And, when the number of sub-swarm is 16, the best standard deviations can be obtained. The best average objective function values means 0.82% cost reduction compared with the original DEEPSO based method. Since annual operational cost of energy plants are large, it is corresponding to a large impact to operation of factories and buildings. Figure 3 shows comparison of the average convergence characteristics by various number of sub-swarms. While, the original DEEPSO based method shows the best results at early iterations, all of the characteristics with more than one sub-swarm show better results at later iterations. When more number of sub-swarms are utilized, the number of agents in one sub-swarm is decreased. Therefore, since convergence is difficult to occur in each sub-swarm, the results by the multi-population based method is inferior to those by the original DEEPSO based method at early iterations. The multi-population model has an advantage which maintains diversification and intensification. Since each sub-swarm searches a different searching area, multi-pupation contributes to diversification. Using the migration function, the whole searching area is expanded and it can contribute to intensification as well. Consequently, the multi-population can improve the solution quality.

Table 1. Comparison of the average objective function value rates and standard deviation rates by various number of sub-swarms.

Num. of sub-swarms	Ave.	Std.
1	100	100
2	99.61	89.95
4	99.52	66.02
8	**99.18**	46.08
16	99.40	**44.56**

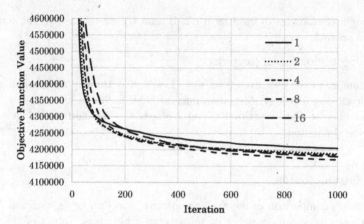

Fig. 3. Comparison of average convergence characteristics by various number of sub-swarms.

5 Conclusions

This paper presents optimal operation planning of energy plants by multi-population DEEPSO. The problem is formulated as a mixed integer nonlinear optimization problem and the proposed method can improve solution quality. Since annual operational cost of energy plants are large, it is corresponding to a large impact to operation of factories and buildings. It is verified that the improvement of solution quality can be realized using the appropriate number of sub-swarms.

As future works, fast computation using parallel and distributed computation and more improvement of the solution quality using other multi-population models are investigated.

References

1. Ministry of Economy, Trade, and Industry of Japan: White paper of annual report on energy 2015 (2016). http://www.enecho.meti.go.jp/about/whitepaper/2015pdf/whitepaper2015pdf_2_1.pdf (in Japanese)
2. Ravn, H., Rygaard, J.M.: Optimal scheduling of coproduction with a storage. J. Eng. **22**, 267–281 (1994)
3. Ito, K., Shiba, T., Yokoyama, R.: Optimal operation of a cogeneration plant in combination with electric heat pumps. J. Energ. Resour. Technol. **116**(1), 56–64 (1994)
4. Yokoyama, R., Ito, K.: A revised decomposition method for MILP problems and its application to operational planning of thermal storage systems. J. Energ. Resour. Technol. **118**, 277–284 (1996)
5. Tsukada, T., Tamura, T., Kitagawa, S., Fukuyama, Y.: Optimal operational planning for cogeneration system using particle swarm optimization. In: The 2003 IEEE Swarm Intelligence Symposium, pp. 138–143 (2003)
6. Wang, J., Zhai, Z.J., Jing, Y., Zhang, C.: Particle swarm optimization for redundant building cooling heating and power system. Appl. Energ. **87**, 3668–3679 (2010)

7. Suzuki, R., Kawai, F., Kitagawa, S., Matsui, T., Matsumoto, K., Xiang, D., Fukuyama, Y.: The ε constrained differential evolution approach for optimal operational planning of energy plants. In: IEEE Congress on Evolutionary Computation, pp. 1–6 (2010)

8. Nishimura, N., Fukuyama, Y., Matsui, T.: Optimal operational planning of energy plants by differential evolutionary particle swarm optimization. In: IEEE International Conference on Power Systems Technology, Wollongong (2016)

9. Committee of new exploration and industrial applications of computational intelligent system.: Optimization benchmark problems for industrial applications. Technical report, IEE of Japan, No. 1287, pp. 9–12 (2013) (in Japanese)

10. Miranda, V., Alves, R.: Differential Evolutionary Particle Swarm Optimization (DEEPSO): a Successful Hybrid. In: The 11th Brazilian Congress on Computational Intelligence, pp. 368–374 (2013)

11. Tang, J., Lim, M.H., Ong, Y.S., Er, M.J.: Study of migration topology in island model parallel hybrid-GA for large scale quadratic assignment problems. In: The Eighth International Conference on Control, Automation, Robotics and Vision, Special Session on Computational Intelligence on the Grid, Kunming, China (2004)

12. Chu, S.C., Pan, J.S.: Intelligent parallel particle swarm optimization alogorithms. Stud. Comput. Intell. (SCI) 22, 159–175 (2006)

13. Wang, D., Wu, C.H., Ip, A., Wang, D., Yan, Y.: Parallel multi-population particle swarm optimization algorithm for the uncapacitated facility location problem using OpenMP. In: IEEE Congress on Evolutionary Computation (CEC) Conference (2008)

14. Lorion, Y., Bogon, T., Timm, I.J., Drobnik, O.: An agent based parallel particle swarm optimization – APPSO. In: Symposium on Swarm Intelligence (SIS), pp. 52–59 (2009)

A Review on Hybridization of Particle Swarm Optimization with Artificial Bee Colony

Bin Xin[1,2,3]($\boxtimes$), Yipeng Wang[1], Lu Chen[1], Tao Cai[1,2], and Wenjie Chen[1,2]

[1] School of Automation, Beijing Institute of Technology,
Beijing 100081, China
brucebin@bit.edu.cn
[2] Key Laboratory of Intelligent Control and Decision of Complex Systems,
Beijing Institute of Technology,
Beijing 100081, China
[3] Beijing Advanced Innovation Center for Intelligent Robots and Systems,
Beijing 100081, China

Abstract. Particle swarm optimization (PSO) and artificial bee colony (ABC) are two formidable population-based optimizers inspired by swarm intelligence(SI). They follow different philosophies and paradigms, and both are successfully and widely applied in scientific and engineering research. The hybridization of PSO and ABC represents a promising way to create more powerful SI-based hybrid optimizers, especially for specific problem solving. In the past decade, numerous hybrids of ABC and PSO have emerged with diverse design ideas from many researchers. This paper is aimed at reviewing the existing hybrids based on PSO and ABC and giving a classification and an analysis of them.

Keywords: Particle swarm optimization (PSO) · Artificial bee colony (ABC) · Hybridization · Review · Swarm intelligence

1 Introduction

The concept of swarm intelligence (SI) derives from the observation of swarms of insects in nature. The swarm refers to a group of individuals that can communicate with each other directly or indirectly, and such swarm can solve complicated problems using decentralized control and self-organization of relatively simple individuals [1,2]. Optimization techniques inspired by SI (i.e. swarm intelligence optimization algorithms, SIOAs), which inherits the features of SI, are population-based stochastic methods mainly for solving combinatorial optimization problems. Besides, SIOAs have become increasingly popular during the last

This work was supported in part by the National Natural Science Foundation of China under Grant 61673058, U1609214 and 61304215, in part by the Foundation for Innovative Research Groups of the National Natural Science Foundation of China under Grant 61321002, and in part by the Research Fund for the Doctoral Program of Higher Education of China under Grant 20131101120033.

© Springer International Publishing AG 2017
Y. Tan et al. (Eds.): ICSI 2017, Part II, LNCS 10386, pp. 242–249, 2017.
DOI: 10.1007/978-3-319-61833-3_25

decade, and plenty of optimizers have been proposed. Particle swarm optimization (PSO), ant colony optimization (ACO), and artificial bee colony (ABC) are among the most representative SIOAs.

In the past decade, hybrid optimizers have attracted persistent attention from scholars that are interested in design of optimizers and their applications [3–8]. As Raidl claimed in his unified view of hybrid meta-heuristics [8], it seems that choosing an adequate hybrid approach is determinant to achieve top performance in solving most difficult problems. A common template for hybridization is provided by memetic algorithms (MAs), which combine the respective advantages of global search and local search (LS) [9]. Due to excellent performance, MAs have been favored by many scholars in their research on different optimization problems [9]. However, MAs only represent a special class in the family of hybrid optimizers. There are manifold possibilities of hybridizing different optimizers, which follow diverse philosophies and paradigms. This paper is aimed at giving a classification and an analysis of various hybrid optimizers based on PSO and ABC by the systematic taxonomy we proposed in a recent work [10].

The rest of this paper is structured as follows. In Sect. 2, we present the simple introduction of PSO and ABC. A systematic taxonomy of hybridization strategies is described briefly in Sect. 3. Section 4 presents different hybrids based on ABC and PSO. Finally, the conclusion is drawn in Sect. 5.

2 Particle Swarm Optimization and Artificial Bee Colony

2.1 Particle Swarm Optimization

The particle swarm optimization algorithm is proposed by Kennedy and Eberhart in 1995, where a swarm stands for the population and the swarm consists of a certain amount of individuals called particles. Originally, PSO was inspired by the social and cognitive behavior of animal groups, such as bird flocks, fish schools and so on. During the past decade, PSO has been successfully and widely applied in the practice of science and engineering, which demonstrates the superiority of this algorithm. In the PSO model, each particle has its own current position $\mathbf{X}_i$ (which represents a solution), current velocity $\mathbf{V}_i$ and the precious best position $\mathbf{pbest}_i$. Particles accumulate their own experiences about the problem space, and learn from each other according to their fitness values as well.

Suppose that the search space of the problem is D-dimensional, then the iteration equations for the velocity and position in standard PSO are given as follows:

$$v_{i,d}^{k+1} = \omega \cdot v_{i,d}^k + c_1 \cdot \text{rand1} \cdot (pbest_{i,d}^k - x_{i,d}^k) + c_2 \cdot \text{rand2} \cdot (gbest_d^k - x_{i,d}^k), \quad (1)$$

$$x_{i,d}^{k+1} = x_{i,d}^k + v_{i,d}^{k+1}, \quad (2)$$

where $\mathbf{X}_i^k = [x_{i,1}^k, x_{i,2}^k, \ldots, x_{i,D}^k]$ and $\mathbf{V}_i^k = [v_{i,1}^k, v_{i,2}^k, \ldots, v_{i,D}^k]$ represent the position and velocity of the ith particle at the kth generation ($i = 1, 2, ..., PS$; PS

is the population size), respectively; $\mathbf{pbest}_i^k = [pbest_{i,1}^k, pbest_{i,2}^k, \ldots, pbest_{i,D}^k]^T$ is the best position that is found so far by the ith particle; $\mathbf{gbest}^k = [gbest_1^k, gbest_2^k, \ldots, gbest_D^k]^T$ is the global best position that is found by particles in the swarm; and $rand1$ and $rand2$ are the random numbers that are uniformly distributed in $[0, 1]$; ω is the so-called inertia weight; c_1 and c_2 are acceleration coefficients, which are also termed as the cognitive factor and the social factor, respectively. The velocity of the particles on each dimension is clamped to the range $[-V_{max}, V_{max}]$.

2.2   Artificial Bee Colony

The artificial bee colony algorithm simulating the foraging behavior of honey bees was developed by Karaboga to solve numerical optimization problems in 2005. The bee colony is a complicated natural society with specialized social divisions, and the ABC just assumes a simplified model composed by three groups of bees: *employed bees*, *onlooker bees* and *scout bees*. Each employed bee is assigned to a food source, and each onlooker bee waits in the hive and chooses a food source depending on the information shared by the employed bees. Besides, the scout bees will search for new food sources surrounding the hive. The food sources are the solutions of the optimization problem and the bees are the variation operators. The exchange of information among bees is the most important occurrence in the formation of the collective knowledge.

Suppose that the solution space of the problem is D-dimensional, then ABC will start with producing food sources randomly, and each food source stands for a candidate solution $\mathbf{X}_i = [x_{i1}, x_{i2}, \ldots, x_{iD}], i \in \{1, 2, \ldots, N_s\}$. N_s is equal to the number of food sources and half the population size.

Employed Bee Phase. Each employed bee generate a new food source v_i by a modification on the position of the old food source, which is shown as follows:

$$v_{i,d} = x_{i,d} + \psi \cdot (x_{i,d} - x_{k,d}), \tag{3}$$

where $k \in \{1, 2, \ldots, N_s\}, k \neq i$, x_k is a food source randomly selected in the neighbor of the ith food source; and ψ is is a random number in the range $[-1, 1]$.

Onlooker Bee Phase. An artificial onlooker bee determines which food source to forage according to the probability value P_i of those food sources shared by the employed bees. The probability value can be calculated by the following equation:

$$P_i = \frac{f_i}{\sum_{k=1}^{N_s} f_k}, \tag{4}$$

where f_i is the fitness value of the ith food source. After the selection, the onlooker bees also generate new food sources as described in Eq. 3.

Scout Bee Phase. If a food source position cannot be further improved within a finite steps, it will be abandoned and the corresponding employed bee will become a scout bee. Then the scout bee will search randomly for new food source position which is obtained as follows:

$$x_{i,d} = x_d^{min} + \text{rand} \cdot (x_d^{max} - x_d^{min}),$$ (5)

where $rand$ is random number selected in the range [0.1]; x_d^{max} and x_d^{min} are the upper and lower borders of the dth dimension of the solution space.

3 Taxonomy on Hybridization Strategies

A favorable taxonomy of hybridization strategies should be capable of differentiating different strategies as well as providing designers a convenient and efficient means of determining a hybridization scheme. In this section, we will introduce the elements of a hybridization strategy (i.e., hybridization factors) and the taxonomy, which we proposed in recent work [10].

Different hybrids can be differentiated by the relationship between parent optimizers (PR), hybridization level (HL), operation order (OO), type of information transfer (TIT), and type of transferred information (TTI). And we will introduce some concise notations to express the candidate classes with respect to each factor as follows.

(1) Parent relationship (PR)
    Collaboration: $\langle C \rangle$, Embedding: $\langle E \rangle$, Assistance: $\langle A \rangle$.
(2) hybridization level (HL)
    Population level: $\langle P \rangle$, Subpopulation level: $\langle S \rangle$, Individual level: $\langle I \rangle$, Component level: $\langle C \rangle$.
(3) Operating order (OO)
    Sequential order: $\langle S \rangle$, Parallel order: $\langle P \rangle$, No order: $\langle N \rangle$.
(4) Type of information transfer (TIT)
    Simplex TIT: $\langle S \rangle$, Duplex TIT: $\langle D \rangle$.
(5) Type of transferred information (TTI).
(6) Solutions: $\langle S \rangle$, Fitness information: $\langle F \rangle$, Solution components: $\langle S_c \rangle$, Auxiliaries: $\langle A \rangle$, Control parameters: $\langle C_p \rangle$, Algorithm-induced in-betweens: $\langle A_i \rangle$.

4 Previous PSOABCs

In this section, we will introduce some typical hybrids based on PSO and ABC in the literature, and give a classification and a simple analysis according to the taxonomy aforementioned. Since the tradeoff between **exploration** and **exploitation** (Tr:Er&Ei) is a core of all kinds of optimizers [11], we primarily divide the hybrids into two parts according to the combination patterns that involve global search (GS) and local search (LS). Besides, it is imperative to implement certain type of global search for solving complex problems, and SIOAs can be used for both GS and LS in solution space. Therefore, the hybrids considered here include two main patterns: "GS⊕GS" and "GS⊕LS".

In order to build a unified nomenclature and differentiate various hybrids with the same parents conveniently, we will name a given hybrid by combining the initial of the last name of each inventor of the hybrid into a suffix that follows the parent algorithms.

4.1   GS⊕GS

Although the tradeoff between exploration and exploitation has been considered in SIOAs, the stagnation phenomenon caused by the weakness of a sole algorithm still could not be prevented. Therefore, the methods with GS ability can be combined with any SIOA to design a new optimizer. A rational motivation behind this hybridization is that the two or more optimizers correspond to different landscapes, which may give birth to some shortcuts to escape from the local optima [12], even though both of them are implemented as GS.

El-Abd proposed a hybrid optimizer combining PSO and ABC in order to gain benefits from their respective strengths [13]. This optimizer is denoted by PSOABC-E, and it incorporates an ABC component into a standard PSO which updates the **pbest** information of the particles in every iteration using the ABC update equation. This component is added to the standard PSO after the main loop. For every particle i in the swarm, the ABC update equation is applied to its personal best $\mathbf{pbest}_i = \{pbest_{i,1}, pbest_{i,2}, \ldots, pbest_{i,D}\}$, which is given as follows:

$$pbest_{ij} = pbest_{ij} + \phi_{ij} \times (pbest_{ij} - pbest_{kj}), \tag{6}$$

where j is a randomly selected number in $[1, D]$ and D is the number of dimensions, and ϕ_{ij} is a random number uniformly distributed in the range $[-1, 1]$, and k is the index of a randomly chosen solution. The new $\mathbf{pbest}_i$ replaces the previous one if it has a better fitness. According to our taxonomy, the type of the hybridization strategy in PSOABC-E is $\langle C, C, S, D \rangle$.

Besides, Sharma *et al.* proposed another PSOABC hybrid [14], and it is termed as PSOABC-SPB. In this PSOABC-SPB, a modified method derived from the velocity update equation of PSO is presented for solution update of the employed as well as onlooker bees, respectively. According to our taxonomy, the type of the hybridization strategy in PSOABC-SPB is $\langle C, P, S, D \rangle$.

Moreover, Shi *et al.* developed a PSOABC hybrid [15], and it is named PSOABC-SLLGWL. The approach is initialized by two sub-systems of PSO and ABC, and then they are executed in parallel. During the two sub-systems are executing, two information exchanging processes are introduced into the system. These processes are called by Information Exchanging Process1 and Information Exchanging Process2, respectively. The first one forwards "better information" from particle swarm to bee colony, and the second is reversed. According to our taxonomy, the type of the hybridization strategy in PSOABC-SLLGWL is $\langle C, P, P, D \rangle$. The hybridization strategies of the hybrids based on 'GS⊕GS' are listed in Table 1.

In addition, there are also various other versions based on PSO and ABC in the literature, and hybridization strategies are similar with those aforementioned. As space is limited, these optimizers will not be covered here.

Table 1. The hybridization strategies of the hybrids based on 'GS⊕GS'

Name	$Parent_A$	$Parent_B$	PR	OL_A	OL_B	HL	OO	TIT	TTI
PSOABC-E	PSO	ABC	C	P	C	C	S	D	S, F, S_c
PSOABC-SPB	PSO	ABC	C	P	P	P	S	D	S, F
PSOABC-SLLGWL	PSO	ABC	C	P	P	P	P	D	S, F
PSOABC-AK [16]	PSO	ABC	C	P	P	P	S	D	S, F
PSOABC-MRR [17]	PSO	ABC	C	I	P	I	S	D	S, F
PSOABC-BM [18]	PSO	ABC	C	P	P	P	S	D	S, F
PSOABC-AKS [19]	PSO	ABC	C	S	S	S	P	D	S, F
PSOABC-VRB [20]	PSO	ABC	C	P	P	P	S	D	S, F

4.2 GS⊕LS

Since the ABC can also be implemented as local search, some PSOABC hybrids are designed follow the concept of MA as well. Li *et al.* proposed a hybrid algorithm denoted by PSOABC-LWYL, which combined the local search phase in PSO with two global search phases in ABC for the global optimum [21]. In the iteration process, the algorithm examines the aging degree of **pbest** for each individual to decide which type of search phase (PSO phase, onlooker bee phase, and modified scout bee phase) to adopt. According to our taxonomy, the type of the hybridization strategy in PSOABC-LWYL is $\langle C, I, S, D \rangle$.

Similarly, Alqattan and Abdullah proposed a PSOABC hybrid optimizer named PSOABC-AA [22]. In this PSOABC-AA, the employed bees are eliminated in the process of ABC. Instead, the particle movement process of PSO is applied to the local search. According to our taxonomy, the type of the hybridization strategy in PSOABC-AA is $\langle C, P, S, D \rangle$. The hybridization strategies of the hybrids based on 'GS⊕LS' are listed in Table 2.

Table 2. The hybridization strategies of hybrids based on 'GS⊕LS'

Name	Parent A	Parent B	PR	OL A	OL B	HL	OO	TIT	TTI
PSOABC-LWYL	ABC	PSO	C	I	I	I	S	D	S, F
PSOABC-AA	ABC	PSO	C	P	P	P	S	D	S, F
PSOABC-WLS [23]	ABC	PSO	E	P	P	P	N	D	S
PSOABC-BDAA [24]	ABC	PSO	C	P	P	P	P	D	S, F
PSOABC-XPZCLZ [25]	ABC	PSO	C	P	P	P	S	D	S, F

5 Conclusion

As can be seen from the existing hybrids based on PSO and ABC, designers have manifold choices to design a hybrid optimizer, and most of the scholars either update onlooker phase or employed bees phase by single update equation of PSO or with other features. Besides, the "GS⊕GS" pattern seems to be more common in the hybridization of PSO and ABC. Nevertheless, more hybridization strategies should be taken into account for the design of a new PSOABC hybrid.

The tradeoff between exploration and exploitation is a core of all kinds of optimizers. Generally, a hybrid follows the hybridization pattern of the MAs can generate a feasible solution to the considered problem. Thus, more PSOABCs which follows the hybridization pattern of the MAs deserve further research.

Besides, the "GS⊕GS" hybridization pattern also contributes to seeking for a more suitable optimizer for a specific problem. Even though the validity of a hybrid optimizer that follows the "GS⊕GS" hybridization pattern cannot be ensured, a rational motivation behind such hybridization is that the two or more optimizers correspond to different landscapes, which can give birth to a new search paradigm and may suppress the stagnation during the iteration process that arises from the limitation of a sole search method [12].

References

1. Dorigo, M., Birattari, M.: Swarm intelligence. Scholarpedia **2**(9), 1462 (2007)
2. Hinchey, M.G., Sterritt, R., Rouff, C.: Swarms and swarm intelligence. Computer **40**(4), 111–113 (2007)
3. Andrea, R., Blesa, M., Blum, C., Michael, S.: Hybrid Metaheuristics-an Emerging Approach to Optimization. Springer, Heidelberg (2008)
4. Voß, S.: Hybridizing metaheuristics: the road to success in problem solving. In: Proceedings of the 8th EU/MEeting on Metaheuristics in the Service Industry, Stuttgart (2007)
5. Talbi, E.-G.: A taxonomy of hybrid metaheuristics. J. Heuristics **8**(5), 541–564 (2002)
6. Gendreau, M., Potvin, J.-Y.: Metaheuristics in combinatorial optimization. Ann. Oper. Res. **140**(1), 189–213 (2005)
7. Blum, C., Roli, A.: Metaheuristics in combinatorial optimization: overview and conceptual comparison. ACM Comput. Surv. (CSUR) **35**(3), 268–308 (2003)
8. Raidl, G.R.: A unified view on hybrid metaheuristics. In: Almeida, F., Blesa Aguilera, M.J., Blum, C., Moreno Vega, J.M., Pérez Pérez, M., Roli, A., Sampels, M. (eds.) HM 2006. LNCS, vol. 4030, pp. 1–12. Springer, Heidelberg (2006). doi:10.1007/11890584_1
9. Krasnogor, N., Smith, J.: A tutorial for competent memetic algorithms: model, taxonomy, and design issues. IEEE Trans. Evol. Comput. **9**(5), 474–488 (2005)
10. Xin, B., Chen, J., Zhang, J., Fang, H., Peng, Z.-H.: Hybridizing differential evolution and particle swarm optimization to design powerful optimisers: a review and taxonomy. IEEE Trans. Syst. Man Cybern. Part C (Appl. Rev.) **42**(5), 744–767 (2012)

11. Chen, J., Xin, B., Peng, Z., Dou, L., Zhang, J.: Optimal contraction theorem for exploration-exploitation tradeoff in search and optimization. IEEE Trans. Syst. Man Cybern. Part A: Syst. Hum. **39**(3), 680–691 (2009)
12. Jones, T.: One operator, one landscape, Santa Fe Institute Technical report, 95-02-025 (1995)
13. El-Abd, M.: A hybrid ABC-SPSO algorithm for continuous function optimization. In: IEEE Symposium on Swarm Intelligence (SIS), pp. 1–6. IEEE (2011)
14. Sharma, T.K., Pant, M., Bhardwaj, T.: PSO ingrained artificial bee colony algorithm for solving continuous optimization problems. In: 2011 IEEE International Conference on Computer Applications and Industrial Electronics (ICCAIE) (2011)
15. Shi, X., Li, Y., Li, H., Guan, R., Wang, L., Liang, Y.: An integrated algorithm based on artificial bee colony and particle swarm optimization. In: 2010 Sixth International Conference on Natural Computation (ICNC), vol. 5, pp. 2586–2590. IEEE (2010)
16. Altun, O., Korkmaz, T.: Particle swarm optimization-artificial bee colony chain (PSOABCC): a hybrid meteahuristic algorithm. In: Scientific Cooperations International Workshops on Electrical and Computer Engineering Subfields, pp. 22–23, August 2014
17. Muthiah, A., Rajkumar, A., Rajkumar, R.: Hybridization of artificial bee colony algorithm with particle swarm optimization algorithm for flexible job shop scheduling. In: Proceedings of 2016 International Conference on Energy Efficient Technologies for Sustainability (ICEETS), pp. 896–903. IEEE (2016)
18. Baktash, N., Meybodi, M.: A new hybridized approach of PSO and ABC for optimization. In: Proceedings of the 2011 International Conference on Measurement and Control Engineering, pp. 69–80 (2011)
19. Amudha, P., Karthik, S., Sivakumari, S.: A hybrid swarm intelligence algorithm for intrusion detection using significant features. Sci. World J. **2015**, 15 (2015). doi:10.1155/2015/574589. Article ID 574589
20. Vitorino, L., Ribeiro, S., Bastos-Filho, C.J.: A mechanism based on artificial bee colony to generate diversity in particle swarm optimization. Neurocomputing **148**, 39–45 (2015)
21. Li, Z., Wang, W., Yan, Y., Li, Z.: PS-ABC: a hybrid algorithm based on particle swarm and artificial bee colony for high-dimensional optimization problems. Expert Syst. Appl. **42**(22), 8881–8895 (2015)
22. Alqattan, Z.N., Abdullah, R.: A hybrid artificial bee colony algorithm for numerical function optimization. Int. J. Mod. Phys. C **26**(10), 1550109 (2015)
23. Chun-Feng, W., Kui, L., Pei-Ping, S.: Hybrid artificial bee colony algorithm and particle swarm search for global optimization. Math. Probl. Eng. **2014**, 8 (2014). doi:10.1155/2014/832949. Article ID 832949
24. Bouaziz, S., Dhahri, H., Alimi, A.M., Abraham, A.: Evolving flexible beta basis function neural tree using extended genetic programming & hybrid artificial bee colony. Appl. Soft Comput. **47**, 653–668 (2016)
25. Xiang, Y., Peng, Y., Zhong, Y., Chen, Z., Lu, X., Zhong, X.: A particle swarm inspired multi-elitist artificial bee colony algorithm for real-parameter optimization. Comput. Optim. Appl. **57**(2), 493–516 (2014)

A Study on Greedy Search to Improve Simulated Annealing for Large-Scale Traveling Salesman Problem

Xiuli Wu[✉] and Dongliang Gao

Department of Logistics Engineering, School of Mechanical Engineering,
University of Science and Technology Beijing, Beijing, China
wuxiuli@ustb.edu.cn, 1144943582@qq.com

Abstract. Traveling salesman problem (TSP) is a typical NP-hard problem. How to design an effective and efficient algorithm to solve TSP within a limited time is of great theoretical significance and practical significance. This paper studies how the greedy search improves simulated annealing algorithm for solving large-scale TSP. First, the TSP formulation is presented. The aim of the TSP is to structure a shortest route for one traveling salesman starting from a certain location, through all the given cities and finally returning to the original city. Second, a simple simulated annealing (SA) algorithm is developed for the TSP. The orthogonal test is employed to optimize the key parameters. Third, a group of benchmark instances are tested to verify the performance of the SA. The experimental results show that for the small-scale and medium-scale instances the simply SA can search the optimal solution easily. Finally, to solve the large-scale instance, we integrate a greedy search to improve SA. A greedy coefficient is proposed to control the balance of the exploration and the exploitation. Different levels of the greedy coefficient are tested and discussed. The results show that the greedy search can improve SA greatly with a suitable greedy coefficient.

Keywords: Simulated annealing algorithm · Greedy search · Traveling salesman problem · Large-scale instances

1 Introduction

With the rapid development of e-commerce in China, the logistics route optimization problem is attracting increasing concerns [1]. As the simplest route optimization problem, the traveling salesman problem (TSP) has remained an interesting problem for a long time. The TSP presents the task of finding an optimum path through a set of given locations (cities), such that each location is visited only once, and the salesman returns to the start location [2, 3]. This problem is a typical NP-hard problem whose computational complexity increases exponentially with the number of cities.

The TSP has been studied for decades. Many heuristics or metaheuristics were proposed to find the optimal or the near-optimal solutions, including Tabu Search (TS) [4], Simulated Annealing algorithm (SA) [5], Genetic Algorithm (GA) [6], Ant Colony Optimization (ACO) [7], Particle Swarm Optimization (PSO) [8], Artificial Immune System (AIS) [9], Artificial Neural Network (ANN) [10], Elastic Net (EN) [3] et al..

© Springer International Publishing AG 2017
Y. Tan et al. (Eds.): ICSI 2017, Part II, LNCS 10386, pp. 250–257, 2017.
DOI: 10.1007/978-3-319-61833-3_26

In this paper, we study the performance of the simulated annealing algorithm for solving the TSP. The main contribution lies in: (1) an improved simulated annealing algorithm integrating greedy algorithm is developed to avoid prematurity for large-scale instances and (2) the improved algorithm takes into account the searching efficiency and the quality of the solution simultaneously.

The rest of this paper is organized as follows. Section 2 formulates the TSP problem. Section 3 proposes a SA to solve TSP. Section 4 reports the experimental results. Section 5 proposes a greedy searchto improve the SA for solving large-scale instance. Section 6 concludes the paper.

2 The Formulation for the TSP

The TSP is to determine a closed tour for the traveling salesman to visit the cities once and only once. It is not hard to compute that the number of the candidate routes equals $(n - 1)!$. When n is large, it becomes huge. The study on the TSP is important because it is not only used to solve the route optimization problem but also employed to many other combinatorial optimization problems such as the flow shop scheduling problem [11].

Let n be the number of cities, $G = (V, E)$ is a directed graph in which $V = \{1, 2, \ldots, n\}$ is the set of vertices and $E = (i,j)|i, j \in V$ is the set of arcs. c_{ij} $(c_{ij} > 0, i \neq j)$ is the distance between city i and city j. K is the number of all the nonempty subset of V. The binary variable x_{ij} determines whether the arc (i,j) is visited by the tour. The TSP can be formulated as follows:

$$\min f = \sum_{i=1}^{n} \sum_{j=1}^{n} c_{ij} x_{ij}. \tag{1}$$

s.t

$$\sum_{j=1}^{n} x_{ij} = 1, i = 1, 2, \ldots, n. \tag{2}$$

$$\sum_{i=1}^{n} x_{ij} = 1, j = 1, 2, \ldots, n. \tag{3}$$

$$x_{i_1 i_2} + x_{i_3 i_4} + \ldots + x_{i_k i_1} \leq K - 1, i_1.i_2, \ldots, i_k = 1, 2, \ldots, n, i_1 \neq i_2 \neq \ldots \neq i_k,$$

$$k = 2, 3, \ldots, n - 1. \tag{4}$$

$$x_{ij} = 0 \text{ or } 1. \tag{5}$$

The Eq. (1) is the objective function to minimize the tour length. The constraint (2) and (3) ensure that each city can be visited only once. The constraint (4) ensures that there doesn't exist a tour that visits only K cities.

3 The Simple Simulated Annealing Algorithm for TSP

3.1 The Simple SA for TSP

The simulated annealing (SA) algorithm is motivated by the metal annealing process [8]. The searching process explores the solution space with the decreasing temperature. At each temperature point, it exploits the neighborhood and accepts the best neighbor according to Metropolis rule. Before we propose the improved algorithm, we first discuss the performance of the simple SA.

The flowchart of the simple SA is as follows.

Step 1. Generate an initial state s and initialize the parameters, including the initial temperature T_s, the terminated temperature T_f, a coefficient α to control the temperature decreasing rate and the Metropolis sampling amount L. let the current temperature $t = T_s$.

Step 2. Repeat the following steps L times.

Step 2.1. Generate a new state s'.

Step 2.2. Decide whether to accept s' with the Metropolis rule. If $f(s) \geq f(s')$, accept the solution s' with 100% probability; Otherwise, accept the solution s' with a probability.

Step 3. Decrease the temperature $t = \alpha t$ and return to Step 2.

Step 4. Terminate when $t < t_f$ and output the final solution as the optimal solution.

3.2 Presentation

Before we employ the simple SA to solve TSP, we need to represent the TSP to bridge the TSP with the SA. For the TSP, we can use a number to represent a city, so a tour is a permutation of n number. Take the example of a TSP with 5 cities, a feasible presentation is "5 3 1 2 4", which means that the tour starts from city 5, visits city 2, 1,3, 4 in turn and finally returns to city 5.

3.3 The Neighbor Generating Function

A neighbor generating function is needed to generate a new state at a certain temperature. Here the swap method is employed. In this method, two randomly selected city indexes are swapped. E.g. the tour in the abovementioned example "5 3 1 2 4", if the position 2 and 3 are chosen randomly, the neighbor is generated to be "5 1 3 2 4".

3.4 The Temperature Decreasing Function

After the full exploitation at one temperature, the temperature decreases with the following decreasing function in Eq. (6).

$$t_{i+1} = \alpha t_i. \tag{6}$$

Where α is a coefficient and $0 \leq \alpha \leq 1$.

3.5   Metropolis Rule

In SA, Metropolis Rule is used to determine whether to accept the new state. Compare the new state with the current state. If the new state dominates the current state, accept it; otherwise, accept the new state with an acceptance probability. The Metropolis rule can be described as follows.

1. If $f(i) \geq f(j)$, the new state j is accepted and the current best state $i = j$;
2. If $f(i) < f(j)$, the new state j will be accepted with a probability. if $\Delta f(x)/t >$ random [0,1], the state j will be accepted and set $i = j$; Otherwise, reject the state j and keep the current best state unchanged.

3.6   The Terminal Condition

At one certain temperature, the search process exploits its neighborhood L times. The whole process will terminate when the temperature decreases to the predefined final temperature.

4   The Experiment Results with SA

To study the performance of the simple SA, we test the benchmark instances from literatures [10]. Before that, the orthogonal test is employed to optimize the parameters first.

4.1   Parameters Setting

There are four parameters in the simple SA: the initial temperature (Ts), the terminal temperature (t_f), the Cooling coefficient (α) and the exploiting times (L) at one certain temperature. We set 3 levels for each parameter so the L9(3^4) orthogonal table [11] (Table 1) can be used. For each setting, we run the SA 10 times to test the Oliver30 instance. The results of the orthogonal experiment are reported in Table 2. From the

Table 1. The orthogonal table

	L	α	T_f	Ts
1	200	0.9	0.1	10000
2	200	0.95	0.01	1000
3	200	0.99	0.001	100
4	500	0.9	0.01	100
5	500	0.95	0.001	10000
6	500	0.99	0.1	1000
7	1000	0.9	0.001	1000
8	1000	0.95	0.1	100
9	1000	0.99	0.01	10000

results, we can see that the 9th group parameter setting is the optimal. Hence we choose them for the simple SA.

4.2   Results for the Benchmark Instances

Five benchmark instances including two small-scale instances (Oliver30 and Att48), two medium-scale instances (Bier127 and Ch130) and one large-scale instance (Fl1577) are tested to verify the performance of the SA. The results are reported in Table 3. It can be concluded that for the small-scale instances (Oliver30 and Att48) SA can easily search the optimal solution within a limited iterations. For the medium-scale instances (Bier127 and Ch130), the error percent increases a little. While for the large-scale instance (Fl1577) the known optimal solution is 22249 and the SA only found a 205098 long tour. The error percent is 862%. There exists many intersected sub-routes in the tour and there also exists some cities which choose a remote city as the next destination. This also shows that the simple SA performs worse and worse with the instances' scale. Hence it's necessary to improve the SA to deal with the large-scale TSP instances.

Table 2. The results of the orthogonal test

	1	2	3	4	5	6	7	8	9	10
1	482.71	490.30	519.20	492.00	480.20	501.00	487.20	472.40	463.70	476.00
2	461.40	466.00	502.00	452.10	434.10	502.50	455.70	443.20	487.80	499.10
3	423.94	449.90	423.94	472.10	423.94	445.97	461.79	424.69	457.90	447.83
4	424.90	495.10	452.16	490.03	452.93	490.80	498.79	488.19	429.38	476.47
5	423.74	425.27	423.74	430.03	425.48	446.99	470.02	429.59	424.90	450.80
6	425.31	424.69	423.94	425.27	450.27	424.69	468.24	425.10	425.10	453.91
7	424.69	478.37	462.75	466.90	425.48	425.51	423.74	462.97	456.27	491.11
8	424.69	423.94	433.09	423.74	451.81	425.48	465.68	424.90	429.38	480.14
9	462.74	423.74	425.50	423.94	425.51	471.96	423.94	423.74	429.83	424.69

5   The Greedy Search to Improve SA for TSP

From the experiment result (Fig. 2), we can see that many city nodes connect to a remote city and thus make a lot of detours. Hence, we consider integrating a greedy search into the SA to avoid the detours. For convenience, we name this hybrid algorithm as g-SA. The greedy strategy is that each city should connect to the nearest city in a tour. The greedy strategy is designed as follows.

Step 1: we first use the simple SA to generate a visit route S. Another route G is generated with the greedy strategy with the same beginning node as in S.
Step2: for each connection, compare the distance between the S and G. Assume the $S(i) = m$ and the $S(i + 1) = n$. Then we locate the node m in G and find its next connection node o. Next, we compare the distance $d(m,n)$ between m and n and the distance $d(m,o)$ between m and o. If the $d(m,n) > fd(m,o)$, swap the city n with the

Table 3. The results for the 5 benchmark instances

Instances	Oliver30	Att48	Bier127	Ch130	Fl1577
1	462.74	34735	141979	8186	213469
2	423.74	34549	148963	7513	211399
3	425.5	35518	141282	7488	216892
4	423.94	34954	138175	8072	205098
5	425.51	34946	141979	7830	220564
6	471.96	35227	138176	7459	210584
7	423.94	35250	139845	7649	213261
8	423.74	34360	141018	7976	218573
9	429.83	34375	141283	8245	217980
10	424.69	36582	145763	7893	212811
average	433.56	35049.6	141846.3	7831.1	214129.1
max	471.96	36582	148963	8245	220564
min	423.74	34360	138175	7459	205098
The optimal	423.74	33522	118282	6110	22249
The error %	2.31%	4.60%	19.90%	28.20%	862%

city o in S. Otherwise, keep the route S unchanged. Here f is a greedy coefficient, which will be optimized later.

To explain the g-SA more clearly, take a TSP with 5 cities as an example. A route S "5 3 1 2 4" is the shortest tour sequence with SA. Another route G "5 1 4 3 2" is generated with the greedy strategy. If $d(5,3) > 2d(5,1)$(here, we assume $f = 2$),swap the "3" and "1" in S. and then compare $d(1,3)$ and $d(1,4)$ and determine whether to swap "1" and "4". Repeat until all the cities in S are all compared.

The greedy coefficient f is a decisive parameter to control whether to adjust the route from the SA. To study its influence, we carry out a group of experiments. The instance Fl1577 is used to test the influence of the parameter. The results are reported in Table 4. First, we set $f = 2$ and run 10 times with SA and the g-SA algorithm. The average result with SA is 214969, while it reduces to 31027 with g-SA. The average optimization is increased by 86.7%.

The results are so inspiring that we determine to continue in this direction. We set $f = 1.5$ and the average result is 28981. It is improved again. So we set $f = 1.3$. The average value decreases to 27829. Does it continue to decrease with a deceasing f value? To answer this question, we do the last test. We set $f = 1.1$, the average value is 27250 and the improvement is little. Therefore, we stop testing due to the long computing time.

On the other hand, we increase the value of the greedy coefficient. We set $f = 3$. The average value is 31181. It becomes worse again. Therefore, it can be concluded that the optimal value for the parameter is near one for large-scale instance. This conclusion can be verified from the route in Fig. 2. There are few detours in the optimized route. By comparing Figs. 1 and 2, we can see that the greedy search improve the performance of the simple SA greatly.

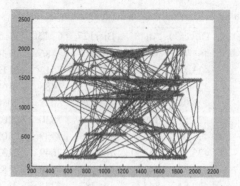

Fig. 1. The route for Fl1577 instance generated with SA

Table 4. The results for different levels of the greedy coefficient

No.	1	2	3	4	5	6	Average
SA	214576	222850	217988	213534	206909	213958	214969
g-SA ($f = 2$)	30064	32705	35014	29664	28778	29937	31027
g-SA ($f = 1.5$)	27697	29253	31633	31676	26996	31709	28981
g-SA ($f = 1.3$)	27117	27233	27389	29382	27250	28601	27829
g-SA ($f = 1.1$)	29742	27170	28887	25298	26824	29516	27250
g-SA ($f = 3$)	33003	34046	32152	28286	33358	33887	31181

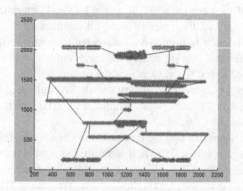

Fig. 2. The route with g-SA ($f = 1.3$)

6 Conclusion

We studied the performance of SA to solve the TSP. To begin with, we found that there is big difference for different scale instances. For small and medium scale instances, the simple SA can search the optimal solution easily. But for the large scale, it is very hard to search the optimal solution within a limited time. Hence, we integrate a greedy search to improve the SA. A greed coefficient is presented to balance the exploration and the exploitation. The experiment results show that the greedy strategy can improve the performance of SA greatly.

Acknowledgement. This paper was partly supported by the National Natural Science Foundation of China under Grant (Grant No.51305024).

References

1. Cui, M., Pan, S., Newell, S., Cui, L.: Strategy, resource orchestration and e-commerce enabled social innovation in Rural China. J. Strateg. Inf. Syst. (2016). http://dx.doi.org/10.1016/j.jsis.2016.10.001
2. Durbin, R.: An analogue approach to the travelling salesman. Nature **326**, 16 (1987)
3. Durbin, R., Szeliski, R., Yuille, A.: An analysis of the elastic net approach to the traveling salesman problem. Neural Comput. **1**(3), 348–358 (1989)
4. Knox, J.: Tabu search performance on the symmetric traveling salesman problem. Comput. Oper. Res. **21**(8), 867–876 (1994)
5. Kirkpatrick, S., Gelatt, C.D., Vecchi, M.P.: Optimization by simulated annealing. Science **220**(4598), 671–680 (1983)
6. Johnson, D.S., McGeoch, L.A.: The traveling salesman problem: A case study in local optim. Local Search Comb. Optim. **1**, 215–310 (1997)
7. Dorigo, M., Gambardella, L.M.: Ant colonies for the travelling salesman problem. BioSystems **43**(2), 73–81 (1997)
8. Shi, X.H., Liang, Y.C., Lee, H.P., Lu, C., Wang, Q.X.: Particle swarm optimization-based algorithms for TSP and generalized TSP. Inform. Process. Lett. **103**(5), 169–176 (2007)
9. Farmer, J.D., Packard, N.H., Perelson, A.S.: The immune system, adaptation, and machine learning. Physica D **22**(1), 187–204 (1986)
10. Jolai, F., Ghanbari, A.: Integrating data transformation techniques with Hopfield neural networks for solving travelling salesman problem. Expert Syst. Appl. **37**(7), 5331–5335 (2010)
11. Cheng, M.Y., Prayogo, D.: Symbiotic organism search: a new metaheuristic optimization. Comput. Struct. **139**, 98–112 (2014)

A Hybrid Swarm Composition
for Chinese Music

Xiaomei Zheng[1,2](✉), Weian Guo[3], Dongyang Li[1],
Lei Wang[1], and Yushan Wang[2]

[1] College of Electronics and Information Engineering,
Tongji University, Shanghai 201804, China
lidongyang0412@163.com, wanglei@tongji.edu.cn
[2] College of Information, Mechanical and Electrical Engineering,
Shanghai Normal University, Shanghai 200234, China
{xmzheng, amywang}@shnu.edu.cn
[3] Sino-German College of Applied Sciences, Tongji University,
Shanghai 201804, China
guoweian@163.com

Abstract. Algorithm composition is an automatic process which uses formalized strategies for music expression and creation. In this paper, we use Particle Swarm Optimization (PSO) and Genetic Algorithm (GA) to simulate the process of Chinese music creation. A hybrid swarm intelligence composition model for Chinese music is proposed. In this model, the PSO is firstly used to generate the initial music material, and then melody is developed by using the evolutionary operators in GA. Finally, a piece of music work is formed. The experimental result shows that the algorithm is feasible, and has better performance than the genetic algorithm composition.

Keywords: Hybrid swarm intelligent composition · PSO · GA · Chinese music

1 Introduction

Since the advent of computers, musicians and computer scientists have been working on computers as a tool for music thinking. For algorithmic composition, it is necessary to consider the formalization of composing behavior and rules, as well as the style of the composing system. Some composing systems can achieve Bach style hymns while some others can be used to produce jazz. So far, the methods used for composition mainly include stochastic process, artificial neural network and GA [1], among which the GA-based method has been studied extensively and achieved rich results. GA is a heuristic search algorithm, which uses a natural evolutionary model. It starts from a population representing the candidate solutions of the problem. According to the survival process of the fittest, the optimal or approximate solutions of the problem are evolved by generations. In the process of evolution, the algorithm uses a series of evolutionary operators including crossover, mutation et al. GA has been widely used in many fields because of its simple concept, easy implementation, strong robustness and suitable for parallel computing. In the field of music creation, elements which make up

© Springer International Publishing AG 2017
Y. Tan et al. (Eds.): ICSI 2017, Part II, LNCS 10386, pp. 258–265, 2017.
DOI: 10.1007/978-3-319-61833-3_27

music, such as pitch and duration, are easily represented by numbers. Composing methods, such as the development of musical material, are similar to the evolutionary operations in genetic algorithms. Therefore, it is feasible to extend GA to the music search space and produce music that conforms to the music rules. In the early 90's of the last century, Horner first applied GA to the field of algorithm composition [2], followed by a large number of computer scholars and musicians having done an useful attempt and exploration. Different evolutionary composing systems are designed and a lot of music creation results have been achieved. Horner completed a computer–aided composing program in which only a simple task of thematic bridging is completed. The famous GenJam [3, 4] is an evolved composing system designed by Biles which produces a 16-bar jazz melody. Ting et al. proposed a phrase imitation-based composing system which uses music theory and imitates the characteristics of melodic progression [5]. Liu proposed an evolutionary composition using the information from music charts in the evaluation criterion [6]. The fitness function is generated by the weighted rules according to the download times from music charts. Prisco achieved an automatic composing system for dodecaphonic music based on genetic algorithm and an evolutionary composer for Bass harmonization [7, 8].

Most of the above systems are for Western music. In this paper we will study the algorithmic composition for Chinese music. In the music creation, Western music focuses on the vertical structure of the works, such as harmony, polyphonic structure; while Chinese music focuses on the horizontal melody structure of the works, such as melody development. One method for Chinese musicians to create a melody is that after a theme is conceived, more music materials can be generated by a variety of melody development techniques and the materials form a complete melody ultimately. The formalization of the method can not be well achieved with a simple GA. Therefore, a hybrid swarm intelligent Chinese music composition algorithm is designed by the combination of PSO and GA. PSO generates the initial music theme which is evolved by GA so as to produce more materials.

2 Algorithm Model

2.1 Chinese Music

Western music has a sophisticated structure, while Chinese music focuses on its intrinsic rhythm. In music composition, melody creation is emphasized and most of the works are monophonic. Melody is the most important characteristic of Chinese music. The raw material to constitute the melody is the scale. The Chinese music mainly uses pentatonic scale "*Gong, Shang, Jue, Zhi* and *Yu*" [9]. With a tone in pentatonic as the keynote, five modes of "*Gong-mode, Shang-mode, Jue-mode, Zhi-mode* and *Yu-mode*" are produced in traditional Chinese music. Notes with five-degree above or five-degree below the keynote are the steady-notes of the keynote. Steady-notes together with keynote constitute the skeleton of mode. Mode is the basis for the melody development. It determines the combination of tones in the music works. In traditional pentatonic modes, the five scales of "*Gong, Shang, Jue, Zhi* and *Yu*" only produce a unison, a major second, a minor third, a perfect fourth, a perfect fifth, a minor sixth, a major sixth, an octave and other intervals.

2.2 Composition Model

According to the creative method of Chinese music, a hybrid swarm intelligent composition model based on PSO and GA is proposed as shown in Fig. 1. The model divides the automatic composition of Chinese music into different stages, i.e., music theme generation, melody development, and final synthesis of works.

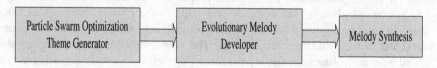

Fig. 1. Composition model of Chinese music

Particle swarm optimization is used for theme generation. PSO is a swarm algorithm proposed in [10]. It simulates the process of bird search for food to solve the optimization problems. PSO considers the individual as a particle in the D-dimensional search space, which flies at a certain speed. The fly speed can be adjusted according to the individual and group's fly experience so as the particle can approximate toward the optimal solution. PSO algorithm of this thought can be used in the process of theme generation in Chinese music. In this paper, the theme is considered as a phrase which is composed of D notes. The location of the particles represents a phrase. By the flight of particles, constantly changing the notes makes the phrase meet the mode regulation and creative rules of Chinese music, and ultimately find a better phrase.

Genetic algorithm is used for the evolution of melody. Unlike most of the previous GA-based composing systems whose initial groups are randomly generated, the theme materials generated by PSO are used as the initial population of GA developer, and the materials are evolved to realize the melody development by GA. This makes the evolutionary process quick and easy.

3 Algorithm Implementation

3.1 PSO Theme Generator.

The PSO theme generator includes two major stages of rhythm controller and PSO theme optimizer.

Rhythm Controller. In [11], the principle of the rhythm decomposition of music is given. The most basic cellular structure which constitutes the rhythm is the smallest odd and even number 1 and 2. A rhythm unit can be broken down into two smallest cell structures. Suppose the shortest note is the sixteenth, whose value is 1, then the eighth unit is 2, the fourth unit is 4, the half unit is 8, and the whole unit is 16, etc. Thus 2 can be broken down into 1 and 1, 4 can be broken down into 2 and 2, or 1 and 3. According to the beat, the number of rhythm unit per bar can be determined, and then rhythm decomposition is made in each bar.

PSO Theme Optimizer. *Particle structure. Particle(i)=Loc[], Vel[], fitness};*

$Loc[\ ] = <n_1, n_2, \cdots, n_j, \cdots, n_D>$ is a location vector, where n_j represents a note, and $Vel[] = <Interval_1, \cdots, Interval_j, \cdots, Interval_D>$ is a velocity vector, where $Interval_j$ represents the speed at which note n_j moves, i.e., the interval between $n_j(t)$ and $n_j(t+1)$.

Fitness. Considering the characteristics of Chinese music, the design of fitness takes into account three factors: f_{mode}, f_{melody} and $f_{tonic}. f_{mode}$ investigates the usage of *keynote* and *steadynotes* of melody, counting the position, frequency and duration of notes they occur. f_{melody} examines the melody pattern. In the creation of melody, wave patterns are mostly usedto avoid a pitch straight up or straight down. Before a big jump, more than four-degree intervals, a reverse movement should be applied. Compared to western music, the tone of Chinese music has the characteristic of calm and tranquil. The tone is mainly reflected in the relationship between intervals. The melody of western music has more jump progression; however, Chinese music is mostly dominated by second-to-third-degree intervals, with few big jump but step progression. f_{tonic} is used to analyze second degree, third degree intervals in melody so as to meet the characteristics of Chinese music.

Algorithm Improvement. The velocity of particle represents the change of interval. The velocity update formula is based on an operation between D-dimensional vectors. According to the pitch relationship of notes, we redefine the operators for velocity and location update formula.

Definition 1. Substraction in velocity update between two particles *Particle(i)* and *Particle(k)* is defined as follow,

$$Particle\,(k).Loc\,[\,] - Particle\,(i).Loc\,[\,] = <n_1^k, \cdots, n_j^k, \cdots, n_D^k> - <n_1^i, \cdots, n_j^i, \cdots, n_D^i>$$
$$= <Interval_{n_1^i}^{n_1^k}, \cdots, Interval_{n_j^i}^{n_j^k}, \cdots, Interval_{n_D^i}^{n_D^k}>$$

Definition 2. Velocity update of particle is defined as follow,

$$Particle\,(i).Vel\,[\,] = \omega \cdot Particle\,(i).Vel\,[\,] + C_1 \cdot rand_1(\,) \cdot (p_{i\_best}.Loc\,[\,] - Particle(i).Loc\,[\,])$$
$$+ C_2 \cdot rand_2(\,) \cdot (p_{best}.Loc\,[\,] - Particle(i).Loc\,[\,])$$

Definition 3. Location update of particle is defined as follow,

$$Particle\,[\,(i).\,Loc\,[\,] = Particle\,(i).\,Loc\,[\,] + Particle\,(i).\,Vel\,[\,]$$
$$= <n_1^i, \cdots, n_j^i, \cdots, n_D^i> + <Interval_1, \cdots, Interval_j, \cdots, Interval_D> = <n_1^{i\prime}, \cdots, n_j^{i\prime}, \cdots, n_D^{i\prime}>$$

3.2 Evolutionary Melody Developer

Melody Development. PSO theme generator produces the theme material of melody development. The theme material of composition, whose size and length may vary according to the work to be created, could be motives, sections or phrases [12]. In this paper, phrases are used as the theme material. These phrases can be used as basic material for melody development, based on which, various developing methods of Chinese melody such as repeating, changing-head, changing-tail, split, and other techniques can be implemented to form new phrases.

Chromosome Expression. The evolutionary individuals are phrases shown as below,

$$Phrase = <Bar^1, Bar^2, \cdots, Bar^i, \cdots, Bar^n >$$
$$= <note_1, note_2, \cdots note_i, \cdots, note_D >$$

Where n is the number of bars that make up the phrase. Here in the paper, we suppose $n = 4$. D is the total number of notes in the phrase. $note_i = <note\_pit_i, note\_val_i >$ represents a note in the bar, where $note\_pit_i$ is the pitch of $note_i$, and $note\_val_i$ is the duration of $note_i$.

Fitness. The fitness is used in the same measure as in the PSO theme generator. The characteristics and rules of Chinese music are used to measure the merits of the chromosome.

Evolutionary Operator. The *selection* operator uses roulette selection.

The *crossover* operators include *note-crossover* and *phrase-crossover*, where *phrase-crossover* operator is an exchange of i bars at the beginning of the phrases t_1 and t_2 to generate new phrases d_1 and d_2, shown as follow,

$$t_1 = <Bar^1_{t_1}, \cdots, Bar^i_{t_1}, Bar^{i+1}_{t_1}, \cdots, Bar^n_{t_1} > ;$$
$$t_2 = <Bar^1_{t_2}, \cdots, Bar^i_{t_2}, Bar^{i+1}_{t_2}, \cdots, Bar^n_{t_2} > ;$$

$$d_1 = <Bar^1_{t_2}, \cdots, Bar^i_{t_2}, Bar^{i+1}_{t_1}, \cdots, Bar^n_{t_1} > ;$$
$$d_2 = <Bar^1_{t_1}, \cdots, Bar^i_{t_1}, Bar^{i+1}_{t_2}, \cdots, Bar^n_{t_2} >$$

The *mutation* operators include the *phrase-ascending* operator, *phrase-descending* operator, *note-variation* operator, etc. For example, *phrase-ascending* operator is a gradual rise of the phrase t on pitch to form a new phrase d,

Let phrase $t = <Bar^1, Bar^2, \cdots, Bar^i, \cdots, Bar^n >$, where

$Bar^i = <note^i_1, note^i_2, \cdots, note^i_j, \cdots, note^i_l >$ and $note^i_j = <note\_pit^i_j, note\_val^i_j >$.

The pitch of each note in phrase t is increased by δ degrees to form a new phrase d as follow, $d = <Bar^{1'}, Bar^{2'}, \cdots, Bar^{i'}, \cdots, Bar^{n'} >$, where $Bar^{i'} = <note^{i'}_1, note^{i'}_2, \cdots, note^{i'}_j, \cdots, note^{i'}_l >$ and $note^{i'}_j = <note\_pit^i_j + \delta, note\_val^i_j >$.

Note-variation operator is a note variation in phrase, which includes pitch and duration variation. Duration variation results in a change of the number of notes. For example, when a whole note changes to a half note, the number of notes in the bar is increased by one. When the j-th note in the i-th bar of phrase t is mutated, a new phrase d is formed. Let phrase $t = <Bar^1, Bar^2, \cdots, Bar^i, \cdots, Bar^n >$, where
$Bar^i = <note_1^i, note_2^i, \cdots, note_j^i, \cdots, note_l^i >$ and $note_j^i = ltnote\_pit_j^i, note\_val_j^i >$.
If only the pitch variation is considered, the new phrase $d = <Bar^1, Bar^2,$ $\cdots Bar^{i'}, \cdots Bar^n >$ is formed, where $Bar^{i'} = <note_1^i, note_2^i, \cdots, note_j^{i'}, \cdots, note_l^i >$ and $note_j^{i'} = <note\_pit_j^{i'}, note\_val_j^i >$.

3.3   Melody Synthesis.

After the implementation of the hybrid swarm algorithm, aseries of melody material are evolved. These materials could be combined to form a complete musicin accordance with Chinese melody development structure. The structure is in the form of a four-phrase in which the melody is composed of four phrases, the relations among which are repetitive, similar or contrast.

The following figure shows the composition results based on *Shang-mode* (Fig. 2).

Fig. 2.   Composition example of *Shang-mode*

4   Analysis and Comparison

Compared with the genetic algorithm composition (GA-CM), the hybrid swarm composition (Hybrid Swarm-CM) can obtain better results. The two algorithms run 20 times respectively, in which the maximum number of iterations is 1000 and the population size is 30. The experimental results are shown in Table 1. As can be seen from Table 1, The best fitness of Hybrid Swarm-CM is better than that of GA-CM, and minimum generation to reach the best fitness in Hybrid Swarm-CM is less than GA-CM. Figure 3 plots the mean of best fitness against generations over 20 runs of the Hybrid Swarm-CM and GA-CM. It shows that the performance of Hybrid Swarm-CMis also better than GA-CM.

Table 1. Best fitness of Hybrid Swarm-CM&GA-CM

No	Hybrid Swarm-CM		GA-CM	
	Best fitness	Generation	Best fitness	Generation
1	1.9074	155	1.9092	213
2	1.8884	218	1.9092	196
3	1.9087	76	1.9036	137
4	1.8995	168	1.9091	463
5	1.8947	105	1.9074	236
6	1.8868	124	1.9035	428
7	1.8917	113	1.9061	193
8	1.8982	203	1.8986	328
9	1.8947	243	1.9108	364
10	1.9022	95	1.9132	51
11	1.9009	226	1.9092	176
12	1.8778	34	1.9005	155
13	1.8975	352	1.912	141
14	1.8817	268	1.9074	213
15	1.8961	82	1.9246	253
16	1.8947	199	1.9101	209
17	1.9024	193	1.9258	375
18	1.8927	208	1.9238	92
19	1.8898	301	1.9035	284
20	1.8986	254	1.9078	247
Avg.	1.8952	181	1.9098	238

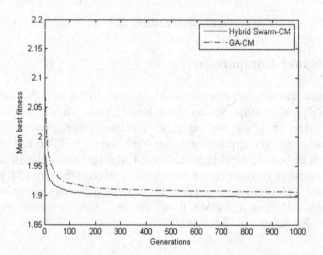

Fig. 3. Comparison of Hybrid Swarm-CM and GA-CM

5   Conclusion

Chinese music has its own creative characteristics. When composing a Chinese music, the algorithm can imitate the musicians' creative process. In this paper, by formalizing the musicians' creation mode, a hybrid swarm intelligent composition model for Chinese melody creation combined with PSO and GA is proposed. PSO algorithm is used as the initial population generator of GA, and the population is evolved by evolutionary operators in GA. Finally a complete melody work is merged through melody synthesis. Compared with the GA-based composition, the hybrid swarm composition algorithm can achieve better fitness and faster performance. However, the algorithm in this paper takes into account the characteristics of Chinese music while the emotions have not been considered. The future work will take the music emotions into account.

Acknowledgments. This work was supported by Shanghai Natural Science Foundation (No. 15ZR1431600), and Shanghai Normal University Research Foundation (No. SK201323)

References

1. Nierhaus, G.: Algorithmic Composition. Springer-verlag/Wien, New York (2009)
2. Horner A., Goldberg D.E.: Genetic algorithm and computer-assisted music composition. In: Proceeding of the 4th International Conference on Genetic Algorithm (1991)
3. Biles, J.A.: GenJam: a genetic algorithm for generating jazz solos. In: Proceedings of the International Computer Music Conference, pp. 131–137 (1994)
4. Biles, J.A.: Autonomous GenJam: eliminating the fitness bottleneck by eliminating fitness. In: Proceedings of the 2001 Genetic and Evolutionary Computation Conference Workshop Program, GECCO, San Francisco (2001)
5. Ting, C.-K., Wu, C.-L., Liu, C.-H.: A novel automatic composition system using evolutionary algorithm and phrase imitation. IEEE Syst. J. **PP**(99), 1–12 (2015)
6. Liu, C.-H., Ting, C.-K.: Evolutionary composition using music theory and charts. In: IEEE Symposium on Computational Intelligence for Creativity and Affective Computing (CICAC), pp. 63–70 (2013)
7. Prisco, R., Zaccagnino, G., Zaccagnino, R.: A genetic algorithm for dodecaphonic compositions. In: Chio, C., et al. (eds.) EvoApplications 2011. LNCS, vol. 6625, pp. 244–253. Springer, Heidelberg (2011). doi:10.1007/978-3-642-20520-0_25
8. Prisco, R., Zaccagnino, R.: An evolutionary music composer algorithm for bass harmonization. In: Giacobini, M., et al. (eds.) EvoWorkshops 2009. LNCS, vol. 5484, pp. 567–572. Springer, Heidelberg (2009). doi:10.1007/978-3-642-01129-0_63
9. Du, Y.X., Qin, D.X.: Chinese Music Theory. Shanghai Conservatory of Music Press, Shanghai (2007)
10. Kennedy, J., Eberhart, R.C.: Particle Swarm Optimization. In: Proceedings of the IEEE International Conference on Neural Networks, pp. 1942–1948 (1995)
11. Zhao, X.S.: Traditional Music Composition. Anhui Literature and Art Publishing House, Hefei (2013)
12. Hankun, S.: Lessons on the Writing of Musical Melodies. Xiamen University Press, Xiamen (2013)

Fuzzy and Swarm Approach

Fuzzy Logic Controller Design for Tuning the Cooperation of Biology-Inspired Algorithms

Shakhnaz Akhmedova[✉], Eugene Semenkin,
Vladimir Stanovov, and Sophia Vishnevskaya

Reshetnev Siberian State University of Science and Technology,
Krasnoyarsky rabochy avenue 31, 660037 Krasnoyarsk, Russia
shahnaz@inbox.ru,
{eugenesemenkin, vladimirstanovov}@yandex.ru,
vishni@ngs.ru

Abstract. Previously, a meta-heuristic approach called Co-Operation of Biology Related Algorithms or COBRA for solving real-parameter optimization problems was introduced and described. COBRA's basic idea consists in a cooperative work of well-known bio-inspired algorithms, which were chosen due to the similarity of their schemes. COBRA's performance was evaluated on a set of test functions and its workability was demonstrated. Thus it was established that the idea of the algorithms' cooperative work is useful. However, it is unclear which bionic algorithms should be included in this cooperation and how many of them. Therefore, the aim of this study was to design a fuzzy logic controller for determining which bio-inspired algorithms should be included in the co-operative work for solving optimization problems using the COBRA approach. The population sizes of the bio-inspired component-algorithms were automatically changed by the obtained controller. The experimental results obtained by the two types of fuzzy-controlled COBRA are presented and their usefulness is demonstrated.

Keywords: Fuzzy controller · Bio-inspired algorithms · Optimization · Cooperation · COBRA

1 Introduction

Co-Operation of Biology Related Algorithms or COBRA is a meta-heuristic approach developed for solving unconstrained real-parameter optimization problems [1]. Its basic idea consists in the cooperative work of different nature-inspired algorithms, which were chosen due to the similarity of their schemes. In the original version of COBRA, five well-known bio-inspired heuristics were used as component-algorithms, namely the Particle Swarm Optimization Algorithm (PSO) [2], the Wolf Pack Search Algorithm (WPS) [3], the Firefly Algorithm (FFA) [4], the Cuckoo Search Algorithm (CSA) [5] and the Bat Algorithm (BA) [6]. Later the Fish School Search Algorithm (FSS) [7] was also added to the cooperation [8].

© Springer International Publishing AG 2017
Y. Tan et al. (Eds.): ICSI 2017, Part II, LNCS 10386, pp. 269–276, 2017.
DOI: 10.1007/978-3-319-61833-3_28

All the mentioned heuristics are similar bio-inspired optimization methods originally developed for continuous variable space. These algorithms mimic the collective behaviour of their corresponding animal groups, thereby allowing the global optima of real-valued functions to be found. For example, PSO and FSS were inspired by bird and fish swarm intelligence respectively, and BA was inspired by the echolocation behaviour of bats. An examination of the efficiency of these six heuristics was conducted on a set of different test functions. Performance analysis showed that all of them are sufficiently effective for solving optimization problems, and their workability has been established.

However, there are various other algorithms which can be used as components for COBRA as well as previously conducted experiments demonstrating that even the bio-inspired algorithms already chosen can be combined in different ways. For example, in [8] five different combinations of the given nature-inspired heuristics for the COBRA algorithm are presented, and their efficiency was examined on test problems from the CEC'2013 competition [9]. It was established that three of them show the best results on test functions depending on the number of variables.

To solve the described problem, controllers based on fuzzy logic [10] were used in this work. Currently researchers frequently use fuzzy rule-based controllers for solving different problems. There are various works in which this method has been used and it has been established that generally it is efficient and works successfully (for example, [11, 12]).

Therefore, in this paper firstly the COBRA meta-heuristic approach and its components are described, and then a description of the fuzzy controller is presented. In the next section, the experimental results obtained by the two types of fuzzy controller are discussed, and after that the implementation of the best obtained fuzzy controller to COBRA and its experimental results are demonstrated. Finally, some conclusions are given in the last section.

2 Co-operation of Biology Related Algorithms

The meta-heuristic approach called Co-Operation of Biology Related Algorithms or COBRA [1] was developed based on five optimization methods, namely Particle Swarm Optimization (PSO) [2], the Wolf Pack Search (WPS) [3], the Firefly Algorithm (FFA) [4], the Cuckoo Search Algorithm (CSA) [5] and the Bat Algorithm (BA) [6] (hereinafter referred to as "component-algorithms"). As was mentioned earlier, the Fish School Search (FSS) [7] was also added as a component-algorithm of COBRA. The main reason for the development of a cooperative meta-heuristic was the inability to say which of the above-listed algorithms is the best one or which algorithm should be used for solving any given optimization problem [1]. Thus the idea was to use the cooperation of these bio-inspired algorithms instead of any attempts to understand which one is the best for the problem in hand.

The originally proposed approach consists in generating one population for each bio-inspired algorithm, therefore six populations, which are then executed in parallel, cooperating with each other. The COBRA algorithm is a self-tuning meta-heuristic, so there is no need to choose the population size for each component-algorithm.

The number of individuals in the population of each algorithm can increase or decrease depending on the fitness values: if the overall fitness value was not improved during a given number of iterations, then the size of each population increased, and vice versa.

There is also one more rule for population size adjustment, whereby a population can "grow" by accepting individuals removed from other populations. The population "grows" only if its average fitness value is better than the average fitness value of all other populations. Therefore, the "winner algorithm" can be determined as an algorithm whose population has the best average fitness value. This can be done at every step.

The main goal of communication between all populations is to prevent their preliminary convergence to their own local optimum. "Communication" was determined in the following way: populations exchange individuals in such a way that a part of the worst individuals of each population is replaced by the best individuals of other populations. Thus, the group performance of all algorithms can be improved.

The performance of the COBRA algorithm was evaluated on a set of benchmark problems with 5, 10 and 30 variables taken from [9] and the experiments showed that COBRA works successfully and is reliable on different benchmarks. Besides, the meta-heuristic COBRA was compared with its component-algorithms and simulations and the comparison showed that COBRA is superior to these bionic algorithms when the dimension grows or when complicated problems are solved.

3 Fuzzy Controller Design

The main idea of using a fuzzy controller is to implement a more flexible tuning method, compared to the original COBRA tuning algorithm. Fuzzy controllers are well known for their ability to generate real-valued outputs using special fuzzification, inference and defuzzification schemes. In this work success rates were used as inputs and population size changes as outputs.

The fuzzy controller had 7 input variables, including 6 success rates, one for each component, and an overall success rate, and 6 output variables, i.e. the number of solutions to be added to or removed from each component. There were two types of success rate evaluation for all input variables except for the last one: the component success rate is either the average fitness value of its population or the best fitness value of its population. The last input variable was determined as the ratio of the number of iterations, during which the best found fitness value was improved, to the given number of iterations, which was a constant period. Thus, the process of population growth was automated by the fuzzy controller.

The Mamdani-type fuzzy inference was used to obtain the output values, and the rules had the following form:

$$R_q : \text{IF } x_1 \text{ is } A_{q1} \text{ and} \dots \text{and } x_n \text{ is } A_{qn} \text{ THEN } y_1 \text{ is } B_{q1} \text{ and} \dots y_k \text{ is } B_{qk} \qquad (1)$$

where R_q is the q-th fuzzy rule, $x = (x_1, \dots, x_n)$ is the set of input values in n-dimensional space ($n = 7$ in this case), $y = (y_1, \dots, y_k)$ is the set of outputs ($k = 6$), A_{qi} is the fuzzy set for the i-th input variable, B_{qj} is the fuzzy set for the j-th output variable. The rule base contained 21 fuzzy rules, which had the following structure:

each 3 rules described the case when one of the components gave better results than the others (as there were 6 components, 18 rules were established); the last 3 rules used the overall success of all components (variable 7) to add or remove solutions from all components, i.e. to regulate the computational resources (Table 1).

Table 1. Part of the rule base

№							
1	IF	X_1 is A_3	X_2-X_6 is A_4	X_7 is DC	THEN	Y_1 is B_3	Y_2-Y_6 is B_1
2	IF	X_1 is A_2	X_2-X_6 is A_4	X_7 is DC	THEN	Y_2 is B_3	Y_2-Y_6 is B_1
3	IF	X_1 is A_1	X_2-X_6 is A_4	X_7 is DC	THEN	Y_3 is B_3	Y_2-Y_6 is B_1
...				...			
19	IF		X_1-X_6 is DC	X_7 is A_1	THEN	Y_1 is B_1	
20	IF		X_1-X_6 is DC	X_7 is A_2	THEN	Y_1 is B_2	
21	IF		X_1-X_6 is DC	X_7 is A_3	THEN	Y_1 is B_3	

The input variables were always in the range $[0, 1]$, and fixed fuzzy terms of triangular shape were used for this case. In addition to the three classical fuzzy sets A_1, A_2 and A_3, the "Don't Care" (DC) condition and the A_4 term with the meaning "larger than 0" (opposite to A_1) were also used to decrease the number of rules and make them simpler.

For the outputs, three fuzzy terms of triangular shape were used. The output fuzzy terms were symmetrical, and the positions and shapes were determined by two values, encoding the left and right position of the central term, as well as the middle position of the side terms in one value, and the left and right positions of the side term in another value. These two values were optimized using the PSO algorithm. The defuzzification procedure was performed by calculating the centre of mass of the shape received by fuzzy inference.

4   Experimental Results

In this study the following 10 benchmark problems taken from [9] were used in experiments: the Rotated Discus Function, the Different Powers Function, the Rotated Rosenbrock's Function, Schwefel's Function, the Rotated Ackley's Function, the Rotated Griewank's Function, the Rotated Katsuura Function, the Rotated Lunacek bi-Rastrigin Function, the Rotated Weierstrass Function and Rastrigin's Function. These benchmark functions were considered to evaluate the robustness of the fuzzy controlled COBRA. However, firstly they were used to determine the best parameters for the two types of fuzzy controllers.

The standard Particle Swarm Optimization algorithm was used for this purpose. Therefore the individuals were each represented as parameters of the fuzzy controlled COBRA, namely the positions of the output fuzzy terms. The objective function optimized by the PSO algorithm was the average best fitness over $T = 10$ runs and all functions. Thus on each iteration all test problems were solved T times by a given fuzzy controlled COBRA and then the obtained results were averaged. Calculations were stopped on each program run if the number of function evaluations exceeded $10000D$. The population size for the PSO algorithm was equal to 50 and the number of iterations was equal to 100; calculations were stopped on the 100-th iteration for the PSO heuristic. When the successfulness of the component-algorithms was evaluated as the best average fitness value, the following parameters for the fuzzy controller were obtained: $[-33; -10; 11; 37]$. Finally when the successfulness of the component-algorithms was evaluated as the best solution found by the component-algorithm, the obtained parameters were $[-12; -2; 0; 19]$.

Figure 1 shows the change of the COBRA component population sizes during the optimization process on two functions, Schwefel's Function (Fig. 1a and c) and the Different Powers Function (Fig. 1b and d) with the best found fuzzy controller parameters. These functions were chosen as they represent two cases: Schwefel's function has many local minima and is quite difficult to optimize, while the Different Powers Function has a more simple landscape. Figure 1a and b demonstrate the standard COBRA tuning procedure behaviour, Fig. 1c and d – the fuzzy controller with the component's success rate as the maximum fitness. The behaviour of these three tuning methods is quite different. The standard COBRA tuning method makes multiple oscillations, but the winning component is changed over time. The method that uses the average fitness (Fig. 1c) with Schwefel's Function gives all resources to one component, which is probably not the best choice, while for the other function it tends to minimize every population size down to 0.

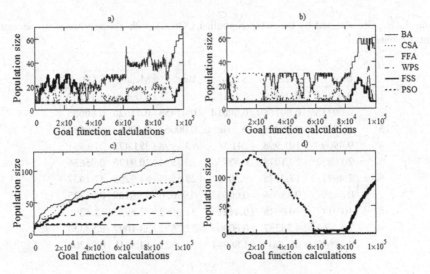

Fig. 1. Graphs of population size change

The method that uses the best fitness (1c and 1d), however, is quite straightforward: for a complicated Schwefel's Function it increases almost every population size in parallel, giving the algorithms an opportunity to work on the problem together. The thing that should be noted here is that, for example, PSO had a minimal resource at the beginning of the optimization until around 45000 calculations, but after this point its resource went up faster than the resource of any other component. By the end of the optimization process, PSO had the second largest resource, which shows that it became particularly effective after some state of the optimization process has been reached. Such cases show why the resource tuning procedure is important.

On the following step the obtained parameters were applied to the fuzzy controlled COBRA and it was tested on the above-listed benchmark functions. There were 51 program runs for each optimization problem; and calculations were stopped if the number of function evaluations became equal to 10000D.

In Tables 2 and 3 the results obtained by the fuzzy controlled COBRA for the two types with the best parameters are presented. The following notations are used: the best found function value (Best), the function value averaged by the number of program runs (Mean) and the standard deviation (STD). Numbers in bold show the values which are better compared to the original COBRA in Table 4, using the Student's t-test for mean comparison. For the case when average fitness values were used for the successfulness evaluation, the best results were obtained for 5 functions out of 10, while for the best values used for successfulness evaluation in Table 3, the best results were received for 7 functions out of 10 (the same number were received for 10D and 30D). Moreover, the values in italics in Table 3 show the cases when the algorithm that uses the best values outperformed the algorithm that uses average values, which was the case for 10D: 6 times out of 10 and for 30D: 8 times out of 10.

For comparison, in Table 4 the results obtained by COBRA with six component-algorithms with the standard tuning method are given.

Table 2. Results obtained by the fuzzy controlled COBRA with average values for successfulness evaluation.

Func	D = 10			D = 30		
	Best	Mean	STD	Best	Mean	STD
1	0.00148	0.03319	**0.11152**	0.03221	0.09689	**0.14292**
2	0.00138	**0.00158**	**0.00014**	0.01453	**0.01526**	**0.00085**
3	6.733e−6	**2.093e−5**	8.304e−6	0.00061	0.00062	6.396e−5
4	9.49093	**101.998**	241.5	139.998	**391.473**	219.351
5	20.0832	**20.1078**	**0.02436**	20.7574	**20.9126**	**0.05656**
6	7.46711	17.2175	9.3068	23.6503	57.9295	**37.1637**
7	0.4889	0.48966	0.00083	0.97626	**0.98138**	**0.01775**
8	10.0177	10.0948	**0.4421**	31.0682	32.5392	12.8296
9	3.62965	**4.36357**	0.60906	32.7502	**33.1097**	**0.53338**
10	4.61672	6.34905	**1.86941**	8.90322	14.8918	6.84208

Table 3. Results obtained by the fuzzy controlled COBRA with best values for successfulness evaluation.

Func	D = 10			D = 30		
	Best	Mean	STD	Best	Mean	STD
1	1.290e−7	8.589e-5	**0.00052**	1.634e-6	0.01862	**0.11131**
2	0	*2.461e−6*	**1.736e−5**	1.416e−6	*0.00072*	**0.00048**
3	0	*2.609e−6*	1.574e−5	4.190e−6	*0.00011*	0.00068
4	23.6667	**83.5599**	81.4823	158.463	*213.352*	116.631
5	20.0035	**20.1159**	**0.13204**	20.8193	*21.0458*	**0.09436**
6	0.07145	*0.09545*	**0.16942**	0.06902	*0.09069*	**0.15303**
7	0.50369	*0.50381*	**0.00013**	0.35558	*0.35605*	**0.00084**
8	10	10.0294	**0.19958**	30.0278	30.1213	**0.21122**
9	1.20414	*1.76373*	0.58829	2.06807	*2.89336*	**0.41837**
10	0.99102	*0.99317*	**0.00423**	1.00182	*1.00887*	**0.01006**

Table 4. Results obtained by COBRA with six component algorithms

Func	D = 10			D = 30		
	Best	Mean	STD	Best	Mean	STD
1	0.00029	0.06666	0.43655	0.01533	0.37206	1.65313
2	0.00422	0.00514	0.00205	0.01581	0.01768	0.00158
3	2.503e−6	9.683e−5	1.082e−6	0.00024	0.00059	4.421e−5
4	228.84	288.457	47.6639	275.284	651.786	114.227
5	20.1164	20.4929	0.20121	20.8419	21.0862	0.10058
6	6.17083	6.61872	1.8081	9.47809	27.0826	44.0647
7	0.29586	0.29657	0.00191	1.08536	1.10279	0.01817
8	10.036	10.263	0.95319	30.6449	33.0245	7.30007
9	6.22935	6.23029	0.0011	35.4404	36.8419	0.76274
10	3.22398	7.11874	9.13148	5.86151	10.2876	3.67494

Thus, the comparison demonstrates that there is no significant difference between the results obtained by either the fuzzy controlled COBRA with the average values for successfulness evaluation or COBRA with six components without a controller. However, the fuzzy controlled COBRA with the best fitness values used for the success rate evaluation outperformed both of them. Therefore, it can be used for solving the optimization problems instead of the versions of the algorithm.

5 Conclusions

In this paper a new modification of the meta-heuristic called COBRA, which was originally based on five nature-inspired algorithms, has been introduced. The proposed modification involves the usage of a fuzzy controller for adjustment of the component-algorithm population sizes and adjustment of the whole collective for

COBRA. Besides, the Fish School Search algorithm was added as a potential component of the COBRA approach. Simulations and comparison showed that the fuzzy controlled algorithm COBRA with the best fitness values used for the success rate evaluation is superior to the other versions. The developed optimization strategy can easily be extended to study constrained and multi-objective optimization problems.

Acknowledgments. Research is performed with the support of the Ministry of Education and Science of Russian Federation within State Assignment project № 2.1680.2017/ПЧ.

References

1. Akhmedova, S., Semenkin, E.: Co-operation of biology related algorithms. In: IEEE Congress on Evolutionary Computation, pp. 2207–2214 (2013)
2. Kennedy, J., Eberhart, R.: Particle swarm optimization. In: IEEE International Conference on Neural networks, vol. IV, pp. 1942–1948 (1995)
3. Yang, C., Tu, X., Chen, J.: Algorithm of marriage in honey bees optimization based on the wolf pack search. In: International Conference on Intelligent Pervasive Computing, pp. 462–467 (2007)
4. Yang, X.S.: Firefly algorithms for multimodal optimization. In: 5th Symposium on Stochastic Algorithms, Foundations and Applications, pp. 169–178 (2009)
5. Yang, X.S., Deb, S.: Cuckoo search via levy flights. In: World Congress on Nature & Biologically Inspired Computing, pp. 210–214. IEEE Publications (2009)
6. Yang, X.S.: A new metaheuristic bat-inspired algorithm. In: González, J.R., Pelta, D.A., Cruz, C., Terrazas, G., Krasnogor, N. (eds.) Nature Inspired Cooperative Strategies for Optimization. SCI, vol. 284, pp. 65–74. Springer, Heidelberg (2010)
7. Bastos, F.C., Lima, N.F.: Fish school search: an overview. In: Chiong, R. (ed.) Nature-Inspired Algorithms for Optimization. SCI, vol. 193, pp. 261–277. Springer, Heidelberg (2009)
8. Akhmedova, S., Semenkin, E.: Investigation into the efficiency of different bionic algorithm combinations for a COBRA meta-heuristic. In: IOP Conference. Materials Science and Engineering, vol. 173 (2016)
9. Liang, J.J., Qu, B. Y., Suganthan P.N., Hernandez-Diaz, A.G.: Problem definitions and evaluation criteria for the CEC 2013 special session on real-parameter optimization. Technical report, Computational Intelligence Laboratory, Zhengzhou University, Zhengzhou China, and Technical report, Nanyang Technological University, Singapore (2012)
10. Lee, C.-C.: Fuzzy logic in control systems: fuzzy logic controller-parts 1 and 2. IEEE Trans. Syst. Man Cybern. **20**(2), 404–435 (1990)
11. Angelov, P., Zhou, X., Klawonn, F.: Evolving fuzzy rule-based classifiers. In: IEEE Symposium on Computational Intelligence in Image and Signal Processing, pp. 220–225 (2007)
12. Valdez, F., Melin, P., Castillo, O.: Fuzzy control of parameters to dynamically adapt the PSO and GA algorithms. In: IEEE International Conference on Fuzzy Systems, pp. 1–8 (2010)

Making Capital Budgeting Decisions for Project Abandonment by Fuzzy Approach

Yu-Hong Liu[1], I-Ming Jiang[2(✉)], and Meng-I Tsai[3]

[1] Graduate Institute of Finance, National Cheng Kung University,
Tainan, Taiwan
yuhong@mail.ncku.edu.tw
[2] Faculty of Finance, College of Management, Yuan Ze University,
Taoyuan, Taiwan
jiangfinance@saturn.yzu.edu.tw
[3] Department of Industrial Engineering and Management, Yuan Ze University,
Taoyuan, Taiwan
tommy0921792267@gmail.com

Abstract. This paper incorporates a fuzzy process into an abandonment option approach, to investigate the effects of managerial optimism and pessimism when a manager considers whether to abandon capital budgeting decisions or not under conditions of irreversible investment and uncertain market environment. We find that managerial optimism would encourage firms to continue the project, while managerial pessimism would encourage them to liquidate it, when the abandonment option value is located in a fuzzy sentiments region.

Keywords: Fuzzy · Abandonment option · Optimism · Pessimism

1 Introduction

Real options are used to evaluate real assets when managers or investors make capital budgeting decisions, such as whether or not to invest in manufacturing plants, adding product lines, and research and development. The valuation of investments is an interesting issue when the firm makes decisions under conditions of irreversibility and uncertainty (McDonald and Siegel [12]; Dixit and Pindyck [5]).

Previous studies that discuss capital budgeting decisions, such as Kaplan and Weisbach [9], document that firm diversification can be motivated by corporate strategy. Lang *et al.* [10] state that divestiture is a source of cash to reduce firm debt. Dyl and Long [6] claim that the optimal time to abandon a project would not be at the beginning, since the firm retains the option to abandon it in the future. Brennan and Schwartz [2] develop abandonment decision rules using a constant salvage value and the price of the underlying commodity. Myers and Majd [14] use a numerical method to evaluate a permanent abandonment option, with dividends paid from the project at any time under a known project lifetime, where the salvage value varies stochastically. However, most projects do not have a predetermined lifespan, and even if they do, this often changes in practice. Comment and Jarrel [3] and John and Ofek [7] highlight the important of financial distress and bankruptcy avoidance, while Kaiser and Stouraitis

© Springer International Publishing AG 2017
Y. Tan et al. (Eds.): ICSI 2017, Part II, LNCS 10386, pp. 277–284, 2017.
DOI: 10.1007/978-3-319-61833-3_29

[8] argue that the presence of agency costs and managerial strategic considerations has a significant impact on the sell-off phenomenon. By considering the sentiments of investors, we could thus forecast the stock market biases and also measure the excess returns by exploiting these. DeBondt and Thaler [4] find that the heuristic of representativeness tends to create extreme predictions or overreactions (optimism), thus leading losers to outperform winners.

The remainder of this paper is organized as follows. Section 2 proposes the fundamental model of the abandonment option by taking instantaneous cash flow into consideration under the conditions of irreversible investment. Section 3 uses a fuzzy approach to measure managerial sentiment, such as optimism and pessimism, in relation to abandonment options. The comparative statistics are shown in Sect. 4. Finally, Sect. 5 presents the conclusions of this work.

2   The Model and the Abandonment Option

In this model, a manager should decide when to abandon the project. Assuming that the investment inputs, I, are known and fixed, then the present value of the investment project's cash inflow per unit time, V_t, which follows geometric Brownian motion, can be modeled as follows(Black and Scholes [1]; Merton [13]).

$$dV_t = \mu_V V_t dt + \sigma_V V_t dz_t, \tag{1}$$

where μ_V denotes the growth rate of the value of the investment and t denotes time index. If the growth rate is positive, then productivity is increasing. The σ_V^2 which represents the uncertainty is the instantaneous variance of the return of project value. The dz_V is the incremental Wiener process, with zero mean and dt variance. After the investment decision has been made, the firm possesses an option to abandon the project at any future time, although the investments are irreversible. The operating cash inflow of the project at time t is equal to $\frac{1}{\rho - \mu_V} V_t - \frac{1}{\rho} C$, where ρ denotes the expected rate of return on the investment project and C is the cost when the project is ongoing.

After imposing several appropriate boundary conditions, the following two roots, β_1 and β_2 are derived from the fundamental quadratic equation,

$$\beta_1 = \frac{\mu_V}{\sigma_V^2} - \frac{1}{2} + \sqrt{\left(\frac{1}{2} - \frac{\mu_V}{\sigma_V^2}\right)^2 + \frac{2\rho}{\sigma_V^2}} \tag{2a}$$

$$\beta_2 = \frac{\mu_V}{\sigma_V^2} - \frac{1}{2} - \sqrt{\left(\frac{1}{2} - \frac{\mu_V}{\sigma_V^2}\right)^2 + \frac{2\rho}{\sigma_V^2}} \tag{2b}$$

where $\beta_1 > 1$ and $\beta_2 < 0$. Equations (2a) and (2b) show that β_1 and β_2 are determined on the growth rate, μ_V, expected rate of return, ρ and variance, σ_V^2.

Proposition 1. Suppose that the manager of one company wants to invest in a project and there exists a series of cash flows and a depreciation rate. The optimal threshold and real option value for the manager to abandon this project are as follows.

$$V^* = \frac{1}{1 + \frac{1}{\rho - \mu_v}} \frac{\beta_2}{\beta_2 - 1} \left[(1 - \phi)I + \frac{1}{\rho}C \right] \tag{3}$$

$$F(V_t) = \begin{cases} \frac{1}{\rho - \mu_v} V_t - \frac{1}{\rho}C + \frac{1}{1 - \beta_2} \left[(1 - \phi)I + \frac{1}{\rho}C \right] \left(\frac{V_t}{V^*} \right)^{\beta_2} & \text{for } V_t > V^* \\ (1 - \phi)I - V^* & \text{for } V_t \leq V^* \end{cases} \tag{4}$$

Equation (3) denotes the optimal threshold value, where ϕ means the depreciation rate and $(1 - \phi)I$ represents the degree of irreversibility, while C is the cost payout when the project is ongoing. Notably, since the underlying assets are non-traded, we use expected rate of return to replace the risk free rate provided by Wong [15]. Equation (4) indicates the binary optimal abandonment rule. A higher value of ϕ implies a higher degree of irreversibility of the investment. $\phi = 0$ and $\phi = 1$ signify full reversibility and complete irreversibility, respectively.

3   Incorporating the Concept of Fuzziness to Evaluate an Abandonment Option

This section uses the real options approach to evaluate the benefits of a firm's abandonment of a project, using a fuzzy approach to measure sentiments (optimism or pessimism) (Zadeh [16]).

Proposition 2. Suppose that the manager of one company whose investment decision-making is affected by his or her own sentiment (we use fuzzy approach to evaluate this effect) wants to invest in a project, and there exists a series of cash flows and the depreciation rates. The optimal threshold and real option value for the manager to abandon this project are as follows.

$$\tilde{V}^{*\pm} = \frac{1}{1 + \frac{1}{(\rho - \mu_v)} \mp \frac{(1 - \alpha)\theta_{\alpha,t}^{\pm}}{\beta_2 - 1}} \frac{\beta_2}{\beta_2 - 1} \left[(1 - \phi)I + \frac{1}{\rho}C \right] \tag{5}$$

$$\tilde{F}^{\pm}(V_t) = \begin{cases} \frac{1}{\rho - \mu_v} V_t - \frac{1}{\rho}C + \frac{[1 \pm (1 - \alpha)\theta_{\alpha,t}^{\pm}] \frac{1}{\rho - \mu_v}}{-1 + \frac{1}{\rho - \mu_v} \mp \frac{(1 - \alpha)\theta_{\alpha,t}^{\pm}}{\beta_2 - 1}} \frac{1}{1 - \beta_2} \left[(1 - \phi)I + \frac{1}{\rho}C \right] \left(\frac{V_t}{\tilde{V}^{*\pm}} \right)^{\beta_2} & \text{for } V_t > \tilde{V}^{*\pm} \\ (1 - \phi)I - \tilde{V}^{*\pm} & \text{for } V_t \leq \tilde{V}^{*\pm} \end{cases} \tag{6}$$

Equation (5) denotes the optimal fuzzy threshold values, $\tilde{V}^{*\pm}$, which represent the rules for an optimal abandonment option. The "$\pm$" denotes the fuzzy random variables which represent the optimal boundary. Different from Eq. (3), Eq. (5) reveals the manager's sentiments, where "+" represents optimism and "−" denotes pessimism. Optimistic managers continue the project when the fuzzy project value is significantly higher than the left-hand side boundary. In contrast, the pessimistic mangers would liquidate the project when $V_t < \tilde{V}^{*+}$. However, if the value is located between the left-hand and right-hand boundaries, i.e. $\tilde{V}^{*-} < V_t < \tilde{V}^{*+}$, the capital budgeting decision becomes ambiguous. Pessimists would exercise their option to abandon while the optimists would not. Different from Eq. (4), Eq. (6) reveals the abandonment project value and the managerial sentiments with irreversible investments and in an uncertain environment. The sentiments can be described as fuzzy regions, which lead a project to have a range of values rather than a single one.

4 Comparative Static Numerical Analysis

To gain more insights into the quantitative effects of the abandonment option on a firm's capital budgeting decisions and firm value, we compute the decision making bias (*DMB*)and treat it as a proxy for strategic considerations, which measures the differences in manager's subjective judgments in order to explore the issue of *DMB*.

$$\text{DMB} = \frac{\text{Ongoing - Abandonment}}{\text{Ongoing}} \qquad (7)$$

Panel A in Table 1 presents the effects of the growth rate (μ_V) of the project on optimal abandonment decisions. An increase in μ_V both increases the value of the abandonment option and DMB. A favorable expectation in the fundamental model shows that managers who enter a higher growth rate would produce both project-related profits and value in relation to the abandonment option'. As for the fuzzy stochastic model, both optimistic and pessimistic managers would strengthen their perceptions about the project due to the rising growth rate. Panel B in Table 1 shows that an increase in the expected rate of return on the investment project (ρ) makes the project less valuable and much more likely to be abandoned. This suggests that managers should abandon the project when the expected rate of return is low. Meanwhile, the fuzzy stochastic model indicates both optimistic and pessimistic managers would strengthen their feelings about the project.

Panel C shows that increasing the variance, (σ_V^2), raises the value of abandonment, ongoing and DMB. The numerical results from the fundamental model are driven by the fact that high volatility refers to high risk, and that firms need more return to offset the potential losses in an uncertain environment. The DMB monotonically increases in σ_V^2, which indicates that the firm should delay the exercise of the abandonment option. In the fuzzy model, an increase in the volatility of the abandonment option also expands the credit region of abandonment option's value, and the possibility for both optimistic and pessimistic manager to continue the project increases. Again, the variance parameter represents an uncertain environment and its payoff function is a convex one which leads both optimistic and pessimistic managers to continue the project and invest more capital in it. Panel D shows that an increasing in operation cost (C) reduces

Table 1. Abandonment, Ongoing, and *DMB* for abandonment option

Panel A: growth rate (μ_v)

	Fundamental model			Fuzzy stochastic model		
	1.0%	1.5%	2.0%	1.0%	1.5%	2.0%
Abandonment	8.6130	8.6500	8.6870	[8.60830,8.6175]	[8.64570,8.6546]	[8.64570,8.6546]
Ongoing	14.227	14.758	15.330	[14.2168,14.238]	[14.7476,14.769]	[15.3194,15.334]
DMB	0.3946	0.4139	0.4333	[0.39450,0.3948]	[0.41380,0.4140]	[0.43560,0.4358]

Panel B: expected interest rate (ρ)

	10%	15%	20%	10%	15%	20%
Abandonment	8.9610	8.6500	8.3550	[8.95850,8.96270]	[8.64570,8.65460]	[8.34790,8.36210]
Ongoing	23.201	14.758	11.005	[23.1972,23.2058]	[14.7476,14.7687]	[10.9835,11.0279]
DMB	0.6140	0.4140	0.2410	[0.61380,0.61378]	[0.41380,0.41400]	[0.23400,0.24170]

Panel C: variance (σ_V^2)

	3.0%	5.0%	7.0%	3.0%	5.0%	7.0%
Abandonment	8.6110	8.6500	8.6770	[8.60730,8.61540]	[8.64570,8.65460]	[8.67220,8.68150]
Ongoing	14.663	14.758	14.835	[14.6563,14.6702]	[14.7476,14.7687]	[14.8218,14.8487]
DMB	0.4130	0.4140	0.4150	[0.41340,0.41272]	[0.41380,0.41340]	[0.41490,0.41530]

Panel D: operating cost (C)

	3.0%	5.0%	7.0%	3.0%	5.0%	7.0%
Abandonment	8.6620	8.6500	8.6390	[8.6573,8.6660]	[8.64570,8.65460]	[8.63410,8.64310]
Ongoing	14.878	14.758	14.639	[14.868,14.888]	[14.7476,14.7687]	[14.6277,14.6498]
DMB	0.4180	0.4140	0.4100	[0.4177,0.4179]	[0.41380,0.41340]	[0.40970,0.41000]

Panel E: depreciation rate (ϕ)

	Fundamental model			Fuzzy stochastic model		
	3.0%	5.0%	7.0%	3.0%	5.0%	7.0%
Abandonment	8.8330	8.6500	8.4670	[8.82830,8.83730]	[8.64570,8.65460]	[8.46310,8.47180]
Ongoing	14.779	14.758	14.738	[14.7679,14.7906]	[14.7476,14.7687]	[14.7283,14.7479]
DMB	0.4020	0.4140	0.4254	[0.40230,0.40250]	[0.41380,0.41400]	[0.42540,0.42560]

Panel F: initial investment inputs (I)

	9	10	11	9	10	11
Abandonment	7.7820	8.6500	9.5180	[7.7783,7.78620]	[8.64570,8.65460]	[9.51320,9.52290]
Ongoing	14.672	14.758	14.869	[14.665,14.6795]	[14.7476,14.7687]	[14.8543,14.8839]
DMB	0.4696	0.4140	0.3600	[0.4696,0.46960]	[0.41380,0.41400]	[0.35957,0.36019]

Panel G: current cash inflow(V_t)

	1.5	2	2.5	1.5	2	2.5
Abandonment	8.6500	8.6500	8.6500	[8.64570,8.6546]	[8.64570,8.65460]	[8.64570,8.65460]
Ongoing	11.372	14.758	18.338	[11.3494,13.272]	[14.7476,14.7687]	[18.3323,18.3439]
DMB	0.2390	0.4140	0.5280	[0.23820,0.3479]	[0.41380,0.41400]	[0.52840,0.52820]

Note: We set the value of each parameter in the base case as follows: Annualized growth rate, μ_V, is equal to 1.5%; the expected rate of return, ρ, is equal to 15%; the annualized variance, σ_V^2, is equal to 5%; the operating cost, C, is 5%; the depreciation rate, ϕ, is 5%; the initial investment inputs, I, is equal to 10; and the current cash inflow, V_t, is equal to 2.

the value of abandonment, ongoing and DMB simultaneously, and thus the manager should exercise the abandonment option at an early stage in order to earn more profit, which implies that the firm would rather hold on cash than continue an uncertain project under the fundamental model. The fuzzy stochastic model also indicates that both optimistic and pessimistic managers would feel anxious about a rise in the operation cost.

Panel E shows that an increase in the depreciation rate (ϕ) reduces both abandonment and ongoing but increases the value of DMB. These results confirm the comparative static analysis results in Table 1. The fundamental model reveals that terminating a project at an early stage would lead to more profits, and would also obtain the potential benefit from real options. More importantly, the depreciation rate is one of the most important factors for pricing the abandonment option under an irreversible investment environment. Again, a higher rate of ϕ implies a higher degree of irreversibility of investment, where $\phi = 0$ signifies full reversibility and $\phi = 1$ signifies complete irreversibility. The fuzzy model suggests that both optimistic and pessimistic managers would feel anxious for the high depreciation of a project and terminate it at the early stage, but would also gain the potential benefits from real options. Again, a higher depreciation rate implies higher irreversibility. In panel F, increasing in the initial investment input (I) enhances both the value of abandonment and ongoing, but decreases the value of DMB. In other words, an increase in I would raise the project's value when abandoning the project at a later stage, but would lose the potential benefit from real options. This concept is the same as diminishing utility. The fuzzy stochastic model shows that increasing the initial investment inputs also widens both the region of the abandonment and ongoing, but the region of DMB is shrinking. Both optimistic and pessimistic managers would strengthen their intentions to continue the project and abandon it at a later stage, but would lose the potential benefit from real options. In panel G, an increase in the current inflow (V_t) enhances the value of ongoing, while the abandonment value remains unchanged and DMB is gradually increasing in V_t. The increase in V_t thus induces the manager to invest in the project and continue holding it in order to obtain more profit at a later stage. Under the fuzzy model, an increase in V_t would widen the region of ongoing while the region of the abandonment value remains unchanged, indicating that the residual values after both optimistic and pessimistic managers abandon the project would not change.

5 Conclusion

This paper adopts a fuzzy stochastic process to evaluate the capital budgeting decisions in relation to an irreversible and uncertain project with an abandonment option, as the heterogeneity of managerial sentiments cannot be modeled by existing models. The fuzzy stochastic process helps us to deal with the ambiguousness of such sentiments under real world conditions, especially when we focus on abandonment decision making, in which the value of the project is assumed to follow the geometric Brownian motion. Since the underlying assets are non-traded in the real world, it is difficult to measure the expected rate of return precisely. We treat the discount rate as the firm's required rate of return which is estimated by using CAPM. Different from previous studies, we also include the instantaneous cash flow, which combines project value (inflow) and operating cost (outflow), into a differential equation.

This paper makes a number of contributions to the literature. First, this paper extends the work of Liu and Jiang's [11] model to consider irreversible investments and an uncertain environment, which can be represented by the depreciation rate and variance of project, respectively. Since the underlying assets are non-tradable, this would lead the project value to be within a region which represents the investment based on managerial sentiment. We thus adopt a fuzzy approach to represent managerial sentiment and use decision making bias (DMB) as a proxy for strategic considerations(Kaiser and Stouraitis [8]). Second, even if the underlying asset of the abandonment option is non-tradable, we can provide a closed-form solution to evaluate the value of the option and the reasonable expected rate of return, and this return is different from the constant risk free rate in Wong's [15] model.

Acknowledgements. The authors would like to thank the anonymous reviewers for their valuable comments and suggestions to improve the quality of the paper. Professor Liu and Jiang acknowledge financially (partially) supported by the Ministry of Science and Technology of the Republic of China, Taiwan (MOST 105-2410-H-006-015 and MOST 105-2632-H-155-001).

References

1. Black, F., Scholes, M.: The pricing of options and corporate liabilities. J. Polit. Econ. **81**(3), 637–654 (1973)
2. Brennan, M., Schwartz, E.: Evaluating natural resource investments. J. Bus. **58**(2), 135–157 (1985)
3. Comment, R., Jarrell, G.A.: Corporate focus and stock returns. J. Financ. Econ. **37**, 67–87 (1995)
4. DeBondt, W., Thaler, R.: Financial decision-making in markets and firms: a behavioral perspective. In: Jarrow, R., et al. (eds.) Handbooks in OR and MS, vol. 9, pp. 385–410. Elsevier, Amsterdam (1995)
5. Dixit, A.K., Pindyck, R.S.: Investment Under Uncertainty. Princeton University Press, Princeton (1994)
6. Dyl, E., Long, H.: Abandonment value and capital budgeting: comment. J. Finance **24**, 88–95 (1969)
7. John, K., Ofek, E.: Asset sales and increases in focus. J. Financ. Econ. **37**, 105–126 (1995)
8. Kaiser, K.M.J., Stouraitis, A.: Agency costs and strategic considerations behind sell-offs: The UK evidence. Eur. Financ. Manage. **7**(3), 319–349 (2001)
9. Kaplan, N.S., Weisbach, M.S.: The success of acquisitions: evidence from divestitures. J. Finance **47**, 107–138 (1992)
10. Lang, L., Poulsen, A., Stulz, R.: Asset sales firm performance, and the agency costs of managerial discretion. J. Financ. Econ. **37**(1), 3–37 (1995)
11. Liu, Y.H., Jiang, I.M.: Influence of investor subjective judgments in investment decision-making. Int. Rev. Econ. Financ. **24**, 129–142 (2012)
12. McDonald, R., Siegel, D.: Investment and the valuation of firms when there is an option to shut down. Int. Econ. Rev. **26**, 331–349 (1985)

13. Merton, R.: Theory of rational option pricing. Bell J. Econ. Manage. Sci. **4**(1), 141–183 (1973)
14. Myers, S.C., Majd, S.: Abandonment value and project life. In: Advances in Futures and Options Research, vol. 4, pp. 1–21 (1990)
15. Wong, K.P.: The effects of abandonment options on investment timing and intensity. Bull. Econ. Res. **64**(3), 305–318 (2012)
16. Zadeh, L.A.: Fuzzy sets. Inf. Control **8**, 338–353 (1965)

An Imputation for Missing Data Features Based on Fuzzy Swarm Approach in Heart Disease Classification

Mohd Najib Mohd Salleh[✉] and Nurul Ashikin Samat

Faculty of Computer Science and Information Technology,
Universiti Tun Hussein Onn Malaysia (UTHM),
Parit Raja, 86400 Batu Pahat, Johor, Malaysia
najib@uthm.edu.my, gi140022@siswa.uthm.edu.my

Abstract. Computational intelligence methods have been broadly applied to define the important features for Heart Disease classification. Nonetheless, imprecise features data such as no values and missing values can affect quality of classification results. Nevertheless, the other complete features are still capable to give information in certain features. Therefore, an imputation approach based on Fuzzy Swarm is developed in preprocessing stage. It will help to fill in the missing values by cluster the complete candidates and optimizes it using Particle Swarm Optimization (PSO). Then, the complete dataset is trained in classification algorithm, Decision Tree. The experiment is trained with Heart Disease dataset and the performance is analyzed using accuracy, precision, and ROC values. Results show that the performance of Decision Tree is increased after the application of Fuzzy Swarm for imputation.

Keywords: Fuzzy C-Means · Particle Swarm Optimization · Imputation · Preprocessing · Decision Tree

1 Introduction

Decision Tree (DT) classifier grow popular and competent classification techniques among the researchers [1–3]. The clear visualization of tree and rules set extracted gives an advantage to the user to make a decision. Motivated by the increasing number of heart disease patients' data around the world [4], researchers have done various studies to help the experts to diagnose early detection of heart disease using DT. However, problems arise when the data collected is not complete thus there will have problems arising in decision making. Real life data especially medical data is hard to collect the accurate, precise, and complete data. Apart from that, incomplete dataset also will affect the accuracy of DT model, might create bias result, and produced inadequate rule set.

According to Rubin [5] missing data can be categorized into three which are Missing Completely at Random (MCAR), Missing at Random (MAR), and Not Missing at Random (NMAR). MCAR category is the missing value has no relationship or dependency towards other data set or variable. While MAR is the missing value that

© Springer International Publishing AG 2017
Y. Tan et al. (Eds.): ICSI 2017, Part II, LNCS 10386, pp. 285–292, 2017.
DOI: 10.1007/978-3-319-61833-3_30

depended on other variables but the missing value can be obtained by estimated other complete variables. The NMAR is the missing value that depended on other missing value, therefore the missing values cannot be estimated from existing data. In this study, MAR data has been investigated.

There are several ways to treat the incomplete dataset such as (1) filter-based (No Imputation) or (2) by filling in the missing values (Imputation). Filter-based treatment means that the missing value is ignored or deleted. According to an investigation by [6], mean imputation is acceptable and suitable if the missing data is in a small number. However, this method will affect the quality of the dataset and it may result to biased data in a bigger number [7, 8]. In recent years, there has been increasing amount of literature on imputation. Mean imputation is the earliest imputation method used by past researchers [9]. Author replaced the missing values with the mean value for the attribute. Although it is easy to use, many researchers have argued that it will affect the relationship between attributes or variables [10]. Meanwhile, k-NN imputation method is a common method to impute the missing data with an actual range of datasets [11]. However, to get a better range of dataset values to substitute, grouping or clustering the data with same similarity features or data will increase the accuracy of imputation values. Therefore, FCM has been proposed and implemented as imputation method in preprocessing stage [8]. Imputation is a way to replace the missing values with appropriate values. Although, there are many imputation ways such as mean imputation, the accuracy for the replaced value can still be improved.

Thus, an alternative imputation method named Fuzzy Swarm (FS) which combined the benefit of Fuzzy C-Means algorithm, known to have good capabilities for their ability to cluster the data [12] and Particle Swarm Optimization algorithm, renowned to have a worthy ability and simple framework in optimization [13] is implemented for training the imputation method in preprocessing stage to overcome the missing data handicaps.

In this work, FS is used as imputation method for Heart Disease dataset in preprocessing stage before trained with Decision Tree. The result of the classification after implemented FS is compared with no imputation, Mean, k-NN, and FCM imputation to examine whether the FS imputation method is capable to improve Decision Tree performance for the task of classification.

This paper is organized as follows: Sect. 2 reviews the related works regarding missing data, FCM and PSO in imputation works, Sect. 3 overviews the proposed methodology, Fuzzy Swarm used in this paper. Section 4 explained about experiments and result analysis. Finally, Sect. 5 provides the conclusion and recommendations.

2   Related Works

In this section, the fundamental scheme of FCM and PSO as well as the proposed fuzzy swarm optimization method are briefly discussed.

2.1 Fuzzy C-Means

Over the last few decades, fuzzy has shown as a great approach for dealing with imprecision and nonlinearity efficiently. Fuzzy C-Means (FCM) algorithm is a clustering tool which overcomes the hard clustering drawback. The clusters are modelled by fuzzy sets assigning each data item of the dataset to each cluster with a membership degree that ranges between 0 and 1. FCM partitions set of n dataset $x = \{x_1, x_2,..., x_n\}$ in R^d dimensional space into fuzzy cluster c, $1 < c < n$ with $r = \{r_1, r_2,..., r_c\}$ cluster centers or centroids. The objective function of FCM algorithm is defined as follows:

$$J_m = \sum_{j=1}^{c} \sum_{i=1}^{n} \mu_{ij}^m \|x_i - r_j\|_2^2 \tag{1}$$

In which $m(m > 1)$ is a scalar termed the weighting exponent and controls the fuzziness of the resulting clusters and $\|x_i - r_j\|_2^2$ stands for the Euclidean distance from dataset, x_i to the cluster center, r_j.

2.2 Particle Swarm Optimization

Particle Swarm Optimization (PSO) algorithm is an optimization tool. The algorithm simulates the natural behavior of bird flocking or fish schooling to find food and was proposed to solve many problems by Eberhart [14]. The PSO algorithm mimics the behavior of birds which fly in a group follows the member that has closest distance to destination. In traditional PSO, population is called the swarm and the candidate of solutions in swarm is called particles while the food is called objective function. Each elements of particle contains parameter; own position, own velocity, and own historical information. Each particle is given random position in search space and random velocity for the particles to fly within the search space. To search for the optimal solution, at each time step, each changes its velocity according to the *pbest* and *gbest* parts according to Eqs. (2) and (3), respectively:

$$v_{id}^t = v_{id}^{t-1} + c_1 r_1 (P_{id}^t - x_{id}^t) + c_2 r_2 (P_{gd}^t - x_{id}^t), d = 1, 2, ..., D \tag{2}$$

$$X_{id}^{t+1} = X_{id}^t + V_{id}^t, d = 1, 2, ..., D \tag{3}$$

Where c_1 indicates the cognition learning rates for individual ability, c_2 indicates the social learning factor and r_1, r_2 are random numbers uniformly distributed in the interval 0 and 1.

3 Fuzzy Swarm Imputation Method

The proposed method is based on the incorporation of FCM and PSO are used to find the optimum value for finding the best value to replace the missing value in the dataset. It is inspired by the efficiency of FCM to group the data with better similarity values and the flexibility and easy to implement of PSO algorithm. The proposed method was expected to overcome the disadvantages caused by missing data in preprocessing stage and improve the classification accuracy.

Step 1:	Normalize the dataset using min-max normalization for the features respectively. Separate the Complete and Incomplete data.
Step 2:	For all Complete dataset,
	(2.1) Calculates the cluster center,

$$r_j = \frac{\sum_{i=1}^{n} \mu_{ij}^m x_i}{\sum_{i=1}^{n} \mu_{ij}^m}$$

(2.2) Computes Euclidean distance $\|x_i - r_j\|_2^2$

(2.3) Updates the membership function using

$$\mu_{ij} = \frac{1}{\sum_{k=1}^{c} \left(\frac{d_{ij}}{d_{ik}}\right)^{\frac{2}{m-1}}}$$

(2.4) If condition is not met, repeat Step 2.

Step 3: For all Incomplete dataset
Calculate the imputation,

$$x_{iMiss} = \sum_{k=1}^{k} U(x_{iCom}, C_k) \bullet (C_k)$$

End for;

Step 4: For each Imputed dataset,
(4.1) Initialize the population size, N = 50; Maximum Iterations, Iters = 100; C1 and C2 = 2; and create a swarm with P particles
(4.2) Calculates the fitness each particle using Equation (6);
(4.3) Calculates the pbest for the particles and gbest for the swarm;
(4.4) Update velocity for using Equation (2) and update the position of matrix using Equation (3);
(4.5) Terminating if condition met otherwise repeat step 4.

Step 5: The optimum value for Imputed dataset is obtained.

4 Experiments, Results and Discussions

4.1 Decision Tree

After the incomplete dataset has been imputed using proposed technique, FS, the complete impute dataset will be trained in Decision Tree algorithm in Waikato Environment for Knowledge Analysis (WEKA) Version 3.6.11. This study used C4.5 algorithm by Quinlan [15]. The data will be partitioned into 60% for training and 40% for testing phase. Training set will build the classification algorithm model and the model will be tested with remaining percentage for testing set.

4.2 Heart Disease Dataset

The Framingham Heart Study is supported by the National Heart Lung and Blood Institute (NHLBI), a part of the National Institutes of Health (NIH). NIH has been committed to classify the collective causes that contribute to Heart Disease. Framingham Heart dataset was got from 5209 men and women between the ages of 32 and 70 from the town of Framingham, Massachusetts. There are 10 attributes in the dataset which are Age, Gender, Systolic blood pressure, Diastolic blood pressure, heart rate, blood Sugar, BMI, cholesterol, smoking and chest pain. The dataset contains missing values in 4 attributes which is Heart Rate, Blood Sugar, BMI, and Cholesterol. In these experiments, there are no artificial missing ratios were inserted into the following dataset.

4.3 Simulations Results

The result of proposed method is compared with No Imputation, Mean Imputation, k-NN Imputation, and FCM Imputation. The performances of the five methods are compared with respect to their corresponding Accuracy, Precision, and ROC values.

The results from Fig. 1 shows the accuracy value of proposed method is higher with 86.3% compared to other imputations. Thus, proves that FS helps the Decision Tree to classify well with better imputation values for missing data. It also shows that missing data can lead to error and confusion in interpreting the data by ignores it. It is disadvantages as the data may contain other important information in the dataset.

Apart from that, although Mean and k-NN Imputation easy to use, the accuracy is lower due to higher number of missing values in the dataset. Hence, it will affect the correlation between the features during classification process. Even FCM shows high accuracy, but with the help of optimization in FS, it gives more accurate classification results.

The results from Fig. 2 show the precision comparison between No Imputation, Mean Imputation, k-NN Imputation, FCM, and FS respectively. From the figure clearly shows that FS gives highest precision value compared to other imputation which means that, it offers better accuracy and more precise reading towards Heart Disease problems.

ROC is a measure tools to evaluate the efficiency of classifiers through the determination of the rates of true positives (elements correctly classified as positive class)

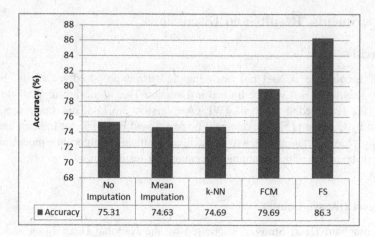

Fig. 1. Accuracy of Decision Tree

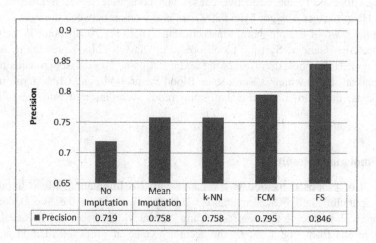

Fig. 2. Precision of Decision Tree

Table 1. ROC values for Decision Tree

	No imputation	Mean imputation	K-NN imputation	FCM imputation	FS imputation
ROC values	0.671	0.690	0.710	0.750	0.830

and false positives (elements incorrectly classified as positive class). From Table 1 it shows that, ROC curve are more to value 1. It shows that, the ability of the test to correctly classify those with and without the Heart Disease.

5 Conclusion and Future Works

The existence of missing values in database is a common fact and can generate problems on the knowledge extractions. Thus, imputation in preprocessing stage will help the data mining algorithm works more accurate. By using the proposed imputation method, FS, it can help to get better imputation by substitute the most plausible values to missing value. The use of PSO which is an optimization algorithm allows FCM to expand the ability to find the best imputation values where the results have led to a significant improving in the classification. The accuracy of Decision Tree results clearly showed that the imputed dataset using FS has improved the classification algorithm analysis compared to no imputation dataset, Mean imputation dataset, k-NN imputation dataset and FCM dataset. To conclude, FS can be considered as a promising imputation method in preprocessing stage for future research.

Acknowledgments. This research was funded by Universiti Tun Hussein Onn Malaysia Research Contract Grant Scheme.

References

1. Tsang, S., Ben, K., Yip, K.Y., Wai-Shing, H., Sau Dan, L.: Decision trees for uncertain data. IEEE Trans. Knowl. Data Eng. **23**, 64–78 (2011)
2. Mohamed, W.N.H.W., Salleh, M.N.M., Omar, A.H.: A comparative study of Reduced Error Pruning method in decision tree algorithms. In: 2012 IEEE International Conference on Control System, Computing and Engineering (ICCSCE), pp. 392–397 (2012)
3. Tomar, D., Agarwal, S.: A survey on Data Mining approaches for Healthcare. Int. J. Bio Sci. Bio Technol. **5**, 241–266 (2013)
4. http://gamapserver.who.int/gho/interactive_charts/ncd/mortality/cvd/atlas.html
5. Rubin, D.B.: Inference and missing data. Biometrika **63**, 581–592 (1976)
6. Batista, G.E., Monard, M.C.: An analysis of four missing data treatment methods for supervised learning. Appl. Artif. Intell. **17**, 519–533 (2003)
7. Dhevi, A.T.S.: Imputing missing values using Inverse Distance Weighted Interpolation for time series data. In: 2014 Sixth International Conference on Advanced Computing (ICoAC), pp. 255–259 (2014)
8. Aydilek, I.B., Arslan, A.: A hybrid method for imputation of missing values using optimized fuzzy c-means with support vector regression and a genetic algorithm. Inf. Sci. **233**, 25–35 (2013)
9. Ravi, V., Krishna, M.: A new online data imputation method based on general regression auto associative neural network. Neurocomputing **138**, 106–113 (2014)
10. Enders, C.K.: A primer on maximum likelihood algorithms available for use with missing data. Struct. Equ. Model. **8**, 128–141 (2001)
11. Jönsson, P., Wohlin, C.: An evaluation of k-nearest neighbour imputation using likert data. In: 10th International Symposium on Software Metrics 2004. Proceedings, pp. 108–118. IEEE (2004)
12. Niu, Q., Huang, X.: An improved fuzzy C-means clustering algorithm based on PSO. J. Software **6**, 873–879 (2011)

13. Mohd Rasip, N., Basari, A.S.H., Ibrahim, N.K., Hussin, B.: Enhancement of nurse scheduling steps using particle swarm optimization. In: Sulaiman, H.A., Othman, M.A., Othman, M.F.I., Rahim, Y.A., Pee, N.C. (eds.) Advanced Computer and Communication Engineering Technology. LNEE, vol. 315, pp. 459–469. Springer, Cham (2015). doi:10.1007/978-3-319-07674-4_45
14. Eberhart, R.C., Kennedy, J.: A new optimizer using particle swarm theory. In: Proceedings of the Sixth International Symposium on Micro Machine and Human Science, New York, NY, pp. 39–43 (1995)
15. Quinlan, J.R.: Induction of decision trees. Mach. Learn. 1, 81–106 (1986)

Clustering and Forecast

Total Optimization of Smart City Using Initial Searching Points Generation Based on k-means Algorithm

Mayuko Sato and Yoshikazu Fukuyama[✉]

Meiji University, Tokyo, Japan
yfukuyama@meiji.ac.jp

Abstract. This paper proposes total optimization of smart city (smart community, SC) using initial searching points generation based on k-means algorithm. In this paper, energy flow models of various sectors in SC are utilized. Namely, energy supply models such as electric power utility, natural gas utility, drinking water treatment plant, and waste water treatment plant, and energy consumption models such as industry, building, residence, and railroad are utilized. Using the SC model, energy costs, actual electric power at peak load hours, and the amount of CO_2 emission of the whole SC is minimized. This paper proposes an initial searching points generation method based on K-means algorithm in order to set the initial searching points in several attractive areas in which quality of solutions are relatively high, without prior information. The proposed method is applied to a model of Toyama city, which is a moderately-sized city in Japan. Optimal operation by the proposed method is compared with that by pseudo-random number generator (PRNG) based initial searching points.

Keywords: k-means · Differential evolutionary particle swarm optimization · Smart city · Total optimization

1 Introduction

SC demonstration projects have been conducted all of the world in order to reduce the CO_2 emission [1, 2]. Using renewable energies such as solar power generation, storage batteries, and recent information technologies, SC can realize a sustainable and low carbon emission community. The importance of SC increases all over the world.

Since it is difficult to evaluate how much the communities can reduce CO_2 emission in actual SC, SC should be evaluated by models. Various models have been developed at each sector such as electric power utility, natural gas utility, drinking water treatment plant, waste water treatment plant, industry, building, residence, and railroad. Static models considering various energy balances and dynamic models considering transient phenomenon have been developed in various sectors [3–5]. However, the models, which can calculate energy consumption and environmental loads among all of the sectors in SC considering communication among various sectors, had not been developed yet. Therefore, a SC model have been developed in Japan so that they can evaluate energy costs or the amount of CO_2 emission of the whole SC [6–8].

© Springer International Publishing AG 2017
Y. Tan et al. (Eds.): ICSI 2017, Part II, LNCS 10386, pp. 295–303, 2017.
DOI: 10.1007/978-3-319-61833-3_31

Using the SC model, the authors have proposed total optimization of whole SC which minimizes energy costs and electric power loads at high load hours namely, peak load shifting by PSO [9], and DE [10]. Reduction of search space considering not only facility characteristics, but also load and cost characteristics and continuity of weekday operation has been also proposed. The problem is a kind of large scale optimization problems and generally, there are two strategies to get high quality solution for the problems. One is generating of appropriate initial searching points and another one is using optimization methods that work efficiently for the problems. Up to now, while the authors have focused on the latter strategy, this paper focuses on the former one.

According to Proximate Optimality Principle (POP) [11], high quality solutions can be found near high quality solutions. Namely, there is a possibility to obtain higher quality solutions by setting the initial searching points in several attractive areas in which quality of solutions are relatively high. Considering above, this paper proposes an initial searching points generation method based on K-means algorithm in order to set the initial searching points in several attractive areas in which quality of solutions are relatively high, without prior information. The proposed method utilizes Differential Evolutionary Particle Swarm Optimization (DEEPSO) for obtaining high quality solutions as well. The proposed method is applied to a model of Toyama city, which is a moderately-sized city in Japan. Optimal operation by the proposed method is compared with that by pseudo-random number generator (PRNG) based initial searching points with promising results.

2 Smart Community Model

2.1 Concept of the Whole Smart Community Models [6]

Japanese experts developed a SC model that can calculate costs and the amount of CO_2 emission in the SC. This model can deal with energy costs and the amount of CO_2 emission. The SC model includes various sectors such as electric power utility, natural gas utility, drinking water treatment plant and waste water treatment plant, industry, building, residence, and railroad. The sector models can be divided into supply-side and demand-side groups. Using the sector models, energy flows among various sectors, the amount of CO_2 emission in various sectors, and the amount of energy supply, demand, and CO_2 emission in the whole SC can be calculated quantitatively.

(a) Energy Supply Models [7]

The supply-side group consists of electric power utility, natural gas utility, drinking water treatment plant, and waste water treatment plant models. The sector models interactively supply electric power, natural gas, and drinking water to the demand-side models. Total supply of each energy in supply-side sectors should be equal to total energy consumption of each energy in demand-side sectors.

(b) Energy Consumption Model [8]

The demand-side group consists of industrial, building, residential and railroad sector models. The models cooperate with energy supply models.

Industrial, building, and residential sector models have energy supply facilities and various energy loads. Primary energies are supplied to the energy supply facilities from the energy supply group and the energy supply facilities supply secondary energies to various energy loads. Therefore, if various hourly load values of one day (24 points) are given, required primary energy values are made clear and the values are supplied from the energy supply group.

3 Problem Formulation of Total Optimization of Whole Smart Community

3.1 Decision Variables

Each decision variables have hourly numbers. (24 decision variables at each) All of decision variables are 648 variables. Therefore, the problem is a large-scale optimization problem. As Example, decision variables of an industrial model are as follows.

The amount of electric power output by a gas turbine generator, the amount of heat output by turbo refrigerators, the amount of heat output by steam refrigerators, and the amount of charged/discharged electric power of a storage battery.

3.2 Objective Function

The problem deals with three objective functions. The first objective function is minimization of 24 h energy costs of sectors except electric power and natural gas utilities. The second one is minimization of actual electric power loads of the whole SC at peak load hours, namely, peak load shifting. The actual electric power load at each hour is summation of the amount of stored electric power, the amount of consumed electric power for converting the electric power to other energies, and the original electric power load at each hour. High load hours of electric power are hours when summation of actual electric power loads of sectors except electric power and natural gas utilities are higher than the average of the summation throughout the day. The peak load hours of actual electric power loads are high load hours when the summation of actual electric power loads are higher than the average of the actual electric loads at high load hours. The third one is minimization of CO_2 emission. The three object functions are collectively arranged in one weighted function as follows:

$$min\{w_1 \sum_{n=1}^{N} \sum_{j=1}^{T} (BuyG_{nj} \times GU_{nj} + BuyE_{nj} \times EU_{nj}) + w_2 \sum_{n=1}^{N} \sum_{m=mps}^{mpe} (GL_{nm})$$
$$+ w_3 \sum_{n=1}^{N} \sum_{j=1}^{T} (BuyG_{nj} \times GC + BuyE_{nj} \times EC)\}$$

$$(1)$$

where, N is the number of sectors except electric power and natural gas, T is the number of hours per day (=24), $BuyG_{nj}$ is the amount of purchased natural gas of sector n at hour j, GU_{nj} is a unit cost of purchased natural gas of sector n at hour j, $BuyE_{nj}$ is the amount of purchased electric power of sector n at hour j, EU_{nj} is a unit cost of purchased electric power of sector n at hour j, GL_{nm} is an actual electric power load of sector n at hour m, mps is the start hour of peak load hours of actual electric power loads, mpe is the final hour of peak load hours of actual electric power loads, GC is a coefficient to change from the amount of purchased natural gas to the amount of CO_2 emission, EC is a coefficient to change from the amount of purchased electric power to the amount of CO_2 emission, w_1, w_2, and w_3 are weighting coefficients ($w_1 + w_2 + w_3 = 1$).

The amounts of electric power and natural gas purchase are calculated after calculation of decision variables as dependent variables. Therefore, penalty function values are added to the objective function values if the calculated dependent variables are out of allowable ranges.

3.3   Constraints

(a) Energy Balances

Electric power, steam, and heat energy balances are treated in each model.

(b) Characteristics of Facilities

Facility characteristics can be expressed with equality constraints, and upper and lower limits of facility outputs can be expressed with inequality constraints. Details of the constraints can be found in [9].

The problem should be formulated as a mixed integer nonlinear optimization problem (MINLP) and various evolutionary computation techniques should be applied to the problem. This paper applies DEEPSO for obtaining high quality solutions.

4   Generation of Initial Searching Points by K-Means Algorithm and DEEPSO for Total Optimization of SC

4.1   Deepso [12]

DEEPSO can generate higher quality solutions than conventional PSOs. DEEPSO utilizes various processes as shown below.

– REPLICATION: each particle is replicated R times.
– MUTATION: the following equations are utilized to mutate weights of all R clones:

$$A^{k+1} = A^k + \tau N(0, 1). \tag{2}$$

$$B^{k+1} = B^k + \tau N(0, 1). \tag{3}$$

$$C^{k+1} = C^k + \tau N(0, 1). \tag{4}$$

$$b_G = b_G + \tau' N(0, 1). \tag{5}$$

where, A^k, B^k, and C^k are weight coefficients at iteration k, τ and τ' are learning parameters, $N(0, 1)$ is a random variable with Gaussian distribution (mean is 0 and variance is 1).

- REPRODUCTION: offspring (original and clones) are generated by the following equations:

$$x_i^{k+1} = x_i^k + v_i^{k+1}. \tag{6}$$

$$v_i^{k+1} = A^{k+1} v_i^k + B^{k+1}\left(x_{r1}^k - x_{r2}^k\right) + P\left[C^{k+1}\left(b_G - x_i^k\right)\right]. \tag{7}$$

where, x_i^k is a current searching point of agent i at iteration k, v_i^k is current velocity of agent i at iteration k, x_{r1}^k, and x_{r2}^k are randomly selected searching points at iteration k, b_G is the best searching point so far found by the agents (*gbest*).

- EVALUATION: objective function values of all offspring are calculated.
- SELECTION: the offspring forming the next generation are selected by stochastic tournament or other selection methods.

In REPRODUCTION, selected x_{r1}^k and x_{r2}^k are ordered so that the following condition is satisfied by calculating object function values of both agents.

$$f\left(x_{r1}^k\right) < f\left(x_{r2}^k\right). \tag{8}$$

P is set to 0 or 1 by probability p as shown below.

$$P = \begin{cases} 1 & (rand(0, 1) \leq p) \\ 0 & (rand(0, 1) > p). \end{cases} \tag{9}$$

where, $rand(0, 1)$ is a uniform random number between 0 and 1.

4.2   The k-means Based Proposed Initial Searching Points Generation

Generally, the objective function of the total optimization of SC problem is multimodal. According to POP, there is a possibility to obtain better solutions by setting initial searching points near attractive objective function value areas. The proposed algorithm for generation of initial searching points is shown below:

Step.1   Generate initial searching points (Kx_{intial}) randomly inside search space.
Step.2   Divide Kx_{intial} into k clusters using k-means method.
Step.3   Rank the divided points in ascending order, based on objective function values at each cluster.
Step.4   Select initial searching points ($x_{initial}$) whose ranks are smaller than ($AgentNum/K$)-th rank at each cluster.

4.3   Total Optimization Algorithm by DEEPSO and k-means

Step.1   Initial searching points of agents for each decision variable are set inside the reduced search space using the proposed k-means based initial searching points generation.

Step.2   Using the cutout transformation function, initial searching points are transformed to operation variables, and an objective function value is calculated at each searching point. If the operation variables are not inside the reduced search space, penalty values are added to the objective function values.

Step.3   The object function value of all agents are calculated with the actual input and output values of the initial searching points. $pbest$ s and b_G are set with the objective function values.

Step.4   REPLICATION: Each agent is replicated R times.

Step.5   MUTATION: Weighting coefficients $(A, B, and C)$ of all R clones and b_G are mutated by (2) to (5).

Step.6   REPRODUCTION: $R + 1$ offspring (the original and clones) are updated by (6) to (9).

Step.7   The searching points of the updated $R + 1$ offspring are transformed by the cutout transformation function to actual input and output values of facilities.

Step.8   EVALUATION: Objective function values of all offspring are calculated with the transformed actual input and output values of facilities.

Step.9   SELECTION: The agent with the best fitness is selected to the next generation among the updated original and cloned agents.

Step.10   If the objective function value of each selected offspring is smaller than a past $pbest$, $pbest$ is updated. b_G is also updated.

Step.11   When the current iteration number reaches the pre-determined maximum iteration number, the procedure can be stopped and go to Step.12. Otherwise, go to Step.4 and repeat the procedures.

Step.12 b_G calculated by DEEPSO is transformed to operation variables using the cutout transformation function. The obtained operational variables and the objective function values are output as a final solution.

5   Simulation

5.1   Simulation Conditions

The proposed method is applied to a typical mid-sized smart community like Toyama city in Japan. The followings are the number of sector models in each sector [13] so that we can compare the calculated operational values by PRNG and the proposed method:

Drinking water treatment plant: 1, Waste water treatment plant: 1, Industry model: 15, Building model: 50, Residential model: 45000, Railroad: 1

The following parameters are utilized for the cutout transformation functions:

The minimum number of parameter $(\alpha_i, i = 1, \ldots, L)$: 0.5, a parameter for divides: -0.05, the maximum number of parameters $(\beta_i, i = 1, \ldots, N)$: 1.05.

The following parameters are utilized for DEEPSO according to pre-simulations: τ is set to 0.2, τ' is set to 0.006, p is set to 0.75, the initial weight coefficients of each term (A, B, and C) are set to 0.5, the number of clones are set to 1.

The number of agents is set to 50. Kx_{intial} is set to 1000 for the proposed method. Therefore, 50 initial searching points are selected within 1000 randomly generated points by the proposed method. 50 initial points are generated randomly within reduced search space by PRNG. The maximum iteration number is set to 1000, and the number of trials is set to 50. Since it is well known that clustering results by k-means method depends on initial clusters, different initial clusters are utilized for each trial. The simulation software has been developed using C language (gcc version 4.92 on Cygwin) on a PC (Intel Core i7 (3.60 GHz)).

5.2 Simulation Results

Table 1 shows comparison of the average objective function values and standard deviations among initial searching points which is determined randomly by PRNG and the proposed method using various values of K. All of values are rates when the average objective function value of the optimal operation by PRNG method is set to 100%. In this simulation, weighting parameters of case 7 (see Table 2) is utilized as one of examples. Since the target of the problem covers energy efficiency and CO_2 reduction of the whole of city, even 1% corresponds to large cost down and large CO_2

Table 1. Comparison of the average objective function values and standard deviations using PRNG and the proposed method with different values of weighting coefficients

Method	K	Objective function values of initial searching points [%]		Objective function values of the optimal operation [%]	
		Ave.	Std.	Ave.	Std.
PRNG	1	4.65E + 13	8.55E + 11	100	0.27
The proposed method	2	1.19E + 13	7.67E + 11	99.15	0.20
	3	8.88E + 12 .	5.60E + 11	99.16	0.20
	4	9.17E + 12	4.58E + 11	99.15	0.21
	5	9.90E + 12	5.62E + 11	**98.82**	**0.13**
	6	1.03E + 13	5.41E + 11	99.07	0.19
	7	1.08E + 13	5.85E + 11	99.12	0.16
	8	1.12E + 13	5.80E + 11	99.14	0.22
	9	1.19E + 13	7.67E + 11	99.12	0.19
	10	1.23E + 13	7.31E + 11	98.92	0.15

Table 2. Comparison of the average optimal objective function values and standard deviations of seven cases using PRNG and the proposed method at different values of weighting coefficients of the objective function

Case		1	2	3	4	5	6	7
w1		1	0	0	0.0001	0	0.5	0.00001
w2		0	1	0	0.9999	0.9	0	0.99998
w3		0	0	1	0	0.1	0.5	0.00001
Conventional method [%]	Ave.	100.00	100.00	100.00	100.00	100.00	100.00	100.00
	Std.	0.77	0.13	0.21	0.51	0.22	0.72	0.27
Proposed method [%]	Ave.	99.64	98.78	98.78	99.59	100.00	96.24	98.72
	Std.	0.61	0.06	0.17	0.42	0.16	0.46	0.21

reduction of the city. It was verified that the objective function values can be reduced using any values of K and standard deviation can be reduced using the proposed method. Especially, the value and the standard deviation can be reduced the most when $K = 5$.

Table 2 shows comparison of the average objective function values and standard deviations using searching points by PRNG and the points by the proposed method with various values of weighting coefficients. Using different values of weighting coefficients, the purpose of the optimization can be arranged. All of values are rates when the objective function value of the optimal operation by the proposed method is set to 100%. In this simulation, K is set to 5. It was verified that objective function values and standard deviations by the proposed method can be reduced at all cases.

For example, SC can reduce about US$100 million per year for energy cost by about 1% reduction. Moreover, CO_2 can be reduced about 1,825,000 Ton-CO_2 per year by 1% reduction. Therefore, the proposed method has a possibility to reduce not only energy cost but also CO2 emission drastically in the SC.

6 Conclusions

This paper presents a generation method of initial searching points for total optimization of smart community (SC) using k-means method. It is verified that the proposed method can work efficiently for the optimization of the large scale problem.

As a future work, other related approaches are investigated to compare with the proposed method. Statistical significant tests are utilized to verify that the proposed method is effective for the problem. In addition, more effective initialization methods for large-scale SC optimization problem are investigated considering uncertainty of renewable energy.

References

1. Xcel Energy, SMARTGRDCITY. http://smartgridcity.xcelenergy.com/
2. Ministry of Economy, Trade, and Industry of Japan, Smart Community. http://www.meti.go.jp/english/policy/energy_environment/smart_community/
3. Marckle, G.: Application of genetic algorithms to pump scheduling for water supply. In: Proceedings of the First International Conference on Genetic Algorithms in Engineering Systems: Innovations and Applications, September 1995
4. Henze, M. (ed.): Activated Sludge Models ASM1, ASM2, ASM2d and ASM3. Scientific and Technical Report No. 9. IWA Publishing, London (2000)
5. Makino, Y., Fujita, H., Lim, Y., Tan, Y.: Development of a smart community simulator with individual emulation) modules for community facilities and houses. In: Proceedings of IEEE 4th Global Conference on Consumer Electronics (GCCE), October 2015
6. Yasuda, K.: Definition and modelling of smart community. In: Proceedings of IEEJ National Conference, 1-H1-2, March 2015. (in Japanese)
7. Yamaguchi, N., et al.: Modelling energy supply systems in smart community. In: Proceedings of IEEJ National Conference, 1-H1-3, March 2015. (in Japanese)
8. Matsui, T., Fukuyama, Y., et al.: Energy consumption models in smart community. In: Proceedings of IEEJ National Conference, 1-H1-4, March 2015. (in Japanese)
9. Sato, M., Fukuyama, Y.: Total optimization of smart community by particle swarm optimization considering reduction of search space. In: Proceedings of the 2016 IEEE International Conference on Power System Technology (POWERCON) (2016)
10. Sato, M., Fukuyama, Y.: Total optimization of smart community by differential evolution considering reduction of search space. In: Proceedings of the IEEE TENCON 2016 (2016)
11. Clover, F., Laguna, M.: Tabu Search. Kluwer Academic Publisher, Norwell (1997)
12. Miranda, V., Alves, R.: Differential Evolutionary Particle Swarm Optimization (DEEPSO): a successful hybrid. In: Proceedings of the 11th Brazilian Congress on Computational Intelligence (BRICS-CCI), Porto de Galinhas, Brazil, September 2013
13. Fukuyama, Y., et al.: Various scenarios and simulation examples using smart community models. In: Proceedings of IEEJ National Conference, 1-H1-5, March 2015. (in Japanese)

Clustering Analysis of ECG Data Streams

Yue Zhang[(⊠)] and Yushuai Liu

The Division of Information Science and Technology,
Graduate School at Shenzhen, Tsinghua University, Shenzhen, China
zhangyue@tsinghua.edu.cn,
liu-ysl4@mails.tsinghua.edu.cn

Abstract. ECG signal is significant for cardiovascular diagnosis. Users may concern about the clustering result of ECG waves in recent time or the whole history. However, most existing stream clustering techniques can't give the two kinds of result at the same time. To tackle this challenge, in this paper, we propose a new stream clustering algorithm, DenstreamD, which can be used to meet the requirement. The core idea of DenstreamD is based on Denstream but to add decay potential core micro-clusters in the online maintenance phase. Comprehensive experiments are conducted using MIT-BIH Long-Term ECG database to demonstrate the effectiveness of proposed algorithms. The experiments show that DenstreamD has better accuracy and efficiency than its original algorithm while obtaining two kinds of clustering results.

Keywords: ECG · Clustering · Data streaming

1 Introduction

ECG signal has been collected for cardiovascular diagnosis in the last century [1]. In traditional analysis process, manual diagnosis is based on the ECG signal records. There are two problems - records cannot be too long limited by device's wearability as well as high cost of labor. With the development of technology, ECG diagnosis has been greatly changed nowadays. First, ECG collection devices tend to be portable and intelligent [2], allowing users to wear for longer time and upload data to a real-time system. Second, many experts and scholars of machine learning, have done a lot of work [3, 4] on the ECG data set, hoping to make intelligence diagnosis.

Here is an ideal scene: a user wears ECG device for 24-hours which can uploads data to a real-time system implementing data analysis automatically. Doctors can quickly know the ECG clustering results of the most recent time or the whole time, which they always be interested in. However, traditional global clustering methods seems to be helpless in this scene: using them on recent few samples leads to poor accuracy, and huge computational time overhead on the whole long-time record. So clustering methods of stream data are concerned in this paper.

A number of techniques for stream clustering have been proposed. CluStream proposed by C.C. Aggarwal et al. [5] uses two-stage framework: online and offline. In online phase, new data points are processed and clustering snapshots are saved for further analysis, while offline phase implements macro-clustering based on snapshots.

© Springer International Publishing AG 2017
Y. Tan et al. (Eds.): ICSI 2017, Part II, LNCS 10386, pp. 304–311, 2017.
DOI: 10.1007/978-3-319-61833-3_32

With the landmark window model, it can present a clustering result of whole time. The two-stage framework has been widely used. DenStream [6] takes DBSCAN as its initial clustering algorithm, overcoming the inherent defects of k-means in Clustream which can only deal with spherical clusters and predefined cluster numbers. By defining density concept to micro-clusters, it differentiates potential core micro-clusters (PCC) from outlier micro-clusters (OC) online, making online maintenance more efficient. We can obtain clustering results of the most recent time for there is a damped window model in it. APStream [7] extends Affinity Propagation (AP) to data steaming. Zhang et al. proposed the Reservoir model, which can effectively reduce the complexity of dealing with outliers.

Based on the DenStream and APStream algorithms, this paper presents an improved algorithm DenStreamD, it takes advantages of both algorithms. The core idea of DenStreamD is to add decay potential core micro-clusters in online maintenance stage. This method allows user obtain the clustering results of ECG waves in recent time or the whole history.

The rest of this paper is organized as follows. Proposed algorithms DenStreamD is discussed in detail in Sect. 2. Section 3 presents the experimental results of DenStreamD on ECG data streams and this paper is concluded in Sect. 4 with some direction for future work.

2 Theory and Steps

The clustering framework of data streaming is composed three steps, and each step involves one key issue of data streaming as follows:

Step 1. Initial Clustering. The initial process of the algorithm, a basic clustering technique is adopted to cluster data points at the very beginning of the data stream and form the initial micro-clusters. This results will be the foundation of subsequent clustering.

Step 2. Online Maintenance. When new data points arrive, the existing clustering results should be adjusted accordingly. This adjustment mechanism should be supported by rigorous theory. The choice of online maintenance mechanism also affects the accuracy as well as the time efficiency of clustering.

Step 3. Offline Stage. This stage is often triggered by the user's request. The results or snapshots in online maintenance stage will be used to form accurate clusters or perform evolution analysis.

Next we will introduce some important concepts and detailed theories which are involved in the proposed algorithm DenStreamD.

2.1 Important Concept

- **Decay Function:** When the user is more concerned about the recent period of data streams, time decay function can be used as a time-tilted technology. In this paper, we choose $f(t) = 2^{-\lambda t}(\lambda \in 0, 1)$ as the time decay function the time decay function, where λ is the decay factor.

- **Density Decay:** The initial weight (which can also be quoted as density) of the data point X at time t_0 is $D_x = 1$ while the weight at time t is $D_x(x, t) = 2^{-\lambda(t-t_0)}$. The density of Micro-cluster c_p consisting of n data points $\{X_1, X_2, \ldots, X_n\}$, can be obtained by the sum of density of the whole data points it has, which is presented as:

$$D_{c_p}(c_p, t) = \sum_{i=1}^{n} D_{x_i}(x_i, t) = \sum_{i=1}^{n} 2^{-\lambda(t-t_{0i})} \tag{1}$$

Note that we do not need to update the density of all micro-clusters when a new data point arrivals, but only the one absorbing a new data point. The new density of the micro-cluster absorbing a new data point can be presented as follows:

$$D_{c_p}(c_p, t) = 2^{-\lambda(t-t_l)}D(c_p, t_l) + 1 \tag{2}$$

where t_l is the last update time.

- **Existing Potential Core Micro-clusters (EPCC):** EPCC, whose densities are greater than threshold σ, represents the distribution of data streams for the most recent time. An eight-dimensional vector $\{C_a, C_w, R_a^2, R_w^2, N, D, t_0, t_l\}$ is used to represent EPCC, where C_a is the absolute center of all points in the cluster, C_w is the weighted center, R_a is the absolute average radius of the cluster, and R_w is the weighted average radius, N denotes the number of all points belong to this micro-cluster, D denotes the last updated weight, t_0 represents the time when this micro-cluster built, and t_l is the last update time. The subscript l means before update, and n means after update. When a new data point X arrivals at current time t, and is absorbed by an EPCC, its C_a, C_w, R_a, R_w can be incrementally update and maintain as follows where $e(C, X)$ represents the square of Euclidean distance between C and X, C is the center of the EPCC:

$$\begin{aligned}
C_{an} &= (C_{al} + X)/(N_l + 1) \\
C_{wn} &= (2^{-\lambda(t-t_l)}C_{wl} + X)/(2^{-\lambda(t-t_l)}D_l + 1) \\
R_{an}^2 &= (R_{al}^2 \cdot N_l + e(C_{al}, X))/(N_l + 1) \\
R_{wn}^2 &= (2^{-\lambda(t-t_l)}R_{wl}^2 + e(C_{wl}, X))/(2^{-\lambda(t-t_l)}D_l + 1)
\end{aligned} \tag{3}$$

- **Decay Potential Core Micro-clusters(DPCC):** if an EPCC does not absorb new data points for a long time, its density gradually decreases. When its density is less than the threshold σ, this EPCC converts to a DPCC. For DPCC does not represent the current distribution, we only adopt a six-dimensional vector $\{C_a, R_a^2, N, D, t_0, t_l\}$ to represent DPCC. Note that the weight D is retained to observe whether the DPCC grows again. If a new data point is absorbed by the DPCC, its C_a, R_a can be updated as same as the EPCC.

- **Reservoir:** When a new data point cannot be absorbed by both EPCC or DPCC, it will be placed into the reservoir, a temporary storage box. There is a restart criterion in the reservoir to discover the new micro-clusters when the trigger condition is satisfied.

2.2 Initial Clustering

Initial clustering is the beginning of the entire process of clustering of data streaming, whose accuracy influences the quality of subsequent clustering directly. To tackle the challenges of lack of domain knowledge and difference between the initial samples, we propose following requirements for the initial clustering algorithm: (a) There is no need to specify the number of clusters and initial cluster centers. (b) It can handle outliers and noise well. (c) It is able to obtain clusters of arbitrary shapes.

We apply DBSCAN to the data set at the very beginning of data streams, with the neighborhood radius ε and minimum density threshold σ_0. A micro-cluster will be put into EPCC buffer if its density is greater than σ_0, and others will be put into the reservoir as outliers.

2.3 Online Maintenance

Once the initial micro-cluster of data streaming can be obtained by DBSCAN, we need to consider how to place new data points which will be discussed in detail in this section. When a new data point arrives:

(1) Find the nearest EPCC through checking the distance between the point and the center of the EPCC. Suppose the point can be absorbed by a certain EPCC and R_{wn} is calculated. If $R_{wn} \leq \varepsilon$, the point is belonged to the EPCC and update the EPCC feature vector.

(2) if $R_{wn} > \varepsilon$ find the nearest DPCC. Suppose it can be absorbed the nearest DPCC and calculate R_{an}. If $R_{an} \leq \varepsilon$, then update the DPCC feature vector. Each time we update a DPCC, if $D_n \geq \delta$, move this DPCC to the EPCC buffer with an initial value $C_w = C_a, R_w = R_a$.

(3) If the new point is not belonged to neither EPCC nor DPCC, throw it to the reservoir waiting for model rebuild.

Note that there are two key issues need careful consideration:

- **Cycle Time for Density Check:** With the continuous evolution of data streams, EPCC may reduce to DPCC, and DPCC may develop into EPCC. The latter case can be identified by checking the density of DPCC after absorbing a new point. While for the former case, a cycle-time density check is necessary. As mentioned by the DenStream, we use the minimum time span in which an EPPC reduce to a DPPC:

$$T_p = \frac{1}{\lambda}\log(\frac{\delta}{\delta - 1})$$ (4)

- **Restart Criterion and Model Rebuild:** Since all possible PCCs have been identified in the past, the data points in the reservoir tend to be the real noise points with more data processed, which is a potential advantage of proposed algorithm. So it is very important to identify a new category in the reservoir. As the reservoir model stems from APStream, we use the same restart criterion and parameter settings, but with DBSCAN as the rebuilding algorithm. When the model rebuilding is triggered, new micro-clusters are added to the EPPC buffer and the rest of data points in reservoir will be delete directly.

2.4   Offline Accurate Clustering

When user's clustering request arrives, an extended DBSCAN is used to obtain the accurate clustering results. Based on the concept of density-connected, all EPCC can give an accurate clustering result in the most recent time, and all PCC (EPCC and DPCC) can give an accurate clustering in the whole time.

2.5   Advantage of DenStreamD

DenStreamD combines the advantages of both DenSttream and APStream, and since DPCC is involved in the maintain stage, DenStreamD has some other advantages: (a) DPPC can be used with EPCC (just PCC in DenStream) to get the whole time clustering results. (b) It overcome two drawbacks of DenStream: mistakenly deletion of outliers and clusters' overlap which will lead to a problem in accuracy. (c) Since DPCC can convert to EPCC at the first time, DenStreamD is more time-sensitive to those data streams who have repeated regular data flow.

3   Experiment and Discussion

We mainly evaluated DenStreamD's performance on ECG data streams. The comparisons among DenStreamD, CluStream as well as APStream were presented. First, DenStreamD was compared with CluStream and APStream on the ECG data streams. CluStream and APStream used a landmark windows model and can only get a clustering result of the whole time. Second, DenStreamD was compared with DenStream for they use the same damped windows model which can get a clustering result of the most recent time. Note that we do not discuss the time frame structure in this paper.

3.1   Environment and Data Sets

Experiment Platform: Core i7-3770 3.4G CPU, 8G memory, windows7, Matlab-2014a.

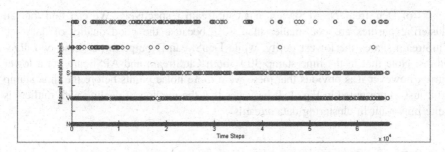

Fig. 1. Distribution of the No.14172 record's category

Data Set: MIT-BIH Long Term Database [8], which contains 7 long-term ECG recordings (14 to 22 h each), with manually reviewed beat annotations. We tested algorithm on all of the 7 recordings and only experimental results of No. 14172 is presented here, for it has more categories (Fig. 1): N means normal beat, V, J, S means different lesion beat, and $\sim$ means the noise. After filtering, segmenting waveform and extracting the characteristics [9], we finally got a data set containing about 66 k sample points with 14-dimensions. These sample points were normalized and marked an integer time stamp in its turn. Then we loaded them into memory, read these data points one by one to simulate the actual ECG data streams.

3.2 Parameter Settings

The algorithm parameters are set as follows: The number of data points involved in the initial clustering is 1000, it is a reasonable value in both computational efficiency and ECG practice. In CluStream, the number of clusters is specified to 40 (ten times of the actual category number, original recommendation), and absorb-radius $\varepsilon = 0.53$, which is set by experience and shared by the other three algorithms. In APStream, the damped coefficient λ_{AP} and parameters of PH test are the same with original value, while the offset parameter $p = 10$, reservoir size $MaxsizeR = 100$. In DenStream and DenStreamD, the minimum density threshold $\sigma_0 = 10$, and time decay factor $\lambda = 0.0016$.

3.3 Experimental Results and Discussions

First, we compared clustering results of a whole time when the time stamps are 5 k, 15 k, 30 k and 60 k respectively. The average purity was used as an indicator of the clustering quality which is calculated as follows:

$$Purity = 100 \times (\sum_{i=1}^{K} \frac{|C_i^d|}{|C_i|}) / K \tag{5}$$

Where K is the number of cluster, $|C_i|$ is the number of data points in cluster i and $|C_i^d|$ is the number of majority class item in cluster i. Figure 2 shows the clustering results of DenStreamD, APStream and CluStream respectively. We can find that all clustering purities are not smaller than 80% because the good quality of data set. CluStream shares the lowest purity, while DenStreamD outperforms these two algorithms. Note that in the time stamp 30 k, both CluStream and APStream get a lower purity, however it is obvious that there were more noise points before the time stamp 30 k just as presented in Fig. 1. It indicates that the method of dealing with outliers is quite important in clustering data streams.

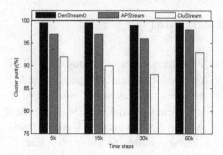

Fig. 2. Clustering purity

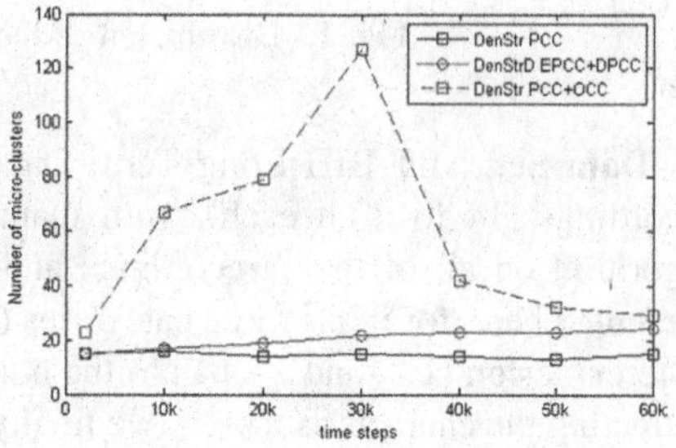

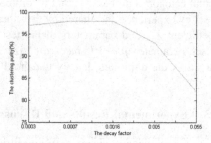

Fig. 3. The number of micro-clusters

Then DenStreamD and DenStream were compared. The clustering purity was almost the same for both of them are based on the DBSCAN. So we only observed the number of micro-clusters in specific times tamps (Fig. 3). More micro-clusters means low efficiency. Obviously, if the data stream contains a lot of noise points, DenStream has to maintain much more micro-clusters than DenStreamD.

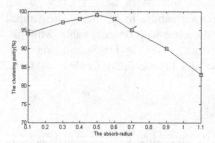

Fig. 4. Influence of the absorb-radius ε

Fig. 5. Influence of decay factor λ

Finally, we discussed the sensitivity of two parameters: the absorb-radius ε and time decay factor λ ε is a key parameter to DBSCAN, if the value is set too large, different natural clusters may be mixed together, while a small value may split natural clusters. In the previous experiment, we tested different values on the initial 1000 sample points to get the clustering purity and finally specified a suitable value ε = 0.53. The test results are shown in Fig. 4. We can see that when ε = 0.1, the clustering purity is also high, but in fact the clustering quality is unacceptable for that we can't get a complete natural cluster. Time decay factor λ controls the importance of historical data to the current cluster, a large value of λ means that only the most recently arrived points determine the clustering results and get a poor clustering quality as Fig. 5 shows. Noted that the discussion of λ is based on the suitable value of ε determined before.

4 Conclusion

The main contribution of this paper is proposing a new algorithm, which combines the advantage of DenStream and APStream, and applying it to the ECG data stream clustering. This new algorithm we called as DenStreamD maintains DPCC online, uses the reservoir model to deal with outlier, can give out two kind of clustering result. Experimental results show that DenStreamD has a good ability to deal with outliers, and can achieve the balance of high precision and efficiency. The time frame structure of DenStreamD is in our next work plan for it can form the clustering result in specified time period, also be used for evolutionary analysis.

References

1. Biel, L., Pettersson, O., Philipson, L., Wide, P.: ECG analysis: a new approach in human identification. IEEE Trans. Instrum. Meas. **50**, N3 (2001)
2. Lobodzinski, S.S.: ECG patch monitors for assessment of cardiac rhythm abnormalities. Progr. Cardiovasc. Dis. **56**(2), 224–229 (2012)
3. Bortola, G., Willems, J.L.: Diagnostic ECG classification based on neural networks. J. Electrocardiol. **26**, 75–79 (1993)
4. Melgani, F., Bazi, Y.: Classification of electrocardiogram signals with support vector machines and particle swarm optimiza. IEEE Trans. Inf Technol. Biomed. **12**(5), 667–677 (1999)
5. Aggarwal, C.C., Han, J.W., Wang, J.Y., Yu, P.S.: A framework for projected clustering of high dimensional data streams. In: Proceedings of the 30th International Conference on Very Large Data Bases (Vol. 30), VLDB Endowment (2004)
6. Cao, F., Ester, M., Qian, W., Zhou, A.Y.: Density-based cluster-ing over an evolving data stream with noise. In: Proceedings of the 2006 SIAM International Conference on Data Mining, Bethesda, USA (2006)
7. Zhang, X., Furtlehner, C., Sebag, M.: Data streaming with affinity propagation. In: Daelemans, W., Goethals, B., Morik, K. (eds.) ECML PKDD 2008. LNCS, vol. 5212, pp. 628–643. Springer, Heidelberg (2008). doi:10.1007/978-3-540-87481-2_41
8. Goldberger, A.L., Amaral, L.A.N., Glass, L., Hausdorff, J.M., Ivanov, P.C., Mark, R.G., Mietus, J.E., Moody, G.B., Peng, C.-K., Stanley, H.E.: PhysioBank, physiotoolkit, and physionet: components of a new research resource for complex physiologic signals. Circulation **101**(23), e215–e220 (2000)
9. Pan, J., Tompkins, W.J.: A real-time QRS detection algorithm. IEEE Trans. Biomed. Eng. **32**(3), 230–236 (1985)

A Novel Multi-cell Multi-Bernoulli Tracking Method Using Local Fractal Feature Estimation

Jihong Zhu, Benlian Xu[✉], Mingli Lu, Jian Shi, and Peiyi Zhu

School of Electrical and Automatic Engineering,
Changshu Institute of Technology, Changshu 215500, China
xu_benlian@cslg.edu.cn

Abstract. A novel multi-cell tracking method based on multi-Bernoulli filter using local fractal feature estimation is proposed in this paper. The Hurst coefficient estimated by the rescaled range analysis method is considered as the local fractal feature. The local fractal feature can offer two advantages for multi-Bernoulli filter. The input of filter is the Hurst coefficient image, the direct effect is that observation can be modeled simply. And the likelihood function can be computed easily using this feature. Experiment results show that our proposed method could achieve an accurate and joint estimate of the number of cells and their individual states especially in the case of the number of cell population varying and the cellular morphology changing. And it shows equivalent accuracy against other tracking methods.

Keywords: Multi-cell tracking · Multi-Bernoulli filter · Local fractal feature · Hurst coefficient · Rescaled range analysis

1 Introduction

The study of analysis of cellular behavior has been attracting more and more attention in the past ten years [1–3]. But cellular motion analysis poses many challenges to those existing techniques due to poor image, irregular cell migration, the varying density of cell populations and changes of cellular morphology. Obviously, the manual analysis of cellular behavior is a tedious process. Sometimes, it becomes impossible for an expert to accurately capture many different events over along sequence, especially when it requires tracking a large number of cells during long period of time in order to obtain robust results. Therefore, the automated tracking methods that eliminate the bias and variability to a certain degree are of great importance. Many efforts have been made in automated cell tracking over past decades [4–10]. Especially, the Mahler's RFS based method [11], an elegant Bayesian formulation of multi-target filtering, which operates on the single-target state space directly and avoids the combinatorial problem that arises from data association, has generated substantial interest in recent years [12], due to the development of the probability hypothesis density (PHD) filter, the cardinalized PHD (CPHD) filter and the multi-target multi-Bernoulli (MeMBer)

© Springer International Publishing AG 2017
Y. Tan et al. (Eds.): ICSI 2017, Part II, LNCS 10386, pp. 312–320, 2017.
DOI: 10.1007/978-3-319-61833-3_33

filter and the realizations by Vo's team through the methods of sequential Monte Carlo (SMC) and Gaussian mixture (GM) [13–16].

In visual tracking, the performance of the RFS based method depends on how to model observation and compute likelihood function. In [11], each target is represented as a rectangular blob, the HSV color histograms of the blob is applied to model the measurement, and the kernel density estimate of a given histogram is applied to compute the likelihood function. In [17], the measurement model and likelihood computation are used kernel density estimate based on background subtraction method. Each frame transformed to a gray-scale image is used as input to the multi-Bernoulli filter, and the gray value of each pixel can be interpreted as the probability of the pixel belonging to the background. So the multi-Bernoulli filter has been proved to enable a tractable solution to the multi-object estimation problem for image data. The idea of our work is derived from the work in [17], a novel multi-cell tracking method based on multi-Bernoulli filter using local fractal feature estimation is proposed. The local fractal feature, i.e., the Hurst coefficient of each pixel, is estimated by the rescaled range analysis method. The estimated result is the input of multi-Bernoulli filter. And the observation model and computation of likelihood function depend this estimated local fractal feature.

The outline of this paper is organized as follows. In Sect. 2, the observation model and likelihood computation based on local fractal feature estimation are discussed. Section 3 briefly introduces the particle multi-Bernoulli filter. The experimental results of the proposed method are given in Sect. 4. And conclusion and possible extensions are presented in Sect. 5.

2 Local Fractal Feature Estimation and Observation Model

So far, the definition of fractal is no widely accepted [18], but the fractal theory and its applications, which offers a new geometry method to describe the natural morphology and simulate the complicated natural bodies on computer easily, becomes an important constituent part of nonlinear science. In recent years, the fractal theory is wildly used in the fields of physics, biology, finance, medical science and so on, especially in field of image processing. The key problem of application of fractal theory is fractal dimension estimation. Currently, the common models to estimate fractal dimension include fractal Brownian motion (FBM) model [19], differential box-counting model [20], blanket covering model [21] and rescaled range (R/S) analysis model [22, 23]. From the texture's point of view, a smaller fractal dimension implies a smoother surface while a larger fractal dimension implies a rougher surface. Since the monitored target or artificial body is smoother than the background surface or natural body in certain images, the fractal dimension can be used to differentiate the monitored target from background surface. With this property, the local fractal feature, namely the Hurst coefficient, is used to build measurement model and compute the likelihood function in this paper, and the R/S method is selected to estimate the Hurst coefficient, which is the unique parameter to describe the fractal dimension, due to its advantages of the easy implementation and relatively small computational cost.

2.1   Local Fractal Feature Estimation

To estimate the Hurst coefficient of each pixel in any image [23], a given $M \times N$ gray image $y(i,j)$ $(i = 1, 2, \cdots, M; j = 1, 2, \cdots, N)$ and a $m \times m$ slide window are required. the pixels in slide window are arranged as a one dimension sequence form according to the order from top to bottom and left to right. For example, in Fig. 1, if p_i is the gray value of the i pixel, the one dimension sequence form of all pixels in 3×3 slide window is given as $p = (p_1, p_2, p_3, p_4, p_5, p_6, p_7, p_8, p_9)$.

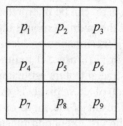

Fig. 1. Pixel in slide window

The mean of pixel in slide window is:

$$\bar{p} = \frac{1}{m \times m} \sum_{i=1}^{m \times m} p_i \tag{1}$$

And the deviation ϕ_i, range R_i and standard deviation S_i of pixel in slide window are computed respectively as follows:

$$\phi_i = |p_i - \bar{p}|, i = 1, 2, \cdots, m \times m \tag{2}$$

$$R_i = \max_{1 \leq i \leq m \times m} (\phi_i) - \min_{1 \leq i \leq m \times m} (\phi_i) \tag{3}$$

$$\bar{\phi}_i = \frac{1}{m \times m} \sum_{i=1}^{m \times m} \phi_i \tag{4}$$

$$S_i = \sqrt{\frac{1}{m \times m} \sum_{i=1}^{m \times m} (p_i - \bar{\phi}_i)^2} \tag{5}$$

The Hurst coefficient H_i of i pixel can be computed as:

$$H_i = \log(R_i + 1) / \log(S_i + 1) \tag{6}$$

Here, it is easily to prove that the range of Hurst coefficient H_i is between 0 and 1.

Note that the modified formula (6) is not calculated by least square fitting method. The advantage of this modification can reduce the amount of computation in practical application. The local fractal feature implies that if the Hurst coefficient of any pixel in image is more close to 0, then the probability of the pixel belonged to the natural bodies is high. Otherwise, the pixel is more likely originated from artificial bodies. Therefore, it is reasonably to use the Hurst coefficient to model observation and compute the likelihood function. Of course, in the practical operation, three constraints are considered as follows:

(1) When the slide window moves in image from left to right and top to bottom, those pixels in some rows and columns of image that do not participate in the calculation are defined as image edge, and the Hurst coefficient of image edge will be set to 0.
(2) The pixel which value of Hurst coefficient is bigger than 1 is defined as the distorted pixel, the Hurst coefficient of distorted pixel will be set to 0.
(3) In order to make better use of the Hurst coefficient to compute likelihood, the Hurst coefficient of some pixels is smaller than a given threshold $\theta(0<\theta<1)$ will be set to 0.

2.2 Observation Model and Likelihood Computation

Without loss of generality, consider the Hurst coefficient image $\mathbf{H}_k = [h_1^k, h_2^k, \cdots, \mathbf{h}_{M \times N}^k]$ in the k frame from a given video sequence, given a multi-target state $\mathbf{X} = \{\mathbf{x}_1, \mathbf{x}_2, \cdots, \mathbf{x}_n\}$, n is the number of targets, each target $\mathbf{x}$ is represented as a fixed rectangular blob $T(\mathbf{x}) = w \times h$, where w, h is the width and high of blob. The likelihood function $g(\mathbf{y}|\mathbf{X})$ which is usually presented in certain visual scenarios is given as [17]:

$$g(\mathbf{y}|\mathbf{X}) = g(\mathbf{y}) \prod_{i=1}^{n} g(\mathbf{y}; \mathbf{x}_i) \tag{7}$$

Where $g(\mathbf{y})$ is the likelihood of background, which is independent from target states, and $g(\mathbf{y}; \mathbf{x}_i)$ is the likelihood that a target is present in the image $\mathbf{y}$ with state $\mathbf{x}_i$. Because the Hurst coefficient of natural background in Hurst coefficient image is usually 0, the likelihood $g(\mathbf{y})$ is equal to 1 or close to 1. So the measurement likelihood function $g(\mathbf{H}_k|\mathbf{X})$ in Hurst coefficient image $\mathbf{H}_k$ with multi-target state $\mathbf{X}$ can be expressed with following form:

$$g(\mathbf{H}_k|\mathbf{X}) = \prod_{i=1}^{n} g(\mathbf{H}_k; \mathbf{x}_i) \tag{8}$$

The observation model (8) is obviously simpler than the model (7), due to the Hurst coefficient of most of background is set to be 0, the observation only considers those nonzero Hurst coefficient originates from targets or small amount of background.

Let N_p is the number of nonzero Hurst coefficient in $T(\mathbf{x})$ and $|T(\mathbf{x})|$ is the number of pixels in $T(\mathbf{x})$. The parameter ε that describes the proportion of the nonzero Hurst coefficient in blob is defined as:

$$\varepsilon = N_p / |T(\mathbf{x})| \tag{9}$$

With this, the likelihood of target $\mathbf{x}_i$ can be computed as follows:

$$g(\mathbf{H}_k; \mathbf{x}_i) = \begin{cases} \delta \times \varepsilon, & \varepsilon \geq \xi \\ 0, & \varepsilon < \xi \end{cases} \tag{10}$$

Where δ is a control parameter, $\xi(0 < \xi < 1)$ is a given judgment threshold. It implies that if the proportion of nonzero Hurst coefficient covered by any blob is bigger than the threshold, this blob will be regarded as a target of interest. It is observed that the likelihood function only depends on the proportion of nonzero Hurst coefficient in blob, the computation of this likelihood becomes easily.

3 The Particle Multi-Bernoulli Filter

The multi-target filtering problem can be cast as a Bayesian filter on the RFS framework. In this framework, the multi-target state is modeled as a RFS whose posterior distribution is propagated forward in time via the multi-target Bayes filter. The multi-Bernoulli filter is a tractable solution to the multi-target Bayes filter which propagates the multi-Bernoulli parameters of the multi-target posterior distribution. The salient features of the multi-Bernoulli filter are that it operates in the single-target state space and it is highly parallelizable and amenable to multiple sensor fusion. The particle multi-Bernoulli filter is a sequential Monte Carlo implementation of multi-Bernoulli filter, which makes Bayes recursion tractable. More details can refer [11, 17].

4 Experiment

In our experiment, targets are modeled by rectangular blobs with constant survival probability $p_S = 0.99$. The number of particles for each Bernoulli target is constrained between $L_{\min} = 200$ and $L_{\max} = 1000$, and the particles resample in each frame of the image sequences. Other parameters are taken as: $\theta = 0.75, m = 3, \delta = 1000$. The size of cell images is 201×201. The parameters of birth model is taken as: $M_\Gamma = 4$, and $r_\Gamma^{(1)} = r_\Gamma^{(2)} = r_\Gamma^{(3)} = r_\Gamma^{(4)} = 0.2$. The range of x-velocity and y-velocity is set to be $[0, 5]$ and $[-5, 5]$ respectively. The experiment is conducted on a Gateway CPU 2.0 GHz processor with 1.84 GB RAM.

The total number of cells is 7 and all cells will move to right. In the first frame, four cells enter at the top and bottom surface, the top cell leaves at the frame 12 and the bottom cells disappear at the frame 17 and frame 26 respectively. The fifth to seventh cell enters in the field of view at frame 7, frame 21 and frame 28 respectively. Some cells change their morphology when they move. The tracking results are shown in Fig. 2 and the corresponding Hurst coefficient images are shown in Figs. 3, 4 and 5 plot the position and velocity estimates of each cell respectively. It can be observed that the proposed method can capture cells entering and leaving the image and track cells of

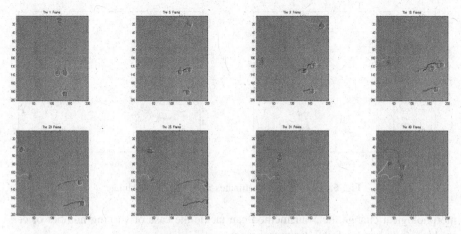

Fig. 2. The tracking results

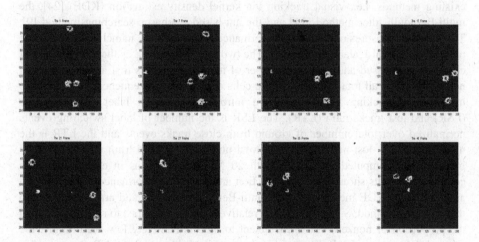

Fig. 3. The corresponding Hurst coefficient image

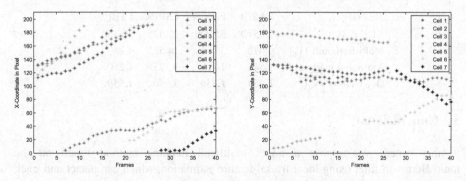

Fig. 4. Cell position estimates in X and Y coordinate

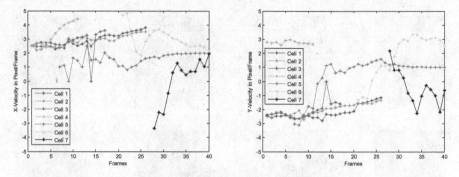

Fig. 5. Cell velocity estimates in X and Y coordinate

morphological change, i.e., our method can tackle the case of varying number of cell tracking and change of cellular morphology.

To get insight into tracking performance, we compare our method with other existing methods, i.e., visual tracking via kernel density estimation (KDE) [24], the multi-Bernoulli filter method [11] and the ant-based stochastic searching method [9]. The first two measures of detection performance are introduced, namely, false negative rate (FNR) and false alarm rate (FAR). The two rates are defined as the total number of cells that are missed and the total number of non-existing cells that are tracked, both normalized over all total number of true cells over all image sequences. Another two measures of tracking performance are introduced, namely, label switching rate (LSR) and lost tracks ratio (LTR). The LSR is the number of label switching events normalized over total number of ground truth close tracks events and the LTR is the number of tracks lost normalized over total number of ground truth tracks. The four measures are computed in Table 1 with 20 Monte Carlo runs in each frame. The comparison results show that both detection and tracking performance of our method are better than KDE method and the multi-Bernoulli method, and are very closed to the ant-based method. Since the cell 5 is relatively small in contrast to regular ones, and the proportion of nonzero Hurst coefficient to blob is relatively low accordingly, the missing detection probably happens during tracking with a higher FNR.

Table 1. Performance comparison results using various methods

Methods	FAR(%)	FNR(%)	LSR(%)	LTR(%)
KDE [24]	23.572	31.729	27.325	10.750
Multi-Bernoulli [11]	4.672	3.461	14.525	4.250
Ant-based method [9]	1.541	1.756	3.175	1.250
Our method	*1.750*	*1.850*	*3.550*	*1.550*

5   Conclusions

In this paper, we propose a novel multi-cell tracking method based on the multi-Bernoulli filter using local fractal feature estimation, which can detect and track directly from image observations without the need for any separate target detection.

The local fractal feature is described by the Hurst coefficient estimated by the rescaled range analysis method. The local fractal feature offers two advantages for multi-Bernoulli filter. The input of filter is the Hurst coefficient image, the direct effect is that observation can be modeled simply. And the Hurst coefficient value of pixel presents the probability belonged to background or target, the likelihood function can be computed easily. According to the experiments, we can find that our proposed method could get an accurate and joint estimate of each cell state and the performance of our method does well to multiple cells, especially, when the number of cell populations varies and the cellular morphology changes. Certainly, we find that the harden computation of likelihood function can't effectively deal with the closed or overlap and small cells. In future work, these shortcomings will be overcome by modeling the robust measurement.

Acknowledgments. This work is supported by national natural science foundation of China (No. 61673075), the natural science fundamental research program of higher education colleges in Jiangsu province (No. 14KJB510001) and the project of talent peak of six industries (DZXX-013).

References

1. Ong, L.L., Dauwels, J., Jr., M.H.A., et al.: A Bayesian filtering approach to incorporate 2D/3D time-lapse confocal images for tracking angiogenic sprouting cells interacting with the gel matrix. Med. Image Anal. **18**(1), 211–227 (2014)
2. Chen, X., Zhou, X., Wong, S.T.: Automated segmentation, classification, and tracking of cancer cell nuclei in time-lapse microscopy. IEEE Trans. Bio-Med. Eng. **53**(4), 762–766 (2006)
3. Dewan, M.A.A., Omair, A.M., Swamy, M.N.S.: Tracking biological cells in time-lapse microscopy: an adaptive technique combining motion and topological features. IEEE Trans. Bio-Med. Eng. **58**(6), 1637–1647 (2011)
4. Xu, C., Prince, J.L.: Snakes, shapes, and gradient vector flow. IEEE Trans. Image Process. **7**(3), 359–369 (1998). A Publication of the IEEE Signal Processing Society
5. Smal, I., Carranza-Herrezuelo, N., Klein, S., et al.: Reversible jump MCMC methods for fully automatic motion analysis in tagged MRI. Med. Image Anal. **16**(1), 301–324 (2012)
6. Zimmer, C., Olivomarin, J.C.: Coupled parametric active contours. IEEE Trans. Pattern Anal. Mach. Intell. **27**(11), 1838–1842 (2005)
7. Shen, H., Nelson, G., Kennedy, S., et al.: Automatic tracking of biological cells and compartments using particle filters and active contours. Chemometr. Intell. Lab. Syst. **82**(1–2), 276–282 (2006)
8. Mukherjee, D.P., Ray, N., Acton, S.T.: Level set analysis for leukocyte detection and tracking. IEEE Trans. Image Process. **13**(4), 562–572 (2004)
9. Xu, B.L., Lu, M.L.: An ant-based stochastic searching behavior parameter estimate algorithm for multiple cells tracking. Eng. Appl. Artif. Intell. **30**, 155–167 (2014)
10. Smal, I., Grigoriev, I., Akhmanova, A., et al.: Microtubule dynamics analysis using kymographs and variable-rate particle filters. IEEE Trans. Image Process. **19**(7), 1861–1876 (2010)

11. Hoseinnezhad, R., Vo, B.N., Vo, B.T., et al.: Visual tracking of numerous targets via multi-Bernoulli filtering of image data. Pattern Recogn. **45**, 3625–3635 (2012)
12. Mahler, R.: Statistical Multisource Multitarget Information Fusion. Artech House, Norwood (2007)
13. Vo, B.N., Singh, S., Doucet, A.: Sequential Monte Carlo methods for multi-target filtering with random finite sets. IEEE Trans. Aerosp. Electron. Syst. **41**(4), 1224–1245 (2005)
14. Vo, B.N., Ma, W.K.: The gaussian mixture probability hypothesis density filter. IEEE Trans. Sig. Process. **54**(11), 4091–4104 (2010)
15. Vo, B.T., Vo, B.N., Cantoni, A.: Analytic implementations of the cardinalized probability hypothesis density filter. IEEE Trans. Sig. Process. **55**(7), 3553–3567 (2007)
16. Vo, B.T., Vo, B.N., Cantoni, A.: The Cardinality balanced multi-target multi-Bernoulli filter and its implementations. IEEE Trans. Sig. Process. **57**(2), 409–423 (2009)
17. Hoseinnezhad, R., Vo, B.N., Vo, B.T.: Visual tracking in background subtracted image sequences via multi-bernoulli filtering. IEEE Trans. Sig. Process. **61**(2), 392–397 (2013)
18. Mandelbrot, B.B.: The fractal geometry of nature. WH freeman, San Francisco (1982)
19. Lin, P.L., Huang, P.W., Lee, C.H., et al.: Automatic classification for solitary pulmonary nodule in CT image by fractal analysis based on fractional Brownian motion model. Pattern Recogn. **46**, 3279–3287 (2013)
20. Sarkar, N., Chaudhuri, B.B.: An efficient approach to estimate fractal dimension of textural image. Pattern Recogn. **23**(9), 1035–1041 (1992)
21. Pentland, A.P.: Fractal-based description of natural scenes. IEEE Trans. Pattern Anal. Mach. Intell. **6**(6), 661–674 (1984)
22. Pilar, G.C.: Empirical evidence of long-range correlations in stock return. Physical **287**, 396–404 (2002)
23. Zhang, F., Zou, H.X., Lei, L.: A CFAR detection algorithm baesd on local fractal dimension. Sig. Process. **28**(1), 105–111 (2012)
24. Elgammal, A., Duraiswami, R., Harwood, D., et al.: Background and foreground modeling using nonparametric kernel density estimation for visual surveillance. Proc. IEEE **90**(7), 1151–1162 (2002)

An Improved Locality Preserving Projection Method for Dimensionality Reduction with Hyperspectral Image

Juan Xiong[1,2(✉)], Sheng Ding[1,2], and Bo Li[1,2]

[1] College of Computer Science and Technology,
Wuhan University of Science and Technology, Wuhan, China
xjuan_wust@163.com
[2] Hubei Province Key Laboratory of Intelligent Information Processing
and Real-Time Industrial System, Wuhan, China

Abstract. Band selection plays a critical role in dimensionality reduction (DR) for hyperspectral image (HSI). In view of the research of the DR method based on manifold, we propose an improved version of the original Locality preserving projection (ILPP), a linear band selection method. The article changes the linear projection constraints of the LPP algorithm by embedding a cluster potential matrix into the Laplacian matrix and forming a projection with two-layer linear structure constructed by a certain rule. The idea is to find a new projection that can preserve the local geometry of the data, enhance the proximity of the similar points and increase the class separability between points that are not similar. Results of experiments on two HSIs confirm that ILPP outperforms several traditional alternatives in the performance of dimensionality reduction.

Keywords: Hyperspectral image (HSI) · Dimensionality reduction (DR) · Manifold learning (ML) · Improved locality preserving projection (ILPP)

1 Introduction

Hyperspectral image (HSI) provides enormous amount of surface information with a very high resolution (VHR), which is now widely used in many fields [1]. However, the high correlation of adjacent bands leads to information redundancy, resulting in a very time-consuming process and "Hughes" phenomenon. Consequently, dimensionality reduction, a preprocessing procedure that strikes out redundant features and preserves valuable information in a low-dimensional representation, becomes an effective strategy to extract information from hyperspectral images efficiently and specifically, and provides new perspectives to the field.

Since there is a nonlinear problem with the multiple scattering between the target and the sensor in hyperspectral remote sensing [2], many nonlinear dimensionality reduction methods have been investigated, among which manifold learning (ML) [3] has become an advanced research hotspot. For example, local linear embedding (LLE) [4], laplacian eigenmap (LE) [5] and their extended methods such as neighborhood preserving embedding (NPE) [6] and local preserving projection (LPP) [7].

© Springer International Publishing AG 2017
Y. Tan et al. (Eds.): ICSI 2017, Part II, LNCS 10386, pp. 321–329, 2017.
DOI: 10.1007/978-3-319-61833-3_34

LPP constructs a graph that excavates local geometries to define a linear projection from the original space to the low-dimensional space, focusing on the local neighborhood relationship in the data set, which is especially sensitive to the noise and the size of the neighborhood. Aimed at this problem, this paper proposes an improved locality preserving projection (ILPP) method.

The objective of the proposed method is to preserve the local geometry of the data from each source domain as well as get similar regions together whilst pushing dissimilar regions apart. It finds the low-dimensional manifold in high-dimensional space and the corresponding embedded mapping by defining a linear projection incorporated both labeled and unlabeled samples. Results show it is beneficial to yield to state-of-the-art performance when applying ILPP to HSI dimensionality reduction.

The rest of this paper is organized as follows. A theoretical introduction to the LPP and the ILPP is separately given in Sects. 2 and 3. Section 4 describes the hyperspectral datasets and setups in the experiment. Section 5 presents the results of the experiments and compares the proposed algorithm with other basic algorithms. Finally, conclusions and further work are summarized in Sect. 6.

2 Locality Preserving Projection

Given a dataset expressed as $X = [x_1, x_2, \ldots, x_n]$, $x_i \in R^d (i = 1, \ldots, n)$, where n is the number of pixels, and d is the dimensionality of the dataset, the purpose is to find a transform f from R^d to R^t, e.g. $y_i = f(x_i)$, $y_i \in R^t$, $t \ll d$.

Locality Preserving Projection (LPP) is a linear projective map that optimally preserves the neighborhood structure of the data set, is also a linear subspace learning method derived from Laplacian Eigenmap. The mathematical theory of LPP can be summarized as follows.

First of all, using the local neighbor structure of the dataset constructs the adjacency matrix W, where w_{ij} is a measurement of the adjacency between x_i and x_j, and $w_{ij} = e^{-\|x_i - x_j\|/\sigma^2}$ if x_i and x_j are connected, otherwise $w_{ij} = 0$. And then we can find a low-dimensional subspace of LPP with W, and the local neighborhood information contained in X can be preserved. The objective function showed as Eq. (1) is used to ensure that if x_i and x_j are adjacent, the corresponding x_j and x_i are also adjacent.

$$\min \sum_{ij} \|y_i - y_j\|^2 w_{ij} = \min \sum_{ij} \|A^T x_i - A^T x_j\|^2 w_{ij}, \quad y_i = A^T x_i \tag{1}$$

According to (1), the minimization problem is equivalent to an eigenvalue decomposition problem defined as:

$$XLX^T a_i = \lambda_i XDX^T a_i \tag{2}$$

where D is a diagonal matrix, $d_{ij} = \sum_{ij} w_{ij}$, $L = D - W$ is the Laplacian matrix, λ_i is the minimum eigenvalue of a_i.

3 Improved Locality Preserving Projection

3.1 Connectivity Graph Construction

It can be seen that the accuracy of the adjacent weight matrix has a direct effect on the performance of dimensionality reduction. To handle both spectral and spatial information mathematically, a manifold point x_i is represented by combining a pixel's spectral information x_i^f and its spatial location x_i^p, i.e., $x_i^T = \begin{bmatrix} x_i^{fT} & x_i^{pT} \end{bmatrix}^T$.

In [8], it describes how to handle graph construction and edge weight definition in a manner that incorporates a penalty on differences in the direction of the spectral information as opposed to a penalty on the norm of their differences.

1. Construct G so that the set of edges ε is defined based on ε-neighborhoods of the spatial locations, i.e., define an edge between x_i and x_j if $\left\| x_i^p - x_j^p \right\|^2 < \varepsilon$.
2. Define edge weights by:

$$W_{i,j} = \begin{cases} \exp(-\cos^{-1}(\frac{\langle x_i^f, x_j^f \rangle}{\|x_i^f\| \cdot \|x_j^f\|}) - \frac{\|x_i^p - x_j^p\|^2}{\sigma_p^2}), & (x_i, x_j) \in \varepsilon \\ 0, & otherwise \end{cases} \quad (3)$$

3.2 New Preserving Projection

The purpose of the proposed method is to matching instances with the same labels, separating instances with different labels and preserving topology of each given domain in an effective way. To meet the goal, the method works by enhancing the connectivity graph in the Laplacian matrix via embedding a cluster potential, i.e., using $L + \beta(L + \alpha V)$ to replace L. That is to say, based on Eq. (2), we propose a new generalized eigenvector problem instead, defined as:

$$X(L + \beta(L + \alpha V))X^T a_i = kXDX^T a_i, \quad (4)$$

where V is a cluster potential matrix, created by defining V to be the sum of non-diagonal matrices $V^{(i,j)}$ defined by: $V_{k,l}^{(i,j)} = \begin{cases} 1, (k,l) \in \{(i,i), (j,j)\} \\ -1, (k,l) \in \{(i,j), (j,i)\} \\ 0, otherwise \end{cases}$, the parameters α and β are introduced to rescale L and V in the manner as multiplication with $\alpha = \partial \cdot tr(L)/tr(V)$, $\beta = \partial \cdot tr(L)/tr(L + \alpha V)$. And $\partial = \frac{1}{C} \frac{n}{(\pi \sigma_m)^{\frac{d+2}{2}}}$, $\sigma_m = 4n^{-\frac{1}{d+2+s}}$, C is a positive constant, $s > 0$.

Finally, the problem is transformed to solve the eigenvalue decomposition of Eq. (4). Supposing the eigenvectors $a_0, a_1 \ldots, a_k$ are the solutions of Eq. (4), and if they are ordered so that their corresponding eigenvalues $0 = \lambda_0 \leq \lambda_1 \leq \ldots \leq \lambda_k$, thus the points $y_1^T, y_2^T, \ldots, y_k^T$ are defined to be the rows of $A = [a_1 a_2 \ldots \ldots a_k]$.

4 Data and Setup

The proposed algorithm is illustrated by two publicly available hyperspectral image datasets in the experiment. The first dataset was captured by the AVIRIS sensor over the rural Indian Pines test site in Northwestern Indiana USA in 1992. The image contains 145×145 pixels with 220 spectral bands (20 channels were discarded due to the noise and water absorption). There are totally 10249 ground truth pixels labeled partially with 16 classes in the dataset. Figure 1 shows the false color image and ground truth image. The second dataset was collected by the ROSIS sensor over the Pavia University, northern Italy in 2003. The image contains 610×340 pixels with 115 spectral bands (12 channels were discarded). A partial set of labels yields 42776 ground truth pixels associated with 9 classes. The false color image and ground truth image are detailed in Fig. 2.

Alfalfa	Corn-mintill
Grass-pasture	Grass-pasture-moved
Otas	Soybean-mintill
Wheat	Bldg-Grass-trees-drives
Corn-notill	Corn
Grass-trees	Hay-windrowed
Soybean-notill	Soybean-clean
Woods	Stone

(a)

Fig. 1. AVIRIS Indian Pines dataset. (a) False color image. (b) Ground truth image. (Color figure online)

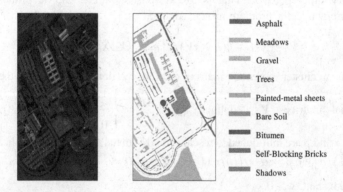

Asphalt
Meadows
Gravel
Trees
Painted-metal sheets
Bare Soil
Bitumen
Self-Blocking Bricks
Shadows

Fig. 2. ROSIS Pavia University dataset. (a) False color image. (b) Ground truth image. (Color figure online)

Table 1. Number of training samples and test samples per-class for the Indian Pins dataset

No	Class	Total	RGB	Training	Test
2	Corn-notill	1428	[255 96 0]	143	1285
3	Corn-mintill	830	[255 191 0]	83	747
5	Grass-pasture	483	[128 255 0]	49	434
6	Grass-trees	730	[32 255 0]	73	657
8	Hay-windrowed	478	[0 255 159]	48	430
10	Soybean-notill	972	[0 159 255]	98	874
11	Soybean-mintill	2455	[0 64 255]	246	2209
12	Soybean-clean	593	[32 0 255]	60	533
14	Woods	1265	[223 0 255]	127	1138
Total		9234		927	8307

Table 2. Number of training samples and test samples per-class for the Pavia University dataset

No	Class	Total	RGB	Training	Test
1	Asphalt	6631	[255 0 0]	332	6299
2	Meadows	18649	[255 170 0]	933	17716
3	Gravel	2099	[170 255 0]	105	1994
4	Trees	3064	[0 255 0]	154	2910
5	Painted metal sheets	1345	[0 255 170]	68	1277
6	Bare Soil	5029	[0 170 255]	252	4777
7	Bitumen	1330	[0 0 255]	67	1263
8	Self-Blocking Bricks	3682	[170 0 255]	185	3497
9	Shadows	947	[255 0 170]	48	899
Total		42776		2144	40632

To make the results convincing, all of the experiments among a dataset for the dimensionality reduction algorithms and classification are performed with the same experimental configuration, detailed as follows.

For the AVIRIS dataset, since only few training samples are available, 7 classes were removed and the remaining 9 classes are used in the experiments, which are detailed in Table 1.

For training samples, we randomly select 10% and 5% of each class from the available ground truth pixels from the Indian Pines and Pavia University images respectively to generate training data and the rest yield the test data. Tables 1 and 2 respectively show the number of training samples and test samples in each class from the Indian Pines and Pavia University.

For dimensionality reduction, some related dimensionality reduction methods including PCA [9], LDA [10], NPE, LLTSA [11], LPP and the proposed ILPP have been implemented to preprocess the dataset. And in the step of graph construction via k-nearest neighbors involved with NPE, LLTSA, LPP and ILPP, we set the neighborhood size $\varepsilon = 12$. For the other parameters in ILPP, we make the initial choice of $\sigma_p = 1$ and adjust when necessary, set α and β range from 0 to 100.

For classification, we use three classifiers, i.e. k-nearest neighbor (KNN) classifier [12], support vector machine (SVM) classifier [13], and extreme learning machine (ELM) classifier [14] to get the accuracy. As for parameters used in the classifiers, we set the neighbors is $k = 5$ for KNN, adopt the genetic algorithm (GA) [15] cross-volidation to get the best value for the penalty parameter C (range: (0,100]) and kernel function parameter γ (range: [0,1000]) in SVM and regularization parameter C (range: Integers in [1, 300]) in ELM, and use the RBF kernel function for SVM.

For comparison, we introduce the overall accuracy (OA) with respect to classification, with which we enable to validate the performance of dimensionality reduction algorithm.

5   Results and Discussion

In this paper, the proposed method is compared with PCA, LDA, NPE, LLTSA and LPP. We choose the reduced dimensionality to be $d = 5, 10, 15, 20, 25, 30, 35, 40, 45, 50$

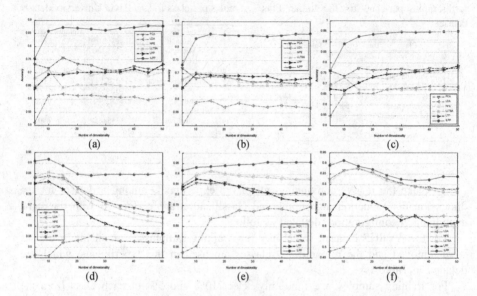

Fig. 3. Overall Accuracy with respect to reduced dimensions with different dataset.
(a) AVIRIS + KNN. (b) AVIRIS + SVM. (c) AVIRIS + ELM. (d) ROSIS + KNN.
(e) ROSIS + SVM. (f) ROSIS + ELM.

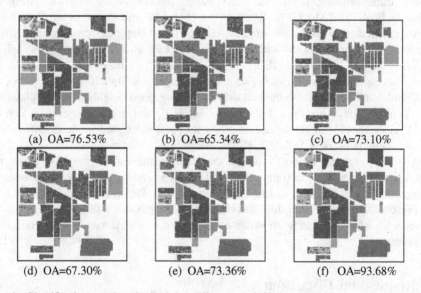

(a) OA=76.53%   (b) OA=65.34%   (c) OA=73.10%

(d) OA=67.30%   (e) OA=73.36%   (f) OA=93.68%

Fig. 4. Classification results obtained by different method for AVIRIS dataset using SVM classifier. (a) PCA. (b) LDA. (c) NPE. (d) LLTSA. (e) LPP. (f) ILPP.

and use the results of classification accuracy obtained respectively from three classifiers to verify the performance of the dimensionality reduction algorithm mentioned above.

Figure 3 separately plots the overall accuracy of different methods with different target dimensions performed among two datasets. As can be seen from the figure, ILPP, as an improved algorithm based on the LPP algorithm, has a significantly superiority over others in each case. And with the increasing of the target dimension, the classification accuracy tends to be close to a specific and good value in ILPP. It is worth noting that even with very small target dimension ($d = 5$), ILPP can also provide good result of $OA = 73.30\%$, 76.19%, 75.31% (AVIRIS), $OA = 90.98\%$, 91.76%, 89.09%(ROSIS) with respect to the KNN, SVM and ELM classifier, while others fail to match it.

Figures 4 and 5 visually depict the classification maps obtained by different methods with the SVM classifier, typically. We produce different target dimension for the two dataset: $d = 20$ (AVIRIS), $d = 10$ (ROSIS). From these maps, we can get an observation that the proposed method ILPP outperforms others in the performance of dimensionality reduction.

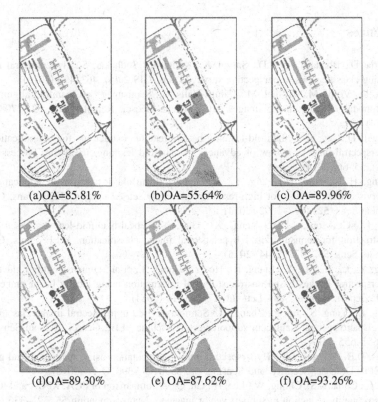

(a)OA=85.81%   (b)OA=55.64%   (c) OA=89.96%

(d)OA=89.30%   (e) OA=87.62%   (f) OA=93.26%

Fig. 5. Classification results obtained by different method for AVIRIS dataset using SVM classifier. (a) PCA. (b) LDA. (c) NPE. (d) LLTSA. (e) LPP. (f) ILPP.

6 Conclusions

In this study, we incorporate the spectral and spatial information of HIS, encourage the cluster potential matrix to form a new projection from the high-dimension to low-dimension, and conduct experiments to evaluate the performance of ILPP. Comparison of the results with those traditional dimensionality reduction algorithms demonstrates that ILPP has a better performance than the prior classical methods. However, the parameter setting in the projection function in ILPP has a serious influence on the performance, and the connectivity graph construction in this paper pays a lot in both time consumption and memory consumption. In the future, we are interested in exploring how to get the parameters in ILPP adaptably and construct the connectivity graph in an efficient way.

Acknowledgments. This work was supported by the National Natural Science Foundation of China (Grant No. 61572381).

References

1. Chutia, D., Bhattacharyya, D., Sarma, K., Kalita, R., Sudhakar, S.: Hyperspectral remote sensing classifications: a perspective survey. Trans. GIS **20**(4), 463–490 (2016)
2. Tuia, D., Volpi, M., Trolliet, M., Camps-Valls, G.: Semisupervised manifold alignment of multimodal remote sensing images. IEEE Trans. Geosci. Remote Sens. **52**, 7708–7720 (2014)
3. Lunga, D., et al.: Manifold-learning-based feature extraction for classification of hyperspectral data: A review of advances in manifold learning. IEEE Sig. Process. Mag. **31**(1), 55–66 (2014)
4. Huang, H.-B., Huo, H., Fang, T.: Hierarchical manifold learning with applications to supervised classification for high-resolution remotely sensed images. IEEE Trans. Geosci. Remote Sens. **52**, 1677–1692 (2014)
5. Ma, L., Crawford, M.M., Yang, X., Guo, Y.: Local-manifold-learning-based graph construction for semisupervised hyperspectral image classification. IEEE Trans. Geosci. Remote Sens. **53**, 2832–2844 (2015)
6. Fang, Y., Li, H., Ma, Y., Liang, K., Hu, Y., Zhang, S., et al.: Dimensionality reduction of hyperspectral images based on robust spatial information using locally linear embedding. IEEE Geosci. Remote Sens. Lett. **11**, 1712–1716 (2014)
7. Yang, L., Yang, S., Jin, P., Zhang, R.: Semi-supervised hyperspectral image classification using spatio-spectral Laplacian support vector machine. IEEE Geosci. Remote Sens. Lett. **11**, 651–655 (2014)
8. Gillis, D.B., Bowles, J.H.: Hyperspectral image segmentation using spatial-spectral graphs. In: SPIE Defense, Security, and Sensing, pp. 83901Q–83901Q-11 (2012)
9. Cao, L., Chua, K.S., Chong, W., Lee, H., Gu, Q.: A comparison of PCA, KPCA and ICA for dimensionality reduction in support vector machine. Neurocomputing **55**, 321–336 (2003)
10. Yu, H., Yang, J.: A direct LDA algorithm for high-dimensional data—with application to face recognition. Pattern Recogn. **34**, 2067–2070 (2001)
11. Zhang, T., Yang, J., Zhao, D., Ge, X.: Linear local tangent space alignment and application to face recognition. Neurocomputing **70**, 1547–1553 (2007)

12. Xiong, L., Chitti, S., Liu, L.: Mining multiple private databases using a knn classifier. In: Proceedings of the 2007 ACM Symposium on Applied Computing, pp. 435–440 (2007)

13. Melgani, F., Bruzzone, L.: Classification of hyperspectral remote sensing images with support vector machines. IEEE Trans. Geosci. Remote Sens. **42**, 1778–1790 (2004)

14. Huang, G.-B., Zhu, Q.-Y., Siew, C.-K.: Extreme learning machine: theory and applications. Neurocomputing **70**, 489–501 (2006)

15. Wu, C.-H., Tzeng, G.-H., Lin, R.-H.: A novel hybrid genetic algorithm for kernel function and parameter optimization in support vector regression. Expert Syst. Appl. **36**, 4725–4735 (2009)

Applying a Classification Model for Selecting Postgraduate Programs

Waraporn Jirapanthong$^{(\boxtimes)}$, Winyu Niranatlamphong,
and Karuna Yampray

College of Creative Design and Entertainment Technology, Dhurakij Pundit
University, Bangkok, Thailand
{waraporn.jir,winyu.nir,karuna.yam}@dpu.ac.th

Abstract. Some people have failed in selection of postgraduate programs. This is due to the lack of potential information to support a making decision. Although a large amount of data based on information systems in academic institutes has been collected for years, the use of the data is still not supporting academic benefits, particularly to the students or applicants. In this work, we present the use of data mining technique, particularly classification technique, to support applicants in selection of postgraduate programs. The paper also presents the study on educational structure in Thailand, and background of data mining concepts and techniques. The details of learning process to built-up the classification model is described and some examples of extracted rules from the classification model are given. We also present the case study and usage of the model.

Keywords: Data mining · Automatic decision · Semi-Intelligent applications · Data utilization

1 Introduction

A key to succeed in an academic life is that an applicant needs to put himself into the right course/program regarding his knowledge, potential skills, and interests. Every applicant is expected to potentially acquire the knowledge according to the course of study chosen. However, the main problem in the selection of study course in postgraduate programs is that the applicants are not supported by the potential and analytical information.

This paper focuses on the study of the influencing factors to the academic success of applicants. The research project thus applies data mining concept and techniques to develop a model for supporting the selection of higher education, particularly postgraduate programs. The model is built by learning an input data set. The learning algorithm is used to discover the model that best fits the relationships between the attribute set and class labels of the input data. The model generated by a learning algorithm should both fit the input data well and correctly predict the class labels of records which it has never seen before. The model is then applied with data set to classify the class labels of applicants. The class label represents the study course which seems to be suitable to an applicant.

© Springer International Publishing AG 2017
Y. Tan et al. (Eds.): ICSI 2017, Part II, LNCS 10386, pp. 330–337, 2017.
DOI: 10.1007/978-3-319-61833-3_35

2 Study on Educational Structure in Thailand

2.1 Higher Education

Many public academic institutes in Thailand are now managed and administered on their own. However, both public and private universities are monitored under the office of the higher education commission. In particular, they set educational standards, approve curriculum, and develop the main institutional and professional accrediting body. In 2010, there are about 300,000 people received Bachelor's degrees and 63,000 people received Master's degree. It shows that 21 percentages of graduates altogether continued their postgraduate studies. In 2006-2010, the number of postgraduate students has been increased from 200,000 to 260,000 students.

The programs and degrees of universities in Thailand can be described as follows:

Stage I: The Bachelor's degree requires four years of full-time study in most fields. However, undergraduate programs in pharmacy, and graphic art requires five years of study. Bachelor's degree programs in medicine, dentistry and veterinary science require six years leading respectively to the Doctor of Medicine, Doctor of Dental Surgery and Doctor of Veterinary Medicine.
Stage II: The Master's degree requires between one and two (usually two) years of full-time study.
Stage III: The Doctorate requires between two and five years of study beyond the master's level.

3 Background on Data Mining

Recently, many research work [1, 4, 6–8] have applied with the data mining techniques. Classification technique is a supervised learning technique that classifies data item into pre-defined class label. This appropriate technique builds model that predict future data trend. There are several algorithms for data classification such as Decision Tree, CART (Classification and Regression Tree) and Back Propagation neural network. Each technique employs a learning algorithm to identify a model that best fits the relationship between the attribute set and class label of the input data. The model generated by a learning algorithm should both fit the input data well and correctly predict the class labels of records it has never seen before. The key objective of the learning algorithm is to build models with good generalization capability.

Moreover, Clustering technique is a division of data item into similar group without training of class labels. There are several clustering algorithms such as K-means, hierarchical agglomerative clustering and Self-Organizing Map. Euclidean distance and two clustering algorithms: Kohonen's Self Organizing Map and K-means applied in this work are described below.

3.1   Euclidean Distance

Euclidean distance is the most popular distance measure to calculate the dissimilarity (similarity) between two data objects from same space is to clustering procedures. Euclidean distance is defined as the Eq. (1).

$$d(i,j) = \sqrt{\left(x_{i1} - x_{j1}\right)^2 + \left(x_{i2} - x_{j2}\right)^2 + \ldots + \left(x_{in} - x_{jn}\right)^2} \tag{1}$$

Where:

$i = (x_{i1}, x_{i2}, \ldots, x_{in})$ are two n- dimensional data objects.
$j = (x_{j1}, x_{j2}, \ldots, x_{jn})$ are two n- dimensional data objects.

3.2   Kohonen's Self Organizing Map Algorithm

The Kohonen's Self Organizing Map (SOM) algorithm is a popular algorithm for unsupervised learning [5]. The concept of SOM is iterative for each data to find weight according number of clusters as follows:

1. Initialize the weights vector of all the output neurons.
2. Determine the output wining neuron m by searching for the shortest normalized Euclidean distance between the input vector and the weight vector of each output neuron, by using the Eq. (1).

$$|X - W_m| = \min_{j=1\ldots M} |X - W_j| \tag{2}$$

Where:

X is the input vector,
W_j is the weight vector of output neuron j, and
M is the total number of output neuron.

3. Let Nm(t) denote a set of indices corresponding to a neighborhood size of the current winner neuron m. The neighborhood size needs to be slowly decreased during the training session. The weights of the weight vector associated with the winner neuron m and its neighborhood neurons are updated by the Eq. (3)

$$\Delta W_j(t) = \alpha(t)\left[X(t) - W_j(t)\right], \quad \text{for} \quad j \in N_m(t) \tag{3}$$

Where α is a positive-valued learning factor from [0, 1]. It needs to be slowly decreased with each training iteration. Thus, the new weight vector is given by the Eq. (4)

$$W_j(t+1) = W_j(t) + \alpha(t)\left[X(t) - W_j(t)\right] \quad \text{or} \quad j \in N_m(t) \tag{4}$$

Steps 2 and 3 are repeated for every exemplar in the training set for a user-defined number of iterations.

3.3   K-Means Algorithm

The k-means algorithm takes the input parameter, k, and partitions a set of n objects into k clusters [5] so that the overall sum of square error is minimized.

A major limitation of k-mean is to determinate the number of k clusters that is to affect the accuracy of the data partition and the efficiency of the clustering processing. As the k parameter increase, the data item will be divided into clusters and the processing will need large computational power. On the contrary, the k parameter is small, the data item will be divided into less clusters and some significant characteristic of data may be lost. The k-means algorithm is described as follows:

1. Determinate the number of k clusters.
2. Randomly selects initial represent point a cluster mean.
3. Assign each object to the closest cluster center, based on similarity measure.
4. Computes the new mean for each cluster. This process iterates until object is no change group (cluster). If object change group go to step 3.

4   Developing a Classification Model to Select Postgraduate Programs

This section describes building-up the classification model for selecting postgraduate programs. Data were divided into two sub-data, one is model building sub-data, and the other is model testing sub-data. By using of decision tree technique, the classification model had been setup. We have described the details of data attributes in Table 1. We considered the data records of applicants who have current accumulative grade from 3.0 since it can be implied that the students are qualified to complete their degree and fair enough to continue further education in the same major or related. Moreover, we applied the 10-fold cross validation method for evaluating the classification model.

In addition, in the research, we declare the terminology as follows:

1. *An instance* – a data record
2. *ConfidenceFactor* – The confidence factor used for pruning (smaller values incur more pruning).
3. *MinNumObj* – The minimum number of instances per leaf.

This dataset is data about applicants' education type and GPA (Grade Point Average) in former degree. The result from this experiment can be used to help applicants in choosing appropriate program for them. There are 7778 records in dataset, in each record there are 5 attributes: Gender (Female, Male), OldMajor (major previously finished), idGPA (previous degree), EdType (Applicant's education type), and Program (that an applicant applies for). We used data mining tool called Weka (http://

www.cs.waikato.ac.nz/ml/weka/) to build decision tree. The algorithm that we used is C4.5 [2], which builds a decision tree by recursively selecting attributes on which to split.

We proposed to build up a classification model to give a guideline to an applicant who is applying for higher education. The technique of decision tree was applied. The tree was developed to determine which major is suitable for an applicant. The data was prepared through process i.e. extraction, cleaning, transformation, loading and refreshing. During the learning process to build up a decision tree, many trees were created and evaluated by the 10-folder cross validation method. The evaluation results of those decision trees were not convincing. Eventually, the final version of the

Table 1. Two test cases

Test case 1	Test case 2
Purpose and Goals	**Purpose and Goals**
• To analyze the accuracy of model in predicting poor academic performance	• To analyze the accuracy of model in predicting poor academic performance
• To use the information/data that are available to suggest the g occur before students enroll	• To use information/data that occur before students enroll
	• To lower cost by screening a selected group of students
Data Used for Model Development	**Data Used for Model Development**
• Juniors and seniors (2005 – 2006) 3,751 cases	• Freshmen (2004 – 2006)
	• 4,652 cases
• Average High School GPA of 2.33	• Average High School GPA of 2.38
56% female and 44% male	• 52% female and 48% male
Variables Used for Model Development	**Variables Used for Model Development**
• Educational status: Third-year and Fourth-year	• Educational status: First-year
• High School Performance	• High School Performance
• High school GPA	• High school GPA
• Old major	• Old major
• Student Demographics	• Student Demographics
• Gender	• Gender
• Type of Feeder High School (Public, Private, Foreign, Out of State)	• Type of Feeder High School (Public, Private, Foreign, Out of State)
Methods	**Methods**
• Apply the decision tree model developed and described in previous section.	• Apply the decision tree model developed and described in previous section.
• Run nine experiments with different confidence and minimum number of object.	• Run nine experiments with different confidence and minimum number of object.
Results	**Results**
According to the nine experiments, the average results of test case 1 show that correctly identified is 87.937% of at-risk students while misclassifying is 47.6%.	According to the nine experiments, the average results of test case 2 show that correctly identified is 83.2% of at-risk students while misclassifying is 55.2%.

decision tree returns the results with high accuracy. Each oval is represented for a factor to be concerned in classification. A rectangle is represented for a major. The classification to find out the most-appropriate major for a student is concerned several factors: *gender, old major in previous program, accumulative GPA in previous degree, type of education,* and *applying program.* The decision tree is generated with the following factors: *ConfidenceFactor* = 0.25; *MinnumObj* = 2; *Evaluating method*: 10-fold cross validation; and *Scheme:* weka.classifiers.trees.J48 -C 0.25 -M 2. The result of generation shows that *Correctly Classified Instances* is 84.1162% and *Incorrectly Classified Instances* is 15.8838%.

A set of rules which are extracted from the decision tree are used to classify and identify the class label for each student. The class label represents the study course of undergraduate programs for a student.

1. If OldMajor = Fundamental-Diploma And Gender = M And OldGPA = 2.50–3.00
   Then Faculty = ICT
2. If OldMajor = Fundamental-Diploma And Gender = M And OldGPA = 3.01–3.49
   Then Faculty = Business-Administration
3. If OldMajor = Fundamental-Diploma And Gender = M And OldGPA = 3.50–4.00
   Then Faculty = Communication-Art

5   Case Study and Usage Analysis

We have created two test cases in order to identifying at risk students. As shown in Table 1, for each of test case we used data set encompassing students' profiles and applied with the model presented in previous section. Test case 1 is aimed to evaluate how the decision tree model can give the useful information to unsuccessful students.

Table 2. Percentage of correctly classified, incorrectly classified, and misclassifying

Test Case		Profile								
		1	2	3	4	5	6	7	8	9
1	Confidence	0.25	0.25	0.25	0.50	0.50	0.50	0.75	0.75	0.75
	Minimum number of instance	2	10	50	10	30	50	5	20	50
	Correctly identifying	91.45	90.42	89.63	88.15	88.02	87.95	86.45	85.42	83.95
	Incorrectly identifying	0.855	09.58	10.36	11.85	11.98	12.05	13.55	14.58	16.05
	Misclassifying	41.02	42.8	47.05	49.28	49.55	50.03	47.85	49.8	51.02
2	Confidence	0.25	0.25	0.25	0.50	0.50	0.50	0.75	0.75	0.75
	Minimum number of instance	2	10	50	10	30	50	5	20	50
	Correctly identifying	89.3	87.25	85.4	84.95	83.6	82.15	79.88	78.82	77.45
	Incorrectly identifying	10.7	12.75	14.6	15.05	16.4	17.85	20.12	21.18	22.55
	Misclassifying	52.1	53.07	53.18	54.28	54.55	55.03	57.35	57.9	59.34

Test case 2 is aimed to analyze the accuracy of model in predicting poor academic performance.

The details of each experiment are shown in Table 2. According to the classification model, we have applied with data records. The percentage of correctly identified is high (87.3%). It implies the model performs the classification of program selection for students with the high performance. The percentage of misclassifying is fairly low (47.6%). It implies the model failed to identify the class label, particularly suitable program, for a student. This is due to incomplete data e.g. missing some attribute values, incorrect data.

6 Conclusions and Future Work

6.1 Conclusions

We applied the technique which is a supervised learning technique that classifies data items composed of several attributes, for example, old major at pre-university, home address, school location, studying faculty, and GPA, into pre-defined class labels. This appropriate technique builds a classifier model that predicts future data trend. The required data is quite simple, and data set is collected from many sources such as university and old schools of students.

According to the research project, it shows that it is a challenge to an academic institute to adopt the techniques of data mining to decrease risk students and improve the quality of education system. A good system including the mining techniques, i.e. classification model, may provide the users with important competitive information. The users here include prospective students, current students, and academic staffs. In addition, the model could be incorporated with available management information system to tackling related issues in education system in Thailand.

6.2 Future Work

At present, the model is used in a semi-automatic way to acquire data from the operational source systems as a data set. The operational database within academic institutes periodically updated in monthly and yearly according to term period. The automatic process is expected to support data extraction, data transformation in data set format, and loading to the data set periodically. So, the researchers who are responding to the system can acquire the updated data for helping to planning and analyzing data which is consequence to rapidly making decision.

Accordingly, in the future we can represent the scatter group of data in scatter plot format to help user forget the overall group of data obviously.

References

1. Agarwal, S., et. al.: Data mining in educational: data classification and decision tree approach. Int. J. e-Educ. e-Bus. e-Manage. e-Learn. 2(2) (2012)
2. Anyanwu, N., et al.: Comparative analysis of serial decision tree classification algorithm. Int. J. Comput. Sci. Secur. (IJCSS) 3(3), 230–240 (2010)
3. Han, J., Kamber, M.: Data Mining Concepts and Techniques, 2nd edn. Morgan Kaufmann publications, San Francisco (2006)
4. Harwati, Sudiya, A.: Application of decision tree approach to student selection model – a case study. In: IOP Conference Series: Materials Science and Engineering, vol. 105 (2016). 012014
5. Kohonen, T.: Self-Organizing Maps. Springer, Heidelberg (2001)
6. Neelamadhab, P., et al.: The survey of data mining applications and feature scope. Int. J. Comput. Sci. Eng. Inf. Technol. (IJCSEIT) 2(3), 1–16 (2012)
7. Tair, A., Halees, E.: Mining Educational Data to Improve Students' Performance: A Case Study, vol. 2(2) (2012). ISSN 2223-4985
8. Umamaheswari, K., Niraimathi, S.: A study on student data analysis using data mining techniques. Int. J. Adv. Res. Comput. Sci. Softw. Eng. 3(8) (2013). ISSN: 128X

University Restaurant Sales Forecast Based on BP Neural Network – In Shanghai Jiao Tong University Case

Liu Xinliang and Sun Dandan[✉]

School of Computer and Information Engineering,
Beijing Technology and Business University, Beijing, China
sundandan07@126.com

Abstract. In recent years, BP (Back Propagation) neural network is widely used in predictive modeling in various fields. But the BP neural network technology which used for university catering service is very few. The article is applied to the data set which is published by the EMC competition of Shanghai Jiao Tong University in 2015. We use BP neural network to analyze and forecast the university restaurant sales, and then through comparing the model with the time series forecasting method. The elements used in the model include the cycle factor, the Baidu index of the network take away, the weather information. The forecasting factors include three aspects of the 11 variables, which is also an innovation of this paper. Finally, we proved that the model we built has a good prediction result and it also has practical availability. This article also explained how the variables impact on university catering service.

Keywords: First keyword · Second keyword · Third keyword

1 Introduction

A very important content for the management of university management is the management for restaurant. But now the restaurant management is still mainly relying on personal experience, there is no data system support [1]. This paper uses BP artificial neural network technology to create a university canteen sales forecast model for the university restaurant management to provide a credible reference data – the sales amount. It is not only helpful for the school to better carry out the logistics management of the canteen, but also helpful to provide a reference for the restaurant to make a more suitable production plan.

In this paper, a pioneering achievement is the use of Baidu index, meteorological factors to predict the restaurant turnover. The prediction model established in this paper can reproduce the restaurant management to a great extent. The results show that the Baidu index of the takeaway site plays a more important role than the rainfall information. Liwen Vaughan, Yue Chen demonstrated the use of Google trends and Baidu index in the availability of data mining [2], and this article will continue to refine its application in the field of prediction. Prediction model based on BP artificial neural network is one of the most effective methods for forecasting problems. In 2008, Wang

© Springer International Publishing AG 2017
Y. Tan et al. (Eds.): ICSI 2017, Part II, LNCS 10386, pp. 338–347, 2017.
DOI: 10.1007/978-3-319-61833-3_36

Wanjun [3] used the gray BP neural network to forecast the sales of commercial housing. In Although the results were satisfactory, the paper did not apply the sales-related factors to model. The model constructed in our research also obtains the predictive variables such as meteorological information and Baidu index. And at the same time, we obtained a more accurate prediction result. In 2014, Song Guofeng and Liang Changyong [4] used the improved BP neural network to forecast the traffic volume of the tourist scenic spot. In the paper, meteorological information and period information were used as the prediction factors, and the satisfactory results were obtained. It can be seen that BP Neural networks have powerful predictive functions for variables with periodic variation rules. Also in 2014, Luo Ronglei et al. [5] used genetic algorithm BP neural network prediction to model the clothing sales, but the predictive variables selected in this paper are expert score which are quite subjective. However, the predictor variables we selected are the objective data that avoid the influence of human subjectivity.

2   The BP Neural Network Algorithm

2.1   Initialization

Input training samples, the input vector X and the expected output vector d are obtained, and the output of each layer is calculated [7].

Net input of hidden layer neuron j:

$$net_j = \sum_{i=0}^{n} v_{ij} * x_i \ j = 1, 2 \ldots \ldots m \tag{1}$$

The net input to the hidden layer neuron j is equal to the sum of initial weight multiplied by the input value.

Output of hidden layer neurons j:

$$y_j = f(net_j) \ j = 1, 2 \ldots \ldots m \tag{2}$$

The output of j in the hidden layer neurons is equal to the activation function acting on the net input value of j.

The net input of the output layer neuron k:

$$net_k = \sum_{j=0}^{m} \omega_{jk} * y_j \tag{3}$$

The net input of the output layer neuron k is equal to the weight of the hidden layer to the output layer multiplied by the output weight of the hidden layer j.

Output the output layer of the neuron j:

$$o_k = f(net_k) \tag{4}$$

The output of k in the output layer neurons is equal to the activation function acting on the net input value of k.

Activation function:

$$f(x) = \frac{1 - e^{-x}}{1 + e^{-x}} \tag{5}$$

2.2 Calculate Network Output Error

The root mean square error is used as the total error of the network:

$$E = \sqrt{\sum_{k=1}^{l} (d_k - o_k)^2} \tag{6}$$

Hidden layer expected output vector [9]:

$$Y'(t) = f(\sum_{i=1}^{n} (W_{ij}(t)X_i(t) + \theta_j(t)) \tag{7}$$

Output node expectation and prediction error:

$$e_j = Y(t) - Y'(t)(1 \leq j \leq k) \tag{8}$$

2.3 Adjust Each Layer Weight

Adjustment of each layer:

$$\Delta\omega_{jk} = \eta(d_k - o_k)o_k(1 - o_k)y_j \tag{9}$$

$$\Delta v_{ij} = \eta\left\{\sum_{k=1}^{l} (d_k - o_k)o_k(1 - o_k)\right\}\omega_{jk}y_j(1 - y_j)x_i \tag{10}$$

η is the learning rate.
Adjusted weights:

$$\omega_j' = \omega_j + \Delta\omega_{jk} \tag{11}$$

$$v_j' = v_j + \Delta v_{ij} \tag{12}$$

Complete training once for all samples, check whether the maximum training time is reached, and if the end condition is satisfied, training is stopped; otherwise, training is continued from step (2) until the end condition is satisfied.

3   Evaluation System of Result

3.1   Network Accuracy

$$\text{Network accuracy} = 1/p \sum\nolimits_{t=1}^{p} \frac{1 - |\text{Predictive value} - \text{Actual value}|}{\text{Maximum predicted value} - \text{Minimum predicted value}} * 100\% \quad (13)$$

Where p represents the total number of test data sets, the greater the network accuracy, the better the quality of the network. Because of the accuracy of the model is calculated for the training set, this parameter is more optimistic than fact.

3.2   Mean Absolute Percentage Error (MAPE)

In order to evaluate the accuracy of the prediction model, this paper uses MAPE as another indicator of model evaluation.

$$\text{MAPE} = \sum \frac{|\text{Predicted value} - \text{Actual value}|}{\text{Actual value}} /p \quad (14)$$

MAPE is less than 10%, indicating that the model has high precision prediction effect: MAPE in 10%–20% shows that the model has a good prediction effect. MAPE between 20% and 50% indicates that the model is feasible. MAPE is greater than 50% indicating that the model is wrong, it doesn't have the function of prediction.

3.3   Importance of Predictive Variables

The importance of the predictor variables indicates the relative importance of each variable in the prediction process, independent of the accuracy of the model predictions. The sum of all predictor variables was 1. The closer the importance value of the predictor variable to 1, the more important it is to establish the model. Here we use the Tchaban algorithm based on network weights to calculate the importance of predictive variables.

The sensitivity coefficient of input variable xi to output variable ok is:

$$Q_{ik} = \frac{X_i(t)}{O'_k(t)} \sum\nolimits_{j=1}^{m} W_{ij}(t) V_{jk}(t) \quad (15)$$

t represents time. For numeric variables, t is selected as 0, 0.23, 0.5, 0.75, and 1. The average value of all the values of all input variables is taken as the input variable.

4 Experiment

The data used in this paper include the students' card record of Minhang campus of Shanghai Jiao Tong University, the information of Minhang campus weather station and the Baidu index in Shanghai area.

4.1 Preliminary Data Processing

First remove the attributes that are not relevant to the purpose of this study from card data set, such as the opening hours of the restaurant and so on, and then complete the missing meteorological data.

Summarize the required variables and add the teaching week, holiday, and Wednesday information according to the Shanghai Jiao Tong University 2014–2015 school calendar, and finally get the data shown in Table 1. "Education Week" represented 2014–2015 year the first few weeks of teaching. Here a negative number indicates before the semester begins, a number more than 18 represent weeks after the end of the winter semester. "Holiday" shows that whether or not the date of the day is a holiday, and the field is divided into five categories: summer vacation (SV), winter vacation (WV), weekends (W), holidays (H), working days (WD). The week is represented by 1–7, 1 here represent Monday.

We get 2014/09/01 to 2015/01/31 in the Shanghai region of the takeaway platform Baidu index from Baidu index search engine. In this article, according to the strength of the demand map in the Shanghai area we selected four network takeaway platform: ele. me, Delivery Hero, Baiduwaimai, Meituanwaimai.

The data were normalized by the students, and the outliers were removed. Finally, 7012 records were obtained.

4.2 Contrast Test

The sales data of four shops (First Floor Restaurant, Sichuan Snack, Xinjiang Restaurant, Chow Mein) were randomly selected by using time series and BP neural network respectively to forecast the sales.

1. Time Series Prediction Model
   Time series algorithm prediction is to use statistical techniques and methods or intelligent algorithms to find out the evolution model from the time series of prediction indicators, then establish mathematical models and make quantitative estimates of the future development trend of forecast indicators [10]. In this paper, we use SPSS Modeler's own expert modeler algorithm for time series modeling, and finally calculate the MAPE value as shown in Table 2.
   As can be seen from Table 2, in addition to the "First Floor Restaurant" outside the use of time series models are built with good prediction accuracy of the forecast model. The error of "First Floor Restaurant" is very large because there are some missing values.

Table 1. The experimental data sample

Date	Merchant name	Amount of transactions	Temperature	Precipitation	Maximum wind speed	Baidu-waimai	Meituan-waimai	Ele. me	Delivery hero	Week	Education week	Holiday
2014/9/2	Grill	3172	27.47	3.7	6.8	222	494	3557	1050	2	−1	SV
2014/9/1	Congee station	3307.9	26.93	22.4	9.7	250	402	2974	839	5	−2	SV
2014/9/2	Chow mein	856.8	27.47	3.7	6.8	222	494	3557	1050	4	−1	SV

Table 2. MAPE of Time series prediction model

Merchant name	MAPE
First floor restaurant	104.71%
Sichuan snack	10.86%
Xinjiang restaurant	18.53%
Chow mein	13.08%

2. BP Neural Network

Applicate of BP neural network on the "First floor restaurant", "Sichuan snack", "Xinjiang restaurant", "Chow Mein" these four restaurant sales data modeling. In order to avoid the over fitting of the model, 70% of the data were randomly selected as the training set, and the remaining 30% were tested. The evaluation index of the model is shown in Table 3.

Table 3. Evaluation index of BP neural network model

Merchant name	Network accuracy	MAPE
First floor restaurant	95.70%	9.58%
Sichuan snack	97.50%	5.82%
Xinjiang restaurant	98.50%	6.68%
Chow mein	92.90%	11.68%

According to the MAPE value, it can be seen that the model based on BP neural network has higher accuracy than the time series prediction model. And for the "Sichuan Snack", "Xinjiang Restaurant" in terms of two restaurants this model achieved high precision of prediction result. The comparison between the predicted results in September 2014 and the actual data in Fig. 1 is drawn for the direct comparison of the two models. It can be seen from the graph that BP neural network has better prediction effect than time series model.

5 Result

Through the comparison experiment, it can be proved that the prediction model based on BP neural network has higher accuracy. We put "Merchant name","Baidu index of Baidu-waimai", "Baidu index of Meituan-waimai", "Baidu index of ele.me","Baidu index of Delivery Hero", "Precipitation", "Maximum wind speed", "Temperature", "Teaching week", "Holidays", "week", a total of eleven variables as input variables to the BP neural network. And then we get a BP neural network model with three layers and 12 hidden layer nodes is established. The accuracy of network prediction can reach 93.7%, and MAPE is 19.74%. The MAPE value is higher than the previous test results. This is because in this model, the business situation of each merchant is different, so for each merchant, whether it is the transaction amount or the importance of the predictive variables have a greater discrepancy. If the total revenue of all restaurants in Minhang

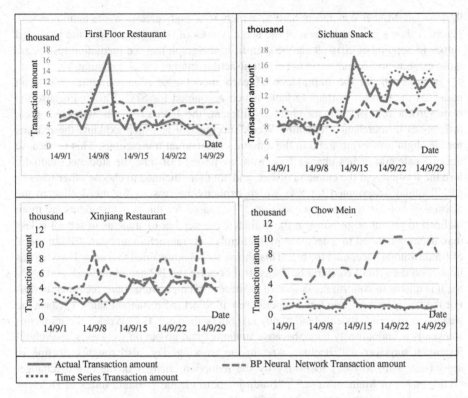

Fig. 1. Comparison of experimental results with real data

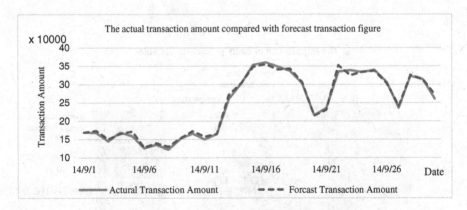

Fig. 2. Comparison of the actual transaction amount and the forecast transaction amount

campus of Shanghai Jiao Tong University as the forecast target, the accuracy of BP neural network is 92.1%, and MAPE can reach to 8.46%, which is a high precision prediction model. If the total income of all restaurants in Minhang campus of Shanghai Jiao Tong University is used as the forecast target, the accuracy of BP neural network is

92.1% and MAPE can reach 8.46%, which is a high precision prediction model. Figure 2 shows the comparison of the actual values of the test data with the predicted values in September 2014. It can be seen from the graph that the prediction results are quite close to the actual results (Total transaction amount for all restaurants).

Figure 3 shows the importance of each model prediction variable. "Merchant name" is the most important variable which accounted for 57%. This is because the management rule of each merchant is different, the forecast results will have great differences in merchant. "Teaching Week" occupied 9%, which explains the reasons for the slight downward trend in the amount of restaurant transactions. This is because that the cafeteria dishes change less, with the passage of time the students gradually lose the freshness of the cafeteria dishes, and are more inclined to choose other ways to eat. "Holiday" accounted for 8%, which explains the reasons for the decline in the amount of holiday restaurant transactions. The reason for this is that students are more inclined to go out or go home in the holidays, the number of students in school is less than usual which led to a decrease in the number of transaction amount. "Baidu index of Baiduwaimai" accounts for 6%. This shows that Baiduwaimai to a certain extent has impact on the school cafeteria business. But because the BP neural network is a black ox, it is unable to confirm the degree of influence. "Temperature" here accounts for 4%. For the "temperature", too low temperature will affect the restaurant sales, because the weather is too cold students tend to buy takeout or dine out. For "Week", the impact of weekends is undoubtedly huge, because students trend go out on weekends, the school restaurant turnover will decline. And in the working day, also roughly decline in accordance with Monday to Friday by the reduction changes in the law of change. "Baidu index of Meituanwaimai", "Baidu index of ele.me", "Baidu index of Delivery Hero" accounts for the same ratio was 3%. These three all have a certain impact on the canteen management, the same as the effect manner of "Baidu index of Baiduwaimai".

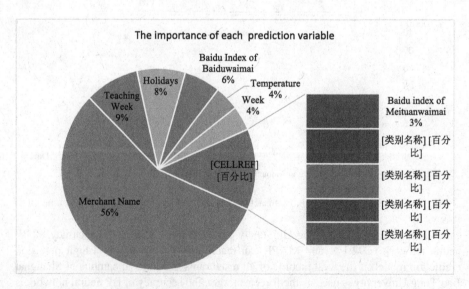

Fig. 3. The importance of each prediction variable

Students are more willing to buy food in the supermarket or order a take-away in the rain and windy weather, so the precipitation and wind speed will also affect the restaurant sales, although its impact is limited.

6 Conclusion

In this paper, the Baidu index of four takeaway websites, meteorological factors and cycle factors are used as predict variables. The choice of variables is also a major innovation in this article. Three-layer BP neural network is used to forecast the restaurant sales of Minhang campus of Shanghai Jiao Tong University. By comparing with the time series modeling method, finally prove that the model established has higher accuracy. Adopt the method of BP neural network model projections for individual businesses, the MAPE index under 20%, network accuracy above 90%. There are some defects, such as the paper chooses variables may not be comprehensive. In the future, we need to complete the variables and improve the algorithm, we need to continue to study.

Acknowledgements. This research is supported by Scientific research plan of Beijing Municipal Commission of Education KM201510011008 and Beijing Key Laboratory of Big Data Technology for Food Safety, School of Computer and Information Engineering, Beijing Technology and Business University.

References

1. Lihui, P.: Research on logistics cost management of logistics in colleges and universities. J. Univ. Logistics Res. **2**, 63–64 (2011)
2. Vaughan, L., Chen, Y.: Data mining from web search queries: a comparison of google trends and baidu index. J. Assoc. Inf. Sci. **66**(1), 13–22 (2015)
3. Wang, W.: The commercial housing consumption forecasting based on grey theory and BP neural network. J. Hebei Polytech. Univ. Soc. Sci. Ed. **8**, 65–67 (2008)
4. Song, G., Liang C., Liang, Y., Zhao, S.: Improved genetic algorithm to optimize the BP neural network to forecast the daily traffic of tourist attractions. J. **35**(9), 2136–2141 (2014)
5. Luo, R., Liu, S., Su, C.: BP neural network based on genetic algorithm clothing sales forecasting method. J. Beijing Univ. Posts Telecommun. **37**(4), 39–43 (2014)
6. Guozheng, Q., Zhenyu, Z., Tong, W.: Modeling the hot deformation behaviors of as-extruded 7075 aluminum alloy by an artificial neural network with back-propagation algorithm. J. High Temp. Mater. Processes **36**(1), 1–13 (2017)
7. Tian, W., Meng, F., et al.: Lifetime prediction for organic coating under alternating hydrostatic pressure by artificial neural network. J. Sci. Rep. **7**, 1–12 (2017)
8. Wang, X., Chen, H., et al.: Study the therapeutic mechanism of amomum compactum in gentamicin-induced acute kidney injury rat based on a back propagation neural network algorithm. J. Chromatogr. B-Anal. Technol. Biomed. Life Sci. **1040**, 81–88 (2017)
9. Jiang, C., Zhang, S.-Q., Zhang, C.: Modeling and predicting of MODIS leaf area index time series based on a hybrid. SARIMA and BP neural network method. Spectral and Spectral Anal. **37**(1), 189–193 (2017)
10. Zhang, Y., Xu, X., Wang, Z.: Support vector machine and time series prediction. J. Comput. Appl. Softw. **27**(12), 127–129 (2010)

Classification and Detection

Swarm ANN/SVR-Based Modeling Method for Warfarin Dose Prediction in Chinese

Yanyun Tao[1] ⓘ, Dan Xiang[3], Yuzhen Zhang[2(✉)], and Bin Jiang[2]

[1] School of Urban Railway Transportation,
Soochow University, Suzhou 215001, China
[2] The First Affiliated Hospital of Soochow University, Suzhou 215006, China
zhangyuzhen@suda.edu.cn
[3] Suzhou Vocational University, Suzhou 215104, China

Abstract. In order to improve the predictive accuracy on warfarin dose, a modeling method using artificial neural network (ANN) and support vector regression (SVR) together with particle swarm optimization (PSO) is developed, which is denoted as PSO-(ANN/SVR). The procedure of PSO-(ANN/SVR) runs a population of ANNs and SVR to develop diverse candidate models, and a PSO is employed as a "shell" to optimize a group of ANNs and SVR by following the current optimum particles(i.e. the best ANN or SVR) in search space. We collected a dataset of 100 Chinese patients provided by The First affiliated Hospital of Soochow University, and divide it into training, validation and test set. (ANN/SVR) models are built on the training and validation set, and finally tested on test set. In the experiment, PSO-(ANN/SVR) is compared with five modeling methods in terms of mean squared error (MSE) and squared correlation (R^2). The experimental results show that the models developed by PSO-(ANN/SVR) present the best MSE and R^2 in these cases. PSO-(ANN/SVR) achieved 11.9–48.9% reduction on MSE over the other methods with different variables in test set. It is noted that PSO-(ANN/SVR) models had a small decrease in R^2 from training set to test set, and obtained a large R^2 on both training and test set. This illustrates that the models of PSO-(ANN/SVR) have a good generalization.

Keywords: Warfarin dose prediction · Machine learning · Neural network · Support vector regression · Particle swarm optimization

1 Introduction

Warfarin is the most popularly used oral anticoagulant for the prevention and treatment of thromboembolic disorders around the world. Despite its effectiveness, treatment with warfarin has several shortcomings. The dose-effect relationship of warfarin is affected by genetic factors and environment factors [1], including cytochrome P450 gene locus mutation, wild-type enzyme CYP2C9, liver enzyme gene polymorphism, patient's weight, height, age and so on. Many commonly used medications interact with warfarin, as do some foods that typically contain large amounts of vitamin K1. In clinical treatment, the international normalized ratio (INR) of warfarin dose-response should be closely monitored by blood testing to ensure an adequate yet safe dose is taken [2]. Due

© Springer International Publishing AG 2017
Y. Tan et al. (Eds.): ICSI 2017, Part II, LNCS 10386, pp. 351–358, 2017.
DOI: 10.1007/978-3-319-61833-3_37

to the influence of a multitude of factors on the warfarin pharmacokinetics and pharmacodynamics, the dose requirements and variability in anticoagulation are usually complicated and unpredictable [3]. Despite of many dose prediction models developed [4], warfarin dose prediction is still mostly depending on the experience of clinicians. Moreover, a patient must undergo a number of adjustments on dose before the does can be finally determined. To mitigate the risk-associated response variability, investigators and clinicians have focused on developing strategies to improve dose prediction with the hopes of improving anticoagulation control with resultant decrease in hemorrhage [5, 6]. Some conventional methods, mechanism analysis and regression analysis, for warfarin dose modeling are explored to provide a basis for dose prediction.

Machine learning (ML) techniques [7], such as Bayesian network (BN), Artificial neural networks (ANNs) and Support vector regression (SVR), have been comprehensively used warfarin dose prediction [8]. BN is utilized to make decision on solving medical problems [9]. ANNs have been widely used in the modeling for warfarin-dose prediction [10, 11]. However, the convergence of global approximation to a complex system is too slow and ANN may easily fall into local optima. For ANNs, a common disadvantage is poor generalization ability. SVR is a modeling method with structured risk minimization, which has been applied in warfarin dose prediction [12]. Despite of achieving many progresses on SVR, the specific kernel functions may be only suitable for a certain kind of problems, and selecting an appropriate kernel function is a rather difficult task for the modeling of warfarin dose prediction.

A single SVR or ANN hardly finds a satisfied solution to warfarin dose prediction problem. The discovered predictive models may over fit the objective function if the only insufficient dataset is available. In our opinion, diversity plays a pivotal role in building a high-accurate and generalized model. This is because the diversity is able to explore a large search space and makes the probability of finding a generalized solution significantly increased. The fact that diverse ensemble members outperform a single such model has also been proved [13]. In this study, multiple ML techniques (i.e. BP, RBF and SVR) are integrated to simultaneously create a variety of candidate models. Particle swarm optimization (PSO) is a global optimization method with a fast convergence, which is very suitable for improving the performance of ANNs [14]. By employing PSO as shell of a group of ANN/SVR, the process of building models with ANN/SVR can be optimized. The group of ANNs/SVRs will follow the ones that yield high-accuracy candidate models so that the better predictive models can probably be achieved in the search space.

2 Methodology

In this study, we propose an evolutionary ensemble modeling methodology, denoted as PSO-(ANN/SVR). ANN and SVR are used as builders of candidate predictive models. When using PSO, ANNs and SVR are regarded as particles and are moved toward the best known positions in the search-space. By this way, the advantages of ANNs and SVR can be compromised in PSO-(ANN/SVR) and the generalization of resulting models can be improved. Figure 1 shows the concept of PSO-ANN/SVR. As Fig. 1, the particles "BP", "RBF" and "SVR" are initially moving in the search space.

When the particles "BP" and "SVR" locate at good position, the other particles gradually follow them around. Finally, the global optimum can be obtained by the particles at overall best position.

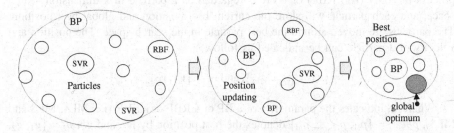

Fig. 1. PSO moving particles "BP, RBF, SVR" to the global optimum

2.1 ANNs, SVR

BP and RBF network are two general ANNs. BP is a global function approximation that can accommodate complex predictors, and it can effectively resist to the noise in sample data. The parameters vector ML1 of BP is shown as follow.

$$ML1 : \{max\_epcho, lr, [S_i], [TF_i], BTF, BLF, \text{network}\}$$

Where *max_epcho*: the maximum number of iterations; $[S_i]$: the number of nodes in layers; *BTF*: training function; *BLF*: weight learning function; $[TF_i]$: transfer function; *lr*: the learning rate; *network:* store the network model;

RBF can approximate to arbitrary complex function at a high accuracy. The parameters vector ML2 of RBF is shown as follow.

$$ML2 : \{max\_neuron, spread, eg, \text{network}\}$$

Where *max_neuron*: the maximum number of neurons; *spread*: spread constant; *e.g.*: training error; *network:* store the network model;

SVR fits a continuous function not only by minimizing the predictive accuracy, but also by reducing the model complexity. However, either of these methods has its disadvantages in mathematical modeling. The parameters vector ML3 of SVR is shown as follow.

$$ML3 : \{C, ker, loss, e, svr\}$$

Where *C*: penalty factor; *ker*: kernel functions of SVR, e.g. 'linear', 'poly', and 'rbf'; *e*: insensitivity; *loss*: loss function; *svr*: store the support regression model;

ML1, ML2 and ML3 are used to represent the position of particles "BP", "RBF", "SVR" in PSO, respectively.

2.2 PSO Optimizer for ANN/SVR

PSO is responsible for improving candidate predictive models with regard to mean squared error (*MSE*) by iteratively optimizing the parameters of ANNs and SVR. Each process of ANN (BP, RBF) or SVR is regarded as a particle in d-dimension search space, and each particle will store the current best position and global best position. The particles are moved around the best particle in the search-space. The position and velocity of a particle can be indicated as follow.

$$x_i = \{x_{i1}, x_{i2}, \ldots, x_{id}\}, \text{ and } v_i = \{v_{i1}, v_{i2}, \ldots, v_{id}\}$$

Where, x_i indicates the parameters of a BP or a RBF or a SVR (i.e. ML1, ML2 and ML3). $pbest_i = \{p_{i1}, p_{i2}, \ldots, p_{id}\}$ denotes the best position by far, and $gbest_i = \{g_1, g_2, \ldots, g_d\}$ indicates the best position of current population. The velocity and position can be updated by (1) and (2).

$$v_{id}^{k+1} = w\,v_{id}^k + c_1 r_1 \left(pbest_{id} - x_{id}^k \right) + c_2 r_2 \left(gbest_{id} - x_{id}^k \right) \tag{1}$$

$$x_{id}^{k+1} = x_{id}^k + v_{id}^{k+1} \tag{2}$$

Mean square error (*MSE*) of the model output and the desired output is used as the evaluation of a predictive model, which can be denoted as (3).

$$mse = \frac{1}{n}\sum_{i=1}^{n} [y(X_i) - Y_i]^2 \tag{3}$$

Where X_i and Y_i indicate the i-th input vector and output value of dataset; $y(X_i)$ denotes the output of model with respect to X_i and n denotes the size of training dataset or validation set.

Fitness function f of a particle is based on the *mse* of models built by corresponding ANN or SVR. Let *train_mse* and *valid_mse* be the *MSE* of the best particle in population on training and validation set, respectively. f is denoted as (4).

$$f = train\_mse + valid\_mse \tag{4}$$

The stopping criterion of PSO is denoted as $f < e$, where e is the accuracy requirement of warfarin dose prediction.

2.3 PSO-(ANN/SVR) Flow

Following, we describe the proposed method as follows.

1. Create a population of $N = 40$ particles, and initialize the particles' position with a random vector x_i and velocity with vector v_i; Generate a set of 40 ANN/SVR built models with the parameters decoded from the particles;

2. Train the set of ANN/SVR models on training set, and calculate *train_mse* of each ANN/SVR built model;
3. Calculate *valid_mse* of each ANN/SVR built model on validation set; Calculate the fitness f (Eq. (4)) of each particle according to *train_mse* and *valid_mse* of ANN/SVR built models;
4. Add the current best ANN/SVR model to *model_pool*;
5. Compare the fitness of particles with the one of *pbest* and the one of *gbest*; if a current particle is better in terms of fitness, *pbest* or *gbest* is updated with the position of this particle; Update the particles' position and velocity by Eqs. (1) and (2);
6. Repeat steps 2 to 6 until the termination criterion of PSO is reached.
7. Output the overall best ANN/SVR model from *model_pool*;

3 Experiment

A dataset that contains 100 Chinese patients of The First affiliated Hospital of Soochow University is collected. The dataset is divided into training set (60), validation set (20) and test set (20). In the dataset, male constitutes 67.3% and female 32.7%; For CYP2C9 genotypes, 84.2% patients have AA and 15.8% ones have AC (a scarce genotype); For another genotype VKORC1, 78.3% AA, 20.7% AG, and 0.9% GG.

PSO-(ANN/SVR) compare with ANN, SVR, conventional regression models, and the tool of "IPWC" in terms of predictive accuracy on the collected dataset. PSO-(ANN/SVR), SVR, RBF and BP are implemented by the toolbox in Matlab. The details of which can refer to the help of matlab. The predictive accuracy is measured by the MSE and R^2, which is the squared correlation between the output of model and the actual data. Table 1 gives four models with different variables and genotypes. M1 is the most popularly used model including five clinical variables. M2 is changed from M1 by adding a variable "Height". M3 includes ten clinical variables except the factors CYP2C9 and VKORC1. A predictive model without genetic factors will apply in these patients. By comparing M1, M2, M3, and M4, we can observe the application effects of different complex models in the modeling of warfarin dose function.

Table 1. The models with different clinical variables and genotypes

	Variables and Genotypes
M1	Age, Weight, CYP2C9, VKORC1, Amiodarone
M2	Age, Weight, Height, CYP2C9, VKORC1, Amiodarone
M3	Age, Weight, Height, Gender, Amiodarone, LA, drinking, ALT, SCr, INR
M4	Age, Weight, Height, Gender, CYP2C9, VKORC1 Amiodarone, LA, drinking, ALT, SCr, INR

4   Results and Discussion

Each method runs ten times to obtain the best model respectively for M1 $\sim$ M4. MSE and R^2 of different methods are shown in Table 2. The bold data indicates the overall best solution. "amiod" indicates the use of Amiodarone, and "no-amiod" means no-use of Amiodarone.

Table 2. MSE(10^{-2})/R^2 of the alternative methods on training and test set

	PSO-ANN/SVR		SVR		RBF		BP		Reg. models	IWPC model
	trn	*tst*	*trn*	*tst*	*trn*	*tst*	*trn*	*tst*	*trn, tst*	*trn, tst*
M1	**1.92**	**1.69**	1.94	1.92	0.61	3.19	1.39	3.31	4.45, 3.89/53.2,	2.25,
	66.5	**73.1**	66.4	72.7	89.7	5.02	77.5	49.8	68.2 (no-amiod)	2.81/52.35,
M2	**1.79**	**1.94**	1.96	1.96	0.89	3.12	1.73	2.36		57.02
	70.1	**70.6**	66.3	70.7	86.2	21.2	71.0	68.2		
M3	**1.51**	**3.16**	2.28	3.08	0.54	3.29	1.74	3.36	3.47, 2.95/53.4,	
	75.4	**53.2**	60.2	50.3	91.8	0.01	70.8	46.9	68.6 (amiod)	
M4	**1.38**	**2.85**	1.46	3.21	1.27	3.14	1.19	3.52		
	78.1	**60.9**	77.7	58.5	87.2	0.00	81.4	46.9		

It is noted that the regression models obtained smaller R^2 values than machine leaning methods. Two conventional regression models [7] gave relatively poor performance in terms of R^2. The method of IWPC based on a large size dataset always obtains a good MSE on both training set and test set we used. The R^2 achieved by IWPC demonstrates that abundant dataset can significantly improve the predictive model.

SVR got the best solution (MSE = 0.0192 and R^2 = 72.7% on test set) with the M1. SVR achieved another good solution by using M2 and the worst solution with M3. BP achieved a solution (MSE = 0.0236, R^2 = 68.2% on test set) with M2, and the solution (MSE = 0.0331, R^2 = 49.8% on test set) with M1 is better than M3. BP got a small decline in R^2 from training to testing set, which shows a commendable generalization in these cases. However, BP discovered models have no significant improvement on MSE as compared to the other methods despite of good generalization. RBF achieved a perfect MSE in training set, but relatively poor MSE in test set. This is attributed to the poor robustness of RBF. It is also noted that RBF models obtained very poor R^2 in test set with different variables despite of a not too bad MSE. This demonstrates that the model is validated only when it has both a large R^2 and a small MSE.

It can be seen that PSO-(ANN/SVR) achieved the overall optimal solution (MSE = 0.0169 and R^2 = 73.1% on test set) with M1. This solution is actual a model obtained by a SVR process. The PSO-(ANN/SVR) solution with M3, which is a model yielded by a BP process, is also the best model as compared to the other methods. This illustrates that PSO-(ANN/SVR) using multiple ML techniques inherit the advantage of these techniques, and is able to achieve a solution to different problems. MSE obtained by PSO-(ANN/SVR) on both training and test set is better than the other methods. This illustrates the good performance of PSO-(ANN/SVR) in terms of the

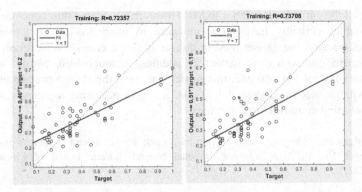

Fig. 2. Regression of models obtained by PSO-(ANN/SVR) (a) with M1 (b) with M3

predictive accuracy. It is also noted that four methods can obtain better solutions with M4 than that with M3, but worse performance than with M1 and M2. PSO-(ANN/SVR) achieved the largest R^2 on test set with different clinical variables and the decline from R^2 from training to testing set is not very large. Compared to other models, PSO-(ANN/SVR) developed the outperforming general models in these cases, even superior to the prediction of famous tool 'IWPC' (4000 patients as training set). Figure 2 gives the regression achieved by PSO-(ANN/SVR) with M1 and M3.

Figure 3 gives comparison of the best MSE and R^2 that each method has achieved in these cases. This can intuitively observe the overall performance of different methods. In general, PSO-(ANN/SVR) performed better than the other machine learning methods, conventional regression methods [7], and a little better than the calculator of IWPC on warfarin-dose prediction.

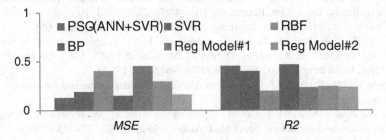

Fig. 3. *MSE* and *R^2* of different methods on test set

5 Conclusion

Overall, the good performance of PSO-(ANN/SVR) demonstrates that drawing diversity into modeling of complex function is beneficial to improve robustness and predictive accuracy. The test results are consistent with the previously proposed points that genotypes are very important to prediction of warfarin-dose. It is noted that M4 including all the variables is inferior to M1. However, we cannot thus conclude that all these variables certainly have no significance to prediction or no influence on dose-effect, because the dataset used in this study is not enough large. Further work must be carried out with more dataset. In addition, although MSE and R^2 with M3 cannot catch that of M1 and M2, this model can solve the dosage prediction on the patients without blood and genetic detection. It is very necessary for some patients in some developing countries such as China

Acknowledgments. This research is support by Natural Science Found of Jiangsu province in China (Grant No. BK20140293) and National Natural Science Found of China (Grant No. 61502327).

References

1. Klein, T.E., Altman, R.B., Eriksson, N., et al.: Estimation of warfarin dose with clinical and pharmacogenetic data. N. Engl. J. Med. **360**, 753–764 (2009)
2. Jonas, D.E., McLeod, H.L.: Genetic and clinical factors relating to warfarin dosing. Trends Pharmacol. Sci. **30**, 375–386 (2009)
3. Niclas, E., Mia, W.: Prediction of warfarin dose: why, when and how? Pharmacogenomics **13**(4), 429–440 (2012)
4. Lai-San, T., Boon-Cher, G., Anne, N., et al.: A warfarin-dosing model in Asians that uses single-nucleotide polymorphisms in vitamin K epoxide reductase complex and cytochrome P450 2C9. Clin. Pharmacol. Ther. **80**(4), 346–355 (2006)
5. Onundarson, P.T., Einarsdottir, K.A., Gudmundsdottir, B.R.: Warfarin anticoagulation intensity in specialist-based and in computer-assisted dosing practice. Int. J. Lab. Hematol. **30**(5), 382–389 (2008)
6. Yang, J., Huang, C., Shen, Z., et al.: Contribution of 1173C > T polymorphism in the VKORC1 gene to warfarin dose requirements in Han Chinese patients receiving anticoagulation. Int. J. Clin. Pharmacol. Ther. **49**(1), 23–29 (2011)
7. Williams, D.R.M.C.: Machine Learning: US, US 20050105712 A1[P] (2005)
8. Martin, B., Filipovic, M., Rennie, L., Shaw, D.: Using machine learning to prescribe warfarin. In: Dicheva, D., Dochev, D. (eds.) AIMSA 2010. LNCS, vol. 6304, pp. 151–160. Springer, Heidelberg (2010). doi:10.1007/978-3-642-15431-7_16
9. Wright, D.F.B., Duffull, S.B.: A bayesian dose-individualization method for warfarin. Clin. Pharmacokinet. **52**(1), 59–68 (2013)
10. Idit, S., Nitsan, M., Gal, C., et al.: Applying an artificial neural network to warfarin maintenance dose prediction. Israel Med. Assoc. J. Imaj. **6**(12), 732–735 (2004)
11. Saleh, M.I., Sameh, A.: Dosage individualization of warfarin using artificial neural networks. Mol. Diagnosis Therapy **18**(3), 371–379 (2014)
12. Erdal, C., Limdi, N.A., Duarte, C.W.: High-dimensional pharmacogenetic prediction of a continuous trait using machine learning techniques with application to warfarin dose prediction in African Americans. Bioinformatics **27**(10), 1384–1389 (2011)
13. Wall, R., Walsh, C.P., Byrne, S.: Explaining the output of ensembles in medical decision support on a case by case basis. Artif. Intell. Med. **28**(2), 191–206 (2003)
14. Chau, K.W.: Application of a PSO-based neural network in analysis of outcomes of construction claims. Autom. Construct. **16**(5), 642–646 (2007)

A Novel HPSOSA for Kernel Function Type and Parameter Optimization of SVR in Rainfall Forecasting

Jiansheng Wu[1,2(✉)]

[1] Department of Mathematics and Computer,
Guangxi Science and Technology, Normal University,
Laibin 546199, Guangxi, China
wjsh2002168@163.com
[2] School of Information Engineering, Wuhan University of Technology,
Wuhan 430070, Hubei, People's Republic of China

Abstract. In this paper, a novel co-evolution algorithm is presented to optimize the type of kernel function and the kernel parameter setting of Support Vector Regression (SVR) for rainfall prediction based on hybrid Particle Swarm Optimization and Simulated Annealing (HPSOSA), namely HPSOSA-SVR. The HPSOSA algorithm is carried out the metropolis process of SA into the movement mechanism and parallel processing of PSO. By combining the two methods, the HPSOSA algorithm has the advantage of both fast calculation and searching in the direction of the global optimum solution, helping PSO jump out of local optima, avoiding into the local optimal solution early and leading to a good solution quality. The developed HPSOSA-SVR model is being applied for monthly rainfall forecasting. Experimental results reveal that the predictions using the proposed approach are consistently better than those obtained using the other methods presented in this study in terms of the same measurements.

1 Introduction

Rainfall forecasting is very important for water resources management because accurate and timely prediction can avoid accidents, such as the life risk, economic losses, etc [1,2]. Developing a rainfall modeling has been a difficult subject in hydrology due to a variety of non-linear factors involved [3,4], such as, rainfall characteristics, watershed morphology, water level and soil moisture, etc [5,6]. Support Vector Regression (SVR) has also been receiving increasing attention to solve nonlinear regression problems due to its many attractive features and promising generalization performance [7,8]. When using SVR, two problems are confronted: how to choose the type of SVR kernel function and how to set the best parameters of SVR kernel function. Proper type of SVR kernel function and the parameters of SVR kernel function can improve the SVR regression accuracy. Inappropriate parameters in SVR lead to over-fitting or under-fitting [9–11]. The effects of SVR applications strongly depend upon operators experience the main

© Springer International Publishing AG 2017
Y. Tan et al. (Eds.): ICSI 2017, Part II, LNCS 10386, pp. 359–370, 2017.
DOI: 10.1007/978-3-319-61833-3_38

reason optimal hyper-parameters are obtained by the trial-and-error. If carelessly used, it can easily learn irrelevant information in the system (over-fitting). Such a model might be doing well in predicting past incidents, but unable to predict future events [12,13]. Recently, several studies have proposed the parameter optimization of Gaussian kernel function by evolutionary optimization, such as Genetic Algorithm (GA) [12], Particle Swarm Optimization (PSO) [11] and Immune Algorithm (IA) [8], but, haven't presented the selection type of SVR kernel function.

The present study proposed a novel and specialized hybrid optimization strategy to simultaneously optimize both the type of SVR kernel function and all the SVR parameters, namely HPSOSA-SVR. Our approach simultaneously determines the appropriate type of kernel function and optimal kernel parameter values for the SVR model. The rainfall data of Guilin, Guangxi, China, is predicted as a case study for our proposed method. An actual case of forecasting daily rainfall is illustrated to show the improvement in predictive accuracy and capability of generalization achieved by HPSOSA-SVR. The rest of this study is organized as follows. Section 2 describes the related works, including the basic SVR concepts. Section 3 describes the SVR HPSOSA, ideas and procedures. For further illustration, different models are used to employ for runoff forecasting analysis in Sect. 4, and conclusions are drawn in the final section.

2   Support Vector Regression and Kernel Parameters

The brief ideas of SVR for the case of regression is introduced. Suppose we are given training data $(x_i, y_i)_{i=1}^N$, where $x_i \in R^n$ is the input vector; y_i is the output value and n is the total number of data dimension. The linear regression function is formulated as follows:

$$f(x) = \omega\phi(x_i) + b, \quad \phi : R^n \to F, \omega \in F \qquad (1)$$

where ω and b are coefficients; $\phi(x)$ denotes the high dimensional feature space, which is nonlinearly mapped from the input space x. The coefficients ε and b can be estimated by minimizing the regularized risk function

$$R(f(x)) = C\frac{1}{n}\sum_{i=1}^n L_\varepsilon(f(x_i) - y_i) + \frac{1}{2}\|\omega\|^2 \qquad (2)$$

$$L(f(x,y)) = \begin{cases} |f(x) - y| - \varepsilon & |f(x) - y| \geq \varepsilon \\ 0 & otherwise \end{cases} \qquad (3)$$

Therefore, the objective of SVR is to include training patterns inside an ε-insensitive tube (ε-tube) while keeping the norm $\|\omega\|^2$ as small as possible. The parameter ε is the difference between actual values and values calculated from the regression function. This difference can be viewed as a tube around the regression function. C denotes a cost function measuring empirical risk; it indicates a parameter determining the trade-off between the empirical risk and the model flatness. According to Ref. [14], the SVR function is given by:

$$f(x, \alpha_i, \alpha_i^*) = \sum_{i=1}^{N} (\alpha_i - \alpha_i^*) K(x, x_i) + b \tag{4}$$

where $K(x, x_j)$ is defined as kernel function. SVR has recently emerged as an alternative and highly effective means of solving the nonlinear regression problem. When designing an effective SVR model, values of the four essential parameters in SVR have to be chosen carefully in advance. Different parameter settings can cause significant difference in performance. The parameters include:

(1) *Kernel function type*: The kernel function is used to construct a nonlinear mapping between the input space and feature space.
(2) *Kernel function parameter*: t represents the intercept of the polynomial kernel, d represents the degree of the polynomial kernel, σ represents the bandwidth of the Guassian kernel function, α and β are parameters of Sigmoid kernel function.
(3) *Regularization parameter* C: determines the trade-off cost between minimizing the training error and minimizing the model's complexity.
(4) *The tube size of ε-insensitive loss function*: It is equivalent to the approximation accuracy placed on the training data points. Types of various kernel function, kernel function parameters and SVR parameters are shown in Table 1.

Table 1. The kernel function type and corresponding parameters of the SVR.

Kernel Type		Kernel	Parameters	SVR	Parameters
Linear	$K(x_i, x_j) = x_i^\gamma x_j$	γ	–	C	ε
Ploy	$K(x_i, x_j) = (t + x_i \cdot x_j)^d$	d	t	C	ε
RBF	$K(x_i, x_j) = exp(-\frac{\|x_i - x_j\|^2}{2\sigma^2})$	C	σ	C	ε
Sigmoid	$K(x_i, x_j) = tanh[\alpha(x_i^T x_j) + \beta]$	α	β	C	ε

3 Hybrid PSOSA Strategy for SVR

3.1 Particle Swarm Optimization and Simulated Annealing

Particle swarm optimization (PSO) was a global optimization technique, which is more powerful, robust and able to provide accurate solution in nonlinear optimization problems [15]. Simulated Annealing (SA) is a stochastic global searching algorithm for optimization problems, which is based on the statistical thermodynamics to simulate the behavior of atomic arrangements in liquid or solid materials during the annealing process. SA includes a rather simple optimization strategy. The SA algorithm simulates this process of slow cooling of molten metal to achieve minimum function value in the minimization problem. The cooling phenomenon is simulated by controlling a temperature like parameter introduced with the concept of Boltzman probability distribution [16].

3.2 Hybrid PSOSA for SVR's Parameters Optimization

To obtain the global or near global optimum solutions, this paper proposed a hybrid Particle Swarm Optimization with Simulated Annealing (HPSOSA) algorithm in order to combine the advantages of SA (ability to jump away from local optimum solutions and converge to the global optimum solution) and the advantages of PSO (fast calculation and easy mechanism). Firstly, particle swarms are randomly created by the PSO structure. Secondly, velocity and position update. The SA metropolis acceptance rule is incorporated into the parallel processing of PSO at this stage. Thirdly, The rule determines whether to accept the new position or recalculate another candidate position according to the fitness function difference between the new and old positions. The conceptual diagram of HPSOSA is shown in Fig. 1.

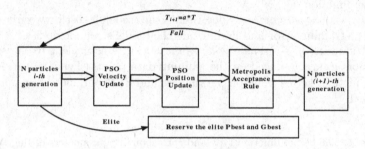

Fig. 1. The HPSOSA conceptual diagram.

3.3 Chromosome Representations

The chromosome is comprised of three parts: Kernel Type (integer-valued), Kernel Parameters and SVR Parameter (real-valued). Chromosome $00, 01, 10, 11$ represents the kernel function type, which the values zero, one, and two denote that the system will choose 'Linear kernel', 'Poly kernel', 'Gaussian (RBF) kernel' and 'Sigmoid function', respectively. Real value chromosome $\{x_{i,1}, x_{i,2}\}$ represents the valued of kernel parameter. Real value chromosome $\{x_{i,3}, x_{i,4}\}$ represents the valued of the penalty parameter and insensitive loss function, respectively.

3.4 HPSOSA-SVR Model

The proper type of kernel function and optimal parameter setting is critical to predict the performance of SVR model. Because kernel type influences kernel parameters, and vice versa, obtaining the proper SVR kernel function type and the optimal SVR parameters must occur simultaneously. Based on the above, HPSOSA algorithm is employed to simultaneously optimize the type of kernel function and the kernel parameter setting of Support Vector Regression (SVR),

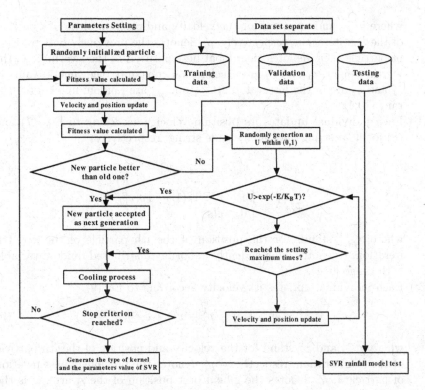

Fig. 2. Flowchart of the HPSOSA optimization SVR.

namely HPSOSA-SVR. Figure 2 illustrates the process of the HPSOSA-SVR algorithm for SVR optimization. Details of our proposed HPSOSA-SVR are described as follows:

1. Particle initialization and PSO parameters setting: Generate initial particles comprised of the SVR's kernel type and parameters. Let $k = 0$, and randomly initialize PSO's position x_i^0, PSO's velocity v_i^0, initial temperature T_0, and cooling rate α.

2. Input training data and calculate the fitness of all particles, which determine G_{best} and P_{best} by a simple comparison of their fitness values according to Eq. (5). The fitness function is defined as follows:

$$F_{fitness}(k) = 1/[1 + \frac{1}{n}\sum_{i=1}^{n}|a_i - p_i|/a_i \times 100\%] \tag{5}$$

where k is the number of iterations, n is the number of training data samples, y_i is the actual value, and p_i is the predicted value.

3. Repeat this step until the stopping criterion is satisfied.

   (a) Perform PSO operators. Update individual velocity according to Eq. (6)

$$v_i^{k+1} = \omega v_i^k + c_1\gamma_1(p_i^k - x_i^k) + c_2\gamma_2(p_g^k - x_i^k) \tag{6}$$

where v_i^{k+1} and x_i^k stand for the velocity and position of the ith particle of the kth iteration, respectively. p_i^k denotes the previously best position of particle i, p_g^k denotes the global best position of the swarm. ω is the inertia weight, c_1 and c_2 are acceleration constants (the general value of c_1 and c_2 are in the interval $[0,2]$), γ_1 and γ_2 are random numbers in the range $[0\ 1]$.

(b) Each individual updates its position velocity according to Eqs. (7) and (8) for float string and binary code string, respectively.

$$x_{id}^{k+1} = x_{id}^k + v_{id}^{k+1} \tag{7}$$

$$x_{id}^{k+1} = \begin{cases} 1 & r < 1/(1 + exp(v_{id}^{k+1})) \\ 0 & else \end{cases} \tag{8}$$

where x_{id}^{k+1} stands for the position of the ith particle of the $k+1$th iteration, r denotes independently uniformly distributed random variable with range $[0, 1]$.

(c) Each individual updates its velocity according to Eq. (9)

$$v_i^{k+1} = \omega v_i^k + c_1\gamma_1(p_i^k - x_i^k) + c_2\gamma_2(p_g^k - x_i^k) \tag{9}$$

where v_i^{k+1} and x_i^k stand for the velocity and position of the ith particle of the kth iteration, respectively. p_i^k denotes the previously best position of particle i, p_g^k denotes the global best position of the swarm. ω is the inertia weight, c_1 and c_2 are acceleration constants (the general value of c_1 and c_2 are in the interval $[0\ 2]$), γ_1 and γ_2 are random numbers in the range $[0\ 1]$.

(d) Evaluate $\Delta Fitness = Fitness(s_i^{k+1}) - Fitness(s_i^k)$ and then randomly select a number $R \in [0, 1]$. If $\Delta Fitness > 0$ meaning that the new position is improved for increasing fitness function, then position s_i^{k+1} is accepted according to the following criterion: $exp(\frac{\Delta Fitness}{T_i}) > R$. Proceed to Step (e) when the velocity of all particles are determined, or return to Step (a) for those particles failing to be accepted, and generate new velocities using the same evaluation process. Too many failures (i.e., 100 in our study) for the same particle will force the last velocity to be accepted, in consideration of CPU time, thereby prevent entering into an endless loop.

(e) Renew each particle to the new position and modify G_{best} and P_{best} by simple comparison their fitness values.

(f) When the evolution process has achieved a satisfactory condition (or maximum evolution number is reached), proceed to Step 4, otherwise, modify the inertia weight ω, and annealing temperature $T_{i+1} = \kappa * T_i$, let $k = k + 1$, and return to Step 3.

4. Once the termination condition is met, output the best solution G_{best} and its fitness value, obtain the appropriate type of kernel and optimal parameter setting for SVR model. Input test samples for the prediction effect of the SVR model.

4   Experiments Analysis and Discussion

The platform adopted to develop the HPSOSA-SVR approach is a PC with the following features: Intel Core i5, 2.3 GHz CPU, 4.0 GB RAM, Windows 10 operating system and the MATLAB R2009a development environment. In this paper, PSO and SA parameters are set as follows: the iteration times are 100; the population is 100; the minimum inertia weight is 0.1; the maximum inertia weight is 0.9 and the learning rate is 2.0. Initial temperature is 5000; termination temperature is 0.9 and temperature reduction factor 0.001.

4.1   Empirical Data

Real-time ground monthly rainfall data have been obtained from January 1951 to December 2015 in Guilin of Guangxi, China. The data set contained 780 data points, whose training data set contained 480 (1951–1990), validation set is 240 (1991–2010) and other 60 (2011–2015) samples is tested modelling. In this paper, first of all, the candidate forecasting factors are selected from the numerical forecast products based on 48 h forecast field, which includes: the 17 conventional meteorological elements and physical elements from the T213 numerical products of China Meteorological Administration, the data cover the latitude from 15^0N to 30^0N, and longitude from 100^0E to 120^0E, with $1^0 \times 1^0$ resolution, altogether there are 336 grid points. We can get 43 variables as the main forecasting factors. This paper used the principal component analysis and obtain 6 variables as SVR's input. The original meteorological data is used as real output.

4.2   Criteria for Evaluating Model Performance

This paper used the following evaluation metric to measure the performance of the proposed model: Root mean square error (RMSE), Mean absolute percentage error (MAPE), Coefficient of efficiency (CE), which can be found in many paper [5].

For the purpose of comparison by the same six input variables, we have also built other two rainfall forecasting models: SVR and pure PSO evolutionary SVR(PSO-SVR). For building SVR rainfall forecasting model, the LIBSVM package proposed by Chang and Lin is adapted for this paper [17], which all SVR parameters are based on the RBF kernel type by the trial-and-error method. The best parameters with the minimum testing RMSE is considered to be optimal. The optimal parameters are ($\sigma = 0.4951, C = 1.3621, \varepsilon = 0.2564$) based on the best testing and validation result (minimum RMSE). For building PSO-SVR rainfall forecasting model, PSO is used to search for optimal type of kernel function and kernel parameter values of SVR for rainfall forecasting by Ming-Wei Li et al. presented [12].

4.3   Results Analysis

Figures 3 and 4 give graphical representations of the validation and forecasting results for rainfall using different models. Table 2 show the training, validation and testing performance of different models in rainfall. In this paper, all rainfall forecasting models are implemented via the same input factors.

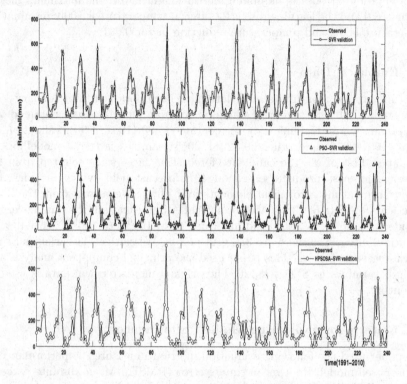

Fig. 3. Comparison of observed and validation for three rainfall models.

From the graphs and table, we can generally see that the forecasting results are very promising for rainfall models under study either where the measurement of forecasting performance is goodness of fit such as RMSE (refer to Table 2) or where the forecasting performance criterion is MAPE (refer to Table 2). These results indicate that the HPSOSA-SVR model are closer to the corresponding observed rainfall values than those of the two other models.

Subsequently, the training, validation and forecasting performance comparisons of various models for the rainfall via RMSE, MAPE, CE and CE_{peak} are reported in Table 2, respectively. As shown in Table 2, for the training data, the RMSE for SVR model is 59.21, PSO-SVR model is 33.47; while for HPSOSA-SVR model, RMSE reaches 8.12. Similarly, for the testing data, the RMSE for the SVR model is 62.56, PSO-SVR model is 35.11; while for the HPSOSA-SVR

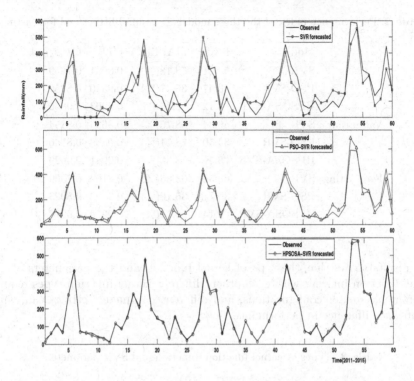

Fig. 4. Comparison of observed and forecasted value for three rainfall models.

model, RMSE reaches 9.14. Focusing on the MAPE indicator of the training case and testing data, the HPSOSA-SVR model is also less than the SVR and PSO-SVR models. These results indicate that the deviations of HPSOSA-SVR model between observed and forecasting value are the smallest. From the experiments presented in this study we can draw the following conclusions that the SVR with Sigmoid kernel function ensemble model can be used as an alternative tool for rainfall forecasting to obtain greater forecasting accuracy and improve the prediction quality further in view of empirical results.

Similarly, for CE and CE_{peak} efficiency index, the HPSOSA-SVR model are maximum in their three models in all stages. These results show the HPSOSA-SVR model have higher correlation relationship with observed rainfall values, it also implies that the HPSOSA-SVR model is capably to capture the average change tendency of monthly rainfall data. To summarize, the HPSOSA-SVR model is superior to the other two models presented here in terms of RMSE, MAPE, CE and CE_{peak} for rainfall prediction under the same input.

Further from Table 3, Sigmoid kernel function is the best choice in HPSOSA-SVR rainfall modeling, the optimal parameters of kernel function α is 0.0813, β is 0.2780, the optimal SVR parameters C is 11.9546, ε is 0.6214, while for SVR-PSO, the best kernel function type is RBF, the optimal parameters of kernel function σ is 1.3013, the optimal SVR parameters C is 34.2674, ε is 0.0847.

Table 2. Performance index of the three models in rainfall fitting and forecasting.

	Model	RMSE	MAPE(%)	CE	CE_{peak}
Training	SVR	59.21	58.18(%)	0.8061	0.7859
	PSO-SVR	33.47	35.47(%)	0.9380	0.8156
	HPSOSA-SVR	8.12	8.40 (%)	0.9963	0.9836
Validation	SVR	56.34	58.16(%)	0.8638	0.8254
	PSO-SVR	34.50	38.51(%)	0.9489	0.8730
	HPSOSA-SVR	6.50	8.30(%)	0.9981	0.9629
Forecasting	SVR	62.56	63.36(%)	0.8102	0.7369
	PSO-SVR	35.11	35.10(%)	0.9402	0.8023
	HPSOSA-SVR	9.14	7.71(%)	0.9959	0.9277

The results shows that the type of kernel function and the parameters of the kernel function interact with each other, different kernel function types require different parameter configurations, and different parameter settings can cause significant difference in performance.

Table 3. Types of kernel function, function and SVR parameters

Model	Kernel type	Kernel parameters			
		α	$\sigma(\beta)$	C	ε
SVR	RBF	–	0.4951	1.3621	0.2564
PSO-SVR	RBF	–	1.3013	34.2674	0.0847
HPSOSA-SVR	Sigmoid	0.0813	0.2780	11.9546	0.6214

Notes: – denotes no parameter needed.

5   Conclusion

The rainfall system is one of the most active dynamic weather systems. In this paper, the advantages and the key issues of the PSO and SA evolve the type and the parameters of SVR kernel function have been presented to model the runoff forecasting. The experiment with real rainfall-runoff data have showed that the HPSOSA-SVR can be of high precision, even better than those of pure SVR, by setting proper values for all parameters (parameter values and type of kernel function) in the SVR model. According to the results obtained in this paper, the HPSOSA-SVR method has the following characteristics:

Remark 1. The HPSOSA algorithm enables the solution to jump out of local optima, and decreases the vibration near the end of locating a solution by incorporating the metropolis acceptance criterion of SA into PSO randomly acceptance rule. The rule determines whether to accept the new position or recalculate another candidate position according to the fitness function difference between the new and old positions. This enables the solution to jump out of local optima, and decreases the vibration near the end of locating a solution. These characteristics improve the quality of the solution and increase the rate of convergence.

Remark 2. The HPSOSA algorithm has the advantage of both fast calculation and searching in the direction of the global optimum solution, leading to a good solution quality by combining the PSO characteristics of movement mechanism and parallel processing into the disturbance mechanism and solution searching procedure of SA.

Remark 3. According to the results obtained in this paper, we can draw the following conclusions that the SVR model with Sigmoid kernel function can be used as an alternative tool for actual rainfall forecasting application to obtain better forecasting accuracy and improve the prediction quality further in view of empirical results.

Acknowledgments. This work was supported the Natural Science Foundation of Guangxi Province under Grant No. 2014GXNSFAA118027, and by the Guangxi Education Department under Grant YB2014467, and by the Key Laboratory for Mixed and Missing Data Statistics of the Education Department of Guangxi Province under Grant No. GXMMSL201405, and by the Key Disciplines for Operational Research and Cybernetics of the Education Department of Guangxi Province.

References

1. Wu, J., Jin, L., Liu, M.: Evolving RBF neural networks for rainfall prediction using hybrid particle swarm optimization and genetic algorithm. Neurocomputing **148**, 136–142 (2015)
2. Wu, J., Jin, L., Liu, M.: A hybrid support vector regression approach for rainfall forecasting using particle swarm optimization and projection pursuit technology. Int. J. Comput. Intell. Appl. **9**(3), 87–104 (2010)
3. Makungo, R., Odiyo, J.O., Ndiritu, J.G.: Rainfall-runoff modelling approach for ungauged catchments: a case study of Nzhelele River sub-quaternary catchment. Phys. Chem. Earth **35**(13–14), 45–62 (2010)
4. Wu, J.: An effective hybrid semi-parametric regression strategy for rainfall forecasting combining linear and nonlinear regression. Int. J. Appl. Evol. Comput. **2**(4), 50–65 (2011)
5. Amin, T., Chua, L.H.C., Quek, C.: A novel application of a neuro-fuzzy computational technique in event-based rainfall-runoff modeling. Expert Syst. Appl. **37**(12), 7456–7468 (2010)
6. Nourani, V., Kisi, Ö., Komasi, M.: Two hybrid artificial intelligence approaches for modeling rainfall-runoff process. J. Hydrol. **402**(1–2), 41–59 (2011)
7. Chi, J.L., Tian, S.L., Chih, C.C.: Financial time series forecasting using independent component analysis and support vector regression. Decis. Support Syst. **47**, 115–125 (2009)
8. Hong, W.-C.: Application of seasonal SVR with chaotic immune algorithm in traffic flow forecasting. Neural Comput. Appl. **21**(3), 583–593 (2012)

9. Huang, C.-L., Dun, J.-F.: A distributed PSO-SVM hybrid system with feature selection and parameter optimization. Appl. Soft Comput. **8**, 1381–1391 (2008)
10. Wu, C.H., Tzeng, G.H., Lin, R.-H.: A novel hybrid genetic algorithm for kernel function and parameter optimization in support vector regression. Expert Syst. Appl. **36**, 4725–4735 (2009)
11. Lin, S.W., Ying, K.C., Chen, S.C.: Particle swarm optimization for parameter determination and feature selection of support vector machines. Expert Syst. Appl. **35**, 1817–1824 (2008)
12. Li, M.-W., Hong, W.-C., Kang, H.-G.: Urban traffic flow forecasting using Gauss-SVR with cat mapping, cloud model and PSO hybrid algorithm. Neurocomputing **99**, 230–240 (2013)
13. Hong, W.-C.: SVR with hybrid chaotic genetic algorithms for tourism demand forecasting. Appl. Soft Comput. **11**, 1881–1890 (2011)
14. Bermolen, P., Rossi, D.: Support vector regression for link load prediction. Comput. Netw. **53**, 191–201 (2009)
15. Rabie, H.M., Ihab, E., Assem, T.: Particle swarm optimization algorithm for the continuous p-median location problems. In: 2014 10th International Computer Engineering Conference (ICENCO), pp. 81–86 (2014)
16. Eglese, R.W.: Simulated annealing: a tool for operation research. Eur. J. Oper. Res. **46**, 271–281 (1990)
17. Chang, C.C., Lin, C.J.: LIBSVM: a library for support vector machines (2001). http://www.csie.ntu.edu.tw/cjlin/libsvm

An Improved Weighted ELM with Krill Herd Algorithm for Imbalanced Learning

Yi-nan Guo[1], Pei Zhang[1], Jian Cheng[1(✉)], Yong Zhang[1], Lingkai Yang[1], Xiaoning Shen[2], and Wei Fang[3]

[1] School of Information and Control Engineering,
China University of Mining and Technology, Xuzhou 221008, Jiangsu, China
chengjian@cumt.edu.cn
[2] School of Information and Control,
Nanjing University of Information Science and Technology,
Ning-Liu Road, Pu-Kou District, Nanjing, China
[3] Department of Computer Science and Technology, Jiangnan University,
Wuxi, China

Abstract. The traditional weighted extreme learning machine chose input weights and hidden biases randomly. This made the algorithm responding slowly and led to worse generalization. In order to overcome the shortcomings, an improved weighted extreme learning machine combining with krill herd algorithm is proposed to solve the class imbalance problems. Krill herd algorithm is adopted to optimize the input weights and hidden biases. The simulation results show that krill herd algorithm can find more suitable input weights and hidden biases, which has better classify accuracy than particle swarm optimization and genetic algorithm. The proposed algorithm also has stable classify accuracy for the data sets with different imbalance ratios.

Keywords: Imbalanced learning · Weight extreme learning machine · Krill herd algorithm

1 Introduction

In many practice problems, such as fault diagnosis and fraud detection, the total number of a class of data (fault data) less than the number of other class of data (normal data). We called them imbalance data set [6]. To classify imbalance data normally had been solved by two kinds of methods, including data-based strategies and algorithm-based strategies. Data-based strategies weakened the imbalance degree of data by over-sampling for minority samples and under-sampling for majority samples. Synthetic minority over-sampling technique (SMOTE) [1] and SMOTE-based improved sampling algorithms were the popular data-based strategies [12]. Algorithm-based strategies enhanced the classification performances by improving the traditional classification algorithms according to the data features, or designing a novel classification algorithm. For instance, cost-sensitive boosting methods [13], in which the classify cost was introduced into

© Springer International Publishing AG 2017
Y. Tan et al. (Eds.): ICSI 2017, Part II, LNCS 10386, pp. 371–378, 2017.
DOI: 10.1007/978-3-319-61833-3_39

the weight updating strategy of AdaBboost, were proposed. Zong et al. [17] proposed a weighted extreme learning machine (weighted-ELM), which assigned an extra weight to each sample so as to strengthen the role of minority class and weaken the impact of majority class.

Compared with other classical machine learning algorithms, such as back propagation(BP) and support vector machine, extreme learning machine(ELM) [9] had attracted more attention due to its fast implementation, less computational cost and better generalization in regression and classification problems [16]. It adopted generalized single hidden layer feed-forward networks (SLFNs) and least-square-based learning algorithm. The input weights and hidden biases of SLFNs were chosen randomly. Corresponding output weights were analytically determined by generalized inverse operation of output matrix in the hidden layer. In order to solve different kinds of problems, some improved ELM algorithms were given. Online sequential ELM adopted additive hidden nodes with radial basis function in SLFNs, which shown better generalization to deal with online data flow or concept-drifting data stream [11]. Incremental ELM [10] randomly added hidden nodes and manually adjusted the output weights linking the hidden layer and the output layer. For weighted regularized ELM [2], there existed more hidden neurons for ELM than traditional SLFN-based learning algorithms. Especially, weighted-SLFN learning algorithms may result in ill-condition problem due to randomly selecting input weights and hidden biases [15]. Non-optimal or redundant input weights and hidden biases also made ELM responding slowly and led to worse generalization [8].

To overcome above disadvantages, some bio-inspired algorithms were introduced to improve the ELM's performances. Genetic algorithm(GA) was applied to select the optimal parameters of the activation function in weighted-ELM [4]. Particle swarm optimization (PSO) was used to find the optimal input weights and hidden biases of ELM [7]. Fruit fly optimization (FOA) was adopted to optimize the input weights and thresholds of hidden layer in ELM [3], and then an optimal ELM prediction model was built. In evolutionary-based ELM, the input weights was selected by differential evolutionary algorithm [16].

In this paper, we focus on the class imbalance problem, in which the majority class with negative label and the minority class with positive label. As we known, an extra weight was assigned in weighted-ELM to fit for the imbalanced data distribution, instead of traditional re-sampling methods. However, the randomly generated input weights and hidden biases may form some useless neuron nodes, which make the networks structure complex. Consequently, krill herd algorithm (KH) is introduced to optimize the input weights and hidden biases of the weighted-ELM and find the optimal number of neuron nodes.

2   Weighted Extreme Learning Machine

In ELM [9], the weights and biases of SLFNs were randomly initialized. Corresponding output matrix of hidden-layer was explicitly calculated based on the output weights by fewer steps. Thus, the computational cost of ELM was low.

Considering a set of N distinct samples (X_i, t_i) with $X_i = [x_{i1}, x_{i2}, ..., x_{in}]^T \in R^n$ and $t_i = [t_{i1}, t_{i2}, ..., t_{im}]^T \in R^m$, an SLFN with L hidden nodes is modeled.

$$\sum_{i=1}^{L} \beta_i g(a_i, b_i, X_j) = t_j, j = 1, ..., N \tag{1}$$

Here, a_i and β_i are the input and output weights. b_i is the biases. $g(\bullet)$ is the activation function. Many nonlinear activation functions can be used in ELM, including sigmoid, sine, radial basis functions, and so on [9]. The error between the estimated outputs o_j and the actual outputs t_i shall be zero once SLFNs can exactly approximate the feature of data. Let $\beta = [\beta_1^T, ..., \beta_L^T]^T$ and $T = [t_1^T, ..., t_N^T]^T$. Above model can be converted to $H\beta = T$. We call H the hidden layer output matrix. H_{ij} represents the output of ith hidden node output corresponding to the input X_i. During the training process, the parameters in the hidden nodes including a_i and b_i are not adjusted. Suppose $H^{\dagger}$ is the MooreCPenrose generalized inverse of H [9]. Corresponding output weights are estimated as follows. The optimal output weights $\hat{\beta}$ can be gotten by minimizing the cost function $\|o - t\|_2$.

$$\hat{\beta} = H^{\dagger}T = \begin{cases} (H^T H)^{-1} H^T T, L < N \\ H^T (HH^T)^{-1} T, L \geq N \end{cases} \tag{2}$$

Weighted-ELM fits for dealing with both balanced and imbalanced data. In weighted-ELM, a diagonal matrix $[W]_{N \times N}$ associated with each training data X_i is defined. If X_i belongs to majority class, the associated weight W_{ii} is relatively smaller. On the contrary, W_{ii} is set the larger value as X_i comes from minority class [17]. The weighting scheme is automatically chosen in terms of the class information. Let $\#(t_i)$ be the number of samples belonging to class t_i and $AVG(t_i)$ be the average number of samples for each class. Two kinds of weighting schemes [17] are defined as follows.

$$W1 : W_{ii} = 1/\#(t_i) \tag{3}$$

$$W2 : W_{ii} = \begin{cases} 0.618/\#(t_i) \ if \#(t_i) > AVG(t_i) \\ 1/\#(t_i) \quad\quad if \#(t_i) \leq AVG(t_i) \end{cases} \tag{4}$$

Subsequently, the output weights matrix for weighted-ELM was calculated.

$$\hat{\beta} = \begin{cases} (H^T W H)^{-1} H^T W T, L < N \\ H^T (W H H^T)^{-1} W T, L \geq N \end{cases} \tag{5}$$

3 The Improved Weighted-ELM with Krill Herd Algorithm

Krill herd algorithm is a novel optimization algorithm derived from the herding behavior of krills [5]. Each krill corresponds to an individual. It moves along the direction, which has the minimum distance from food and highest density of the herd. Three main actions of krill individuals include the motion induced by other krill denoted as N_i, the foraging motion F_i and the physical diffusion D_i. The Lagrangian model is generalized to an n dimensional decision space [5].

$$\frac{dX_i}{dt} = N_i + F_i + D_i \tag{6}$$

Assuming that $N^{\max}$ be the maximum induced speed. $\omega_n \in [0, 1]$ is the inertia weight of the motion. $N_i(t-1)$ is the last motion influenced by other krill. The target and local effect denoted by α_i determines the moving direction of krill. For ith krill, $N_i(t)$ is the sum of $N^{\max}\alpha_i$ and $\omega_n N_i(t-1)$.

Suppose V_f is the foraging speed. $\omega_f \in [0, 1]$ is the inertia weight of the foraging motion. $F_i(t-1)$ is the last foraging motion. For ith krill, the foraging motion $F_i(t) = V_f \beta_i + \omega_f F_i(t-1)$, where $\beta_i = \beta_i^{food} + \beta_i^{best}$.

For ith krill, the third motion is essentially a random process defined as $D_i = D^{\max}\delta$. $D^{\max}$ is maximum diffusion speed and $\delta \in [-1, 1]$ is a random directional vector.

Let LB_j and UB_j are lower and upper bounds of jth variables. $C_t \in [0, 2]$ is a constant. Suppose $\Delta t = C_t \sum_{j=1}^{n} (UB_j - LB_j)$. Obviously, Δt changes with C_t and shall be regulated in terms of specific problems because it can be regarded as a scale factor of the speed vector. According to above three actions, the position of ith krill during the time interval $[t, t + \Delta t]$ can be formulated as follows. In order to enhance the algorithm performance, genetic reproduction mechanism is introduced. More details about regular KH algorithm can be referred as [5].

$$X_i(t + \Delta t) = X_i(t) + \Delta t \frac{dX_i}{dt} \tag{7}$$

Based on krill herd algorithm, an improved weighted extreme learning machine (KHW-ELM) is proposed to deal with the class imbalance problems. In traditional weighted-ELM, the fixed weights assigned to the samples were proportional to the size of class corresponding to the samples. Being different from it, the reasonable input weights and the hidden biases of weighted-ELM are found by KH algorithm so as to improve the generalization and simplify the network. In KHW-ELM, each individual is composed of input weights and hidden biases, denoted by $P = [a_{11}, ..., a_{1N}, ..., a_{L1}, ..., a_{LN}, b_1, b_2, ..., b_L]$. Corresponding output weights are calculated according to Eq. 2. The fitness of each individuals is defined as the root mean square error (RMSE) on the training set X_{train}.

$$fitness = \sqrt{\frac{\sum_{j=1}^{N} \left\| \sum_{i=1}^{L} \hat{\beta}_i g(a_i, b_i, X_j) - t_j \right\|_2^2}{N}} \tag{8}$$

The detailed steps of the proposed method are shown as follows.

Step 1: The imbalanced data set $X = \{(x_i, t_i), i = 1, 2, ...N\}$ is divided into the training set X_{train} and test set X_{test}.

Step 2: The initial population P is randomly generated based on the defined lower and upper bounds of variables. The length of each individual is $L(N+1)$.

Step 3: Evaluate each krill individuals fitness according to its position.

Step 4: Three main actions including motion induced by the presence of other individuals, foraging motion and physical diffusion are done. Subsequently, the position of each individual is updated.

Step 5: If the stop criteria does not satisfy, jump to Step 3. Otherwise, the optimal input weights and hidden biases of weighted-ELM are output.

4   Experimental Results and Discussion

In order to fully analyze the performances of the proposed method, 9 binary-class data sets chosen from UCI are tested in the experiments. Detailed information about these data sets are provided in Table 1. The imbalanced ratio of data sets varies from 0.1 to 0.4. Especially, WPBC and Iris miss some attribute values.

Table 1. Detailed information about binary-class data sets from UCI

Dataset	Number of attributes	Number of classes	Number of training data	Number of test data	Imbalance ratio
Wine1	13	2	100	78	0.3333
Wine 2	12	2	2898	2000	0.2461
Adult	14	2	28842	20000	0.25
WPBC	34	2	100	98	0.2533
WDBC	32	2	300	317	0.4
Dota2	116	2	50000	52944	0.1
Iris	4	2	75	75	0.3333
Car	6	2	1000	728	0.2997
Poker hand	11	2	25010	100000	0.2382

For imbalanced data, we give more insight into the accuracy obtained within each class instead of the accuracy of all samples. Suppose TP, TN, FP, FN stand for the number of data with true positive, true negative, false positive and false negative. Let $\frac{TP}{TP+FN}$ be the minority accuracy and $\frac{TN}{TN+FP}$ be the majority accuracy. $G-mean = \sqrt{\frac{TP}{TP+FN} \times \frac{TN}{TN+FP}}$ is used to measure the classify accuracy [14].

For weighted-ELM with GA, PSO and KH, the terminal iteration is 200 and population size is 20. There are 150 hidden neurons. For GA, the crossover probability is 0.9 and the mutation probability is 0.01. The generation gap is 0.9. For PSO, the learning factors $c_1 = 1.5, c_2 = 1.7$ and $w = 1$. For KH, maximum induced speed $N^{\max} = 0.01$, the foraging speed $V_f = 0.02$, maximum diffusion speed $D^{\max} = 0.005$ and $C_t = 1$.

4.1   Comparison of the Performances Among Weighted-ELM Combining with GA, PSO and KH

In order to improve the generalization of weighted-ELM, some intelligent optimization algorithms were introduced. In this paper, three kinds of bio-inspired algorithms including GA, PSO and KH are used to optimize the input weights and hidden biases in weighted-ELM. The algorithm performances of three improved weighted-ELM under 20 running times are compared in Table 2.

Table 2. Compare the performances of weighted-ELM with GA, PSO and KH

Data set	Weighted-ELM with GA			Weighted-ELM with PSO			Weighted-ELM with KH		
	Minority accuracy	Majority accuracy	G-mean	Minority accuracy	Majority accuracy	G-mean	Minority accuracy	Majority accuracy	G-mean
Wine	0.9811	0.841	0.9084	1	0.9129	0.9555	1	**0.9663**	**0.983**
Wine 2	0.9349	**0.9779**	0.9562	0.9424	0.951	0.9467	**0.9787**	0.9687	**0.9737**
Adult	0.942	0.9001	0.9208	**0.9542**	0.8451	0.898	0.9447	**0.9151**	**0.9298**
WPBC	**0.9524**	0.6105	0.7625	0.9048	0.7895	0.8452	0.9048	**0.853**	**0.8785**
WDBC	0.913	0.9311	0.922	0.882	**0.9427**	0.9118	**0.9291**	0.9342	**0.9316**
Dota2	**0.9379**	0.8535	0.8947	0.9157	**0.9042**	0.9099	0.9365	**0.9199**	**0.9282**
Iris	0.8571	0.8449	0.851	0.914	0.8594	0.8863	**0.9374**	**0.9145**	**0.9259**
Car	0.8515	0.7885	0.8194	**0.9214**	0.5508	0.7124	0.917	**0.8858**	**0.9013**
Poker hand	0.865	0.8003	0.832	**0.9524**	**0.8751**	**0.9129**	0.9114	0.8527	0.8816

Weighted-ELM with KH algorithm obtains the maximum G-mean for 8 data sets from UCI. Although the minority and majority accuracy gotten by KHW-ELM are not optimal for all data sets, they are better enough.

4.2   The Effect of the Imbalance Ratio on the Algorithm Performance

For above nine data sets, we change the imbalanced ratios from 0.1 to 0.4. The algorithm performances under different imbalanced ratios are shown in Fig. 1 by the boxplot of G-mean and Fig. 2 by the trend of G-mean. Both figures indicate that no matter the imbalance ratios change or not, weighted-ELM with KH algorithm has the stable classification performance.

4.3   The Effect of Key Parameters in KHW-ELM on the Algorithm Performance

Larger L means there are ineffective neuron nodes in the network and the computational complexity becomes larger. Thus, to choose a suitable L is essential for any ELM algorithms. Taking wine and WDBC as examples, as L varies from 10 to 300, G-mean gotten from KHW-ELM is shown in Fig. 3. Obviously, the algorithm performance becomes better with the increasing of L. Especially, when $L \geq 150$, G-mean trends to stable. Considering less computational complexity and better algorithm performance, we choose L from [150, 200].

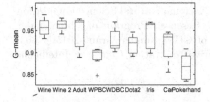

Fig. 1. The boxplot of G-mean

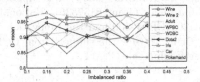

Fig. 2. G-mean under different imbalanced ratios

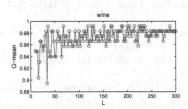
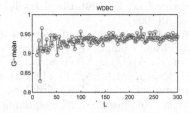

Fig. 3. G-mean under different L in KHW-ELM

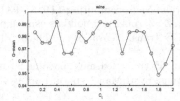
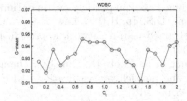

Fig. 4. G-mean under different C_t in KHW-ELM

For KH algorithm, time interval decided by C_t has a direct impact on the krills behavior [5]. The algorithm performance as $C_t \in [0, 2]$ is plotted in Fig. 4. It is obvious that when C_t is close to 1, G-mean is quite good and stable. Thus, we choose $C_t = 1$ so as to find the optimal input weights and hidden biases.

5 Conclusions

Aiming at improving the classify accuracy, an improved weighted-ELM combining with KH algorithm was proposed to deal with the class imbalance problems. KH algorithm was used to optimize the input weights and hidden biases in weighted-ELM. The experimental results show that weighted-ELM with KH algorithm had better performance than PSO and GA. It also had stable classify accuracy for the data sets with different imbalance ratio. By analyzing the effect of the key parameters on the algorithm performance, the most reasonable network was constructed. How to find the optimal number of neuron nodes and their weights simultaneously is our future works.

Acknowledgments. This work is supported by National Natural Science Foundation of China under Grant 61573361, National Basic Research Program of China under Grant 2014CB046300 and the Innovation Team of CUMT under Grant 2015QN003. Also, thank you for the support from Collaborative Innovation Center of Intelligent Mining Equipment, CUMT.

References

1. Chawla, N.V., Bowyer, K.W., Hall, L.O., Kegelmeyer, W.P.: Smote: synthetic minority over-sampling technique. J. Artif. Intell. Res. **16**(1), 321–357 (2002)
2. Deng, W., Zheng, Q., Chen, L.: Regularized extreme learning machine. In: IEEE Symposium on Computational Intelligence and Data Mining (CIDM 2009), pp. 389–395 (2009)
3. Dong, L.I.: Stock price prediction based on extreme learning machine and fruit fly of algorithm. Comput. Eng. Appl. **50**(18), 14–18 (2014)
4. Ertam, F., Avci, E.: A new approach for internet traffic classification: GA-WK-ELM. Measurement **95**, 135–142 (2017)
5. Gandomi, A.H., Alavi, A.H.: Krill herd: a new bio-inspired optimization algorithm. Commun. Nonlinear Sci. Numer. Simul. **17**(12), 4831–4845 (2012)
6. HaiboHe, Y.: Imbalanced Learning: Foundations, Algorithms, and Applications. John Wiley & Sons, Inc., Hoboken (2013)
7. Han, F., Yao, H.F., Ling, Q.H.: An improved evolutionary extreme learning machine based on particle swarm optimization. Neurocomputing **116**, 87–93 (2013)
8. Huang, D.S., Ip, H.H.S., Law, K.C.K., Chi, Z.: Zeroing polynomials using modified constrained neural network approach. IEEE Trans. Neural Netw. **16**(3), 721–732 (2005)
9. Huang, G.B., Zhou, H., Ding, X., Zhang, R.: Extreme learning machine for regression and multiclass classification. IEEE Trans. Syst. Man Cybern. Part B **42**(42), 513–529 (2012)
10. Huang, G.B., Chen, L., Siew, C.K.: Universal approximation using incremental constructive feedforward networks with random hidden nodes. IEEE Trans. Neural Netw. **17**(4), 879–892 (2006)
11. Liang, N.Y., Huang, G.B., Saratchandran, P., Sundararajan, N.: A fast and accurate online sequential learning algorithm for feedforward networks. IEEE Trans. Neural Netw. **17**(6), 1411–1423 (2006)
12. Ramentol, E., Caballero, Y., Bello, R., Herrera, F.: Smote-RSB*: a hybrid preprocessing approach based on oversampling and undersampling for high imbalanced data-sets using smote and rough sets theory. Knowl. Inf. Syst. **33**(2), 245–265 (2012)
13. Sun, Y., Kamel, M.S., Wong, A.K.C., Wang, Y.: Cost-sensitive boosting for classification of imbalanced data. Pattern Recognit. **40**(12), 3358–3378 (2007)
14. Yu, H., Sun, C., Yang, X., Yang, W., Shen, J., Qi, Y.: ODOC-ELM: optimal decision outputs compensation-based extreme learning machine for classifying imbalanced data. Knowl. Based Syst. **92**, 55–70 (2015)
15. Zhao, G., Shen, Z., Miao, C., Man, Z.: On improving the conditioning of extreme learning machine: a linear case. In: 7th International Conference on Information, Communications and Signal Processing (ICICS 2009), pp. 1–5 (2009)
16. Zhu, Q.Y., Qin, A.K., Suganthan, P.N., Huang, G.B.: Evolutionary extreme learning machine. Pattern Recognit. **38**(10), 1759–1763 (2005)
17. Zong, W., Huang, G.B., Chen, Y.: Weighted extreme learning machine for imbalance learning. Neurocomputing **101**(3), 229–242 (2013)

Fast Pseudo Random Forest
Using Discrimination Hyperspace

Tojiro Kaneko[✉], Hidehisa Akiyma, and Shigeto Aramaki

Fukuoka University, Fukuoka, Japan
Td156501@cis.fukuoka-u.ac.jp

Abstract. In recent years, machine learning technique has been applied to various problems. The improvement of computational power enables the processing of large scale data in a practical time and brought the success of machine learning technique. However, the processing speed of current machine learning models still have a potential to be improved. We are trying to improve the processing speed of Random Forest, which is known as a fast and reliable classification model. In this study, we propose a Discrimination HyperSpace called DHS, which realized a pseudo Random Forest. Experiment results show our method runs much faster than original Random Forest without losing classification performance.

Keywords: Random forest · Decision tree · Classification · Speeding up · Discrimination hyperspace · DHS

1 Introduction

In recent years, machine learning technique have been applied to various problems [3–5]. And these achieve high performance. The improvement of computational power brought the success of machine learning because large scale data can be processed in a practical time. On the other hand, the execution speed of the model constructed by machine learning is rarely discussed [5].

Execution speed is one of the most important aspect in actual applications. The model acquired by the current machine learning method tends to increase in size as the complicated and highly accurate processing is required. In order to process them in real time, there are solutions at the hardware level such as parallel processing by GPU [3]. However, it is difficult to apply these approaches on the small portable device like smart phones which hardware resources are limited. Therefore, there is great significance in saving computation resources required by machine learning models.

2 Related Research

Random Forest (hereinafter referred to as "RF") [1] is known as a machine learning method that has high execution speed. This method is a kind of the ensemble learning and realize complex classification using Decision Tree that is a simple discriminator.

© Springer International Publishing AG 2017
Y. Tan et al. (Eds.): ICSI 2017, Part II, LNCS 10386, pp. 379–386, 2017.
DOI: 10.1007/978-3-319-61833-3_40

In recent years, RF has been applied to a wide variety of fields because of its high classification accuracy and execution speed [2, 3, 5].

However, the execution speed of RF is still needed to be improved. For example, if RF increase the number of tree for expressing more complex phenomenon, execution speed decreases due to increasing blanch process of Decision Trees. On the other hand, there are study cases that generates a huge Decision Tree for reducing the total branches process while expressing complex phenomenon [3]. However, the huge tree needs to have many branch in itself for express complex phenomenon. This may cause a reduction in processing speed per processed unit in parallelization.

3   Proposed Method

In this paper, we propose a solution space named DHS (Discrimination HyperSpace) that can process a classification faster than RF without losing classification capability. DHS is created using a classification model obtained by RF.

3.1   Random Forest

In this section, we explain the characteristic of RF first in order to need understand our proposal method. Detailed principle of RF can be found in [1, 2].

RF is constructed by Decision Tree. Decision Tree is a simple branching tree model as shown in Fig. 1. Each node represented by circle in Fig. 1 has a branch condition for classification. Each leaf represented by square in Fig. 1 has a classification result. When the root of Decision Tree takes input data, the process flows according to the branch condition and outputs the result from the reached leaf. Data used for learning and execution of Decision Trees is composed of several feature values. The feature used as branching condition in the node is called "explanatory variable", and that stored as result in leaf is called "response variable". Decision Tree also has an ability to select the most suitable explanatory variable.

RF uses randomness to create multiple Decision Trees that are different from each other. As a result, each Decision Tree has different branch conditions each other. RF gives same input data to each Decision Tree when the classification is performed,

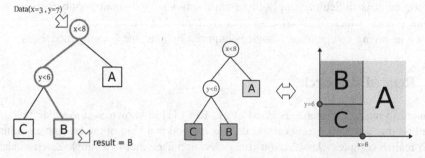

Fig. 1. Example of decision tree   **Fig. 2.** Compatibility of decision tree and solution space

and each Decision Tree outputs its result. RF uses arithmetic mean of these output as the classification result. In other words, execution speed of RF will decrease as increase the number of Decision Tree, because that increases necessary process inside it.

3.2 Decision Tree as Solution Space

The Decision Trees that compose RF can be considered as a model of dividing solution space (Fig. 2). In Fig. 2, axes of solution space correspond to each explanatory variable of data. In the solution space, each leaf of Decision Tree can be represented as a region in the solution space, and each conditional branch means border lines among regions. The classification process using Decision Tree can be considered as a search process to find a region corresponding with input values.

On the other hand, if all dimensions are discrete and all grid cells are filled by the result value of leaf, the result value at the input vector can be referred directly from solution space by indexing the input vector. This referring process would be much faster than Decision Tree which needs many branch condition. We propose DHS based on this view point.

3.3 Representing RF by DHS

In this paper, we realized DHS as a multidimensional discrete space using an array on the memory (Fig. 3). Each element of the array stores the classification result corresponded to the value of explanatory variable. In order to refer the solution space, we just need to refer the element of array using an index correspond to the value of explanatory variable.

RF is a strong classifier bundling a large number of Decision Trees. Each Decision Tree is independent of each other, and the result of whole RF is obtained by collecting their classification result. Usually, arithmetic mean value of results from all trees is used as whole result. Since the solution space of the Decision Tree can be represented by DHS as described above, in order to represent RF, it is only necessary to replace each Decision Tree with DHS.

However, because of the following reasons, it is enough to generate only one DHS. Figure 4 shows two Decision Trees represented in solution space. The solution spaces

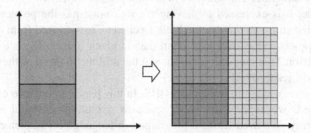

Fig. 3. Expressed solution space by DHS on memory

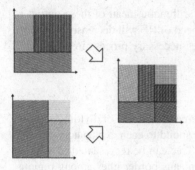

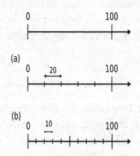

Fig. 4. Overlap of solution spaces   **Fig. 5.** Example of resolution value on the axis

are 2D plane regions taking the same two explanatory variables. Because both solution spaces take same explanatory variables, it is possible to superimpose both spaces as shown in Fig. 4. The new region formed by the overlap of both space has an arithmetic mean value of classification results. This superimposing can be performed independent of the number of Decision Trees and the number of explanatory variables constituting solution space. Even if each Decision Tree has different explanatory variables, it is possible to superimpose them by increasing additional dimension axes in DHS.

From these reasons, solution spaces of several Decision Trees can be represented by one DHS. This means that RF can be represented by one DHS. In contrast to RF which needs a large number of conditional branching process, the solution space of DHS can be referred by a unique address on the memory. Therefore, the processing speed for DHS drastically outperforms the one for RF representing same solution space.

4 Implementation

4.1 Implementation of RF and DHS

In this chapter, we will describe the information on the implementation necessary for the experiment. We Implemented RF and DHS using the same programming language.

We used C# as a programming language for implementing RF. The Decision Tree is a general tree structure. The structure to become a node expresses by having a pointer of a child node. It is expressed by having the root node has the pointer of the child node. RF is a list structure that has multiple Decision Trees created from object. When classification processing is executed, input data is given to each Decision Tree in the list. Each Decision Tree outputs the result, and the arithmetic mean of these is taken as the classification result of RF.

There are several ways to implement DHS. In this paper, in order to emphasize the execution speed, we express the super space as a multidimensional array by one dimensional array. We used C# as a programming language for implementing DHS.

Since RF can represent a solution space with real numbers by a conditional branch of a Decision Tree, there is no theoretical upper limit in the expression range. On the other hand, DHS is represented as an array on the memory, and its solution space is represented with discrete value. Therefore, expressiveness of DHS depends on the size of the array. Therefore, we introduce the concept of "resolution" as an indicator of the expressiveness of DHS. This is equivalent to the number of elements of the array representing each axis of the DHS solution space. Figure 5 shows an example of it.

In Fig. 5, one axis of the hyperspace takes real values in the range 0 to 100. When the resolution of DHS is set to 5, the axis is represented by an array with 5 elements as shown in Fig. 5(a). In this time, one element of the array corresponds the real number range of 20. Likewise, when the resolution is set to 10, one element of the array corresponds to the real number range of 10 as shown in Fig. 5(b).

4.2 Generating DHS

In order to actually use DHS, it is necessary to replace the solution space of the generated RF to that of DHS. Several methods are conceivable for this, but this time we chose the simplest method.

As described above, the index value of the array expressing the DHS corresponds to the value of each explanatory variable of the data to be discriminated. Therefore, when converting the index value of DHS to explanatory variable and giving it as input value to RF, RF outputs the classification result for that. This result is the value of RF solution space corresponding to the element of the solution space on DHS indicated by the index used for input. By performing this processing on all elements of the array expressing DHS, it is possible to transfer the solution space acquired by RF to the solution space of DHS.

In this way, DHS is generated based on RF. As a result, generation time of DHS is required in addition to RF generation time, and the total learning time is increased compared with the case of only RF. However, this process is independent of the classification process and does not affect the execution speed in the classification process.

5 Experiment

In this section, we actually generate DHS based on RF, and perform classification processing for the data set. We compare its accuracy and execution speed by giving the same data to both RF and DHS.

5.1 Data Set to Be Used

In this experiments, we used Iris dataset [6] for classification processing. This data consists of four explanatory variables and one objective variable. The Iris data set has 150 instances and can be categorized with high precision, so it is frequently used for testing classification processing.

Since the explanatory variables are used as an index of the array in DHS, it is necessary to convert the value of each explanatory variable to an integer value. We solved this problem by calculation that adapts the minimum and maximum values of data to the range of values that the axes of DHS takes.

Assume that the minimum value of explanatory variable is "min", the maximum value is "max", and the resolution of DHS is "d". At this time, in order to convert the certain explanatory variable "n" to the index value "Ind", the following formula (1) is calculated.

$$Ind = n/((max - min)/d) \tag{1}$$

In the case of the Iris data used this time, the value that each explanatory variable can take is within the range from 0.0 to 10.0. If this is scaled with a resolution of 10, the each explanatory variable will be converted to an integer value from 0 to 9 by the above equation. This scaling process was performed every time access to the DHS occurred, for not to affect the data set.

In this experiment, we will generate DHS that represents the solution space of RF. And then we will measure classification accuracy and execution speed of both. Classification accuracy is measured by cross validation. We randomly divided the Iris data set into two groups. The one is used as training data, and the other as test data. The each data have 75 instances. The same test data is given to RF and DHS. Numerical values such as RF generation parameters and DHS resolution are specified for each experiment.

5.2   Comparative Experiment of Accuracy

We describe about comparative experiments of classification accuracy. In order to compare, the parameter of RF used in this experiment was fixed to the following value. The RF to be generated has 100 Decision Trees, and each tree has subsets configured from 20 data. Limit of maximum depth of tree is set to 10. DHS is generated from this RF with gradually changing the resolution. In the experiment, we observe the relationship between the resolution of DHS and the classification accuracy. The RF and the DHS are regenerated for each measurement.

Table 1 shows the results of this experiment. The accuracy of the model generated in RF frequently changes depending on the randomness. To alleviate this, we measured three times for each condition and used the average value. From Table 1 it can be seen that the classification accuracy of DHS is approaching that of RF as the resolution of DHS increases. This shows that the classification accuracy of DHS depends on the resolution and that of RF.

5.3   Comparative Experiment of Execution Speed

Next, we described a comparative experiment of execution speed. In this experiment, we generate models while changing the RF setting parameter and the resolution of DHS to various values. And then, we give the same test data to both generated models

Table 1. Result of experiment of accuracy

RF				DHS		
Number of Tree	Limit number of layer	Number of data in subset	Accuracy rate of classification	Resolution value	Time of generation(s)	Accuracy rate of classification
100	10	20	0.92	10	2.19	0.83
100	10	20	0.92	20	32.54	0.87
100	10	20	0.89	30	167.19	0.87
100	10	20	0.92	40	554.17	0.91
100	10	20	0.91	50	1,366.42	0.89
100	10	20	0.89	100	24,600.00	0.87

Table 2. Result of experiment of execution speed

Number of processed data	RF				DHS	
	Number of Tree	Limit number of layer	Number of data in subset	Total processing time(ms)	Resolution value	Total processing time(ms)
10,000	100	5	100	2,678	10	7
10,000	100	5	100	2,849	20	6
10,000	100	5	100	2,742	50	6
10,000	10	5	100	261	10	6
10,000	500	5	100	15,184	10	6
10,000	1,000	5	100	37,218	10	6
10,000	5	10	10,000	347	10	6
10,000	5	20	10,000	165	10	6
100,000	100	5	100	26,844	10	36
1,000,000	100	5	100	298,554	10	335
10,000,000	100	5	100	2,930,230	10	3,246

as input data for classification process. In this experiment, we generate models while changing the RF setting parameter and the resolution of DHS to various values. And then, we measure both execution speed of classification process with giving the same test data to both generated models. Since RF and DHS are both high speed, it is desirable to give a large amount of data to lengthen the classification process time, in order to measure with sufficient accuracy. Therefore, this is realized by repeatedly giving the same test data a plurality of times. All the experiments were executed on the CPU (Intel(R) Core(TM) i7-6700 K CPU @4.00 Hz 4.00). Table 2 show the results.

From Table 2, it can be read that the RF execution speed changes depending on the generation parameters and the number of input data. On the other hand, the execution speed of DHS depends only on the number of input data. Regarding the execution speed, the DHS has been several tens to several hundred times as fast as the RF in this experiment.

6   Discussion

From the experimental results, we consider the relationship between the DHS and RF and the characteristic. First, we found that the classification accuracy of DHS greatly depends on RF. This is because the solution space expressed by DHS is the solution space of RF simply replaced. For the same reason, accuracy of DHS is possible to approach that of RF by increase the resolution of DHS. Second, the execution speed of the DHS was overwhelmingly faster than the RF, and it was always stable. This is because DHS always only requires a simple processing of accessing memory, whereas RF requires a large amount of branching process which varies depending on setting parameters. From this, we can conclude that DHS can execute faster than RF.

7   Future Work

As problems of DHS, there are several things caused by generating super space on the memory. For example, DHS imitate the solution space of RF with higher accuracy as the resolution is higher, but at the same time, the number of elements in the super space increases too. Therefore, high resolution consumes a large amount of memory and increases generation time of DHS exponentially.

Also, since DHS has all the explanatory variables of the data as axes, there is a problem that physically memory becomes insufficient as the number of explanatory variables increases. For this reason, our DHS at present can only be applied to simple data sets such as Iris data. In order to solve these problems, we need realize more lightweight expression of DHS while maintaining high speed.

References

1. Breiman, L.: Random forests. Mach. Learn. **45**, 5–32 (2001)
2. Habe, H.: Random forests. In: CVIM, vol. 182, no. 31, pp. 1–8 (2012)
3. Shotton, J., Fitzgibbon, A., Cook, M., Sharp, T., Finocchio, M., Moore, R., Kipman, A., Blake, A.: Real-time human pose recognition in parts for single depth images. Commun. ACM **56**(1), 116–124 (2013)
4. Toshev, A., Szegedy, C.: DeepPose: human pose estimation via deep neural networks. In: The IEEE Conference on Computer Vision and Pattern Recognition (CVPR), pp. 1653–1660 (2014)
5. Kato, M.F.N, Qi, W.: Automatic image annotation by variational random forests. In: IEICE Technical report, vol. 110, no. 414, PRMU2010-209 (2011)
6. UCI Machine Learning Repository, iris. https://archive.ics.uci.edu/ml/machine-learning-databases/iris/iris.data

A Fast Video Vehicle Detection Approach Based on Improved Adaboost Classifier

Tao Jiang[1], Mingdai Cai[1]([⊠]), Yuan Zhang[1], and Xiaodong Zhao[2]

[1] School of Mechanical Engineering,
Tongji University, Shanghai 201804, China
dl129442045@163.com
[2] State Grid Zhengzhou Power Supply Company, Zhengzhou 400006, China

Abstract. Aiming at the problem that traditional vehicle detection methods often fail to ensure the accuracy and speed simultaneously, a vehicle detection method based on background difference and improved Gentle Adaboost classifier is proposed. Firstly, the foreground region is obtained by using the background difference method, and the morphological processing is applied properly to get better candidate foreground regions. Then, the cascaded Adaboost classifiers are used to detect multi-scale vehicles in these regions. In this paper, we adopt effective search strategy, which can greatly reduce the number of search windows, and further improve the detection speed. The experimental results show that the proposed method not only can obtain high accuracy, but also has strong real-time performance. Precisely, the highest accuracy reaches to 96.0% and the highest detection speed reaches to 51.4 FPS.

Keywords: Fast vehicle detection · Traffic video · Background subtraction · Gentle adaboost · Search strategy

1 Introduction

In recent years, with the rapid development of computer vision technology, image-based vehicle detection technology has become a research hotspot. In general, vehicle detection approaches can be divided into two categories, motion-based approaches (such as frame difference method [1], background difference method [2–4] and optical flow method [5]) and appearance-based approaches [6, 7]. Motion-based approaches use the motion information of a video sequence, which can quickly detect moving targets. But the information is relatively simple, while ignores the other important information, so it is sensitive to noise and easy to cause the high false alarm rate. Relatively speaking, appearance-based approaches may get higher detection accuracy and better robustness, because it can make full use of the information contained in the images. But it generally needs to scan the entire image by adopting multi-scale detection strategy, which is really difficult to get real-time performance [8].

Many researchers adopt one of the approaches mentioned above. J. Song et al. [1] proposed a frame difference method by using the parallel affine transformation, which can be applied in a simple scene to detect moving vehicles. S. Li et al. [2] proposed an adaptive background subtraction method, dynamically updating the background and

© Springer International Publishing AG 2017
Y. Tan et al. (Eds.): ICSI 2017, Part II, LNCS 10386, pp. 387–394, 2017.
DOI: 10.1007/978-3-319-61833-3_41

choosing the optimal threshold for segmentation of motion regions. The method achieved good accuracy of vehicle counting, but it may easily fail when there exists vehicle overlap problem. Q. Wu et al. [5] proposed an optical flow method by utilizing the edge information, which is not easy to meet the real-time requirements. Van Pham et al. [6] proposed an approach of car hypothesis based on car windshield appearance and obtained high accuracy under the condition of high contrast of the windows and other areas.

A few researchers try to combine the motion information and feature classification together. Firstly, use motion information to narrow the range of detection, then use the classifiers to search for vehicles in the ROI (Region of Interest), which can improve the performance to a certain extent. X. Zhuang et al. [8] proposed a vehicle detection algorithm that is based on the Haar-like features and combines motion detection with a cascade of classifiers. The detection accuracy and speed are considerable, but it is not robust enough. For example, it may fail in dark condition. JIANG Xinhua et al. [9] presented a vehicle detection method based on a semi-supervised support vector machine (SVM) classification algorithm which is combined with three-frame difference. The method has good adaptability, but the speed needs to be improved.

In order to get better accuracy and speed simultaneously, we proposed a vehicle detection method based on background difference and Gentle Adaboost classifier. In general, the main contributions of this paper are as follows:

1. The motion information is utilized to narrow the range of detection which can accelerate the detection for real-time application.
2. A special search strategy for classifiers is proposed that can greatly improve the detection speed without reducing the accuracy.
3. Training samples largely determine the performance of detection. Many special but useful positive and negative samples are added to the training samples and the results demonstrate the superiority of the strategy.

The layout of this paper is as follows: Sect. 2 describes the vehicle detection approach, including background subtraction method, cascaded Gentle Adaboost classifiers and our special search strategy. In Sect. 3, we design the experiments to evaluate the performance of the proposed algorithm. Specifically, we test the algorithm in challenging conditions, including but not limited to, the changing illumination condition, multiple moving vehicles, shadows and vehicle overlap. The conclusions are presented in Sect. 4.

2 The Proposed Approach Based on Improved Adaboost Classifier

The proposed approach combines the superiority of motion-based methods and appearance-based method to enhance the performance. Firstly, the background subtraction method is used to remove most of the background regions, and the foreground regions are processed by morphological processing properly, so we can get better foreground regions. Then, the Adaboost classifier is used to detect multi-scale vehicles in these regions. Different from the traditional search strategy, in this paper, we use the prior knowledge and adopt effective search strategy, which can greatly reduce the number of search windows, and further improve the detection speed. Figure 1 shows the flow chart of the proposed approach.

2.1   Fast Background Subtraction

The background difference method is widely used in real-time vehicle detection due to its high efficiency which is easy to implement. First, the initial background is obtained by calculating the average value of each pixel from a collection of images. We can get a "variation image" through the subtraction of the background image from the current image. Then select an appropriate threshold to transform the "variation image" into a binary image, in which morphological processing (dilation and erosion) is applied to get a better foreground image. Typically, in order to make the algorithm faster, we do not use complex background modeling methods. We simply update the background by using the weighted average of the current frame and the current background to adapt to the environmental changes. In this paper, the size of dilation's element is a little larger than that of erosion's element to ensure all vehicles not to be removed in the foreground regions and the whole vehicle has greater probability to be remained. Therefore, this measure can help to gain higher accuracy combined with the further detection. Figure 2 shows the result of background subtraction.

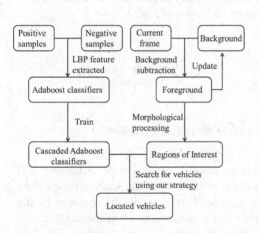

Fig. 1. The flow chart of the proposed approach

Fig. 2. The result of background subtraction

2.2 Feature Extraction Method

In this paper, Local Binary Pattern (LBP) is used as the feature extraction method rather than Haar-like features, because Haar-like features are too simple to recognize those samples that are difficult for classification. Relatively, LBP is more robust and still ensures the speed of detection. LBP image is gained by using the LBP operator to scan the entire gray image. Then select a number of regions to extract LBP histograms, which can be used as a feature vector for classification.

2.3 Cascaded Adaboost Classifiers

Many classification methods have been used in two-class problem, such as artificial neural network method, support vector machine (SVM), Adaboost etc. Considering the efficiency of the methods, in this paper, Adaboost method is chosen to solve the classification problem. P. Viola et al. [10] proposed a real-time face detection method based on cascaded Adaboost classifiers, which is widely used in the field of object detection nowadays. The Adaboost algorithm is used to train a number of strong classifiers, and these strong classifiers are cascaded from simple to complex, each of which has high detection rate and tolerable error detection rate. Given a sub-window of the frame to be detected, each classifier will classify whether the sub-window is a vehicle. Only if the sub-window is classified positive by all classifiers, it will finally be regarded as a vehicle. Otherwise, if the sub-window is classified negative by one classifier of them, it will be denied directly without the need to let the other non-used classifier to make decision. Actually, most of the sub-windows is negative and only a few of them are positive, so this strategy can greatly accelerate the detection. The schematic diagram of the cascaded classifiers is shown in Fig. 3.

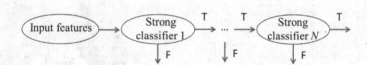

Fig. 3. The schematic diagram of the cascaded classifiers

There are several Adaboost algorithms, among which Gentle Adaboost [11] algorithm is relatively more efficient than the original Discrete Adaboost algorithm. The output of Discrete Adaboost classifier is either 1 or -1, while the output of Gentle Adaboost can be a real number between -1 and 1, which is more realistic. Moreover, Discrete Adaboost classifier focuses too much on outlier samples, so it can easily lead to over-fitting problem, while Gentle Adaboost can effectively avoid the problem by updating the weights smoothly during the training stage. Considering the high efficiency, Gentle Adaboost algorithm is adopted as the proposed classification method.

2.4 Search Strategy

Traditional appearance-based methods adopt overall search strategy to search for multi-scale vehicles by scanning the entire image. It is really time-consuming since the vehicles often occupy a few space of the image. In this paper, motion information is exploited to narrow the range of detection. After morphological processing, the foreground image will be more qualified for the follow-on searching and there will be several regions where vehicles are likely to locate. Here comes a problem that it will take much time to search for vehicles adopting overall search strategy in each foreground region when the traffic is not very congested and most of the regions only contain a vehicle. Considering this situation, we proposed a special search strategy to greatly accelerate the detection. Figure 4 illustrates the proposed search strategy. The foreground regions consist of several connected domains that are marked as Region 1, Region 2, etc. The red rectangles are a little larger than the minimum enclosing rectangles of connected domains in case that some parts of vehicles are not in the regions. Taking Region 3 as an example, the blue square is the biggest square that the red rectangle contains and the center of the blue square coincides with the center of the red rectangle. The proposed strategy is as follows:

For each region, take the image of the blue square as the input of the classifiers. If it is classified positive, it will be regarded as a vehicle directly and there is no need to search for the rest windows of the region. Otherwise, in the blue square, a few windows a bit smaller than the blue square are classified positive or negative. If one of them is classified positive, it will be regarded as a vehicle and the searching of the region will stop. Otherwise, use the traditional overall search strategy to search for vehicles in the red rectangle of the region.

Adopting the proposed strategy, use a few windows of Region 1, 2, 4, the vehicles can be found quickly. As for Region 3, the blue square and a few windows a bit smaller than it are all classified negative, so the traditional strategy is used to search for vehicles in the red rectangle of the Region 3 to find out the two vehicles.

The proposed strategy can greatly improve the detection efficiency. For example, given a window whose size is 100×100 (assume that the minimum detection window is 24×24, the scale factor is 1.1, the minimum step size is 4), the traditional overall search strategy based classifiers need to classify more than 1000 windows, meanwhile, adopting the proposed strategy, classifiers may only have to use one window or a few

Fig. 4. The illustration of proposed search strategy

Fig. 5. The examples of positive samples

windows if it only contains a vehicle. Moreover, using the proposed strategy often gets better vehicle locations rather than the result by using the traditional strategy because it needs to merge a lot of positive windows.

3 Experiment Results

3.1 Training Samples

Training sample plays an important role in classification which largely determines the performance of detection. Good training samples can help to design classifiers with high performance. We collect the positive samples from urban roads and highways in different weather conditions. And the positive samples also vary in type, size and color to ensure the greater variety. Negative samples are from the traffic images that do not contain vehicles and other non-vehicle images on the Internet. Both positive and negative samples are normalized to a uniform size of 24×24. Occluded vehicles are added to the positive training samples to handle occlusion. Moreover, in order to make the classifiers with better performance, we add some negative samples that contain two or more vehicles as well as a small parts of a vehicle. This measure can reduce the possibility that the classifiers mistakenly recognize two or more neighboring vehicles or a small part of a vehicle as a vehicle. Figure 5 shows the examples of positive samples.

3.2 Experiment Datasets

The experiment datasets come from the real traffic videos shot in Shanghai. The datasets contain various interfering factors, such as the changing illumination condition, multiple moving vehicles, shadows and vehicle overlap problem. The first dataset (dataset I) was collected in a cloudy day with low illumination and the second dataset (dataset II) was collected in a sunny day with strong sunlight.

3.3 Result and Analysis

The methods used in experiments are as follows: traditional Adaboost-based method, method based on traditional search strategy and the proposed method in this paper. They all used the same Adaboost classifiers cascaded with 22 strong Adaboost classifiers trained by more than 10000 samples and worked on 400×340 videos and were tested on a laptop with the equipment of 2.53 GHz Core i3 CPU, 4 GB RAM and implemented through C++ programming. The differences among them are as follows: traditional Adaboost-based method adopted overall search strategy to search for vehicles in the whole image. Method based on traditional search strategy adopted overall search strategy to search for vehicles in the foreground regions. The proposed method adopted the proposed search strategy to search for vehicles in the foreground regions.

A comparison of the three methods is shown in Tables 1 and 2. Generally, from the tables, we can see that the proposed method are comparable to the other two methods in TPR (true positive rate) and FDR (false discovery rate), but is much faster than other two methods which strongly demonstrates its real-time performance. As shown in

Table 1, on dataset I, the TPR of traditional Adaboost-based method reaches to 96.2% which is approximately equal to 96.0% the proposed method reaches to. And the FDR of the proposed method is slightly superior to the FDR of traditional Adaboost-based method. However, the detection speed of the proposed method reaches to 51.4 FPS (frames per second), which is more than 3 times the detection speed of traditional Adaboost-based method. At the same time, compared to the traditional search strategy, adopting our special search strategy really makes a difference in improving the detection speed. Table 2 shows a similar result, though the contrast is not as strong as that in Table 1, for the reason that the traffic of dataset II is more congested with a lot of shadows. From the experiment results, it is proved that the proposed method is really efficient and robust in handling different weather conditions, multiple vehicles and partial vehicle occlusion. Figure 6 shows some examples of detection results.

Table 1. Experimental results on dataset I

Methods	TPR/%	FDR/%	Speed/FPS
Traditional Adaboost-based method	96.2	3.8	14.8
Method based on traditional search strategy	96.0	3.3	33.9
Proposed method	96.0	3.3	**51.4**

Table 2. Experimental results on dataset II

Methods	TPR/%	FDR/%	Speed/FPS
Traditional Adaboost-based method	91.6	1.6	18.7
Method based on traditional search strategy	93.2	2.7	35.1
Proposed method	93.2	2.7	**41.5**

Fig. 6. Some examples of detection results

4 Conclusion

In this paper, we proposed a vehicle detection method based on improved Adaboost classifier. Firstly, use the background difference method to obtain foreground regions. Then, the cascaded Adaboost classifiers are used to detect multi-scale vehicles in these regions. Specially, in this paper, we adopt effective search strategy, which can greatly reduce the number of search windows, and further improve the detection speed. The experimental results show that the proposed method not only can obtain high accuracy, but also has strong real-time performance in different weather conditions. Moreover, the proposed method is able to handle partial vehicle occlusion.

References

1. Song, J., Song, H., Wang, W.: An accurate vehicle counting approach based on block background modeling and updating. In: 2014 7th International Congress on Image 2 and Signal Processing (CISP), Dalian, pp. 16–21 (2014)
2. Li, S., Yu, H., Zhang, J., Yang, K., Bin, R.: Video-based traffic data collection system for multiple vehicle types. IET Intell. Transport Syst. 8(2), 164–174 (2014)
3. Wójcikowski, M., Zaglewski, R., Pankiewicz, B.: FPGA-based real-time implementation of detection algorithm for automatic traffic surveillance. Sensor Network J. Sig. Process. Syst. 68(1), 1–18 (2012)
4. Unzueta, L., Nieto, M., Cortes, A., Barandiaran, J., Otaegui, O., Sanchez, P.: Adaptive multicue background subtraction for robust vehicle counting and classification. IEEE Trans. Intell. Transp. Syst. 13(2), 527–540 (2012)
5. Chen, Y., Wu, Q.: Moving vehicle detection based on optical flow estimation of edge. In: 2015 11th International Conference on Natural Computation (ICNC), Zhangjiajie, pp. 754–758 (2015)
6. Van Pham, H., Lee, B.-R.: Front-view car detection and counting with occlusion in dense traffic flow. Int. J. Control Autom. Syst. 13, 1150–1160 (2015)
7. Li, Y., Er, M.J., Shen, D.: A novel approach for vehicle detection using an AND–OR-graph-based multiscale model. IEEE Trans. Intell. Transp. Syst. 16(4), 2284–2289 (2015)
8. Zhuang, X., Kang, W., Wu, Q.: Real-time vehicle detection with foreground-based cascade classifier. IET Image Process. 10(4), 289–296 (2016)
9. Jiang, X., Gao, S., Liao, L., et al.: Traffic video vehicle detection based on semi-supervised SVM classification algorithm. CAAI Trans. Intell. Syst. 10(5), 690–698 (2015)
10. Viola, P., Jones, M.: Rapid object detection using a boosted cascade of simple features. In: Proceedings of the 2001 IEEE Computer Society Conference on Computer Vision and Pattern Recognition, 2001. CVPR 2001, vol. 1, pp. I-511–I-518 (2001)
11. Friedman, J., Hastie, T., Tibshirani, R.: Additive logistic regression: a statistical view of boosting. Ann. Stat. 28(2), 337–407 (2000)

Detection of Repetitive Forex Chart Patterns

Yoke Leng Yong[✉], David C.L. Ngo, and Yunli Lee

Department of Computing and Information Systems,
Sunway University, Bandar Sunway, Malaysia
14071856@imail.sunway.edu.my, dclngo@ieee.com,
yunlil@sunway.edu.my

Abstract. Throughout the years, numerous methods have been proposed for FOREX trading analysis and forecasting. As analysts/traders prefer to work with historical trading data, technical analysis based methods are often used. This paper presents an in-depth examination of technical analysis methods with an emphasis on charting/pattern-based analysis. Our findings indicate how to overcome the subjectivity often associated with identification and extraction of patterns within FOREX historical data. Based on historical facts that FOREX chart patterns repeat over time, the proposed method improves the approach towards identification of chart patterns as well as prediction of their recurrence regardless of the time warping effect affecting their formation.

Keywords: Pattern recognition · Forex forecasting · Linear regression line · Piecewise Linear Regression · Dynamic Time Warping

1 Introduction

The continuous evolution of the finance market has contributed to its increasingly complex nature with a myriad of financial investment opportunities offered to the traders. However, the FOREX market has proven to be trader's preferred choice with an average volume surpassing the trillion dollar mark [1, 2]. It is precisely due to the dynamic and volatile nature of the market and the emphasis on profitability that necessitates quick analysis and forecasting of the price fluctuation. Therefore, extensive work has been done to dissect the FOREX data with the expectation to unearth underlying relations that could be used for forecasting market trends and making trading decisions. Numerous methods of technical analysis, as well as econometrics and computational finance, have been conceptualised and incorporated into the market over the years.

While the advancement in FOREX market research provides additional insight to the traders, analysts are often faced with the element of subjectivity as heuristics and trading rules remains an important part of analysis. Following the three fundamental precepts introduced in technical analysis, the proposed method focuses on charting based methods and attempts to reduce the subjectivity in identifying the repeating chart pattern within the FOREX trading data. This can be achieved by simplifying the predefined chart patterns used for recognition. Subsequent sections will delve into more details on the intricacies of the Forex market in conjunction with the analysis methods used at present.

© Springer International Publishing AG 2017
Y. Tan et al. (Eds.): ICSI 2017, Part II, LNCS 10386, pp. 395–402, 2017.
DOI: 10.1007/978-3-319-61833-3_42

2   FOREX Analysis

2.1   Technical Analysis

Analysis of FOREX data is often an overwhelming task driven by various external factors and constraints such as information availability, selection of currency exchange pair and the frequency of the data (time interval) are often dependent on the trader's knowledge and prior experience. Although its effectiveness and predictive power are often viewed with scepticism, technical analysis has always been an important trading technique used by 90% of the market participants [3, 4] with researchers such as Hsu et al. [5] obtaining positive results from the studies conducted. Based heavily upon the notion that FOREX price move in trends and that history often repeats itself [3, 4, 6], technical analysis encompasses two main approaches, namely charting and mechanical rules. Technical analysis methods have also been used in conjunction with Artificial Intelligence (AI) methods such as Artificial Neural Network (ANN) [7, 8] and Genetic Algorithm [9] for forecasting.

Mechanical analysis revolves around the use of trading rules which are derived from mathematical functions by using past and present exchange rate to provide a mathematical justification and theoretical background. There are currently six (6) well-known groups of indicators, viz: trend, momentum, volume, volatility, cycle and Bill Williams' indicator [10, 11]. On the other hand, charting based technical analysis revolves around the interpretation of predefined patterns, which develops over time. Potential repeatability of FOREX trends and patterns contributes to the development of a set of predefined chart patterns, which are extensively documented in the Encyclopedia of Chart Patterns [12] and are religiously followed by chartist to analyse and forecast market fluctuations.

2.2   FOREX Chart Patterns

The underlying chart patterns identified within the Forex historical trading data are important to chartist as it often provides invaluable information when thoroughly scrutinised for market analysis and forecasting. Depending on the viewpoint adopted (short or long term data analysis), the recurrence of a chart pattern derived from the past often reflects the current market activity and volatility. The original investigation by Ito et al. [13] considers the effects of seasonality using intraday trading information such as the number of deals, price change, return volatility, and bid-ask spread for both the USD/JPY and EUR/USD exchange pair. Findings from the study eventually reveal the existence of trading patterns that occur not only daily but adapts itself across seasons.

On the other hand, the use of documented chart patterns allows for a different type of analysis that emphasises on pattern matching. When developing patterns are detected based on predefined patterns, these patterns are not only useful as a forecasting mechanism when correctly matched but they could signal one of the following trends: reversal, continuation or bilateral movement of the trend. The FOREX trends investigated are often formed using the peaks and troughs identified from FOREX charts and connecting two major tops or bottoms together [12, 14, 15]. From the trends identified whether uptrend, downtrend or sideway trend, it provides a global overview of the market direction.

3   Proposed Research

3.1   Detection of Underlying Trend

The emphasis of the proposed study is to identify and extract underlying FOREX trends and chart patterns that are simple enough to provide adequate features for classification. The proposed method was built on few technical analysis algorithms to offer the benefits of analysed data. Methods such as Linear Regression (LR) and Piecewise Linear Regression (PLR) could potentially be used to identify the relationship between fluctuating currency price over time. However, PLR implementation such as proposed by [16, 17] offers a more flexible analysis whereby multiple 'breakpoints' are used for representing the dynamic nature of the financial data as indicated in Fig. 1 (left) below. The PLR lines detected will determine how the data is segmented. As a single pass using PLR on the dataset might not result in an optimal detection and extraction of trends within the dataset, iterative detection is introduced whereby the segments identified in the first iteration is further processed by iterative PLR calculation process such as shown in Fig. 1 (right) to ensure optimal trend identification.

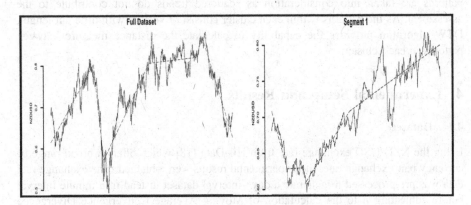

Fig. 1. PLR implementation: (Left) full dataset; (Right) first segment.

3.2   Extraction of Hidden Patterns

Once the underlying trends have been successfully detected and segmented, the subsequent step is to detect possible repeating patterns within each segment. Herein the proposed algorithm, the patterns extracted are not the common predefined patterns as documented by Bulkowski [12] but reduced to the basic trigonometric graph pattern as shown in Fig. 2 whereby the left plot represents the downtrend chart pattern and the corresponding right plot represents uptrend chart pattern. In an attempt to automate the identification and extraction of patterns from the Forex historical data, the PLR lines act as a baseline whereby the patterns are extracted based on the crossover above and below the PLR lines. Here, the proposed approach attempts to compensate for the main shortcoming of subjectivity that is often associated with charting method as it reduces the need for human interpretation of the chart patterns.

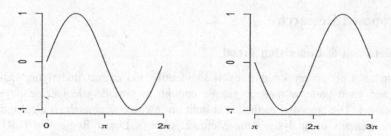

Fig. 2. Archetypes of chart patterns: (Left) downtrend; (Right) uptrend.

3.3 Detection of Repeating Patterns

Agglomerative Hierarchical Clustering using Dynamic Time Warping (DTW) plays a crucial role in establishing the repeatability of patterns as well as a method to represent the similarities between the patterns extracted. Once the patterns have been extracted, the hierarchical clustering performed gives an overview of how the trends compared against each other. At this stage, only clusters containing a significant amount of trend patterns are taken into consideration as scattered trends do not contribute to the assessment. As the patterns within each cluster consist of vectors with different length, DTW algorithm provides the capability to calculate the distance measure between patterns in each cluster.

4 Experimental Setup and Results

4.1 Dataset

Using the NZD/USD exchange rate from HistData [18] which offers a broad range of currency pair exchange rate, the experimental results were obtained. The exchange rate data was pre-processed to generate a day –interval dataset instead of a minute interval before subjecting it to the calculation of Moving Average Convergence Divergence (MACD) technical indicator. A full description of the dataset used is summarised in Table 1 below with the corresponding plot of the dataset shown previously in Fig. 1.

Table 1. Dataset information.

Currency dataset	NZD/USD (date/time, open, high low and close)
Total time frame	2 Jan 2006 to 31 December 2015
Number of data points	3119
Time interval	Daily interval
Technical indicator	MACD (12, 26, 9)

4.2 Algorithm Implementation

The proposed algorithm works to simplify the analysis process as it bypasses the requirement to have a pre-defined set of chart patterns for comparison. With the chart

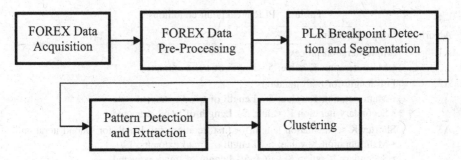

Fig. 3. Proposed algorithm.

patterns as depicted in Fig. 2 successfully extracted, it is consequently subjected to multiple iterations of hierarchical clustering process that enables the identification of similar patterns. The clustering process ensures that similar patterns will eventually be grouped together in the same cluster. Elimination of irrelevant trends is then performed by only taking into consideration clusters that contain a significant amount of trend patterns than scattered trends. The overall algorithm is highlighted in Fig. 3.

Iterative trend detection processes are also introduced whereby PLR calculations are used to identify uptrend and downtrend segments within the dataset. The preliminary segments identified could be taken as is or further processed by iterative PLR

```
While (stopping condition is not true)
   Perform PLR calculation and obtain breakpoints.
   Calculate Mean Squared Error (MSE) value for each
   segment.
End

Select PLR for the main iterations with the minimum
Mean Squared Error (MSE) value for segmentation.

For (each segment detected)
   While (stopping condition is not true)
      Perform PLR calculation and obtain breakpoints.
      Calculate Mean Squared Error (MSE) value for each
      segment for segmentation.
   End
End

Select PLR for the secondary iteration with the min-
imum Mean Squared Error (MSE) value.
```

Fig. 4. Breakpoint detection and segmentation pseudocode.

Table 2. PLR breakpoint conditions.

Algorithm	Looping conditions
A1	Single Iteration process • Main iteration K value: 5 - length of trend extracted
A2	Full length for both iterations • Main iteration K value: 5 - Length of trend extracted • Secondary iteration K value: 5 - Length of trend extracted
A3	Shorter K segmentation length for first iteration; full length for second iteration • Main iteration K value: 5 - Length of trend extracted/120 • Secondary iteration K value: 5 - Length of trend extracted
A4	Shorter K segmentation length for both first and second iteration • Main iteration K value: 5 - Length of trend extracted/120 • Secondary iteration K value: 5 - Length of trend extracted/15

calculation process to ensure that the best PLR adaptation to represent the trend patterns are extracted depending on the algorithm selected from four different ones. Subsequent implementation for the PLR breakpoint detection and segmentation module as depicted by the pseudocode in Fig. 4 are performed using R statistics software with the controlling variable being the number of quantiles chosen to represent the initial breakpoints (denoted as K). The proposed implementation offers four different algorithms implemented with the number of quantiles as listed in Table 2. The stopping condition follows Table 2 unless the PLR algorithm encounters an error or failed to converge for five (5) consecutive K value.

4.3 Results

The Forex trend clustering results obtained from the algorithm previously discussed in Sect. 4.2 clearly shows that hidden repetitive patterns do exist within the FOREX historical dataset. Using the pre-processed data as input, clusters of similar patterns are

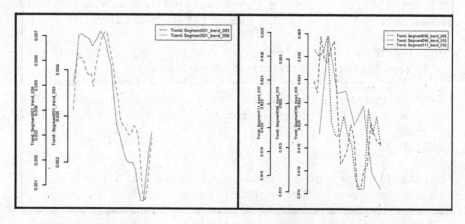

Fig. 5. An example of clustering result: (Left) A3 algorithm, (Right) A4 algorithm.

detected using the extracted trends. As there are four different variations of the algorithm, Table 3 denotes the findings using each of the variations and Fig. 5 illustrates an example of the clusters detected.

Table 3. Dataset information.

Algorithm	Total no. of trends detected	Total no. of clusters	No. of rejected trends	No. of clustered trends
A1	43	8	19	24
A2	334	31	241	93
A3	280	36	185	95
A4	136	22	84	52

From the results tabulated, it is clear that the FOREX historical data does not fluctuate at random and that the assumptions held by technical analyst hold true. FOREX price does indeed move in trend with the detected chart patterns repeating over time albeit with slight difference caused by time warping effect. However, the issue can be resolved easily with the use of DTW method which ensures comparison could be carried out seamlessly.

5 Conclusion

In conclusion, the proposed method explored an alternative towards automated identification and extraction repeating patterns for analysis. This will eventually pave the way towards removing the element of subjectivity from the patterns recognition process as current analysis still relies upon the analysis by traders. The results obtained are encouraging and solidifies the theory that patterns could potentially be used for FOREX forecasting. Further research options utilising the clustering results not only involves process optimisation but the chart patterns extracted could be exploited for forecasting purposes using the DTW mapping of the patterns.

References

1. King, M.R., Mallo, C.: A user's guide to the Triennial Central Bank Survey of foreign exchange market activity. BIS Q. Rev., 71–83 (2010)
2. Bank of International Settlements: Triennial Central Bank Survey - Foreign Exchange Turnover in April 2013: Preliminary Global Result (2013)
3. Schulmeister, S.: Components of the profitability of technical currency trading. Appl. Financ. Econ. **18**(11), 917–930 (2008)
4. Neely, C.J., Weller, P.A.: Technical analysis in the foreign exchange market. Federal Reserve Bank of St. Louis Working Paper Series 2011-001B (2011)
5. Hsu, P.H., Taylor, M.P., Wang, Z.: Technical trading: is it still beating the foreign exchange market? J. Int. Econ. **102**, 188–208 (2016)

6. Bagheri, A., Peyhani, H.M., Akbari, M.: Financial forecasting using ANFIS networks with quantum-behaved particle swarm optimization. Expert Syst. Appl. **41**(14), 6235–6250 (2014)
7. Gerlein, E.A., McGinnity, M., Belatreche, A., Coleman, S.: Evaluating machine learning classification for financial trading: an empirical approach. Expert Syst. Appl. **54**, 193–207 (2016)
8. Czekalski, P., Niezabitowski, M., Styblinski, R.: ANN for FOREX forecasting and trading. In: 20th International Conference on Control Systems and Computer Science (CSCS 2015), pp. 322–328 (2015)
9. Ozturk, M., Toroslu, I.H., Fidan, G.: Heuristic based trading system on Forex data using technical indicator rules. Appl. Soft Comput. **43**, 170–186 (2016)
10. Forex indicators: The best guide to indicator's world. http://forex-indicators.net/list
11. Gallo, C.: The forex market in practice: a computing approach for automated trading strategies. Int. J. Econ. Manag. Sci. **3**(169), 1–9 (2014)
12. Bulkowski, T.N.: Encyclopedia of Chart Patterns, 2nd edn. Wiley, New York (2005)
13. Ito, T., Hashimoto, Y.: Intraday seasonality in activities of the foreign exchange markets: evidence from the electronic broking system. J. Jpn. Int. Econ. **20**(4), 637–664 (2006)
14. Person, J.L.: A Complete Guide to Technical Trading Tactics: How to Profit Using Pivot Points, Candlesticks & Other Indicators. Wiley, New York (2004)
15. Talebi, H., Hoang, W., Gavrilova, M.L.: Multi-scale foreign exchange rates ensemble for classification of trends in forex market. Procedia Comput. Sci. **29**, 2065–2075 (2014)
16. Muggeo, V.M.R.: Estimating regression models with unknown break-points. Stat. Med. **22**(19), 3055–3071 (2003)
17. Wood, S.N.: Minimizing model fitting objectives that contain spurious local minima by bootstrap restarting. Biometrics **57**(1), 240–244 (2001)
18. HistData. http://www.histdata.com/download-free-forex-data

Damage Estimation from Cues of Image Change

Hang Pan[1], Yi Ning[2], Jinlong Chen[3,4], Xianjun Chen[1,3(✉)],
Yongsong Zhan[5], and Minghao Yang[6]

[1] Guangxi Key Laboratory of Trusted Software,
Guilin University of Electronic Technology, Guilin 541004, Guangxi, China
hingini@126.com
[2] Guangxi Colleges and Universities Key Laboratory
of Intelligent Processing of Computer Image and Graphics,
Guilin University of Electronic Technology, Guilin 541004, Guangxi, China
[3] Guangxi Key Laboratory of Cryptography and Information Security, Guilin
University of Electronic Technology, Guilin 541004, Guangxi, China
[4] Guangxi Cooperative Innovation Center of Cloud Computing and Big Data,
Guilin University of Electronic Technology, Guilin 541004, Guangxi, China
[5] Guangxi Experiment Center of Information Science,
Guilin University of Electronic Technology, Guilin 541004, Guangxi, China
[6] Chinese Academy of Sciences, Beijing 100190, China

Abstract. This paper proposes a damage estimation algorithm from cues of image changes. We get the feature map of damage area through comparing the Haar feature matrix and the LBP feature matrix by two images before and after the change. We then take the offset comparison method for fusion comparison results of different migration. At last, we get accurate location of damage detection by Gaussian filter and image morphology processing. Experimental results show that the algorithm can accurately detect the image damage area effectively, and is not too sensitive for the changes of light and color temperature. Furthermore this method does not need to establish different damage detection and evaluation models for different targets, and it can adapt to a variety of conditions of damage detection.

Keywords: Feature extraction · Damage detection · Damage estimation

1 Introduction

Since the middle of the last century, especially after the gulf war, the damage detection of the battlefield oriented is becoming more and more important for the combat troops. And the army uses it as an important link in the missile troops. When the missile aiming at target and the successful launch, the army will pay attention to some problems, For example, Whether the missile hit the target, launch the damage area where how mutilate hit area, etc. At the same time, as the key factor that affects the operation decision making, the detection of the damage target is very important.

In the existing damage detection algorithm, some damage detection and estimation methods are soft change detection. For example, the Bayesian network inference model

© Springer International Publishing AG 2017
Y. Tan et al. (Eds.): ICSI 2017, Part II, LNCS 10386, pp. 403–411, 2017.
DOI: 10.1007/978-3-319-61833-3_43

is established based on the attributes of the weapon and the target, and the detection and evaluation method is independent of the image [1, 2, 7]. However, this kind of soft change detection method needs to know in advance the data of weapon attribute, target attribute, hitting parameter and so on [1, 2, 8]. At present, most of the damage detection and estimation methods are hard detection or soft and hard combination detection method [3–5]. But the image acquisition before and after the damage mainly depends on aerial remote sensing and UAV detection technology, this technology has made rapid development in the last 30 years, every day can be obtained with TB as the unit of remote sensing data [6]. These data make a lot of fine quickly and accurately obtain high resolution images of possible target area, the traditional hard change damage detection method based on image analysis mainly rely on artificial interpretation, in the face of massive data for this method is slow, but also by the individual's subjective factors, so use computer to replace artificial damage on the target detection is very necessary.

The traditional methods of detecting the damage against fire based on image analysis, difference method and ratio method has strict requirements on image registration and accurate registration of images is often a difficult thing, which limits the applicability of this method [10]. Some methods are based on feature points matching, the construction of multiple sets of feature vectors, the calculation of the same kind of vector Euclidean distance to achieve the target detection and damage assessment [9, 10]. The method requires that the image has obvious edge features or angular features, and it is difficult to extract features of the image which is too smooth. Some methods need to know the target to be detected in advance, and different detection models are established for the damage detection of specific targets, such as airports, roads and so on. This kind of method has better detection effect, but the versatility is poor.

In this paper, a new damage detection method is proposed, which is based on the features of LBP and Haar, which is not sensitive to light and color, and then the 3D feature matrix is constructed based on the pixel level LBP and Haar features; and due to the impact of battlefield artillery and other factors making it very difficult to select markers for accurate registration of images before and after damage. So in this paper, we obtain the feature map according to the comparison of the migration. Because it can reduce the dependence on image registration to a certain extent, the damage detection based on image change analysis can be carried out under rough registration.

Because the algorithm is based on the change of the image pixel level to achieve the detection of damage, and this algorithm has nothing to do with the specific damage targets, has a universal. Therefore, this algorithm not only can be used to detect artillery fire damage targets, but also can be applied to other areas (such as natural disasters, urban expansion, etc.).

2 Method Principle and Step

In this method, we input two images of before and after damage. In the time span, these two images are not required, and are available at the right time according to actual combat conditions. Then in the space span, as long as the main area to be detected is the main area of the image on the line, and these two areas can be the same block in different positions of different angles of the two images. In imaging media, it can be used as optical, SAR, multispectral, hyper spectral, and other heterologous image, but requires the two images must be the same class. As illustrated in Fig. 1, the proposed algorithm includes three modules: image preprocessing, feature extraction, damage detection and estimation.

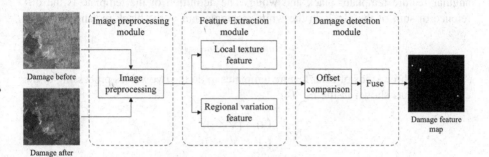

Fig. 1. Change detection flow chart

2.1 Calculation of Features Matrix Based on LBP Algorithm

LBP (Local Binary Patterns) is a non parametric operator for describing the local spatial structure of an image. Because the LBP eigenvalue response is the contrast between the pixel values of the local pixels, the LBP algorithm has strong robustness to illumination and other environmental change factors.

First we calculate the LBP feature matrix, the specific steps are as follows:

We traverse all pixels in the original image and compare the nuclear center point P with eight points (P0, P1, P2, $\cdots$, P7), and the results act as one of the eigenvalues. Below is Traditional LBP eigenvalue calculation formula:

$$LBP_P = \sum_{n=0}^{7} b_n 2^n. \tag{1}$$

Where LBP_P is the LBP eigenvalue of the ranges from 0 to 255. On the basis of the traditional LBP feature extraction, a set of vectors are used to represent the LBP eigenvalue, and the calculation method of the vector element value is the Eq. (2).

$$b_i = \begin{cases} 0 & (p \le p_i) \\ 1 & (p > p_i) \end{cases}. \tag{2}$$

Where $\{b0, b1, b2 \cdots b7\}$ is the LBP eigenvalue of the P. We use the kernel to traverse the entire sub image to obtain the LBP feature matrix of the image. According to the difference of the image resolution and the number of channels, the matrix is a three-dimensional matrix.

2.2   Calculation of Features Matrix Based on Haar Feature

The Haar feature is a simple rectangle features introduced by Viola et al. in the face detection system, which is defined as the difference the sum of the gray value in an image region adjacent area. The rectangular features can reflect the local features of the gray level change of detection object. There are two kinds of rectangles in the rectangular feature template: black and white. The definition of the template is the difference of sum of the pixels in the white rectangle and in the black rectangle:

$$V = Sum_{white} - Sum_{black} \qquad (3)$$

We use three kinds of Haar feature templates to do the two values processing of the eigen value, as shown in the Eq. (4):

$$b_i = \begin{cases} 0 & (V \le 0) \\ 1 & (V > 0) \end{cases}. \qquad (4)$$

Based on the characteristics shown above, we can change the size and location of the feature template to generate a large number of eigenvalues. There are four kinds of template sizes: 5 * 5, 7 * 7, 9 * 9, 11 * 11. We use four feature templates to traverse all the pixels in the image. For each pixel, there are 3 kinds of features, 4 sizes of the feature templates, a total of 3 * 4 = 12 features. So for each pixel, the feature value is a set of vectors with a length of 12 elements with the value of 0 or 1. For one images, the feature matrix is a rows × cols × 12 three-dimensional matrix.

2.3   Damage Detection Based on LBP and Haar Features

The process of damage detection is to generate damage feature map. We use the method of comparing the results of the feature map. First, we specify the offset $m_1, m_2 \cdots m_n$. Then according to the feature matrix, we use the coordinate of point p in matrix 1 to compare with coordinates $p + m_i$ in matrix 2, contrast rules as shown in Eq. (5):

$$R_i = \begin{cases} 0 & (e_{p,p+m_i} < th) \\ 255 & (e_{p,p+m_i} \ge th) \end{cases}. \qquad (5)$$

$$S_i = \frac{l - e_{p,p+m_i}}{l} \times 100. \qquad (6)$$

Where R_i is the value of the comparison of point p in matrix l and $p + m_i$ in matrix 2. $e_{p,p+m_i}$ is the number of points p that equals to the point $p + m_i$ in their feature vector elements. th represents threshold. S_i means the score of the evaluation, l is the length of eigenvector. After the comparison, the final eigenvalue and score of points p are:

$$R = M(R_1, R_2, R_3 \ldots R_n). \tag{7}$$

$$S = \frac{1}{n} \sum_{i=0}^{n} S_i. \tag{8}$$

where M is a feature fusion algorithm, which includes not only the fusion of different offset features, but also the fusion of different feature matrix, which can be used as "and" operation and "or" operation.

3 Experimental Analyses

In order to verify the effectiveness of the proposed algorithm, we have carried out experiments in a number of sets of data. The experimental diagram as shown below: the first set of image (Fig. 2(a)) for Pristina military airport facilities before and after the bombing effect; the second are the images (Fig. 2(b)) before and after a domestic regional damage.

(a)   (b)

Fig. 2. Comparison of damage area

3.1 Experimental Result

This algorithm was used to detect the damage area of the two groups respectively. The experimental images where the system environment for the Windows 7 64 operating system, Inter Core i3-2130 processor, 4G ram and Visual Studio 2013 programming environment. The experimental parameters and the experimental results are as follows (Table 1):

Table 1. Comparison of experimental parameters of damage detection between two groups

Parameter	First group	Second groups
Size	520 KB	1.65 MB
Resolution	280 * 360	981 * 588
Damage detection time	0.768 s	1.772 s
Memory	96 MB	257 MB

Because the offset acquisition results fusion feature map, so the offset choice is particularly important, the offset appropriate for obtaining accurate results have important influence characteristics of offset map. The offset direction represented by the selected offset $m_n (1 \leq n \leq 14)$ is as follows (Table 2):

Table 2. The migration direction represented by different offsets

n	1	2	3	4	5	6	7
Offset direction	LT_1	T_1	RT_1	R_1	RB_1	B_1	LB_1
n	8	9	10	11	12	13	14
Offset direction	L_1	T_2	B_2	LT_2	RT_2	RB_2	LB_2

The offset in the table is represented by D_x, which D represents the direction of the offset, respectively: T(Top), B(Bottom), L(Left), R(Right); x representing the offset size in pixels.

In order to verify the influence of different direction migration comparison on the results, we choose three different fusion offsets. The first selects are m_1–m_8 offset to get which are used to get the fused feature map, and the second selects the offset of m_9–m_{10} are used to get the feature map, and the third selects the offset of m_{11}–m_{14} are used to get the feature map. The results of the two groups of images of the image fusion of the three kinds of different offsets, as shown in Fig. 3:

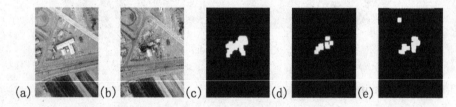

(a)   (b)   (c)   (d)   (e)

Fig. 3. the results of the first group of images with different offsets; (a) Image before damage; (b) Image after damage; (c) The results obtained from the first set of offset fusion; (d) The results obtained from the second set of offset fusion; (e) The results obtained from the third set of offset fusion

Through the comparison of the experimental results of the two groups of images, we can see that the fusion of the third kinds of offsets can be compared with the predicted results. At the same time, the experimental results obtained by the image features of the image, and other scholars of the first group of image damage detection after the comparison of the characteristics of the figure, as shown in Fig. 4:

We can see that this method described in the available range of the damage area, can get very good effect, and have a good description of the scope on damage region. As shown in Fig. 4(c) This method can be used to determine the extent of damage area that white area marked by red rectangle. However, the description of the method in the damage region boundary is relatively rough. Because the results obtained in this paper is the result of the comparison of the image migration in the image registration.

Fig. 4. Comparison between the method and other methods in damage detection. (a) Image before damage; (b) Image after damage; (c) The feature map of the first group is obtained by this algorithm; (d) The feature map of the first group is obtained by literature [10]; (Color figure online)

Compared with the results obtained in literature [10] in Fig. 4(d), the results obtained by using the traditional ratio method after the precise registration of the image are finally obtained. Although it can be used to describe the damage boundary, but there are many holes in the results. At the same time, this method needs to be further improved in the removal of the boundary pixels.

In conclusion, the advantages of this method are that the image registration accuracy is not high, and does not need to specify how much damage area in image. Therefore, this method has strong versatility, but the detection effect is not good.

3.2 Damage Estimation

At the same time, this paper uses the method of changing the number of pixels and the degree of change to evaluate the damage area, The results are shown in the following figure:

(a)   (b)

Fig. 5. Estimation results of damage degree of damage zone

Figure 5(a), For the true damage zone, the estimation method is 61, 70 and 74, After manual evaluation, the image of the middle of the two regions of the damage is more serious damage is similar, it is reasonable to score 70 and 74. Figure 5(b), the three damage zones were scored from left to right for 70, 65 and 66 respectively. The top of the upper part of the higher, after the manual assessment, the damage level of the two rightmost zones does not differ significantly, it is difficult to determine artificially the lower part of the terrain due to the difference between the right and the larger of the two zones.

In conclusion, according to the experimental results, the damage assessment method can take a simple scoring way to assess the damage area of image pixels based on the change of general evaluation results can explain the damage, to a certain extent, but not for the different targets of different environment were established the model of damage assessment, damage assessment results are also too simple. It is not accurate enough, only to be used as a reference manual assessment.

4   Conclusion

In this study, we propose a new method for damage estimation. We use the LBP features and Haar features that were not very sensitive to light and color. We then through compare the results of fusion multiple sets of migration making the registration precision is lower. Furthermore, we use the spatial filtering and morphology processing operations to feature map that improving the detection accuracy and reducing the false alarm rate. According to the pixel level changes in the degree of damage degree we had a simple estimation score. This method did not want to registration strictly; it needed to registration only roughly. Compared to most of traditional methods, this study not only has stronger commonality, but also fewer restrictions. The method in this paper is effectiveness by through experimental results. Our next step work is mainly to further improve the detection precision and delve into the target damage estimation system.

Acknowledgments. This research work is supported by the grant of Guangxi science and technology development project (No: 1598018-6, AC16380124), the grant of Guangxi Key Laboratory of Cryptography & Information Security of Guilin University of Electronic Technology (No: GCIS201604), the grant of Guangxi Cooperative Innovation Center of Cloud Computing and Big Data of Guilin University of Electronic Technology (No: YD16E11), the grant of Guangxi Key Laboratory of Trusted Software of Guilin University of Electronic Technology (No: KX201513), the grant of Guangxi Colleges and Universities Key Laboratory of Intelligent Processing of Computer Image and Graphics of Guilin University of Electronic Technology (No: GIIP201403), the grant of Guangxi Experiment Center of Information Science of Guilin University of Electronic Technology (No: 20140208).

References

1. Luo, W.: Research on the Methods of Change Detection and Image Annotation for Remote Sensing Images. University of Electronic Science and Technology of China (2012)
2. Xu, Fi., Zhu, Y.B.: Semi-automatic road centerline extraction from high-resolution remote sensing by image utilizing dynamic programming. J. Geomat. Sci. Technol. **32**(6), 615–618 (2015)
3. Zhang, Z., Jian, F.Q.: Change detection based on aerial image of urban area. Geomat. Inf. Sci. Wuhan Univ. **3**, 240–244 (1997)

4. Biscainho, L.W.P., Freeland, F.P., Diniz, P.S.R.: Using inter-positional transfer functions in 3D-sound. In: IEEE International Conference on Acoustics, Speech, and Signal Processing, pp. II-1961–II-1964 (2002)

5. Huo, C.L., Lu, J., Han, C.Q.: Object-level change detection based on multiscale fusion. Acta Automatica Sinica **34**(3), 251–257 (2008)

6. Li, D.R.: Change detection from remote sensing images. Geomat. Inf. Sci. Wuhan Univ. **28**(S1), 12–17 (2003)

7. Franzen, D.W.: A Bayesian Decision Model for Battle Damage Assessment (1999)

8. Jensen, J.R., Toll, D.L.: Detecting residential land-use development at the urban fringe. Photogrammetric Eng. Remote Sensing **48**(4), 629–643 (1982)

9. Wang, M.: Change detection using high spatial resolution remotely sensed imagery by combining evidence theory and structural similarity. J. Remote Sensing **14**(3), 558–570 (2010)

10. Yang, Y.P.: Study of battle damage effect assessment based on image change detection. Xidian University (2013)

Identifying Deceptive Review Comments with Rumor and Lie Theories

Chia Hsun Lin[1(✉)], Ping Yu Hsu[1], Ming Shien Cheng[2],
Hong Tsuen Lei[1], and Ming Chia Hsu[1]

[1] Department of Business Administration, National Central University,
No. 300, Jhongda Road, Jhongli City 32001, Taoyuan County, Taiwan (R.O.C.)
984401019@cc.ncu.edu.tw
[2] Department of Industrial Engineering and Management,
Ming Chi University of Technology, No. 84, Gongzhuan Road,
Taishan District, New Taipei City 24301, Taiwan (R.O.C.)
mscheng@mail.mcut.edu.tw

Abstract. The survey data showed that consumers trusted online reviews growing year by year, in which deceptive reviews had much more influences on consumer decisions. Unfortunately, due to the rapid transfer and enormous influence were typical of online reviews, many organizations began to deliberately exaggerate their own products or fabricated negative comments to attack competitors in order to derive benefit. Studies had shown that, it had much more influence by online reviews were tourism and hotel industry. This study discussed the negatively truthful review and the deceptive reviews from top twenty famous hotels in Chicago, including the true reviews taking from six famous review sites and the comparison group deceptive reviews on Amazon Mechanical Turk10. On the basis of the rumors and lies theories, the method created six attributes, key words of hotel, vague words personal pronoun negative words pronouns and pleonasm. By using text mining combined classification algorithm to forecast outcome and apply to build models. In this model showed that the mathematical operations not only worked more efficiently but kept the accuracy reasonably, so it could distinguish true or deceptive reviews well.

Keywords: Rumor · Lie · Deceptive review · Negative review · Text mining

1 Introduction

User review sites and webpages that have continued to increase (such as TripAdvisor and Yelp) have given ill-intentioned people the incentive to release fake reviews on the Internet to recommend or discredit target products and businesses. Deliberately falsified and fictitious fake reviews are said to have the potential to benefit certain businesses. Over the past few years, fake review or junk mail related problems have continued to spread, with many infamous cases reported. For instance, Amazon filed a lawsuit against more than 1,000 fake review writers in October 2015 [17], allegedly for releasing falsified fake reviews on the Amazon platform that not only misled consumers but also damaged Amazon's reputation. Today, businesses, distributors,

© Springer International Publishing AG 2017
Y. Tan et al. (Eds.): ICSI 2017, Part II, LNCS 10386, pp. 412–420, 2017.
DOI: 10.1007/978-3-319-61833-3_44

publishers, and the like no longer underestimate the impact of word-of-mouth, direct or indirect, on consumers' final decisions. Review information in an online review written by a former consume is said to be helpful in eliminating a new consumer's distrust towards a product. Conversely, if distributors, publishers, and the like can falsify reviews written by real consumers, online reviews' function in eliminating the consumer's distrust towards a product will be lost altogether, leading to consumers' incorrect product knowledge. In that case, consumers will be completely misled (e.g. product quality and supplier reputation) by online reviews [2].

From previous studies on the identification of real and fake reviews, it was found that when selecting characteristics and establishing attributes from raw data, fake or falsified review related theories were not used as the basis for setting the basis for determining fake reviews. Only a few classification algorithms were used to carry out assessment experiments. When determining the effectiveness of the classification model, only accuracy rates rather than computing efficiency were considered. This study explores the difference between real reviews and fake reviews for the purpose of constructing a method for identifying real and fake reviews. Based on fake review related theories, attributes or characteristic words applicable for identifying real and fake reviews were compiled. However, in addition to taking the effectiveness of the classification model into account, the computing efficiency of the classification model is also taken into consideration in this study. Therefore, the data dimension reduction method was adopted to further enhance the computing efficiency of the classification model, thereby achieving the goals of effectiveness and efficiency.

This paper is organized as follow: (1) Introduction: This part explains the research background, motivation and purpose; (2) Literature review: This part contains a review of scholars' researches on rumors, lies, negative and false comments; (3) System design: This part describes the content of the model design in this study; (4) Empirical analysis: This part describes the research data acquisition process. Numerical values and diagrams are also used to present the experimental results; (5) Conclusion and future research: The contribution of the model developed in this study is proposed, and possible future research direction is discussed.

2 Related Work

2.1 Rumors

Scholar Koenig pointed out in this research that rumors have three basic constituents [7]: 1. the target: It may be a person, matter, location, or thing; 2. the charge or allegation; 3. the source: An authoritative and reliable source of symbolism that enhances the reliability of a rumor. Shibutani (1966) [16] pointed out that a rumor is accepted and believed because the disseminated event has certain importance to the receiver although the event cannot be verified by the receiver through sufficient knowledge or information. This viewpoint coincides with the concept formula researched and developed by Allport and Postmann in 1947 [1]. Kapferer's research results [6] also proved the multiplication relationship between the formula factors. The formula is:

$$\text{Rumors} = \text{Importance of events} \times \text{Ambiguity of events} \qquad (1)$$

If the importance equals zero, or if the event itself is not ambiguous, a rumor will not exist.

2.2 Characteristics of Lies (Fabricating Stories)

Eduator Noah Zandan [18] has analyzed the characteristics of lies through linguistic and psychological perspectives. He has pointed out that "On the psychological level, we tend to self-optimize, casting people we worship on ourselves, rather than the original self". The study shows that it takes time to fabricate stories, in different language forms. The language form of fabricated stories possesses the following four characteristics:

1. Deliberately Ignoring Self: It refers to the liar knows a message has been deliberately fabricated, thus the reluctance to be liable for the lie. Therefore, the liars will deliberately be ignoring self to focus on someone else and detach from the lie itself.
2. Negative Words: They are uttered at the psychological level, because a story fabricator subconsciously feels regret for a lie told. Thus, the liar is usually more negative.
3. Simplified thinking: It is expressed from simplified thinking, because a story fabricator has to focus on making up a lie, the liar sometimes subconsciously simplifies matters. For example, US President Bill Clinton at the time of denying the affair said: I did not have sexual relations with that woman", rather than "I have no sexual relationship with Miss Monica Lewinsky".
4. Wordiness: A story fabricator often beats around the bush, inserting meaningless words and sentences and trivial words that appear plausible to complement a lie. In other words, the liar will connect many straightforward terms and mix them into a pile of roundabout sentences.

3 System Design

This study adopted two methods. First, the TF-IDF (Term Frequency–Inverse Document Frequency) method was adopted to evaluate the importance of every word in an article or a dataset. The weights of the words served as important bases for review determination. Subsequently, the dataset was divided into training data and testing data. The training data was used to carry out classifier training using SVM algorithm. SVM method is that it is capable of processing high-dimensional data., while other algorithms such as neural network, Bayesian classification, decision tree, and other methods may give e encounter calculation efficiency problems when faced with large amounts of data. However, Kantardzic in 2011 [5] mentioned in the book titled Data Mining that data dimension reduction could achieve computing efficiency while maintaining reasonable accuracy.

This study adopted the definition of rumor proposed by Allport and Postman [1] and the linguistic characteristics of lies (fabricating stories) as the conceptual basis for inferring possible characteristics of fake reviews. Six attributes were extended as the bases for determining real and fake online reviews. All the reviews and the words contained in the six attributes underwent word frequency calculation using the thesaurus matching method. After compiling the frequencies of words that appear in reviews, different classification algorithms were used in the first stage to train multiple assessment models. Targeting negative reviews in the review dataset, whether fabricated fake reviews were present were determined. The research processes promoted in this study are as shown in Fig. 1.

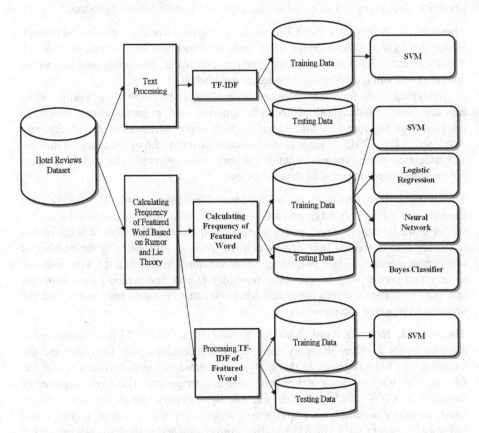

Fig. 1. Research process

3.1 Data Dimension Reduction

Based on the literature-based definition of rumors and the linguistic characteristics of lies (fabricating stories) proposed in this study, six characteristics for determining real and fake reviews were developed. Based on these six characteristics, thesauruses related to the six attributes were constructed. During the key word extraction phase, the thesaurus matching method is adopted to calculate the total number of key words that appear in every review.

According to the rumor concept equation inferred by Allport and Postman [1] (i.e. A rumor = importance of a message (event) * ambiguity of a message (event), possible reasons leading to rumors include the ambiguity of a message and the importance of a message.

Thesaurus 1. Important Hotel Attribute Thesaurus: The six indicators listed on hotel review website TripAdvisor are: price, hotel class, style, hotel brand, location, and amenities. After obtaining 38 relevant words from the respective indicators, WordNet Search 3.1 was adopted for synonym extension. Finally, 150 important hotel attribute keywords were compiled to establish the important hotel attribute thesaurus.

Thesaurus 2. Ambiguous Word Thesaurus: Ambiguous words were collected through Hedge Words search. Ambiguous words from three websites were compiled, 57 in all. WordNet Search 3.1 was adopted for synonym expansion, 166 ambiguous key words compiled and ambiguous word thesauruses established.

According to the four major characteristics of lies (fabricating stories) proposed by educator Noah Zandan [18]: deliberately ignoring self, negative words, simplified thinking, and wordiness, words under the four major attributes were subsequently developed. LIWC 2015 Thesaurus was adopted as the matching thesaurus. From the 125 categories of the thesaurus, suitable categories were selected as the matching words for the respective attributes, as detailed below:

Thesaurus 3. First-person Pronoun Thesaurus: First characteristic of a lie: The liar knows it is a deliberately fabricated message. Dare not taking responsibility for the lie, the liar will deliberately ignore self in the fabricated content. Therefore, it is inferred in this study that fabricated fake review and real reviews will differ in the number of first-person pronouns. Thus, two categories of thesaurus in LIWC 2015 were selected, namely first-person singular pronouns (category I) and first-person plural pronouns (category we). The 36 words contained in the two categories were used to establish the first-person pronoun the thesaurus.

Thesaurus 4. Negative Word Thesaurus: Second characteristic of a lie: The concept of negative words involves the story fabricator's subconscious guilt for lying, and the emotions tend to be negative. In this study, it is inferred that in the datasets of real and fake reviews vary in terms of negative word use frequency. The five categories of thesaurus in LIWC 2015 were selected, namely, category swear, negative emotion words (category negemo), anxiety (category anx), angry words (category anger), and sad words (category sad). The 831 negative words contained in the five categories were used to establish a negative word thesaurus.

Thesaurus 5. Simplified Thinking Word Thesaurus: Third characteristic of a lie: According to simplified thinking, the liar tends to subconsciously simplify matters. Therefore, it is inferred in this study that fabricated review contents often replace names or specific terms with pronouns. Thus, the four pronoun category thesauruses in LIWC 2015 were selected, namely, personal pronouns (category ppron), third-person singular pronoun (category shehe), third-person plural pronoun (category they), and indefinite personal pronoun (category ipron). After establishing a thesaurus using the 148 words contained in the four categories as the key words.

Thesaurus 6. Redundant Word Thesaurus: Fourth characteristic of a lie: A reviewer that fabricates contents will fall into wordiness. Therefore, non-fluent (category nonfl), filler (category filler) in LIWC 2015 Thesaurus were selected in this study. The 34 wordiness words in the two categories were selected to establish the thesaurus.

3.2   Establishment of Assessment Models

According to the literatures, the Support Vector Machine, SVM and Bayes Classifier achieved sound performance in related studies [4, 12, 19]. According to the literatures, it has been pointed out that regression analysis and neural network can be used as bases for predicting future situations through the analysis of previous data characteristics or attributes. Therefore, four classification prediction methods in data mining classification algorithm, namely SVM, Bayes Classifier, regression analysis, and neural network were used for research. Furthermore, in view of regression analysis, different types of regression analysis were adopted based on the different variable sizes. Since the purpose of this study is to explore whether the reviews are real or fake, the independent variables fall under two category variable types, thus the adoption of binary logistic regression.

4   Empirical Analysis

4.1   Data Collection

In this study, the hotel review datasets collected by Ott et al. [14] were adopted as the research samples. The review data includes the top 20 most-well-known hotels in Chicago on international travel review website TripAdvisor, and the positive and negative fake review datasets were collected through Amazon Mechanical Turk10 (AMT).

4.2   Model Results

The research samples include 20 negative reviews and 20 negative fake reviews from the 20 targeted hotels, 800 negative review datasets in all. The 800 reviews and the words from the six attributes collected then underwent matching calculation. The classification method involves finding fake review clues from the review content characteristics. These characteristics can be used in training classifiers to obtain the best classification results. Statistical software (StatSoft) Statistica 12 was subsequently adopted to carry out research analyses using four classification methods, namely, SVM, logistic regression, neural network, and Bayes Classifier. The classifiers' precision, recall, accuracy, and F-measure are as shown in Table 1.

Table 1. Single classifier experimental data

Classifier	Precision	Recall	Accuracy	F-measure
(1) TF-IDF + SVM	**81.1**	**82.5**	**81.7**	**81.8**
(2) FRL + SVM	67.2	68.3	67.5	67.8
(3) FRL + LR	66.0	77.5	68.8	71.3
(4) FRL + ANN	74.4	77.5	75.4	75.9
(5) FRL + NB	66.2	75.0	68.3	70.3
(6) FRL + TF-IDF + SVM	69.6	65.0	68.3	67.2

Note: FRL (Features of Rumor and Lie theory); SVM (Support Vector
Machine); LR (Logistic Regression); ANN (Artificial Neural Network);
NB (Naïve Bayes Classifier)

The experimental data in Table 1 shows that when using TF-IDF (Term Frequency
– Inverse Document Frequency) to directly carry outward weighting calculation on
preprocessed data. The SVM training classifier has the best performance among the six
classifiers developed in this study. However, the excessive words led to the reduced
calculation efficiency of the classification algorithm. This study adopted the six attributes extended from rumor and lie theories as the bases for data dimension reduction.
Then, the four classification algorithms, namely, SVM, logistic regression, neural
network, and Bayes Classifier were adopted to train multiple assessment models.
The SVM classification algorithm was conjunctively used to construct the review
classification assessment model. As shown in the data in Table 1, the classifiers
developed using the six attributes are the modest ideal in terms of three indicators:

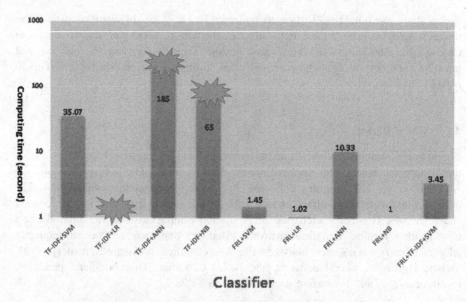

Fig. 2. Efficiency comparison of various classifier

precision, accuracy, and F-measure. As for the recall part, both the logistic regression and neural network method reached 77.5%.

In this study, the calculation efficiency of the respective classifiers was also explored. When training single classifiers, the computing time each needed as shown in Fig. 2. However, under reasonable accuracy rates, the calculation efficiency was improved.

5   Conclusion and Future Research

The analysis results show that that valid characteristic words for identifying hotel real and fake reviews were found through the six attributes extended from rumor and lie theories. With this as the basis for determining reviews, ideal accuracy was achieved. The characteristics were also used in data dimension reduction, making the review identification model not only able to maintain correct determination rates but also achieve calculation efficiency.

The research limitations and future research recommendations will discuss below:

1. Exploration of Hotel Reviews from Different Aspects: This study only targets negative hotel reviews for classification and analysis. Literatures have also pointed out that many fabricated reviews are intended to promote one's own products in order to create a product and corporate image. Therefore, it is suggested that whether positive reviews vary in characteristics or applicable methods.
2. Adoption of Review Information from Different Fields: It is recommended extended studies on other travel related reviews, such as tourist attractions, food services, etc. be carried out. Extended studies on fields outside the travel industry, such as reviews of books, brands, home appliances, cosmetics, etc.

References

1. Allport, G.W., Postman, L.: The psychology of rumor (1947)
2. Hu, N., Liu, L., Sambamurthy, V.: Fraud detection in online consumer reviews. Decis. Support Syst. **50**(3), 614–626 (2011)
3. Jindal, N., Liu, B.: Opinion spam and analysis. In: Proceedings of the 2008 International Conference on Web Search and Data Mining, pp. 219–230. ACM (2008)
4. Kantardzic, M.: Data Mining: Concepts, Models, Methods, and Algorithms. Wiley, New York (2011)
5. Kapferer, J.N.: Rumors: Uses, Interpretations, and Images. Transaction Publishers (2013)
6. Koenig, F.: Rumor in the Marketplace. Dover, Auburn House (1985)
7. Mihalcea, R., Strapparava, C.: The lie detector: explorations in the automatic recognition of deceptive language. In: Proceedings of the ACL-IJCNLP 2009 Conference Short Papers, pp. 309–312. Association for Computational Linguistics (2009)

420   C.H. Lin et al.

8. Ott, M., Choi, Y., Cardie, C., Hancock, J.T.: Finding deceptive opinion spam by any stretch of the imagination. In: Proceedings of the 49th Annual Meeting of the Association for Computational Linguistics: Human Language Technologies, vol. 1, pp. 309–319. Association for Computational Linguistics (2011)
9. Shibutani, T.: Improvised News: A Sociological Study of Rumor. Ardent Media (1966)
10. Techcrunch: Amazon Files Suit Against Individuals Offering Fake Product Reviews on Fiverr.com (2015).http://techcrunch.com/2015/10/16/amazon-files-suit-against-individuals-offering-fake-product-reviews-on-fiverr-com/
11. TEDxTaipei (2015). http://tedxtaipei.com/articles/the_language_of_lying/
12. Zhou, L., Sung, Y.W.: Cues to deception in online Chinese groups. In: Hawaii International Conference on System Sciences, Proceedings of the 41st Annual, p. 146. IEEE (2008)

Identifying Fake Review Comments
for Hostel Industry

Mei Yu Lin[1], Ping Yu Hsu[1], Ming Shien Cheng[2(✉)],
Hong Tsuen Lei[1], and Ming Chia Hsu[1]

[1] Department of Business Administration, National Central University,
No. 300, Jhongda Road, Jhongli 32001, Taoyuan, Taiwan (R.O.C.)
984401019@cc.ncu.edu.tw
[2] Department of Industrial Engineering and Management,
Ming Chi University of Technology, No. 84, Gongzhuan Road, Taishan District,
New Taipei City 24301, Taiwan (R.O.C.)
mscheng@mail.mcut.edu.tw

Abstract. Nowadays, consumers are inclined to issue their opinions for merchandise in the era of Web 2.0. As a result, numerous review comments about different products or services are accumulated on various websites every day. It has been found that to manipulate customer opinions, some dealers created the review comments in order to exaggerate the advantages of their own products or defame rival's reputation. This study strived to identify the negative fake review comments which were falsely created and aimed at attacking targeted products. The method created three word banks, namely, vagueness, and positive and negative attacks. The number of these words appearing in each review comments were calculated and applied to build logistic regression models. The experiment was conducted with true hostel review comments taking from "TripAdvisor" and the comparison group "Fake reviews" on Amazon Mechanical Turk. In the case where the ratio of fake and true review comments are10% in the training data, the proposed method reached 100%, 51.5% and 3% of precision, accuracy and recall, respectively. When the ratio is 50%, the method could reach 64%, 64%, 64% of precision, accuracy and recall respectively. The performance is better than the benchmark method which based on LIWC and SVM.

Keywords: Rumor · Fake review · Text mining · Logistic regression

1 Introduction

Rumors on the Internet, whether in terms of degree of complexity or validity, are far more powerful than rumors conventionally spread through word of mouth. In addition, rumors rapidly disseminated through the Internet pipeline are potentially devastating and fatal. Hu [2] found in his study that when consumers search product comments on the web, they tend to trust comments released by other users. Businesses also know this fact to be true; therefore, they maliciously fabricate comments to attempt to affect consumers' purchase decisions.

© Springer International Publishing AG 2017
Y. Tan et al. (Eds.): ICSI 2017, Part II, LNCS 10386, pp. 421–429, 2017.
DOI: 10.1007/978-3-319-61833-3_45

Targeting negative comments, fake comment qualities and text mining technology in data mining was the starting point for exploring the key attributes of offensive comments to establish a classification model that can accurately identify whether or not comments possess offensive characteristics under the actual data environment where different ratios of fake comments are inputted, thereby proposing a new form of classification method used to analyze which actual Internet comments are more likely to be false offensive comments.

This paper is organized as follow: (1) Introduction: This part explains the research background, motivation and purpose; (2) Literature review: This part contains a review of scholars' researches on rumors, negative and false comments, and logistic regression; (3) System design: This part describes the content of the model design in this study; (4) Empirical analysis: This part describes the research data acquisition process. Numerical values and diagrams are also used to present the experimental results; (5) Conclusion and future research: The contribution of the model developed in this study is proposed, and possible future research direction is discussed.

2   Related Work

2.1   Rumors

Scholar Koenig pointed out in this research that rumors have three basic constituents [5]: 1. the target: It may be a person, matter, location, or thing; 2. the charge or allegation; 3. the source: An authoritative and reliable source of symbolism that enhances the reliability of a rumor. Shibutani [10] pointed out that a rumor is accepted and believed because the disseminated event has certain importance to the receiver although the event cannot be verified by the receiver through sufficient knowledge or information. This viewpoint coincides with the concept formula researched and developed by Allport & Postmann in 1947 [1]. Kapferer's research results [4] also proved the multiplication relationship between the formula factors. The formula is:

$$Rumors = Importance\ of\ events \times Ambiguity\ of\ events \qquad (1)$$

If the importance equals zero, or if the event itself is not ambiguous, a rumor will not result.

2.2   Negative and False Comments

Lee [6] stated that while comments can be classified into positive and negative comments, it is highly important to accurately determine which comments are false comments. Negative comments devastate businesses and jeopardize companies' long-term interests and development. In a market environment of vicious competition, negative and false comments are even more likely. Ravid et al. [9] pointed out in their study that negative comments significantly reduce box-office revenues. Thus, the identification and filter of negative and false comments have substantial significance and theoretical values. Through the relevant studies on false comments conducted by Losiewicz [7] and Jindal [3], one type of false comments is deceptive review which deliberately provide positive reviews to facilitate the selection of a certain product or business entity; deliberately provide negative reviews to lower the reputation of a product or business entity in order to affect sales. The "deceptive" comments, coincides with the viewpoint of Shibutani [10] on the characteristics of rumors. This type of false comments is characterized by secrecy and diversity. Hence, it is extremely difficult for consumers to identify this kind of matters. This study is intended to identify more offensive negative comments in the first type, "deceptive" comments.

3 System Design

In this study, the text mining technology was adopted to carryout pre-processing of original comment contents. Then the keyword matching was used to extract keywords that are relevant to comments, which are a basis for classifying Internet comments. In addition, according to the concept formula put forth by Allport and Postman [1], with the concept basis of rumor = importance of an event * fuzziness of an event, coupled with the legalistic regression equation, a research model was developed. Targeting negative comments, whether or not they are by fake nature was identified.

In the study process, the review related vocabulary and synonyms were first collected. Then, the content of the original review and the vocabulary underwent pre-processing. After cross matching the keywords, the vocabulary not mentioned in the review were deleted. The t-test was then conducted to screen suitable vocabulary for classification and thesaurus establishment. The thesaurus establishment process is as shown in Fig. 1.

Fig. 1. Thesaurus establishment process

The total number of relevant vocabulary from the respective thesauruses that appear in reviews was first calculated. Logistic regression was then used to find relevant vocabulary under different ratios of fake review data environment. The predictor model of the experimental data sets was used to identify fake reviews. To determine reviews in the future, this model can be used for initial filtering, so as to screen dubious reviews. The review determination process is as shown in Fig. 2.

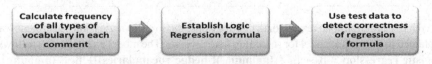

Fig. 2. Review service website determination process

When extracting keywords, thesaurus cross matching was selected. Three thesauruses were established, namely, hedge vocabulary thesaurus, importance vocabulary thesaurus in fake reviews, and reality review importance vocabulary thesaurus. By extracting basic words from three thesauruses, the study adopted the rumor theory as the basis. Using "hedge word" on the Internet, three websites were found to compile related basic hedge words. A total of 56 basic hedge words were extracted. Targeting the 56 basic hedge words and using WorldNet Search 3.1., all possible synonyms were found to expand the vocabulary. After integration, 133 hedge words were found.

When establishing importance vocabulary thesauruses, based on the related screening indicators for hotels listed on the TripAdvisor hotel review website, six features were classified: price, hotel class, style, hotel brand, location, and amenities. Among them, the indicators contained in hotel brand and location were prerequisites travelers chose when engaged in a hotel review, and thus they were excluded. A total of six features were extracted, which contain 38 basic importance words. In targeting the 38 basic importance words, WordNet Search 3.1 was used to find all possible synonyms to expand the vocabulary. After integration, a total of 159 importance words were found.

After the initial filtering, the relevant number vocabulary in every review underwent the t-test, population mean test, and inequality of variance analysis to screen hedge vocabulary and importance vocabulary that reach significant difference and relative suitability. In particular, the importance vocabularies, based on the mean obtained using the t-test, are divided into the larger of the reality and fake variables, based on which fake review importance vocabulary and reality review importance vocabulary can be classified. The final results of the hedge vocabulary show in Table 1. The 12 relevant fake and importance vocabularies, as shown in Table 2. The final results of the reality review importance vocabulary show in Table 3.

Table 1. Hedge words vocabularies

1	about	2	actually	3	almost	4	apparently	5	appear
6	around	7	as if	8	barely	9	could	10	evidently
11	fairly	12	fake	13	genuinely	14	guess	15	hardly
16	hopefully	17	imagine	18	in general	19	just about	20	kind of
21	like	22	likely	23	look	24	might	25	most
26	mostly	27	must	28	nearly	29	obviously	30	or so
31	overall	32	possibly	33	presumably	34	pretty	35	primarily
36	probably	37	rather	38	roughly	39	say	40	seem
41	seem a bit	42	seemingly	43	should	44	some	45	sometimes
46	somewhat	47	sort of	48	suppose	49	supposedly	50	think
51	truly	52	would						

Under different fake review data distributions, the logistic regression coefficient was obtained to develop a model for screening fake reviews. The model equation is as follows:

logistic regression = $\beta_0 + \beta_1$ * (amount of hedge vocabularies) + β_2 * (amount of

Table 2. Fake and importance vocabularies

1	clerk	2	experience	3	food	4	hotel	5	luxury
6	money	7	motel	8	reserve	9	room	10	site
11	suite	12	vacation	13	website					

Table 3. Reality and importance vocabularies

1	bar	2	bathroom	3	bed	4	business	5	charge
6	comfort	7	free breakfast	8	lobby	9	location	10	lounge
11	manager	12	manner	13	quiet	14	review	15	tv
16	wall								

importance vocabularies of fake comment) + β_3 * (amount of importance vocabularies of reality comment) *Evaluating value* $= e^{\log icregression}/(1 + e^{\log icregression})$.

4 Empirical Analysis

4.1 Data Collection

This study adopted the review data collection collected by scholars Ott et al. [8], which is the only marked data collection for public use. The data collection targets the 20 most popular hotels in Chicago listed on the world's largest travel website TripAdivsor [11]. Fake reviews mainly came from the Amazon Mechanical Turk (AMT). Through the platform, a specific task was announced, 400 positive fake reviews and 400 negative fake reviews were collected in batches, and this study identified only the offensive reviews.

4.2 Model Results

Targeting the negative 800 reality and fake reviews, keyword cross matching was conducted in this study to calculate the total frequency the relevant vocabulary appeared. Then, 75% of the reviews were randomly extracted as training data, 300 negative reality reviews and 300 negative fake reviews in total; 25% were used as testing data, containing 100 negative reality reviews and 100 negative fake reviews. In addition, since reality and fake data distribution in the real world is often completely out of balance, the SAS Enterprise Guide 5.1 software was used. The 600 training data entries in this study were divided into ten experimental data sets. That is, every set had fixed input of reality reviews that totaled 300. Fake reviews in different ratios of 10%, 20%, 30%, 40%, 50%, 60%, 70%, 80%, 90%, and 100% were gradually added to the 300 fake reviews. Finally, the logistic regression coefficient was obtained. As for the results, the hedge vocabulary only had insignificant effect at 10%, and thus it was not included in the model; the effect of relevant vocabulary for the rest of the ratios reached significance. The results are as follows:

logistic regression$_{10\%}$ = $-2.8292 + 0.2236$ * (amount of importance vocabularies of fake comment) $- 0.343$ * (amount of importance vocabularies of reality comment)

logistic regression$_{20\%}$ = $-2.0428 + 0.1549$ * (amount of hedge vocabularies) $+ 0.1697$ * (amount of importance vocabularies of fake comment) $- 0.5171$ * (amount of importance vocabularies of reality comment)

logistic regression$_{30\%}$ = $-1.458 + 0.1407$ * (amount of hedge vocabularies) $+ 0.1575$* (amount of importance vocabularies of fake comment) $- 0.5214$ * (amount of importance vocabularies of reality comment)

logistic regression$_{40\%}$ = $-2.231 + 0.1638$ * (amount of hedge vocabularies) $+ 0.149$ * (amount of importance vocabularies of fake comment) $- 0.5138$ * (amount of importance vocabularies of reality comment)

logistic regression$_{50\%}$ = $-2.0918 + 0.1499$ * (amount of hedge vocabularies) $+ 0.1775$* (amount of importance vocabularies of fake comment) $- 0.5293$ * (amount of importance vocabularies of reality comment)

logistic regression$_{60\%}$ = $-0.8756 + 0.1604$ * (amount of hedge vocabularies) $+ 0.1693$ * (amount of importance vocabularies of fake comment) $- 0.5377$ * (amount of importance vocabularies of reality comment)

logistic regression$_{70\%}$ = $-0.7602 + 0.1559$ * (amount of hedge vocabularies) $+ 0.1755$ * (amount of importance vocabularies of fake comment) $- 0.5176$ * (amount of importance vocabularies of reality comment)

logistic regression$_{80\%}$ = $-0.6117 + 0.1459$ * (amount of hedge vocabularies) + * (amount of importance vocabularies of fake comment) $- 0.5111$ * (amount of importance vocabularies of reality comment)

logistic regression$_{90\%}$ = $-0.5168 + 0.1574$ * (amount of hedge vocabularies) $+ 0.1751$ * (amount of importance vocabularies of fake comment) $- 0.5176$ * (amount of importance vocabularies of reality comment)

logistic regression$_{100\%}$ = $-0.4202 + 0.1529$ * (amount of hedge vocabularies) $+ 0.1767$ * (amount of importance vocabularies of fake comment) $- 0.5123$ * (amount of importance vocabularies of reality comment)

The experimental results in this study show that the precision and recall were a growing curve, while the precision was a declining curve, as shown in Fig. 3.

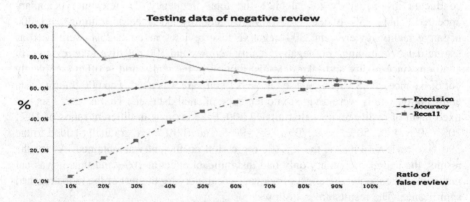

Fig. 3. Experimental result of negative reviews involved variant percentage of fake reviews

4.3   Control Experiment

According to the research of scholars Ott et al. [8], LIWC was also selected to screen relevant vocabulary. The Support Vector Machine (SVM), regarded as one of the best models for classifying algorithms among the classification methods, was used as the control model. The 12 vocabulary for identifying fake reviews selected by LIWC were imported into the R statistical software RStudio. The vocabulary selected by LIWC is as shown in Table 4. Negative reviews (25%) as those used in this research experiment were adopted as testing data, a total of 200 reviews. The experimental results of the control experiment show that under different ratios, the accuracy slightly increased, and the increase gradually stabilized after 50% of fake review input as shown in Fig. 4.

Table 4. Selected vocabularies of LIWC

1	I	2	Family	3	See	4	Pronoun
5	Leisure	6	Exclam	7	Sixltr	8	Posemo
9	Comma	10	Cause	11	Future	12	Feel

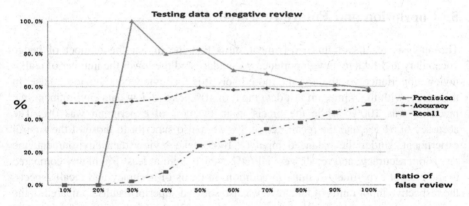

Fig. 4. Results of negative reviews involved variant percentage of fake reviews (SVM)

4.4   Analysis and Comparison of Experimental Results

Targeting the same 20 review data for testing, classification and prediction were carried out. In terms of precision, accuracy, and recall, the numerical experimental results of the two experiments are as shown in Tables 5, 6, and 7. Overall, the experimental model in this research proves to be more effective in terms of classification and prediction.

Table 5. Experimental comparison result of precision

Ratio of false review	10%	20%	30%	40%	50%	60%	70%	80%	90%	100%
Our study	100%	78.9%	81.3%	79.2%	72.6%	70.8%	67.1%	67%	66%	64%
LIWC + SVM	0%	0%	100%	80%	82.8%	70%	68%	62.3%	61.1%	59%

Table 6. Experimental comparison result of accuracy

Ratio of false review	10%	20%	30%	40%	50%	60%	70%	80%	90%	100%
Our study	51.5%	55.5%	60%	64%	64%	65%	64%	65%	65%	64%
LIWC + SVM	50%	50%	51%	53%	59.5%	58%	59%	57.5%	58%	57.5%

Table 7. Experimental comparison result of recall

Ratio of false review	10%	20%	30%	40%	50%	60%	70%	80%	90%	100%
Our study	3%	15%	26%	38%	45%	51%	55%	59%	62%	64%
LIWC + SVM	0%	0%	2%	8%	24%	28%	34%	38%	44%	49%

5 Conclusion and Future Research

The analysis results of this experiment show that the higher the number of hedge vocabulary and fake review importance vocabulary and the lower the number of reality review importance vocabulary, the more likely it is for review contents to be false. In this study, with the minimum test data input of 10% fake review data environment, the precision was 100%, while the precision in the control experiment was 0%. The accuracy of 51.5% and the recall rate of 3% were also superior to those of the control experiment. Under the balanced input of 100% fake review data environment, the precision, accuracy, and recall were all 64%, which were at least 5% higher compared to the control experiment results. In addition, in terms of accuracy and recall aspects, the respective input ratios of fake reviews also showed superior results compared to the control experiment, specifically the accuracy was 6.45% higher than that of the control experiment, and the recall average was 19.1% higher than the control experiment. Obviously, the keywords selected in this study and the use of the model to determine offensive comments had certain effectiveness.

The review website service management team may use the model in this study to filter contents in the review area and delete reviews that had no factual basis or affected ratings in order to enable consumers to obtain more correct product rating information, when engaging in purchase decision-making and enable consumers to regain trust for web reviews provided by review website service suppliers.

References

1. Allport, G.W., Postman, L.: The Psychology of Rumor. Henry Holt, New York (1947)
2. Hu, N., Liu, L., Zhang, J.(Jennifer).: Do online reviews affect product sales? The role of reviewer characteristics and temporal effects. Inf. Technol. Manage. **9**, 201–214 (2008)
3. Jindal, N., Liu, B.: Analyzing and detecting review spam. In: Proceedings of the 7th International Conference on Data Mining, Washington, DC, USA, pp. 547–552. IEEE Computer Society (2007)
4. Kapferer, J.N.: Rumors - Uses, Interpretations, and Images. Transaction Publishers, New Brunswick (1990)
5. Koenig, F.: Rumor in the Marketplace. Dover, Auburn House (1985)
6. Lee, D., Kim, H.S., Kim, J.K.: The role of self-construal in consumers' electronic word of mouth (eWOM) in social networking sites: a social cognitive approach. Comput. Hum. Behav. **28**, 1054–1062 (2012)
7. Losiewicz, P., Oard, D.W., Kostoff, R.N.: Textual data mining to support science and technology management (2000). http://www.onr.navy.mil/sci_tech/special/technowatch/textmine.htm
8. Ott, M., Yejin, C., Claire, C., Jeffrey, T.H.: Finding deceptive opinion spam by any stretch of the imagination. In: Proceedings of the 49th Annual Meeting of the Association for Computational Linguistics: Human Language Technologies, Stroudsburg, and Portland, USA, pp. 309–319 (2011)
9. Ravid, S.A., Basuroy, S., Chatterjee, S.: How critical are critical reviews? The box office effects of film critics, star power, and budgets. J. Mark. **67**(4), 103 (2003)
10. Shibutani, T.: Improvised News: A Sociological Study of Rumor. Bobbs Merrill, Indianapolis (1966)
11. TripAdvisor. 20 Best Chicago hotels on TripAdvisor - prices & reviews for the top rated accommodation in Chicago (2015). http://www.tripadvisor.com/Hotels-g35805-Chicago_Illinois-Hotels.html

Planning and Routing Problems

Multi-UAV Cooperative Path Planning for Sensor Placement Using Cooperative Coevolving Genetic Strategy

Jon-Vegard Sørli[1]([✉]), Olaf Hallan Graven[1], and Jan Dyre Bjerknes[2]

[1] University College of Southeast Norway, Kongsberg, Norway
Jon-Vegard.Sorli@usn.no
[2] Kongsberg Defence Systems, Kongsberg, Norway

Abstract. With the continuing increase in use of UAVs (Unmanned Aerial Vehicles) in various applications, much effort is directed towards creating fully autonomous UAV systems to handle tasks independently of human operators. One such task is the monitoring of an area, e.g. by deploying sensors in this area utilizing a system of multiple UAVs to autonomously create an efficient dynamic WSN (Wireless Sensor Network). The locations, order and which UAV to deal with deployment of individual sensors is a complex problem which in any real life problem is deemed to be hard to solve using brute force methods. A method is proposed for multi-UAV cooperative path planning by allocation of sensor placement tasks between UAVs, using a cooperative coevolving genetic algorithm as a basis for the solution to the described challenge. Algorithms have been implemented and preliminary tested in order to show proof of concept.

Keywords: Cooperative coevolution · Multi-UAV · Genetic strategy

1 Introduction

Research on the use of UAVs has increased massively the last few years. Today most UAVs are controlled either manually or have a simple pre-programmed path/pattern. Research is continuously progressing in various fields such as path planning [8], collision avoidance [9], mission planning [4], with the aim of reaching UAVs that can navigate and perform various tasks with true autonomy. This paper is a sub-part of a bigger project where the aim is to aid the first responders to an incident by gathering information about the area. The aim of this part of the project is to use a swarm of autonomous UAVs to cooperate in the task of placing and recovering sensors in the area.

A project was completed to create a prototype mechanism to be attached under a UAV for deployment and retrieval of sensor packages. A device was created that can carry up to four sensor packages in a revolving magazine which easily can be filled by an operator or by the UAV and mechanism. The sensor packages can be deployed one by one by rotating the magazine, and picked

© Springer International Publishing AG 2017
Y. Tan et al. (Eds.): ICSI 2017, Part II, LNCS 10386, pp. 433–444, 2017.
DOI: 10.1007/978-3-319-61833-3_46

Fig. 1. Sensor deployment and retrieval mechanism

up by moving the three arms automatically on top of the sensor package. The mechanism and sensor packages are light weight to not exceed UAV payload limits. The mechanism design is scalable in size to enable the carrying of more sensor packages. It has a generic interface enabling it to connect and be carried by any UAV like a quadcopter. The finished design can be seen in Fig. 1.

1.1   Problem Statement

Most research on path planning involves only one UAV flying from a start position to a goal position within a restricted area [8]. This paper proposes a solution to the issue of cooperative placement problem, CPP. Multiple UAVs shall cooperate efficiently to create a WSN by placing sensors in an area. This is a combined problem of:

1. Finding an optimized allocation of sensor placement tasks between the particular UAVs. Similar to a vehicle routing problem (VRP).
2. Finding optimized flight paths for each UAV in a real life environment.

These problems are strongly connected since the efficiency of flight paths will directly affect the allocation of sensors between UAVs and cannot be solved separately. Evolving paths for multi UAVs in parallel by using coevolution is not a new concept. The novelty in our approach lies in tackling the very specific concept of combined task allocation and path planning as a combined problem that needs to be solved simultaneously. The CPP is an optimization problem that is not feasible to solve using a normal exhaustive search method due to the time it would take, hence we use a heuristic. The hypothesis is that the described problem can be solved efficiently using a special implementation of an algorithm based on the Cooperative Coevolving Genetic Algorithm (CCGA) [7] in combination with a multi objective genetic algorithm (MOGA) for path planning. The proposed method enables us to optimally cover an area using a number of UAVs handling the tasks by finding a balance in the workload of each UAV. By "optimal" in this article we mean an approximation to the optimal solution. The research is divided into two major parts:

First which is the main focus of the work (and preliminary results) described in this article is the creation of a new method based on CCGA [7] to handle allocation of tasks optimally. Its purpose is to have multiple UAVs cooperating in planning their paths and allocating the tasks of placing sensors in an optimized way, by evolving in parallel into a joint optimized "complete solution" which in this context is the fastest way (shortest maximum path for a single UAV assuming constant airspeed) to allocate all the sensors. See Fig. 2 for examples of complete solutions where two and three UAVs cooperate in deploying six, nine and twelve sensors.

Secondly is the creation of a MOGA for creating flyable trajectories between sensor waypoints. It will be integrated in parallel with the coevolutionary algorithm feeding it with path costs for use during fitness calculations. The problem of UAV flight path must be solved using multi objective optimization due to the many factors affecting the efficiency of a flight path, e.g. an UAV's fuel consumption. Using a multi objective optimization method is both difficult and time consuming which is why we want it done in the most computationally efficient way as possible. We want to calculate several consecutive paths (trajectories) between waypoints where the goal of one planned path is the start of the next path. Due to the time constraint the algorithm cannot be too computationally expensive, so we have to make a trade-off in performance versus level of optimality. Having a computationally efficient algorithm is even more important in finding optimal paths given the real-time constraints and its ability to re-plan during a mission (run algorithm again based on a new layout for sensor placement if there are changes in the scenario or WSN layout). Another aspect of reaching high efficiency is the algorithm's suitability for parallel implementation.

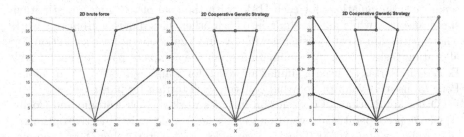

Fig. 2. Examples of complete solutions, test cases 1, 2, 3

For simplicity during the work presented in this paper we simplified the problem by using straight lines for the paths between sensors placement points and restricting them to 2D instead of using MOGA, hence saving time in evaluating the approach taken. This simplification is deemed to be a usable approximation due to the fact that it will not affect the algorithm presented in this paper, only the cost of each path in the algorithm. This limits the cost to distance. In a real scenario this path cost would be based on objectives such as distance and fuel/energy limitations of UAVs.

2   Related Work

The first major part of the algorithm in this work is inspired by the first work on cooperative coevolution presented by M. Potter and K. Jong in [7], CCGA which was used to optimize functions. They presented a solution to some of the traditional GA's weaknesses, giving opportunity to optimize more complex structures, as well as neural networks by having "coevolution of cooperative species" [7]. The big idea was that to evolve complex structures, modularity was needed, i.e. individual parts of an organism can evolve independently in order to provide the possibility of new complex solutions by how the different parts interact. Figure 3 gives an overview on the basic difference between CCGA and normal GA. The biggest difference is that we have multiple populations known as species evolving independently, but have their fitness evaluated together as a complete solution, i.e. the fitness of an individual is determined by how well it cooperate with individuals from the other species, together forming a complete solution to the problem, i.e. the complete solution represents the cooperative path plan for all UAVs to place all sensors, see Fig. 2 for example. A problem (vehicle routing problem) similar to the simplified CPP focused on in this article has been solved before using CCGA [3]. In their approach one subpopulation contain individuals representing permutations of complete solutions with size equal to the number of destinations. Another subpopulation contains information on how many destinations each vehicle should visit. The combined set is decoded to each vehicle's complete route.

The second major part of the work is related to Zheng [14], where they used cooperative coevolution to plan paths for multiple UAVs by evolving paths in separate subpopulations for each UAV, with the goal of reaching a destination simultaneously while not crashing into each other. It is different from this work where paths are created using a MOGA and the cooperative coevolution for allocating tasks, which in combination gives the complete path for each UAV. E. Besada-Portas [2] used a similar approach for planning multi UAV missions. Peng [6] uses coevolution to avoid collision with terrain for multi UAVs.

Other applications of coevolution includes work such as Antonio and Coello [1] who uses cooperative coevolution to solve large scale multi objective optimization problems of up to 5000 decision variables. It also states that there is empirical evidence indicating that the efficiency of multi objective meta-heuristics decreases as the number of variables increase for normal genetic algorithms. Evidence shows that cooperative coevolution is effective in solving large scale problems. The reported approach is that, subpopulations represents subset of all decision variables instead of only one and one, in order to handle the problem with non-separable functions, i.e. vectors of decision variables of that kind interact with each other and are not independent. H. Omid [5] uses it in evolving computer chess games. Wang [12] coevolves sensor placement and fuzzy logic controller to facilitate navigation and obstacle avoidance. Takahama [11] uses cooperative coevolution for structural learning of neural networks as a better alternative than GA. Sim [10] uses cooperative coevolution to optimize refinery scheduling.

3   Cooperative Coevolutionary Multi-UAV Path Planning

Figure 3 shows how the coevolutionary genetic algorithm GA differs from normal genetic algorithm as shown on the left most column of Fig. 3. From the top we can see that just like the normal GA, it is initialized by creating a random population. However, this is done one time for each species, i.e. for each subpopulation. In our algorithm each species represent a population of complete paths of a single UAV, and we can have up to N number of species where N represent the number of drones used to solve the task. Each individual in a species represents a subcomponent of the complete solution which is the cooperative path plan for all UAVs to place all sensors. We obtain a complete solution by combining single individuals from each species. Their fitness, i.e. the fitness of each individual is determined by how well it performs together with individuals of the other species as a complete solution to solve the problem. The combined fitness of a specific individual from each species is the complete fitness of the solution to

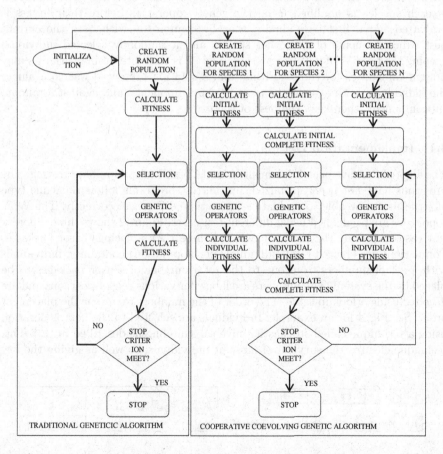

Fig. 3. Comparison standard generic algorithm vs coevolving cooperative genetic algorithm

the problem. They have an individual component of the fitness, the cost of a path, i.e. cost of trajectories planned by a MOGA. This is not the case for the simplified version used in this paper that assumes perfect straight lines.

In theory, one UAV may be able to deploy all the sensors, but this will probably not be an optimal solution, nor feasible considering limitations such as maximum payload and distances. Each species evolve independently using genetic operators, evolving each UAV's flight path for a set of sensor placements. A component of the original CCGA not used in the algorithm is the evolution of number of species (death and birth of species). This could be incorporated as a way of minimizing the number of UAVs needed to complete a mission successfully, or adding UAVs to complete a mission faster, or even run the algorithm faster.

After the initialization of the populations, an initial fitness is determined for each individual in each species. This is done by combining each individual of a species with a random individual from all the other species. Then individuals are selected for the next generation based on their fitness. Then each individual of each species is modified using the genetic operators before their fitness is evaluated again, first individually, then by comparison with both the current best subcomponents of the other species and a randomly selected individual keeping the best. A hypothesis presented by [7] is that interacting variables may present difficulties for the algorithm, and a solution presented was to evaluate the individuals both against the most fit subcomponents, and against a random subcomponent, then keep the best of the two.

3.1 Implementation Details

The algorithms have been implemented in MATLAB for preliminary testing. The programs have not been optimized, but all test programs follow the same type of structure to be able to evaluate relative time differences correctly. The WSN topologies are predefined using different test cases. Figure 2 shows three different test cases depicting the paths of two and three UAVs placing sensor packages. Populations are created (100 individuals per population) containing individuals with a set of numbers from one to the total number of sensor packages to be placed by the system. The numbers are labels for various sensor positions and are decoded using a lookup table. The order of the numbers represent the placement order. See Fig. 4 for an example. Individuals are selected to the next generation using a technique called stochastic universal sampling as described in [13]. This technique ensures the survival of most fit individuals, as well as giving the low

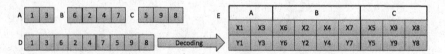

Fig. 4. Genotype/phenotype decoding of a complete solution D formed by three individuals A, B and C

fitness individuals a chance of survival, hence avoiding local minima problems. No crossover is used hence calling it genetic strategy. Algorithm 1 shows the general structure of the coevolutionary algorithm.

Algorithm 1. Coevolutionary coevolution pseudo code

```
PositionTable ←Sensor position data
SetupGAparameters();
GenerateRandomPopulation([P₁...Pₙ]);
for population[P₁...Pₙ] do
    for each individual pick individuals for comparison do
        CalculateFitness();
        if fitness(current) < BestSolution then
            BestSolution ←current
        end if
    end for
end for
while stop criteria not met do
    for population[P₁...Pₙ] do
        for each individual pick individuals for comparison do
            CalculateFitness();
            if fitness(current) < BestSolution then
                BestSolution ←current
            end if
            Selection();
            Mutation();
            Replacement();
            UpdatePositionTable();
        end for
    end for
end while
return BestSolution
```

3.2 Fitness Calculations

Algorithm 2 gives an overview of the fitness calculation. Weights are used to set the impact from coverage and duplication on the combined fitness (found by experimentation).

The fitness is calculated from two parts:

1. Individual fitness is a calculation of the path cost, i.e. total distance of individual UAV path.
2. Combined fitness is calculated by comparison to individuals from the other species.

The combined set of individuals are evaluated on two criteria:

1. Coverage: Check if all sensor locations are covered by at least one UAV.
2. Duplicate coverage: Check if more than one UAV covers a sensor package.
3. The cost of the maximum individual distance.

Algorithm 2. Fitness calculation

Random ←getRandomIndividuals();
Best ←getBestIndividuals();
SetIndividualStartPositions();
FillIndividualCalculationArrays();
CalculateIndividualDistances();
BufferCompleteSolution();
CheckSensorCoverage();
CheckDuplication();
Length = MaxOf(Distances);
Fitness = Length + coverage*200 + duplication*100
return Fitness

Individual Selection. The selection of individuals for combined fitness evaluation is done in three different ways:

1. Selection of one random individual from each other species.
2. Selection of one random and one individual with the best fitness from each of the other species, keeping the one giving the best combined fitness.
3. Comparison to all individuals in all other species, i.e. exhaustive fitness evaluation.

3.3   Mutation Strategies

Four different mutation strategies have been used in this implementation:

Random point mutation: A random single number from an individual is mutated into a new number.

Switch point mutation: Two numbers in an individual are switched with each other, e.g. the route order is changed.

Addition mutation: A new random number is added to the individual.

Deletion mutation: A single random number is removed from the individual.

Mutation Rates. Mutation rates used for all tests are 0,5 for random and switch mutation, and 0, 1 for deletion and addition.

3.4   Setup Limitations

The algorithm described in this article is based on a real life application using multiple UAVs to deliver and recover sensor packages in an area. A prototype has been made for the handling of sensor packages as described in Sect. 1 which introduces a limitation of four sensors per UAV.

A limit of one is set to the minimum number of sensor packages the individuals can have when experiencing deletion mutation since we want the problem solved quickly.

We assume the topology of the wireless sensor network is known a prior and will not be changed during runtime in the initial testing.

4   Results

Preliminary test results are shown in Table 1. Tests were conducted on three different test cases as shown in Fig. 2. Abbreviation used are: R for random comparison. R & B for random and best fitness comparison. Ex for exhaustive comparison. BF for brute force method. See Sects. 3.2 and 4.2 for more details. Average times and spread (range) to reach acceptable results are shown. For test seven and nine, a solution is deemed acceptable if its value is within a ten percent of the fitness of the perfect solution. For the other test cases, the acceptable result is the perfect solution.

Table 1. Results

Nr	Type	UAVs	Sensors	Avg time	Nr tests	Testcase	Range
1	R	2	6	1,25	30	1	4,95
2	R & B	2	6	0,64	30	1	1,53
3	Ex	2	6	4,57	30	1	7,39
4	BF	2	6	0,06	1	1	
5	BF	3	6	0,03	1	1	
6	R	3	6	0,58	30	1	1,11
7	R	3	9	6,76	30	2	16,1
8	BF	3	9	504,93	1	2	
9	R	3	12	86,89	30	3	75,1

The average time of test one and two shows how comparing to both a random individual and a best individual during each fitness evaluation performs better than only comparing to a random individual. One reason why the random comparison used longer time on average is that it showed a tendency of sometimes getting stuck around non-optimal solutions giving some longer runs, where as the R & B have less spread in the results and does not get stuck in local minima.

Test number six perform better than test number one both with the same number of sensors. This indicates how decreasing the number of sensors per UAV increases the speed of the algorithm as the number of combinations becomes smaller. This may suggest adding extra UAVs to handle complex tasks faster.

Comparing tests six, seven and nine, all done by the comparison to random individuals, where three UAVs cooperate in placing six, nine and twelve sensors, an interesting find is that the increase in time going from six to nine sensors (11,65 times longer) is almost the same as from nine to twelve sensors (12,85 times longer). When not considering the increase in amount of partitions going from two to four sensors per UAV, we have an increase from 6! permutations for test six, 9! permutations for test seven and 12! permutations for test nine, i.e. an 50300% increase in permutations from test six to seven and a 1319620% increase from test seven to nine. These results tells us that the algorithm runtime does not exponentially increase even with exponential increase in combinations.

4.1 Exhaustive Fitness Evaluation

The method of exhaustive fitness evaluation was discarded quickly because of poor results as shown in test number three. For the two UAV, six sensor test case the average time used was 4,57 s which is high considering 0,64 s for the best and random comparison. For problems with few combinations, e.g. three UAV and six sensors, using a large population often give the solution in one iteration, but considering the increase in combinations as shown in Sect. 4.2 the size of the populations would become too large very quickly.

4.2 Comparison to Brute Force

It is easy to think that the problem faced in this article can be solved easily using a brute force method, i.e. using raw computing power to test the fitness of all possible solutions. A simple brute force method was implemented to run though all legal solutions to find the point when brute force method is no longer a viable option. The fitness is evaluated the same way as in the other algorithms. Tests number four and five in Table 1 shows good results for the simple cases with only six sensors but the limit for acceptable execution time is breached already in test number eight, at three UAVs placing out a total of nine sensors using approximately 500 s to run. The number possible solutions to the problem is shown in Eq. 2 when considering the ten different partitions below (each UAV can deploy n_i sensors from one to four, n1 + n2 + n3 = 9) and confirms we are dealing with a NP-Hard/Complete problem depending on variant used.

$$\{(1,4,4)\}, \{(2,4,3),(2,3,4)\}, \{(3,4,2),(3,2,4),(3,3,3)\},$$
$$\{(4,4,1),(4,3,2),(4,2,3),(4,1,4)\}$$

$$C_{(n1,n2,n3)} = \frac{9!}{(9-n1)!} \cdot \frac{(9-n1)!}{(9-n1-n2)!} \cdot (9-n1-n2)! = 9! \tag{1}$$

$$N(9,3,4) = 10 \cdot 9! = 10! = 3628800 \tag{2}$$

4.3 Conclusion

This article presented the preliminary work for a proposed cooperative coevolving genetic strategy for solving the challenge of multi-UAV cooperative path planning for sensor placement. The results show that the concept works. Further work will show its efficiency when applied to bigger scale test cases, and with the implementation of parallelized use of multi objective genetic algorithm for path calculations. Limitations included predefined wireless sensor network layout for placement of sensor packages, straight line paths between sensor packages for simplicity and sensor package payload limits of UAVs. The placement of sensors, i.e. where to place each sensor in the environment to build up an efficient sensor network to gather information about the area is a complex problem that can be solved using multi objective optimization. In a real world scenario the environment is dynamic so the system would have to adjust to changes in the environment and re-plan real-time, or close to real time. This can be done by making changes to the sensor position table of the algorithm during run time. Results show that the number of sensors per drone has the most impact on the efficiency of the algorithm. The system has found close to perfect solutions when using the limit of four sensors per UAV, so the algorithm is a viable method with the limitations used.

Further Work. Replace distance calculation inside the fitness calculation with a look up table of path costs between waypoints to speed up the algorithm. This table can be continuously updated by the MOGA running in parallel, which also will account for changes to the environment. Expand the algorithm for a high number of sensors and UAVs, i.e. higher than the limitations used in this work. Investigate other new methods for selecting individuals for comparison. Further work includes refinement/completion of the algorithm presented in this article. A part of this work is the refinement of a MOGA for UAV path planning in 3D environments, optimizing on shortest collision free paths and fuel/energy consumption of UAVs which will replace the straight line costs. Further work in this area considers the development of a "Coevolving Crisis Algorithm", CCA which solves both the WSN placement problem and cooperative path planning using a coevolving genetic algorithm, i.e. we use a coevolving genetic algorithm to solve the sensor placement problem together with the system presented in this paper to solve the problem of dynamic sensor placement. The cooperative coevolving path planning method bases its sensor placement inputs on the results of the sensor placement GA which will continually evolve with changes to the environment. Other related work in progress includes a sensor placement planning algorithm for WSN in 3D environments utilizing multi-objective genetic algorithm. Where the above mentioned CCA focuses on the dynamic sensor placement problem, this work focuses on the creation of WSN layout by considering both terrain and sensor models. This includes terrain analysis to find the best viable sensor placement points, sensor limitations and wireless communication. Optimization is performed to achieve the best coverage, longest lifetime and optimal conditions for communication. Work is also done on mutation strategies for sensor placement, i.e. comparison of different GA mutation strategies for moving sensors around in a 3D environment.

References

1. Antonio, L.M., Coello, C.A.C.: Use of cooperative coevolution for solving large scale multiobjective optimization problems. In: IEEE Congress on Evolutionary Computation, CEC 2013, vol. 2, pp. 2758–2765 (2013)
2. Besada-Portas, E., De La Torre, L., De La Cruz, J.M., De Andrés-Toro, B.: Evolutionary trajectory planner for multiple UAVs in realistic scenarios. IEEE Trans. Robot. **26**(4), 619–634 (2010)

3. Machado, P., Tavares, J., Pereira, F., Costa, E.: Vehicle routing problem: doing it the evolutionary way. In: Proceedings of the Genetic and Evolutionary Computation Conference (January), p. 690 (2002)

4. Meng, B.B., Gao, X., Wang, Y.: Multi-mission path re-planning for multiple unmanned aerial vehicles based on unexpected events. In: International Conference on Intelligent Human-Machine Systems and Cybernetics, IHMSC 2009, vol. 1, pp. 423–426 (2009)

5. David, O.E., van den Herik, H.J., Koppel, M., Netanyahu, N.S.: Genetic algorithms for evolving computer chess programs. ACM Comput. Surv. **18**(5), 779–789 (2014)

6. Peng, Z.H., Wu, J.P., Chen, J.: Three-dimensional multi-constraint route planning of unmanned aerial vehicle low-altitude penetration based on coevolutionary multi-agent genetic algorithm. J. Central South Univ. Technol. **18**(5), 1502 (2011). doi:10.1007/s11771-011-0866-4

7. Potter, M.A., Jong, K.A.: A cooperative coevolutionary approach to function optimization. In: Davidor, Y., Schwefel, H.-P., Männer, R. (eds.) PPSN 1994. LNCS, vol. 866, pp. 249–257. Springer, Heidelberg (1994). doi:10.1007/3-540-58484-6_269

8. Roberge, V., Tarbouchi, M., Labonte, G.: Comparison of parallel genetic algorithm and particle swarm optimization for real-time UAV path planning. IEEE Trans. Industr. Inform. **9**(1), 132–141 (2013). http://ieeexplore.ieee.org/lpdocs/epic03/wrapper.htm?arnumber=6198334

9. Shanmugavel, M., Tsourdos, A., White, B.A.: Collision avoidance and path planning of multiple UAVs using flyable paths in 3D. In: 15th International Conference on Methods and Models in Automation and Robotics, MMAR 2010, pp. 218–222 (2010)

10. Sim, L., Dias, D., Pacheco, M.: Refinery scheduling optimization using genetic algorithms and cooperative coevolution. In: IEEE Symposium on Computational Intelligence in Scheduling, pp. 151–158 (2007)

11. Takahama, T., Sakai, S.: Structural learning of neural networks by coevolutionary genetic algorithm with degeneration. In: Conference Proceedings - IEEE International Conference on Systems, Man and Cybernetics, vol. 4(3), pp. 3507–3512 (2004)

12. Wang, X., Yang, S.X., Shi, W., Meng, M.H.: A co-evolution approach to sensor placement and control design for robot obstacle avoidance. In: International Conference on Information Acquisition, pp. 107–112 (2004)

13. Yu, X., Gen, M.: Introduction to Evolutionary Algorithms. Springer, London (2010). http://www.springer.com/gp/book/9781849961288

14. Zheng, C., Ding, M., Zhou, C., Li, L.: Coevolving and cooperating path planner for multiple unmanned air vehicles. Eng. Appl. Artif. Intell. **17**(8), 887–896 (2004)

Optimal Micro-siting Planning Considering Long-Term Electricity Demand

Peng-Yeng Yin[✉], Ching-Hui Chao, Tsai-Hung Wu,
and Ping-Yi Hsu

National Chi Nan University, Puli, Nantou, Taiwan
pyyin@ncnu.edu.tw

Abstract. Wind farm micro-siting is to determine the optimal placement for the wind turbines such that the cost of energy (COE) is minimal. The problem is clearly a long-term decision one, once the micro-siting was constructed, it is extremely costly to reconfigure the layout. Long-term electricity demand forecasting is a necessity for formation of governmental energy policy. We anticipate that the two problems should be considered simultaneously to create potential benefits because they have resembling properties and close supply-and-demand relationship. This paper proposes a demand-aware micro-siting system to COE minimization. The system is a holistic integration of long-term electricity demand forecasting and optimal micro-siting. A case study in central Taiwan area is conducted to validate the feasibility of the proposed system.

Keywords: Micro-siting · Demand forecasting · Regression · Optimization

1 Introduction

Sufficient energy supply plays an important role in economic development of industry countries. However, energy is considered as limited and expensive resources. The traditional fossil fuels will be exhausted by the end of this century [1], while the renewable energy is expensive due to current production technology. As of the end of 2014, the production of wind and solar energy contributes about 4% and 6% in global energy production (REN21. http://www.ren21.net/). On the other hand, the global awareness for restraining the emission of greenhouse gases has urged industry countries to expedite the reformatting process of energy structure towards renewable energy. It thus becomes increasingly important for a government to develop an effective energy policy considering long-term electricity demand. Taiwan has abundant wind capacity due to its long coast line. However, 97.84% of Taiwan's energy supply in 2015 is imported and only 0.16% is generated from wind power (MOEA. http://www.moeaboe.gov.tw/). The wind capacity is under-explored and the development of new wind farms should consider locational electricity demand since the generated wind power decays with transmission distance and the storage of excess wind power is expensive.

For the wind energy sectors, it is crucial to select a potential site for constructing the wind farm and to determine the optimal placement for the wind turbines such that

© Springer International Publishing AG 2017
Y. Tan et al. (Eds.): ICSI 2017, Part II, LNCS 10386, pp. 445–453, 2017.
DOI: 10.1007/978-3-319-61833-3_47

the cost of energy (COE) is minimal. The optimal wind turbine placement (OWTP) problem, referred to as *micro-siting* in the literature, is clearly considered on a long-term decision basis. Once the micro-siting was installed, the construction is permanent and it is extremely costly to reconfigure the layout. The OWTP problem is region-dependent and its influence factors include wind conditions, land surface roughness and environmental impacts. Analogously, electricity demand landscape obtained from long-term forecasting is a necessity for formation of governmental energy policy. Long-term demand prediction is also region-dependent and is highly influenced by the number and sales of profit-seeking enterprises, type of regional industries, local temperature and human population. We anticipate that the two problems (micro-siting and demand forecasting) should be considered simultaneously to create potential benefits because they have resembling properties (long-term and regional) and close supply-and-demand relationship.

Most of the state-of-the-art methods for solving the two problems employ some form of metaheuristic algorithms, whose advantages include generality, easy implementation, and fast computation. For the demand forecasting problem, genetic algorithm (GA) [2] and particle swarm optimization (PSO) [3] have been chosen to predict the energy demand in Turkey. A PSO-GA approach is proposed in [4] for predicting the demand in China. Askarzadeh [5] compared several PSO variants on electricity demand estimation of Iran. For the micro-siting problem, GA was applied in [6] to determine the optimal turbine positions in an array of grids. Yin and Wang [7] proposed a hybrid algorithm combining GRASP and VNS for finding the optimal layout of wind turbines.

Given the discussions above, we feel that the demand forecasting and the micro-siting problems should be simultaneously considered from a holistic point of view and the best solution method should contain metaheuristics as its core algorithms. This paper proposes a holistic integration of a demand-aware micro-siting system to COE minimization. An effective metaheuristic algorithm is developed for both problems.

2 Literature Review

2.1 Electricity Demand Forecasting

From the perspective of lead time forecasting range and the managerial objective, the electricity demand forecasting can be classified as short-term, medium-term, and long-term forecasts. The short-term prediction anticipates the next immediate demand from a few seconds to hours. The purpose is to regulate the power production and enhance the efficiency. The medium-term prediction forecasts the electricity load in days to months which is useful for pricing and production plan setting. The long-term prediction estimates the future power demand in years. The government sectors rely on this sort of forecasts for formation of energy policies.

On the other hand, the technology for electricity demand forecasting is mainly classified into three types. (1) Econometric models [8] use economy indicators as input and apply statistical approaches to find the optimal model fitting to the economy data.

(2) Soft computing approaches [9] have been broadly used for prediction applications. Researchers have applied artificial neural networks, fuzzy logic, and grey theory to electricity demand prediction. (3) An emerging class is metaheuristic algorithms [2–5], most of which are bio-inspired and easily implemented. Among others, GA, PSO, and ACO are most contemplated for electricity demand forecasting.

2.2 Micro-siting Problem

Given a potential site for capturing the wind energy, the micro-siting problem seeks to optimize the turbine placement considering the wake effect. Formally, the wind farm is divided into d × d grids, at the center of each grid can it be placed a turbine. Mosetti's wake model [6] is broadly adopted for calculating the downstream wind speed. Let W_0 be the free-flow wind speed. The downstream wind speed W_i at the position with l upstream turbines is in a ratio to W_0, and W_i is estimated by

$$\frac{W_i}{W_0} = 1 - \sqrt{\sum_{j=1}^{l} \left(\frac{2a}{1 + \alpha(h_j/r)} \right)^2} \tag{1}$$

where a is the axial induction factor, α is the entrainment constant, h_j is the distance to the jth upstream turbine, and r is the wake radius. The parameters a, α, and r are related to the turbine specification. Let the wind farm be placed with N turbines which are in operation for a time period T. The power production e is estimated as follows.

$$e = \sum_{i \in S} \sum_{j \in D} \sum_{k=1}^{N} t_{ij}(\xi W_k^3) \tag{2}$$

where S and D are the set of wind speeds and directions, t_{ij} is the time occurrence distribution in T for observing wind having speed i and direction j, and ξ is the efficiency coefficient for the turbine power generator.

3 Proposed Methods

The proposed optimal micro-siting planning system considering long-term electricity demand is conceptualized in Fig. 1. The proposed system has three functional models as indicated by the boxes with a boldface heading.

3.1 Demand Forecasting Model

As evidenced from [2–5], the quadratic regression with econometric indicators is very effective in forecasting electricity demand. We employ a quadratic regression equation which uses past electricity demand and governmental statistics as input parameters. From our preliminary experiments, we found the indicators, namely, the number and

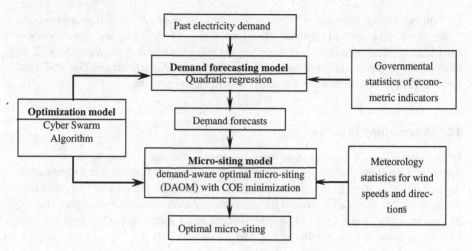

Fig. 1. Flowchart of the proposed demand-aware micro-siting planning system.

sales of profit-seeking enterprises, local temperature and population, are the most influential ones. Let us denote the four indicators by y_1, y_2, y_3, and y_4, respectively. The quadratic regression equation for the predicted demand $\hat{d}$ is formulated by

$$
\begin{aligned}
\hat{d} = {} & w_1 y_1^2 + w_2 y_2^2 + w_3 y_3^2 + w_4 y_4^2 + w_5 y_1 y_2 + w_6 y_1 y_3 + w_7 y_1 y_4 + w_8 y_2 y_3 \\
& + w_9 y_2 y_4 + w_{10} y_3 y_4 + w_{11} y_1 + w_{12} y_2 + w_{13} y_3 + w_{14} y_4 + w_{15}
\end{aligned} \tag{3}
$$

where w_i is the coefficient for the best-fit quadratic regression equation. The task for electricity demand forecasting is now reduced to the learning problem for finding the optimal value of w_i which minimizes the error between the predicted electricity demand and the actual ones from a given training set. The error can be measured in a number of ways. In this paper, we adopt the mean absolute error (MAE) as follows.

$$
\mathrm{MAE} = \sum_{i=1}^{n} \left| \hat{d}_i - d_i \right| \tag{4}
$$

where d_i is the actual electricity demand and n is the number of time interval instances in the training horizon.

3.2   Micro-siting Model

The electricity demand predicted by the demand forecasting model covers a planning horizon consisting of m time intervals into the future. A demand-aware wind farm system should alleviate power-undersupply situations. Let us denote the predicted demand and the generated power from wind turbines at time interval i by $\hat{d}_i$ and e_i, respectively. We specify an upper bound for the estimated power deficiency in the planning time horizon. For those time instances where the generated power is undersupplying the demands, the deficiency is filled by purchasing power from other energy sources. The classic unconstrained COE model is infeasible for the studied context and needs significant modifications. Therefore, we propose a demand-aware optimal micro-siting (DAOM) model to COE minimization as follows.

$$\text{Minimize } \left(C_{\text{WT}} + C_{\text{Electricity}}\right) \Big/ \left(\sum_{i=n+1}^{n+m} \left(e_i + \max\{0, \hat{d}_i - e_i\}\right) \right) \tag{5}$$

$$\text{subject to } \sum_{i=n+1}^{n+m} \max\{0, \hat{d}_i - e_i\} \leq U \tag{6}$$

where C_{WT} and $C_{\text{Electricity}}$ are the cost for installed wind turbines and purchased electricity from other energy sources. The two types of cost are calculated as follows.

$$C_{\text{WT}} = N\left(\frac{2}{3} + \left(\frac{1}{3}\right)\left(e^{-0.00174N}\right)\right) \tag{7}$$

$$C_{\text{Electricity}} = \sum_{i=n+1}^{n+m} p_i \max\{0, \hat{d}_i - e_i\} \tag{8}$$

where p_i is the purchase price for 1 kW in operation in time interval i and N is the number of installed turbines. The cost C_{WT} is designed for seducing a purchase of more turbines and the maximal discount is one third off, while $C_{\text{Electricity}}$ is the sum of product of power deficiency and the varying price determined by the energy market.

3.3 Optimization Model

We customize the cyber swarm algorithm (CSA) [10] for establishing the optimization model for solving both the optimization tasks incurred in the demand forecasting model and the micro-siting model. In the CSA, particle represents a candidate solution to the underlying problem and the particle's fitness is an assessment value indicating the performance obtained by the corresponding solution. Clearly, the design for particle representation and fitness assessment is problem-dependent. For our first optimization task, we need to find the optimal value of the fifteen coefficients (w_i) for the quadratic regression equation. Therefore, the particle is naturally perceived as a linear array containing fifteen real numbers. The fitness of a particle is designed as the reciprocal of the MAE over the training set,

$$fitness = \left(\sum_{i=1}^{n} |\hat{d}_i - d_i| \right)^{-1} \tag{9}$$

For the optimization task of 10×10 micro-siting, a feasible solution can be described by a binary matrix where entry value one indicates a wind turbine is placed in the grid, and zero otherwise. Each row in the matrix is a binary string having 10 bits and can be converted to an integer value in the range of [0, 1023]. We design the particle representation as a linear array containing ten real numbers. The fitness of a particle is designed as follows.

$$fitness = \left\{ (C_{WT} + C_{Electricity}) \middle/ \left(\sum_{i=n+1}^{n+m} (e_i + \max\{0, \hat{d}_i - e_i\}) \right) + \max\left\{0, \left(\sum_{i=n+1}^{n+m} \max\{0, \hat{d}_i - e_i\} - U \right) \right\} \right\}^{-1} \quad (10)$$

Referring to our micro-siting model (see Eqs. (5) and (6)), it is seen that our fitness function consists of the objective value (the first term in Eq. (10)) for assessing the performance and the penalty value (the second term in Eq. (10) for accessing the feasibility violation of the power deficiency constraint.

The CSA is an effective form of PSO and the former adds three salient features, reference set learning, dynamic social network, and responsive strategies, to the latter. The three features operate in a higher level and they are problem-independent. We omit the description for saving the space. The reader is referred to [11].

4   Experimental Results

4.1   Datasets

The first dataset is used for electricity demand forecasts and it contains the electricity consumption for Taichung city in central Taiwan from 2001 to 2014, where the data falling in the first thirteen years are used as training set, and the remaining data as test set. The econometric indicators are collected from the governmental statistics databases maintained by the Ministry of Finance and the Ministry of the Interior. The second dataset is prepared for optimal micro-siting planning and the dataset is collected from the Central Weather Bureau from 2011 to 2013. The dataset contains hourly wind speeds and directions. We integrate the data on a weekly basis to form a wind rose distribution, indicating various speed-direction occurrences. Fifty-two wind roses are constructed to simulate the wind conditions.

4.2   Simulations

We first conduct the simulations with the electricity demand forecasting using the past demand records and the four econometric indicators. With 30 repetitive runs of CSA, the best value for the coefficients of the quadratic regression equation is obtained by referring to the best run which outputs the minimum MAE. Figure 2 shows the convergence curve of the MAE obtained in the best run as the number of fitness evaluations increases. It is seen that, at the early stage of evolution, the MAE value decreases drastically until reaching a salient local optimum. At the middle and final stage of evolution, the reduction of MAE gradually converges.

Our second experiment conducts the simulations with the DAOM micro-siting model which seeks to minimize the COE while respecting the upper bound for the estimated power deficiency in the planning time horizon. Figure 3(a) shows the

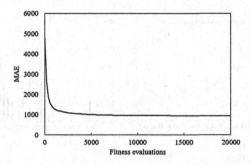

Fig. 2. The convergence of the MAE as the number of fitness evaluations increases.

comparison between the weekly demand forecasts and the produced power by simulations with the DAOM micro-siting model. It is interesting to note that the most intensive electricity consumption appears around summer, however, the strongest winds are observed in winter. The situation raises the difficulty for managing optimal wind farm layouts considering long term electricity load. Our DAOM model circumvents the difficulty by constraining on an upper bound for the whole year production deficiency and filling the gap by purchasing power from other energy sources. Under this criterion, the optimal COE solution is approached. It is seen in Fig. 3(a) that the produced wind power goes along with the whole year demand instead of merely satisfying the consumption peaks. The benefit of our DAOM model is twofold. *First*, if the wind farm layout is organized by satisfying the high-demand period, more turbines need to be installed (high cost), certainly incurring severe wake effect (low efficiency). It is likely that much produced power in the low-demand period will be wasted or stored with an expensive cost. *Secondly*, if we optimize the layout by the classic COE model with no demand information, the wind turbines will be positioned in the low-wake zone facing to the strongest wind direction. Therefore, much more power will be produced in winter than in summer. In other words, the classic COE model is not demand-aware and if it is adopted, the problem of power waste and deficiency is more significant. Figure 3(b) shows the weekly amount of purchased power as the DAOW model is adopted. The whole year sum of power deficiency is 1279 kW which is about 18% of the total electricity demand (7047 kW). This is an effective micro-siting decision because an economic COE value could be obtained by transferring the inefficient power production to multi-source energy integration from a holistic perspective. If we break down the COE value into each week, clearly the COE value varies depending on the wind conditions. The variation of weekly COE would be very significant and dynamic due to the wind uncertainty. On contrary, the COE value of other energy sources, such as fossil fuel and hydropower, is relatively stable. Our DAOW model facilitates different combinations of wind power and other forms of energy to reach an effective COE overall.

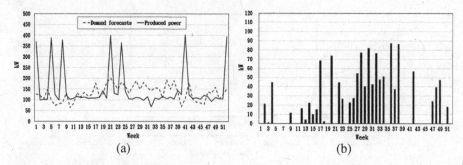

Fig. 3. (a) Comparison between the demand forecasts and the produced power. (b) Amount of purchased power as the DAOW model is adopted.

5 Conclusions

In this paper, we have proposed a demand-aware micro-siting system to COE minimization. The system is a holistic integration of long-term electricity demand forecasting and optimal micro-siting. The CSA is customized to develop an optimization model for both problems. A case study in central Taiwan area is conducted to validate the feasibility of the proposed system.

Acknowledgments. This research is partially supported by Ministry of Science and Technology of ROC, under Grant MOST 105-2410-H-260-018-MY2.

References

1. Shafiee, S., Topal, E.: When will fossil fuel reserves be diminished. Energy Policy **37**, 181–189 (2009)
2. Ceylan, H., Ozturk, H.K.: Estimating energy demand of Turkey based on economic indicators using genetic algorithm approach. Energy Convers. Manage. **45**, 2525–2537 (2004)
3. Unler, A.: Improvement of energy demand forecasts using swarm intelligence: the case of Turkey with projections to 2025. Energy Policy **36**, 1937–1944 (2008)
4. Yu, S., Zhu, K., Zhang, X.: Energy demand projection of China using a path-coefficient analysis and PSO-GA approach. Energy Convers. Manage. **58**, 142–153 (2012)
5. Askarzadeh, A.: Comparison of particle swarm optimization and other metaheuristics on electricity demand estimation: a case study of Iran. Energy **72**, 484–491 (2014)
6. Mosetti, G., Poloni, C., Diviacco, B.: Optimization of wind turbine positioning in large wind farms by means of a genetic algorithm. Wind Eng. Ind. Aerodyn. **51**, 105–116 (1994)
7. Yin, P.Y., Wang, T.Y.: GRASP-VNS algorithm for optimal wind turbine placement in wind farms. Renew. Energy **48**, 489–498 (2012)
8. Limanond, T., Jomnonkwao, S., Srikaew, A.: Projection of future transport energy demand of Thailand. Energy Policy **39**, 2754–2763 (2011)

9. Ekonomou, L.: Greek long-term energy consumption prediction using artificial neural networks. Energy **35**, 512–517 (2010)
10. Yin, P.Y., Glover, F., Laguna, M., Zhu, J.X.: Cyber swarm algorithms – improving particle swarm optimization using adaptive memory strategies. Eur. J. Oper. Res. **201**, 377–389 (2010)
11. Glover, F., Laguna, M., Marti, R.: Fundamentals of scatter search and path relinking. Contr. Cybern. **29**, 653–684 (2000)

A Hyper-Heuristic Method for UAV Search Planning

Yue Wang, Min-Xia Zhang, and Yu-Jun Zheng[✉]

College of Computer Science and Technology, Zhejiang University of Technology,
Hangzhou 310023, China
wyzidu.2015@outlook.com, zmx@zjut.edu.cn, yujun.zheng@computer.org

Abstract. Motivated by the wide use of unmanned aerial vehicles (UAV) in search-and-rescue operations, we consider a problem of planning the search sequence and search modes of UAV, the aim of which is to maximize the probability of finding the target in a complex environment with probabilistic belief of target location. We design five meta-heuristic algorithm for solving the complex problem, but find that none of them can always obtain satisfactory solutions on a variety of instances. To overcome this obstacle, we integrate these meta-heuristics into a hyper-heuristic framework, which adaptively manage the low-level heuristics (LLH) by using feedback of their real-time performance in problem solving, and thus can find the most suitable LLH or their combination that can outperform any single LLH on each given instance. Experiments show that the overall performance of the hyper-heuristic is significantly better than any individual heuristic on the test instances.

Keywords: Hyper-heuristic · Meta-heuristic · Search-and-rescue · Unmanned aerial vehicles (UAV)

1 Introduction

Unmanned aerial vehicles (UAV) have now been widely adopted by the emergency response community by providing critical extensions of responders, especially in environments that are inaccessible or dangerous to humans. The potential of using UAV for search-and-rescue operations has also been supported by a number of studies in the areas of target recognition, task scheduling, and path planning [3,6,10,14].

Primitive methods for UAV search planning [1,4,5,8] typically discretize a search area into cells and assume that the UAV can measure the state of an entire cell at a constant altitude. However, such an assumption is often inappropriate for today's UAV that keep evolving to lightweight and agile [17]. Considering that a UAV equipped with downward-looking cameras can change its flying altitude, the higher the altitude, the larger the area it covers (as shown in Fig. 1) and the less search time it needs, but the precision of observation and the consequent probability of finding the target decrease. Since the search time is often very

© Springer International Publishing AG 2017
Y. Tan et al. (Eds.): ICSI 2017, Part II, LNCS 10386, pp. 454–464, 2017.
DOI: 10.1007/978-3-319-61833-3_48

limited, it is reasonable to perform more refined search on the cells with higher probabilities of target presentation. However, when the total search range (and the number of cells) is large, determining the UAV's search path as well as the altitude at each cell is a highly complex combinatorial optimization problem.

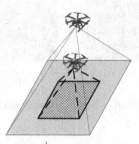

Fig. 1. Illustration of the search of UAV at different altitudes.

In this paper we formulate a UAV search problem which is to maximize the total weighted probability of finding a target in a complex environment with probabilistic belief of target location. First we use some popular meta-heuristics, including genetic algorithm (GA) [7], particle swarm optimization (PSO) [9], biogeography-based optimization (BBO) [13], fireworks algorithm (FWA) [16], and water wave optimization (WWO) [21], to solve the problem. However, experiments on various test instances show that none of them can always obtain satisfactory solutions. To overcome the obstacle, we integrate the meta-heuristics into a hyper-heuristic [2] framework, which performs a stochastic search on the low-level heuristics (LLH) by using feedback of their performance during problem-solving and yields a high performance on the whole test set.

In the remainder of the paper, we formulate the problem in Sect. 2, propose the five meta-heuristic methods in Sect. 3, and present the hyper-heuristic method in Sect. 4. Section 5 depicts the experiments, and Sect. 6 concludes.

2 Problem Description

We consider the problem of searching a target (e.g., a victim) in a search area which is discretized into m cells, as illustrated in Fig. 2. Each cell i is assigned with a probability $p(i)$ that the target is present in it ($1 \leq i \leq m$). That is, we assume that we have a prior probability distribution function that describes the initial belief of the target location. This can be a coarse estimate of the target location depending on the victim's information such as the time and place last seen and environmental features such as rivers or roads; if no prior information is known, we assume a uniform distribution [18].

Ideally, a UAV can fly at any altitude. Here, for simplicity, in our problem the UAV is limited to conduct search at one of K different heights, defined as $h(1), h(2), ..., h(K)$, above the surface of each cell. If the UAV uses pattern k

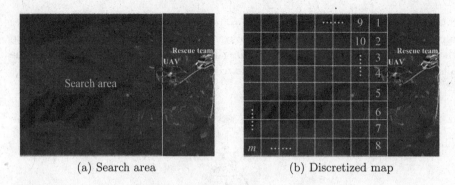

(a) Search area   (b) Discretized map

Fig. 2. An example of the problem, where the search area is divided into a set of cells.

to search cell i, then the expected search time is $t_s(i, k)$, and the probability of finding the target is $p_s(i, k)$ if the target is located at the cell $(1 \leq i \leq m; 1 \leq k \leq K)$. The values of $t_s(i, k)$ and $p_s(i, k)$ can also be estimated based on environmental features.

Let $l(i)$ be the (average) altitude of the terrain of cell i, if the search pattern on the cell is k, then the (average) altitude at which the UAV search the cell is $\alpha(i) = l(i) + h(k)$. Let d_{ij} denote the horizontal distance between cell i and cell j, the time needed for the UAV to fly from cell i at altitude $\alpha(i)$ to cell j at altitude $\alpha(j)$ is estimated as:

$$
t_f(i, j, \alpha(i), \alpha(j)) = \begin{cases} \frac{d^2(i,j,\alpha(i),\alpha(j))F_{\max}}{(d(i,j,\alpha(i),\alpha(j))F_{\max}+d_{ij}mg)v_{\max}}, & \text{if } \alpha(i) \geq \alpha(j) \\[2ex] \frac{d^2(i,j,\alpha(i),\alpha(j))F_{\max}+d(i,j,\alpha(i),\alpha(j))d_{ij}mg}{d_{ij}F_{\max}v_{\max}}, & \text{else} \end{cases}
$$

(1)

where $d(i, j, \alpha(i), \alpha(j)) = \sqrt{d_{ij}^2 + (\alpha(i) - \alpha(j))^2}$, $F_{\max}$ is the maximum thrust force of the UAV, $v_{\max}$ is the maximum horizontal speed of the UAV, m is the weight of the UAV, and g is the gravitational acceleration. As a special case, let d_{0i} denote the horizontal distance between the starting position of the UAV and cell i and $l(0)$ be the altitude of the starting position, the time needed for the UAV flies from the starting position to cell i at altitude $\alpha(i)$ is:

$$
t_f(0, i, l(0), \alpha(i)) = \begin{cases} \frac{d^2(0,i,l(0),\alpha(i))F_{\max}}{(d(0,i,l(0),\alpha(i))F_{\max}+d_{0i}mg)v_{\max}}, & \text{if } l(0) \geq \alpha(i) \\[2ex] \frac{d^2(0,i,l(0),\alpha(i))F_{\max}+d(0,i,l(0),\alpha(i))d_{0i}mg}{d_{0i}F_{\max}v_{\max}}, & \text{else} \end{cases}
$$

The problem is to decide the sequence $\mathbf{x} = (x_1, x_2, ..., x_m)$ of the cells searched by the UAV and the corresponding search patterns $\mathbf{y} = (y_1, y_2, ..., y_m)$, such that the sum of the time weighted probability of finding the target is maximized:

$$
\max f(\mathbf{x}, \mathbf{y}) = \sum_{i=1}^{m} \frac{t(x_m) - t(x_i) + 1}{t(x_m) + 1} p_s(x_i, y_i)p(x_i)
$$

(2)

where $(x_1, x_2, ..., x_m)$ is a permutation of $(1, 2, ..., m)$, $1 \leq y_i \leq K$, and $t(x_i)$ is the time that the UAV completes the search of cell x_i. The terms $(t(x_m) - t(x_i) + 1)/(t(x_m) + 1)$ and $p_s(x_i, y_i)$ in the right hand of Eq. (2) indicate that, the higher the cell's probability that the target is present in it, the earlier we expect the UAV to search it with a more refined pattern.

For the first cell x_1 we have:

$$t(x_1) = t_f(0, i, l(0), \alpha(x_1)) + t_s(x_1, y_1) \tag{3}$$

By fixing $x_0 = 0$, $\alpha_{x_0} = l(0)$, and $t(0) = 0$, the completion time of the search on cell x_i can be written as follows $(1 \leq i \leq m)$:

$$t(x_i) = t(x_{i-1}) + t_f(i - 1, i, \alpha(x_{i-1}), \alpha(x_i)) + t_s(x_i, y_i)$$
$$= \sum_{i'=1}^{i} \left(t_f(i' - 1, i', \alpha(x_{i'-1}), \alpha(x_{i'})) + t_s(x_{i'}, y_{i'}) \right) \tag{4}$$

Noted that although the dimension of $\mathbf{x}$ is m, in the real implementation of a solution, whenever the UAV finds the target in an intermediate cell x_i, the search is terminated and the remaining $(m - i)$ cells will not be further investigated.

A feasible solution to the problem also needs to satisfy the following two constraints, i.e., the total search time cannot exceed an upper time limit, and the total probability of finding the target cannot be below a lower limit

$$t(x_m) \leq \widehat{t} \tag{5}$$

$$\sum_{i=1}^{m} p(x_i)p_s(x_i, y_i) \geq \widehat{p} \tag{6}$$

From the formulation we can see that the problem is of highly complexity: Its subproblem for determining the search sequence $\mathbf{x}$ can be regarded as an extension of the traveling salesman problem (TSP) which is known to be *NP*-hard; moreover, our problem needs to determine a search pattern y_i for each x_i, and its objective function is much more computationally expensive than TSP. Thus it is necessary to design efficient heuristic methods for the problem.

3   Some Meta-Heuristic Based Methods for the Problem

First we use some popular meta-heuristics to solve the considered problem. Unless otherwise stated, all the algorithms use the penalty method to handle the constraints of the problem, i.e., the objective value of an infeasible solution will be subtracted by a large positive penalty value.

3.1   Genetic Algorithm

GA is one of the most famous meta-heuristics for combinatorial optimization problems including TSP. It maintains a set of solutions called chromosomes in a population, and uses selection, crossover and mutation operators to mimic the process of natural evolution to search the solution space.

The selection operator determines which solutions will have offspring in the population of the next generation. Here we use the standard roulette-wheel selection method, so that the selection probability of a solution $\mathbf{z} = (\mathbf{x}, \mathbf{y})$ is proportional to its fitness defined as follows:

$$fit(\mathbf{z}) = \frac{f(\mathbf{z}) - f_{\min} + \epsilon}{f_{\max} - f_{\min} + \epsilon} \tag{7}$$

where $f_{\max}$ and $f_{\min}$ are the maximum and minimum objective function values among the population, respectively, and ϵ is a small positive number to avoid division-by-zero.

After testing a number of permutation-based crossover operators, we choose the position-based crossover (PBX) [15] which first selects a subset of positions in the first parent, copies the components at these positions to the offspring at the same positions, and then fills the other positions with the remaining components in the same order as in the second parent. For our problem, when performing crossover on two solutions $(\mathbf{x_1}, \mathbf{y_1})$ and $(\mathbf{x_2}, \mathbf{y_2})$, we use PBX to produce an offspring $\mathbf{x}$ from $\mathbf{x_1}$ and $\mathbf{x_2}$, and make the search pattern of each cell in the offspring as the same as the corresponding cell in the parent, as shown in Fig. 3.

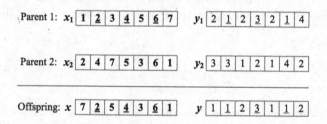

Fig. 3. Illustration of the PBX operation for the considered problem.

We define two mutation operators for our problem. The first mutates a solution $(\mathbf{x}, \mathbf{y})$ by randomly reversing a subsequence of $\mathbf{x}$ while keeping each cell's search pattern unchanged, and the second randomly chooses a small number of cells and changes their search patterns.

3.2 Particle Swarm Optimization

In the proposed PSO algorithm, the position of each particle, denoted by $\mathbf{z}'$, also consists of two vectors $= (\mathbf{x}', \mathbf{y})$, where $\mathbf{x}' = (x_1', x_2', ..., x_m')$ is a real-value vector that is mapped to a search sequence $\mathbf{x} = (x_1, x_2, ..., x_m)$ by setting each component x_i as the order of x_i' in $\mathbf{x}'$. Rather than the standard PSO [12], here we use a comprehensive learning strategy [11,20] where every particle can learn from a different exemplar *cbest* at each dimension i as follows:

$$v_i = wv_i + cr(cbest_i - z_i') \tag{8}$$

where w is the inertial weight, c is the acceleration constant, r is a random number uniformly distributed in $[0, 1]$), and *cbest* is determined as the better one between two randomly selected solutions other than $\mathbf{z}'$ from the population.

The position of the particle is then updated as follows (the components of the integer-value pattern vector are always rounded to the nearest integers):

$$x_i' = x_i' + v_i \tag{9}$$
$$y_i' = \min(\max(y_i' + v_i, 1), K) \tag{10}$$

3.3 Biogeography-Based Optimization

BBO assigns each solution $\mathbf{z}$ with an immigration rate $\lambda(\mathbf{z})$ and an emigration rate $\mu(\mathbf{z})$ as follows ($f_{\max}$, $f_{\min}$, and ϵ have the same meanings as Eq. (7)) [13]:

$$\lambda(\mathbf{z}) = \frac{f_{\max} - f(\mathbf{z}) + \epsilon}{f_{\max} - f_{\min} + \epsilon} \tag{11}$$

$$\mu(\mathbf{z}) = \frac{f(\mathbf{z}) - f_{\min} + \epsilon}{f_{\max} - f_{\min} + \epsilon} \tag{12}$$

At each iteration, each cell x_i and its search pattern y_i of each solution $\mathbf{z} = (\mathbf{x}, \mathbf{y})$ has a probability of $\lambda(\mathbf{z})$ of being "immigrated" and, if so, an emigrating solution $\mathbf{z}' = (\mathbf{x}', \mathbf{y}')$ is selected with a probability proportional to $\mu(\mathbf{z}')$, and the migration operation is performed by moving (x_i, y_i) to the corresponding position of (x_i', y_i') in $\mathbf{z}$, and then filling the empty position of $\mathbf{z}$ with (x_i', y_i').

The BBO also uses the same two mutation operators as GA in Sect. 3.1.

3.4 Fireworks Algorithm

FWA [16] is a relatively new meta-heuristic that mimics the explosion process of fireworks, such that each solution (firework) $\mathbf{z}$ can generate multiple offspring (sparks). The number s of sparks and the explosion amplitude A are calculated as follows (where P is the population, and M and $\widehat{A}$ are two control parameters):

$$s = M \cdot \frac{f(\mathbf{z}) - f_{\min} + \epsilon}{\sum_{\mathbf{z}' \in P}(f_{\max} - f(\mathbf{z}')) + \epsilon} \tag{13}$$

$$A = \widehat{A} \cdot \frac{f_{\max} - f(\mathbf{z}) + \epsilon}{\sum_{\mathbf{z}' \in P}(f(\mathbf{z}') - f_{\min}) + \epsilon} \tag{14}$$

For the considered problem, when performing an explosion with an amplitude A, we generate a random integer k uniformly distributed in $[1, A]$, and then swap two randomly selected adjacent cells together with their search patterns for k times, such that a larger A will result sparks more deviated from the firework.

3.5 Water Wave Optimization

WWO [21] is also a novel meta-heuristic, where each solution (wave) $\mathbf{z}$ is assigned with a wavelength λ representing the search range of the solution. For the considered problem, we propose a new wavelength model as follows:

$$\lambda(\mathbf{z}) = \frac{\left(\sum_{z' \in P} f(\mathbf{z}')\right) - f(\mathbf{z})}{\sum_{z' \in P} f(\mathbf{z}')} \tag{15}$$

When propagating a wave $\mathbf{z} = (\mathbf{x}, \mathbf{y})$, for each $i \in [1, m]$, we have a probability of $\lambda(\mathbf{z})$ of reversing the subsequence $(x_i, x_{i+1}, ...x_{i+l})$ and the corresponding part of the search pattern vector $(y_i, y_{i+1}, ...y_{i+l})$, where l is a random integer uniformly distributed in $[1, m - i]$. Thus, the larger the wavelength, the higher the degree to which the solution will be changed.

At each iteration, the best wave in the population will be broken into k_{max} solitary waves (where k_{max} is a control parameter), each of which is obtained by randomly setting a small number of y_i to $(y_i + 1)$ or $(y_i - 1)$. Moreover, if a wave cannot be propagated to a better position for h_{max}, it will be replaced by a new randomly generated solution (where h_{max} is another control parameter).

4 A Hyper-Heuristic Method for the Problem

Using the above different meta-heuristics as the low-level heuristics (LLH), we propose a hyper-heuristic method (HHM) which maintains a pool of LLH operators that compete with each other to become the active operator to generate new solutions. Currently our HHM uses eight LLH operators:

- The genetic OBX operator.
- The genetic mutation operator on search sequences.
- The genetic mutation operator on search patterns.
- The PSO learning operator.
- The BBO migration operator.
- The FWA explosion operator.
- The WWO migration operator.
- The WWO breaking operator.

Each operator is assigned with an integer-valued suitable index $\rho(op)$ which represents the probability of applying the operator in the HHM. All the indices are initially set to 100 and updated as follows:

- If the operator generates a better solution than its parent(s), set $\rho(op) = \rho(op) + 1$;
- If the operator generates new best solution of the population, set $\rho(op) = \rho(op) + 2$;
- Otherwise, keep $\rho(op)$ unchanged.

Algorithm 1 presents the pseudo-code of the HHM.

Algorithm 1. The hyper-heuristic for the UAV search problem.

1   Randomly initialize a population P of solutions;
2   Set $\rho(op) = 100$ for all the operators;
3   **while** stop criterion is not satisfied **do**
4   Record the best and worst solutions, and update the LLH parameters;
5   Create an empty population P' for the next generation;
6   **for each** solution $\mathbf{z} \in P$ **do**
7   Select an op with a probability proportional to $\rho(op)$;
8   **if** op is FWA explosion or WWO breaking **then**
9   Apply op on $\mathbf{z}$ to generate a set of offsprings and select the best as $\mathbf{z}'$;
10   **else** Apply op on $\mathbf{z}$ to generate an offspring $\mathbf{z}'$;
11   **end if.**
12   **if** $f(\mathbf{z}') > f(\mathbf{z})$ **then**
13   Add $\mathbf{z}'$ to P';
14           $\rho(op) \leftarrow \rho(op) + 1$;
15   **if** $f(\mathbf{z}') > f_{\max}$ **then** $\rho(op) \leftarrow \rho(op) + 2$; **end if.**
16   **else** Add $\mathbf{z}$ to P';
17   **end if.**
18   **end for.**
19     $P \leftarrow P'$;
20 **end while.**
21 **return** the best solution found so far.

5   Computational Experiment

We use a test set of 15 problem instances which are all constructed based on real-world terrains, and compare our HHM with the five meta-heuristic algorithms described in Sect. 3. For HHM, the population size $|P|$ is set to 150. For GA we set $|P| = 50$, and set the mutation rate of search sequences to 0.06, while tune the mutation rate of search patterns to linearly increase from 0.04 to 0.25 with iteration. For PSO we set $|P| = 30$, $c = 1.98$, and w linearly decrease from 0.9 to 0.4. For BBO we set $|P| = 50$ and set the mutation rates as GA. For FWA we set $|P| = 20$, $M = 30$, and $\hat{A} = 40$. For WWO we set $|P| = 40$, $k_{\max} = 6$, and $h_{\max} = 6$. To ensure a fair comparison, for all the algorithms the termination criterion is that the number of objective function evaluations reaches 500 m. The experiment is conducted on a computer with Intel i7-6700 CPU, 8 GB memory, and the Windows 7 system. Every algorithm is run for 30 times on each instance.

Table 1 shows the experimental results including the maximum, minimum, median and standard deviation of the resulting objective function values among the 30 runs, where the best maximum and median values among the six algorithms are shown in boldface. The instances #1–#15 are ordered roughly by their difficulties. On the first two simple instances, all the six algorithms achieve the exact same solution (no any deviation among the 30 runs). Among the five meta-heuristic algorithms, none of them can overwhelm all the others on all the remaining instances. In general, BBO exhibits good performance on relatively small size instances, as it can achieve the best maximum objective values on

Table 1. Comparative results of HHM and the five meta-heuristics on test instances.

#	Metrics	GA	PSO	BBO	FWA	WWO	HHM
1	Max	**0.434**	**0.434**	**0.434**	**0.434**	**0.434**	**0.434**
	Min	0.434	0.434	0.434	0.434	0.434	0.434
	Median	**0.434**	**0.434**	**0.434**	**0.434**	**0.434**	**0.434**
	Std	0.000	0.000	0.000	0.000	0.000	0.000
2	Max	**0.398**	**0.398**	**0.398**	**0.398**	**0.398**	**0.398**
	Min	0.398	0.398	0.398	0.398	0.398	0.398
	Median	**0.398**	**0.398**	**0.398**	**0.398**	**0.398**	**0.398**
	Std	0.000	0.000	0.000	0.000	0.000	0.000
3	Max	**0.410**	0.399	**0.410**	**0.410**	**0.410**	**0.410**
	Min	0.355	0.301	0.343	0.377	0.410	0.410
	Median	0.383	0.373	0.399	**0.410**	**0.410**	**0.410**
	Std	0.030	0.034	0.022	0.007	0.000	0.000
4	Max	**0.353**	**0.353**	**0.353**	**0.353**	**0.353**	**0.353**
	Min	0.282	0.308	0.288	0.289	0.314	0.328
	Median	0.316	0.337	0.314	0.324	0.328	**0.339**
	Std	0.034	0.026	0.031	0.032	0.023	0.014
5	Max	**0.320**	0.296	**0.320**	0.278	0.292	**0.320**
	Min	0.268	0.239	0.268	0.223	0.244	0.272
	Median	0.298	0.273	**0.301**	0.262	0.268	**0.301**
	Std	0.023	0.032	0.021	0.035	0.023	0.027
6	Max	0.308	0.275	**0.318**	0.279	0.297	**0.318**
	Min	0.224	0.209	0.239	0.193	0.209	0.258
	Median	0.279	0.239	0.275	0.251	0.261	**0.299**
	Std	0.045	0.037	0.043	0.052	0.037	0.032
7	Max	0.299	**0.330**	**0.330**	0.273	0.307	**0.330**
	Min	0.223	0.201	0.238	0.147	0.216	0.216
	Median	**0.275**	0.266	**0.275**	0.255	0.263	**0.275**
	Std	0.036	0.057	0.039	0.060	0.038	0.051
8	Max	0.232	0.254	0.273	**0.289**	0.281	**0.289**
	Min	0.133	0.164	0.133	0.146	0.147	0.176
	Median	0.199	0.218	0.190	0.195	0.218	**0.227**
	Std	0.044	0.041	0.068	0.062	0.051	0.044
9	Max	0.177	0.188	0.175	**0.217**	0.199	**0.217**
	Min	0.128	0.151	0.123	0.145	0.147	0.158
	Median	0.158	0.172	0.143	**0.186**	0.180	**0.186**
	Std	0.029	0.013	0.029	0.032	0.025	0.025
10	Max	0.237	**0.241**	0.219	0.205	0.227	**0.241**
	Min	0.166	0.152	0.133	0.127	0.166	0.159
	Median	**0.224**	0.202	0.180	0.180	0.202	**0.224**
	Std	0.036	0.029	0.046	0.034	0.031	0.036
11	Max	**0.199**	0.190	0.195	0.188	**0.199**	**0.199**
	Min	0.159	0.153	0.153	0.127	0.153	0.153
	Median	**0.179**	0.174	0.172	0.172	0.177	**0.179**
	Std	0.020	0.020	0.019	0.020	0.020	0.019
12	Max	0.189	0.196	0.179	0.189	0.211	**0.218**
	Min	0.152	0.096	0.117	0.109	0.145	0.152
	Median	0.178	0.174	0.148	0.153	**0.196**	**0.196**
	Std	0.015	0.023	0.027	0.030	0.015	0.016
13	Max	0.157	**0.182**	0.140	0.149	**0.182**	**0.182**
	Min	0.109	0.150	0.113	0.099	0.130	0.150
	Median	0.133	0.166	0.129	0.125	0.165	**0.175**
	Std	0.022	0.018	0.014	0.021	0.019	0.012
14	Max	0.080	0.103	0.071	0.080	0.091	**0.131**
	Min	0.053	0.058	0.053	0.048	0.058	0.084
	Median	0.071	0.072	0.065	0.068	0.079	**0.109**
	Std	0.013	0.021	0.020	0.015	0.013	0.019
15	Max	0.062	0.073	0.056	0.055	0.080	**0.111**
	Min	0.037	0.042	0.030	0.028	0.052	0.075
	Median	0.056	0.062	0.048	0.042	0.067	**0.090**
	Std	0.008	0.013	0.009	0.012	0.013	0.012

#3–#7 and the best median values on #5 and #7; However, with the increase of instance difficulty, the performance of BBO decreases dramatically, as it always achieves the worst or the second worst median values on the remaining instances. FWA achieves the best results among the five meta-heuristics on medium-size instances #8 and #9. On relatively large instances, GA and WWO exhibit more desirable performance. The performance of PSO, however, varies greatly among the different instances. In summary, the search mechanism of each meta-heuristic can be suitable for some instances but unsuitable for others, and thus it is hard to select the most suitable meta-heuristic for a variety of instances, particularly under the emergency conditions of search-and-rescue [22].

The proposed HHM, by using the operators of the five meta-heuristics as LLH, exhibits the best performance (in terms of both the maximum values and the median values) on all the instances. This is because, by adaptively managing the LLH operators according to their real-time performance, HHM can not only decide which LLH is most suitable for each stage of the search on each given instance, and which LLH is most, but also determine how to combine different LLH to maximize the search performance if such a combination could outperform any single LLH. As we can see, on most small and medium size instances, HHM and some meta-heuristics simultaneously achieve the best results, which indicates that HHM successfully find the most suitable LLH on the instances. On the last (and the largest) two instances, HHM uniquely achieves the best results by effectively coordinating multiple LLH such that they can exchange information with each other to promote the search. As a result, the performance of HHM is significantly better than any individual meta-heuristic on the whole test set.

6   Conclusion

The paper proposes five meta-heuristic algorithms and a hyper-heuristic that integrates the meta-heuristics for a UAV search planning problem. Experimental results show that, by adaptively managing the LLH based on their real-time performance during problem-solving, the hyper-heuristic outperforms any individual meta-heuristic on a variety of test instances. Our future work will consider the cooperation of one or more UAV swarms [19] to the search problem.

Acknowledgements. This work is supported by National Natural Science Foundation (Grant No. 61473263) and Zhejiang Provincial Natural Science Foundation (Grant No. LY14F030011) of China.

References

1. Bourgault, F., Göktogan, A., Furukawa, T., Durrant-Whyte, H.F.: Coordinated search for a lost target in a bayesian world. Adv. Robot. **18**(10), 979–1000 (2004)
2. Burke, E., Hart, E., Kendall, G., Newall, J., Ross, P., Schulenburg, S.: Hyper-heuristics: an emerging direction in modern research technolology. In: Glover, F., Kochenberger, G. (eds.) Handbook of Metaheuristics, pp. 457–474. Kluwer, Dordrecht (2003)

3. Doherty, P., Rudol, P.: A UAV search and rescue scenario with human body detection and geolocalization. In: Orgun, M.A., Thornton, J. (eds.) AI 2007. LNCS (LNAI), vol. 4830, pp. 1–13. Springer, Heidelberg (2007). doi:10.1007/978-3-540-76928-6_1

4. Elfes, A.: Using occupancy grids for mobile robot perception and navigation. Computer **22**(6), 46–57 (1989)

5. Gemeinder, M., Gerke, M.: GA-based path planning for mobile robot systems employing an active search algorithm. Appl. Soft Comput. **3**(2), 149–158 (2003)

6. Goodrich, M.A., Morse, B.S., Gerhardt, D., Cooper, J.L., Quigley, M., Adams, J.A., Humphrey, C.: Supporting wilderness search and rescue using a camera-equipped mini UAV. J. Field Robot. **25**(1–2), 89–110 (2008)

7. Holland, J.H.: Adaptation in Natural and Artificial Systems: An Introductory Analysis with Applications to Biology, Control and Artificial Intelligence. MIT Press, Cambridge (1975)

8. Jin, Y., Liao, Y., Minai, A.A., Polycarpou, M.M.: Balancing search and target response in cooperative unmanned aerial vehicle (UAV) teams. IEEE Trans. Syst. Man Cybern. Part B **36**(3), 571–587 (2005)

9. Kennedy, J., Eberhart, R.: Particle swarm optimization. In: IEEE International Conference on Neural Networks, vol. 4, pp. 1942–1948 (1995)

10. Li, C., Duan, H.: Target detection approach for UAVs via improved pigeon-inspired optimization and edge potential function. Aeros. Sci. Techn. **39**, 352–360 (2014)

11. Liang, J.J., Qin, A.K., Suganthan, P., Baskar, S.: Comprehensive learning particle swarm optimizer for global optimization of multimodal functions. IEEE Trans. Evol. Comput. **10**(3), 281–295 (2006)

12. Shi, Y., Eberhart, R.C.: A modified particle swarm optimizer. In: IEEE Congress on Evolutionary Computation, pp. 69–73 (1998)

13. Simon, D.: Biogeography-based optimization. IEEE Trans. Evol. Comput. **12**(6), 702–713 (2008)

14. Symington, A., Waharte, S., Julier, S., Trigoni, N.: Probabilistic target detection by camera-equipped UAVs. In: 2010 IEEE International Conference on Robotics and Automation, pp. 4076–4081 (2010)

15. Syswerda, G.: Schedule optimization using genetic algorithms. In: Davis, L. (ed.) Handbook of Genetic Algorithms. Van Nostrand Reinhold, New York (1991)

16. Tan, Y., Zhu, Y.: Fireworks algorithm for optimization. In: Tan, Y., Shi, Y., Tan, K.C. (eds.) ICSI 2010. LNCS, vol. 6145, pp. 355–364. Springer, Heidelberg (2010). doi:10.1007/978-3-642-13495-1_44

17. Waharte, S., Symington, A., Trigoni, N.: Probabilistic search with agile UAVs. In: 2010 IEEE International Conference on Robotics and Automation, pp. 2840–2845 (2010)

18. Waharte, S., Trigoni, N.: Supporting search and rescue operations with UAVs. In: 2010 International Conference on Emerging Security Technologies, pp. 142–147 (2010)

19. Zheng, Y.J., Chen, S.Y.: Cooperative particle swarm optimization for multiobjective transportation planning. Appl. Intell. **39**, 202–216 (2013)

20. Zheng, Y.J., Ling, H.F., Xue, J.Y., Chen, S.Y.: Population classification in fire evacuation: a multiobjective particle swarm optimization approach. IEEE Trans. Evol. Comput. **18**(1), 70–81 (2014)

21. Zheng, Y.J.: Water wave optimization: a new nature-inspired metaheuristic. Comput. Oper. Res. **55**(1), 1–11 (2015)

22. Zheng, Y.J., Chen, S.Y., Ling, H.F.: Evolutionary optimization for disaster relief operations: a survey. Appl. Soft Comput. **27**, 553–566 (2015)

An Efficient MVMO-SH Method for Optimal Capacitor Allocation in Electric Power Distribution Systems

Hiroyuki Mori[(⊠)] and Hiromitsu Ikegami

Department of Network Design, Meiji University, Nakano, Japan
hmori@meiji.ac.jp

Abstract. This paper proposes an efficient method for optimal capacitor allocation in electric power distribution systems. The proposed method is based on a new metaheuristic method, MVMO (Mean Variance Mapping Optimization) that makes use of information on the mean and variance of archives through the mapping function. The optimal capacitor allocation is aimed at minimizing the active power network losses under some constraints with capacitor banks. The mathematical formulation results in a combinatorial optimization problem. In this paper, MVMO-SH is proposed to evaluate better solutions. It introduces swarm intelligence into MVMO that finds out better solutions with the mean and variance obtained from the archives. It improves the performance of MVMO with the multi-point search and multi-parent crossover. The proposed method is successfully applied to the IEEE 69-bus and 119-bus electric power distribution systems.

Keywords: Capacitor allocation · Network loss minimization · Meta-heuristics · MVMO · MVMO-SH

1 Introduction

This paper proposes a metaheuristic method for electric power distribution network capacitor allocation. The proposed method is based on MVMO-SH that makes use of the mean and variance of agents in a swarm. The objective of distribution capacitor allocation is to minimize the network loss with capacitor banks of discrete variables [1–9]. As a result, the capacitor allocation problem results in one of combinatorial optimization problems. The conventional methods for the capacitor allocation may be classified into the following: (i) Mathematical Programming [1–5], and (ii) Metaheuristics [6–9]. Method (i) has a drawback to require a lot of computational time or evaluate low quality solutions due to the approximation of continuous variables. On the other hand, Method (ii) is defined as an optimization method that repeatedly uses simple rules and heuristics to evaluate highly approximate solutions to a global optimum within given time. Thus, Method (b) is the mainstream in the distribution system capacitor allocation. As the typical metaheuristics, the following methods have been applied to the capacitor allocation: Simulated Annealing (SA) [6], Genetic Algorithm (GA) [7], Tabu Search (TS) [8], Particle Swarm Optimization (PSO) [9], *etc.*

© Springer International Publishing AG 2017
Y. Tan et al. (Eds.): ICSI 2017, Part II, LNCS 10386, pp. 465–474, 2017.
DOI: 10.1007/978-3-319-61833-3_49

In this paper, an efficient MVMO method is proposed to deal with the optimal capacitor allocation. MVMO developed by Erlich *et al.* is a new metaheuristic method for evaluating solutions with the archives that memorize good solutions found in the past, their mean values and variance, and mapping functions [10]. The studies on the applications of MVMO to electric power transmission systems have been conducted [11–13]. In 2011, Erlich *et al.* applied MVMO to the optimal dispatch of reactive power sources in wind farms [11]. MVMO showed good results and the robustness of the solutions in a sense that the obtained solutions were not affected by the initial solutions. Also Worawat *et al.* solved the optimal reactive power dispatch by MVMO in the IEEE 57-bus and118-bus electric power transmission systems [12]. The convergence characteristics of MVMO were fast in both systems, and demonstrated better results especially in the 118-bus transmission system. As the application to the electric power distribution system, Rueda *et al.* presented an MVMO method for the distribution network reconfiguration problem [13]. In the 33-bus distribution system, it indicated that MVMO was superior in terms of solution accuracy and computational time. Although the applications of MVMO to electric power transmission systems have been conducted, there are a few examples applied to distribution power systems. From the standpoint, this paper proposes an optimal capacitor allocation method with MVMO-SH proposed by Rueda and Erlich [14] that applied swarm intelligence to MVMO.

2   Mathematical Formulation

This section describes the mathematical formulation of electric power distribution system capacitor allocation. It may be expressed as one of combinatorial optimization problems with some constraints. The objective is to minimize the distribution network losses with capacitor banks of discrete variables. Also, there exist some constraints on the power flow equation, the upper and lower bounds of nodal voltage magnitudes and active line flows. Therefore, the mathematical formulation may be written as

Cost function :
$$f = P_{loss} \rightarrow \min \tag{1}$$

Constraints :
$$y = k(z), \tag{2}$$

$$V_i^m \leq V_i \leq V_i^M, \tag{3}$$

$$p_{ij} \leq \left| p_{ij}^M \right|. \tag{4}$$

where f is the cost function, P_{loss} is the active power loss, y is the nodal specified values of the power flow equation, z is the power flow solution, V_i is the voltage magnitude at node I, $V_i^m(V_i^M)$ is the lower (upper) bound of V_i, p_{ij} is the active line flow between nodes i and j, p_{ij}^M is the upper bound of p_{ij} and $|\cdot|$ is the absolute value of $\cdot$.

Equation (1) shows the cost function to be minimized which means the minimization of the active power network loss in the distribution system. Equation (2) gives the equality constraint on the power flow equation that guarantees the existence power flow of the solution for the conditions of changed capacitor banks. Equation (3) indicates that the nodal voltage magnitude of the obtained solution is maintained within the specified upper and lower bounds. Equation(4) means that active line flows are set up not to exceed the thermal limitation of distribution line. To solve (1) – (4), the penalty function method is introduced to minimize the cost function.

$$g = f + \sum_{k=1}^{3} \alpha_k c_k \rightarrow \min. \tag{5}$$

where g is the new cost function, c_k is the penalty of constraint k and α_k is the weight of constraint k.

3 MVMO-SH

3.1 Algorithm of MVMO

In this paragraph, MVMO developed by Erlich et al. [10] is explained. It makes use of the mean and variance, construct the mapping function and evaluate better solutions. MVMO has the following features:

1. All the variables are normalized to the variables of interval [0, 1] expect cases of cost function calculation and local search.
2. Archives play a key role to keep better solution obtained by the past search and determines the search direction.
3. The mapping function has a function of mutation that makes solution candidates more diverse.
4. MVMO has a rule that one descendant.

In item 1 above, the created solution candidates are not allowed to exceed the interval of [0, 1]. Item 2 implies that a group of better solutions are maintained and the mean for the solutions are used to create the descendant and find out better solutions in the neighborhood. In item 3, the mapping function seems similar to the Gaussian one, but is peculiar to MVMO. Item 4 means that the number of iteration is equal to the number of the cost function calculation, i.e., one iteration requires one cost function calculation.

Next, consider the updating formula of MVMO. It creates the descendant by carrying out the crossover operator between the parent corresponding to the mean in the archives and mutated solution candidates by the mapping function. The updating formula may be written as

$$x_i = h_x + (1 - h_1 + h_0) \cdot x_i^* - h_0, \tag{6}$$

$$h(\bar{x}_i, s_1, s_2, x) = \bar{x}_i \cdot (1 - e^{-x \cdot s_{i1}}) + (1 - \bar{x}_i) \cdot e^{-(1-x) \cdot s_{i2}}, \tag{7}$$

$$h_x = h(x = x_i^*), h_0 = h(x = 0), h_1 = h(x = 1). \tag{8}$$

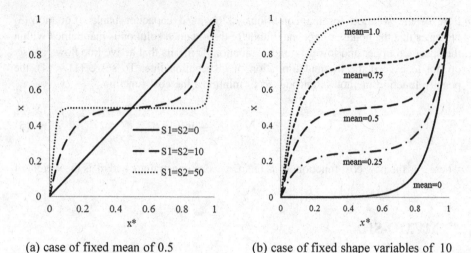

(a) case of fixed mean of 0.5   (b) case of fixed shape variables of 10

Fig. 1. The effects of s_i and mean on the mapping function [10]

where, x_i is variable i, x_i^* is the uniform random variable of [0, 1], $h(\cdot)$ is the function h, $\bar{x}_i$ is the mean of variable x_i and s_1 and s_2 are the shape variables.

Equation(6) consists of (7) and (8) with uniform random number x_i^* and determines the variables to be mutated. The mapping function is adjusted by the shape variables and mean of the archives so that it varies from 0 to 1. The variance obtained from the archives determines the shape variables. They become lager as the variance is smaller. Figure 1 shows the transition of the shape variables and mean for given parameters. Figure 1 (a) gives a case where the mean of 0.5 is fixed and shape variables s_1 and s_2 change in synchronization. It can be seen than the mapping function becomes flat as the shape variables are lager. Figure 1 (b) shows a case where the shape variables is fixed to 10 and the mean is changed. It can be observed that the mapping function becomes a flat function after a sharp rise as the mean approaches unity. It can be confirmed that MVMO finds out a solution by changing the mapping function with the mean and variance.

3.2 MVMO-SH

This section describes MVMO-SH proposed by Rueda and Erlich [14]. It is a new metaheuristic method that introduces swarm intelligence and multi-parent crossover strategies into MVMO. The use of MVMO-SH allows to improve the performance of MVMO and carry out global search. Now, let us define Np and ΔFE as the number of particles and cost function calculations, respectively. Each particle updates the archives after finishing the calculation of the cost function or the local search. At the initial stage of search, each particle independently creates and evolves the descendants. At this stage, the best solution for each particle (the top of the archive) is chosen as the parent for the next descendant. After going through independent stages, the whole particles are classified into two types, *i.e.*, good and bad ones by the best solution of the particle.

Information between particles corresponding to good and bad particles through the crossover and mutation are exchanged to create the descendants. The selection of different parents, crossover and mutation for each particle may be described as follows:

For simplicity, let us define GP and Np-Gp as the particles clarified into good and bad solutions, respectively. The parent selection is done for GP by the following formula:

$$x_k^{parent} = x_{GB}^{best}, \tag{9}$$

$$\bar{x}_k = \bar{x}_{RG}^{best}. \tag{10}$$

where, x_i^{parent} is parent of particle k belonging to GP, x_{GB}^{best} is the best solution in GP, $\bar{x}_i$ is mean of particle k and $\bar{x}_{RG}^{best}$ is the best solution selected from GP randomly.

The current best solution is chosen as the parent by (9), and the mean value is randomly selected from GP by (10). For the shape variable, its own information is used.

$$x_i^{parent} = x_{RG}^{best} + \beta\left(x_{GB}^{best} - x_{LG}^{best}\right), \tag{11}$$

$$\bar{x}_i = x_i^{parent}. \tag{12}$$

where, x_i^{parent} is particle i belonging to Np-GP, β is a certain random large number proportional to the number of iterations and x_{LG}^{best} is the best solution with the worst cost function in GP.

The multi-parent crossover is performed by (11), and the value of the parent is used as the mean value in Np-GP. As a result, Np-GP evaluates a search region different from the conventional one, and the search becomes more active. Also, regarding the shape variables, their own information is used. If MVMO-SH has one particle, the algorithm is the same as MVMO. In other words, MVMO may be regarded as MVMO-SH with one particle.

4 Proposed Method

This paper proposes an efficient MVMO-SH [14, 15] method for the optimal capacitor placement in electric power distribution systems. Particles using archives that store solutions with good cost function values create descendants by multi-parent crossover and mapping functions although MVMO does not consider local search due to speed-up of computational time [15]. The algorithm may be written as

Step 1:   Set the initial conditions
Step 2:   Calculate the cost function
Step 3:   Update the archive of particles if better solution are obtained and calculate the mean value, variance, and shape variable
Step 4:   Classify the particles into GP and Np-GP according to the cost function values

Step 5: If the particle belongs to the GP, select the parent by (9) and (10). Otherwise, select parents by (11) and (12)

Step 6: Create the next generation particles by carrying out the mutation of selected variables with the mapping function and combining solution candidates

Step 7: Stop if the termination condition is satisfied. Otherwise return to Step 3.

The proposed method evaluates solutions by memories of good solutions in the past and the mapping function and multi-parent crossover. Unlike MVMO, the particles independently evolve at the beginning of the search, and then evaluate better solutions by creating descendants with multi-parent crossover.

5 Simulation

5.1 Simulation Conditions

(i) The proposed method were applied to the IEEE 69-bus [16] and the IEEE 119-bus [17] distribution systems. Equation (13) shows the upper and lower limits of the nodal voltage magnitude at each bus, and the upper limit of the line current flow is individually set from data of the original system.

$$0.9 \leq V \leq 1.1[p.u.]. \tag{13}$$

In addition, it was assumed that four types of capacitor banks of 50, 100, 200, and 400 [kVar] could be arranged at all the buses.

(ii) To show the effectiveness of the proposed method of MVMO-SH, this paper made a comparison between the proposed and the conventional methods of PSO [18] and EPSO [19]. PSO is famous for swarm intelligence as a comparative method, and EPSO is an improved PSO that introduces Evolutionary Strategy into PSO. Also, to deal with discrete variables by these two methods above, this paper made use of the sigmoid function to make solution condition more diverse [20]. For convenience, the following methods are defined:

Method A: Discrete PSO (DPSO)
Method B: DEPSO (Discrete EPSO)
Method C: MVMO
Method D: MVMO-SH (proposed method)

Table 1 shows the parameters of each method that were determined by preliminary simulations. This paper used the Very Fast Decoupled method [21] for the flow calculation due to the computational efficiency.

(iii) The coding of 4 bits were prepared to express switching on/off for 4 kinds of shunt capacitor banks at each bus. The size of the capacitor bank combinations results in 1.21×10^{83} and 1.22×10^{142} for the 69-bus and 119-bus systems, respectively.

(iv) To investigate the influence of the initial solutions on the final one, a hundred of random solutions were prepared. All the calculations were performed on

Table 1. Parameters of each method

Methods	Parameters	Values	
		69-bus syst.	119-bus syst.
A	No. of agents	100	200
	w_1, w_2	1.5,1.5	1.8,1.5
	V_{max}	5	5
	No. of iterations	4000	15000
B	No. of agents	100	200
	Replication rate	3	2
	τ, τ'	0.01,0.01	0.01,0.02
	V_{max}	5	6
	No. of iterations	2500	7500
C	Archive size	5	5
	m_{ini}, m_{final}	20,1	20,1
	Δd_0	0.2	0.2
	fs_{ini}, fs_{final}	1.0,2.0	1.0,2.0
	dini	2	2
	No. of iterations	100000	200000
D	Archive size	5	5
	No. of agents	20	30
	GP_{ini}, GP_{final}	0.7,0.2	0.7,0.2
	m_{ini}, m_{final}	20,1	20,1
	Δd_0	0.2	0.2
	fs_{ini}, fs_{final}	1.0,2.0	1.0,2.0
	dini	1	1
	No. of iteration	100000	200000

Fujitsu PRIMEGY RC350 S7 (CPU: Xeon E5-2680 2.70 GHz × 2, Memory: 256 GB, SPECint2006: 55.8, SPECfp2006: 83.7).

5.2 Simulation Results

Table 2 shows simulation results in the 69-bus distribution system, where the original system has the network loss of 20.87 [kW]. In the table, the best, the worst and the average cost functions as well as the standard deviation (SD) of the cost functions were given for one hundred trials. Also the average CPU means average computational time for one hundred trials. Looking at the best cost function, Methods A-D succeeded in evaluating the same best function of 13.2782 [kW] that corresponded to about 64% of the original system network loss. Regarding the worst cost function, although Method B was a little worse than Method A, there were not significant difference of the results for Methods A-D. In particular, it was noteworthy that Method D outperformed other methods in terms of SD significantly. It succeeded in reducing SD of Method A by 83%, which implied that Method D was more robust to the initial values. It was

Table 2. Simulation results of each method

No. of buses	Methods	Cost functions [kW]				Ave. CPU [s]
		Best	Worst	Ave.	SD	
69	A	13.27819	13.30686	13.29058	0.0084 (1.00)	72.4 (1.00)
	B	13.27819	13.30756	13.28771	0.0080 (0.95)	161.47 (2.23)
	C	13.27819	13.29874	13.28104	0.0038 (0.45)	16.85 (0.23)
	D	13.27819	13.28457	13.27920	0.0014 (0.17)	16.86 (0.23)
119	A	775.7017	776.1856	775.9190	0.1041 (1.00)	808.22 (1.00)
	B	775.6690	776.2798	775.9495	0.1113 (1.07)	947.76 (1.17)
	C	775.5294	775.5889	775.5464	0.0120 (0.115)	47.98 (0.059)
	D	775.5288	775.5621	775.5401	0.0070 (0.067)	49.97 (0.062)

observed that Methods C and D of MVMO variants outperformed Methods A and B of PSO ones. That was because MVMO had advantage to create one descendant and reduce computational effort. Next, regarding the average CPU time, Method D had almost the same average CPU time as Method C because Method D of the multi-point search has the same number of the cost function calculations as Method C. Methods C and D was about 4.3-times and 9.7-times faster than Methods A and B, respectively. Figure 2 shows the convergence characteristics of each method in the 69-bus system. The vertical axis indicate the average of the network losses [kW] while the horizontal one gave the number of the cost function calculations. It can be seen that Methods C and D converge to the final solution efficiently. Therefore, the simulation results have shown that Method D provided better solution than others.

Table 2 shows simulation results in the 119-bus system with the original network losses of 1297.4 [kW]. The simulation results had the same trend as those in the 69-bus system. Namely, the performance of optimization was better in order of Methods D, C, B, and A. Method D evaluated the best network loss of 775.529 that corresponded to 60% of the original system network loss. The difference between results in the 69-bus and 119-bus systems was that Methods C and D outperformed Methods A and B significantly. For example it was a great surprise that Method D reduced SD of Method A by 99.33%. Although there were no significant difference of the best functions, the worst and the average cost functions were improved in Methods C and D in comparison with Methods A and B. Regarding average CPU time, Method D reduced average CPU time of Methods A and B by 93.3% and 94.72%, respectively. Method D was a little inferior to Method C because Method D had additional exchange calculations of multi-point search.

It can be observed that Methods C and D had excellent computational performance in comparison with the results in the 69-bus system. Thus, it can be concluded that the proposed method is much better than other methods in terms of solution quality and computational time.

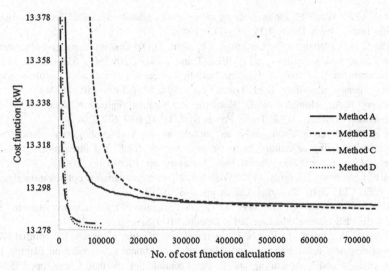

Fig. 2. Convergence characteristics of each method in 69-bus distribution system

6 Conclusion

This paper has proposed an efficient MVMO-SH based method for optimal capacitor placement in electric power distribution systems. MVMO-SH has excellent search process with the mapping function and multi-parent crossover in a way that swarm intelligence is introduced into MVMO. The proposed method was successfully applied to the IEEE 69-bus and 119-bus distribution systems. The simulation results have shown that the proposed method outperforms PSO, EPSO and MVMO in terms of solution quality and computational time significantly. Therefore, the proposed method allows distribution system operators to smooth distribution systems with the optimal capacitor placement efficiently.

References

1. Duran, H.: Optimum number, location and size of shunt capacitors in radial distribution feeders a dynamic programming approach. IEEE Trans. Power Apparatus Syst. **PAS-87**(9), 1769–1774 (1968)
2. Grainger, J.J., Lee, S.H.: Optimum size and location of shunt capacitors for reduction of losses on distribution feeders. IEEE Trans. Power Apparatus Syst. **PAS-100**(3), 1105–1118 (1981)
3. Grainger, J.J., Civanlar, S.: Volt/Var control on distribution systems with lateral branches using shunt capacitors and voltage regulators parts I, II and III. IEEE Trans. Power Apparatus Syst. **PAS-104**(11), 3278–3297 (1985)
4. Baran, M.E., Wu, F.F.: Optimal capacitor placement on radial distribution systems. IEEE Trans. on Power Delivery **4**(1), 725–734 (1989)

5. Baran, M.E., Wu, F.F.: Optimal sizing of capacitors placed on a radial distribution system. IEEE Trans. Power Deliv. **4**(1), 735–743 (1989)
6. Chiang, H.D., Wang, J.C., Cockings, O., Shin, H.D.: Optimal capacitor placements in distribution systems: parts I and II. IEEE Trans. Power Deliv. **5**(2), 634–649 (1990)
7. Sundhararajan, S., Pahwa, A.: Optimal selection of capacitors for radial distribution system using a genetic algorithm. IEEE Trans. Power Syst. **9**(3), 1499–1507 (1994)
8. Gallego, R.A., Monticell, A.J., Romero, R.: Optimal capacitor placement in radial distribution networks. IEEE Trans. Power Syst. **16**(4), 630–637 (2001)
9. Oo, N.W.: A comparison study on particle swarm and evolutionary particle swarm optimization using capacitor placement problem. In: IEEE 2nd International Power and Energy Conference 2008, Johor Baharu, Malaysia, pp. 1208–1211 (2008)
10. Erlich, I., Venayagamoorthy, G.K., Worawat, N.: A mean-variance optimization algorithm. In: IEEE CEC 2010, Shanghai, China, pp. 1–6 (2010)
11. Erlich, I., Nakawiro, W., Martinez, M.: Optimal dispatch of reactive sources in wind farms. In: IEEE PES General Meeting 2011, Detroit, MI, USA, pp. 1–7 (2010)
12. Nakawiro, W., Erlich, I., Rueda, L.: A novel optimization algorithm for optimal reactive power dispatch: a comparative study. In: 4th International Conference on Electric Utility Deregulation and Restructuring and Power Technologies, Weihai, China, pp. 1555–1561 (2011)
13. Rueda, J.L., Loor, R., Erlich, I.: MVMO for optimal reconfiguration in smart distribution systems. In: 9th IFAC Symposium on Control of Power and Energy Systems, New Delhi, India, pp. 276–281 (2015)
14. Rueda, J.L., Erlich, I.: Hybrid mean-variance mapping optimization for solving the IEEE-CEC 2013 competition problems. In: IEEE CEC 2013, Cancun, Mexico, pp. 1664–1671 (2013)
15. Rueda, J.L., Erlich, I.: Testing MVMO on learning-based real-parameter single objective benchmark optimization problems. In: IEEE CEC 2015, Sendai, Japan, pp. 1025–1032 (2015)
16. Chiang, H.D., Jumeau, R.J.: Optimal network reconfigurations in distribution systems: parts I and II. IEEE Trans. Power Deliv. **5**(4), 1902–1909 (1990)
17. Zhang, D., Fu, Z., Zhang, L.: An improved TS algorithm for loss-minimum reconfiguration in large-scale distribution systems. Electr. Power Syst. Res. **77**, 685–694 (2007)
18. Kennedy, J., Eberhart, R.: Particle swarm optimization. In: IEEE IJCNN 1995, Perth, Australia, vol. 4, pp. 1942–1948 (1995)
19. Miranda, V., Fonseca, N.: EPSO-best-of-two-words meta-heuristic applied to power system problems. In: IEEE CEC 2003, Newport Beach, CA, USA, vol. 2, pp. 1080–1085 (2003)
20. Kennedy, J., Eberhart, R.C.: A discrete binary version of the particle swarm algorithm. In: IEEE SMC 1997, Orlando, FL, USA, vol. 5, pp. 4104–4108 (1997)
21. Chiang, H.D.: A decoupled load flow method for distribution power network: algorithms, analysis and convergence study. Electr. Power Energy Syst. **13**(3), 130–138 (1991)

A Capacity Aware-Based Method of Accurately Accepting Tasks for New Workers

Dunwei Gong$^{(\boxtimes)}$ and Chao Peng

School of Information and Control Engineering,
University of Mining and Technology, Xuzhou 221116, China
dwgong@vip.163.com

Abstract. The lack of a new worker's capacity of accepting tasks seriously affects his/her incomes obtained by fulfilling the tasks issued by requesters. We propose a capability aware-based method of accurately accepting tasks for a new worker in this paper. In the proposed method, the problem of accepting tasks is first formulated as a constraint optimization problem with an unknown parameter. Then, the time consumption is estimated based on information provided by similar workers and tasks in the crowdsourcing platform. Finally, the strategy of accepting tasks is generated by solving the optimization problem using a genetic algorithm. We evaluate the proposed method based on data provided by Taskcn and compare the results obtained by the proposed method with the actual earnings of workers in the platform. The results show that the proposed method can be accurately aware of a new worker's capacity of accepting tasks.

Keywords: Crowdsourcing · New worker · Accurate acceptance of tasks · Capability aware · Genetic algorithm

1 Introduction

Crowdsourcing is a manner of efficiently solving a problem at a low cost through an open platform of Internet [1]. In view of a close relation between capacity aware, accurate acceptance of tasks and the earnings of a worker [5], it has become one of focuses for scholars and businessmen to study appropriate methods of being aware of capacity of a worker and accurately accepting tasks.

To this end, we present a capacity aware-based method of accurately accepting tasks for a new worker. The processes of this paper are organized as follows. The optimization problem of accepting tasks is formulated in Sect. 2. Section 3 describes the method of estimating the capacity of a new worker in accepting tasks in detail. The strategy of accepting tasks is generated using a genetic algorithm in Sect. 4. Section 5 evaluates the proposed method. Finally, Sect. 6 concludes the whole paper and points out topics to investigate in the future.

© Springer International Publishing AG 2017
Y. Tan et al. (Eds.): ICSI 2017, Part II, LNCS 10386, pp. 475–480, 2017.
DOI: 10.1007/978-3-319-61833-3_50

2   Formulation of the Problem of Accepting Tasks

The optimization problem of accepting tasks can be described as follows. The level of a task is denoted as $i = 1, 2, \cdots, I$, with m_i the amount of tasks in the i-th level. Besides, the flag of whether or not the worker selects and fulfills the j-th task in the i-th level is denoted as x_{ij}, and if the worker selects and fulfills the task, $x_{ij} = 1$; otherwise, $x_{ij} = 0$. When $x_{ij} = 1$, b_{ij} is denoted as the reward of the corresponding task gained by the worker. The time consumption is required for a worker to fulfill a task. Let t_{ij} be the time consumption of the worker in fulfilling the j-th task in the i-th level. Besides, the maximum time available for the worker is denoted as T. If we denote the number of tasks available as M, the number of tasks selected by the worker should be less than or equal to M.

Based on the above analysis, the optimization problem of accepting tasks for a worker can be formulated as follows:

$$
\begin{aligned}
\max \quad & F(x) = \sum_{i=1}^{I} \sum_{j=1}^{m_i} b_{ij} x_{ij} \\
\text{s.t.} \quad & \sum_{i=1}^{I} \sum_{j=1}^{m_i} t_{ij} x_{ij} \leq T \\
& \sum_{i=1}^{I} \sum_{j=1}^{m_i} x_{ij} \leq M
\end{aligned}
\tag{1}
$$

However, t_{ij} is unknown since he/she has not accepted any task in the platform.

3   Estimation of the Capacity of a New Worker in Accepting Tasks

3.1   Idea of the Proposed Estimation Method

We first seek a number of workers similar to a new worker based on browsing behaviors of previous workers. Then, we search for published tasks similar with those fulfilled by similar workers via the attributes of tasks, such as the domain, the reward and the expected completion time, in the platform. Finally, we estimate the time consumption of a task fulfilled by the new worker with information of the similar workers and their time consumption in fulfilling similar published tasks. Figure 1 depicts the process of estimating the value of t_{ij} using the above idea.

3.2   Seeking Similar Workers

Let U be a set composed of workers, and $U = U^N \cup U^O$. The k-th worker in U^O is denoted as u^k. The number of clicking the j-th task in the i-th level is denoted as r_{ij}^k. Let u^0 be a new worker and his/her number of clicking that task be r_{ij}^0. The similarity between u^0 and u^k in clicking tasks can be calculated using the

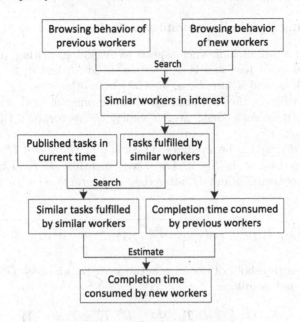

Fig. 1. Estimated process of t_{ij}

Pearson correlation [3] with tasks in the platform, denoted as $s_r(u^0, u^k)$, with the following expression:

$$s_r(u^0, u^k) = \frac{\sum\limits_{i=1}^{I} \sum\limits_{j=1}^{m_i} r_{ij}^0 \cdot r_{ij}^k}{\sqrt{\sum\limits_{i=1}^{I} \sum\limits_{j=1}^{m_i} \left(r_{ij}^0\right)^2} \cdot \sqrt{\sum\limits_{i=1}^{I} \sum\limits_{j=1}^{m_i} \left(r_{ij}^k\right)^2}} \tag{2}$$

We denote the similarity in the time consumption of browsing tasks as $s_t(u^0, u^k)$, with the following expression:

$$s_t(u^0, u^k) = \frac{\sum\limits_{i=1}^{I} \sum\limits_{j=1}^{m_i} t_{ij}'^0 \cdot t_{ij}'^k}{\sqrt{\sum\limits_{i=1}^{I} \sum\limits_{j=1}^{m_i} \left(t_{ij}'^0\right)^2} \cdot \sqrt{\sum\limits_{i=1}^{I} \sum\limits_{j=1}^{m_i} \left(t_{ij}'^k\right)^2}} \tag{3}$$

Since we consider only two aspects, the weight of $s_t(u^0, u^k)$ is $1-\alpha$ when assigning that of $s_r(u^0, u^k)$ to α, and $0 \leq \alpha \leq 1$. Let the similarity between u^0 and u^k be $s(u^0, u^k)$, we have

$$s(u^0, u^k) = \alpha \cdot s_r(u^0, u^k) + 1 - \alpha \cdot s_t(u^0, u^k) \tag{4}$$

Let s^0 be a threshold reflecting the similarities between each pair of workers, and $0 \leq s^0 \leq 1$. In this way,

$$S(u^0) = \left\{ u^k | s(u^0, u^k) \geq s^0, u^k \in U^O \right\} \tag{5}$$

3.3   Searching for Similar Completed Tasks

The domain of a task includes such aspects as designing, writing, programming, multimedia, to say a few, and is denoted as d. Let d_i and d_j be the domains of two tasks. If d_i and d_j are same, $|d_i - d_j| = 0$; otherwise, $|d_i - d_j| = 1$. To conveniently facilitate, they are normalized in the range of $[0, 1]$, and denoted as b and $\tilde{t}$, respectively. As a result, we can employ the vector, $(d, b, \tilde{t})$, to represent a published task.

Let $T_{ij} = (d_{ij}, b_{ij}, \tilde{t}_{ij})$ be a published task in a platform and $T_l^k = (d_l^k, b_l^k, \tilde{t}_l^k)$ a task completed by u^k in $S(u^0)$. Then, we can utilize $D(T_{ij}, T_l^k)$ to indicate the distance between T_{ij} and T_l^k after they are normalized with the following expression:

$$D(T_{ij}, T_l^k) = \frac{1}{\sqrt{3}} \sqrt{\left(d_{ij} - d_l^k\right)^2 + \left(b_{ij} - b_l^k\right)^2 + \left(\tilde{t}_{ij} - \tilde{t}_l^k\right)^2} \qquad (6)$$

Let D^0 be a threshold of distance between tasks, and $0 \leq D^0 \leq 1$. $S(T_{ij})$ can be formulated as follows:

$$S(T_{ij}) = \left\{ T_l^k | D(T_{ij}, T_l^k) \leq D^0, T_l^k \in T(S(u^0)) \right\} \qquad (7)$$

where tasks completed by all the workers in $S(u^0)$ form a set, denoted as $T(S(u^0))$.

3.4   Estimation of Time in Completing a New Task

The estimation of the time consumption, t_{ij}, of u^0 in completing T_{ij}, denoted as $\hat{t}_{ij}$, can be calculated as follows:

$$\hat{t}_{ij} = \frac{1}{\sum S(T_{ij})} \sum_{T_l^k \in T(S(u^0))} (1 - D(T_{ij}, T_l^k)) \cdot t_l^k \qquad (8)$$

4   Problem Solving Based on Genetic Algorithms

Since genetic algorithms [2,4] are a kind of efficient methods of solving combinatorial optimization problems, we employ them to solve the problem in this paper.

The fitness of the individual, X, denoted as $Fit(X)$, can be calculated as follows:

$$Fit(X) = \sum_{i=1}^{I} \sum_{j=1}^{m_i} b_{ij} x_{ij} - \gamma_1 \left(\sum_{i=1}^{I} \sum_{j=1}^{m_i} t_{ij} x_{ij} - T \right) - \gamma_2 \left(\sum_{i=1}^{I} \sum_{j=1}^{m_i} x_{ij} - M \right) \qquad (9)$$

where γ_1 and γ_2 are two penalty factors reflecting the constraints related to the time consumption and the number of tasks [6].

5   Experiments

We run 30 times under the following condition in Table 1 and take the average of their results as the basis of performance evaluation.

Here, we denote the E as the income per time unit and utilize it to evaluate a method, with its expression as follows:

$$E = \frac{TR}{TM} \times 100\% \tag{10}$$

where the TR reflects the total incomes earned by the new worker and the TM indicates his/her the total time consume in the platform.

Figure 2 shows the values of E of new workers when utilizing different methods, where the abscissa represents the identifier of a new worker, and the ordinate expresses the value of E. Figure 2 tells that, the value of E of each new worker is larger when using the proposed method than that without the proposed method, meaning that a new worker can earn more using the proposed method. Through the above experimental results and analysis, we can obtain the

Table 1. Parameters setting

Parameter	Symbol	Value
The weight of browsing behaviors	α	0.9
Threshold of similar workers	s^0	0.5
Threshold of similar tasks	D^0	0.7
The weight of penalty factors of the time consumption	γ_1	70
The weight of penalty factors of the number of tasks	γ_2	15
Probability of crossover	p_c	0.3
Probability of mutation	p_m	0.5
The number of iterations	n	550

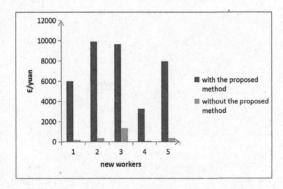

Fig. 2. The values of E of a new worker in different methods

following conclusions: the proposed method can accurately estimate the capacity of a new worker, which is beneficial to his/her incomes when he/she anticipates in a crowdsourcing platform.

6   Conclusion

We have applied the proposed method to the optimization problem of a new worker in accepting tasks, and compared the gains obtained by the proposed method with those provided in Taskcn. If the threshold related to the similarity of workers is too large, or that of tasks is too small, we will have difficulties in seeking similar workers or similar tasks in a crowdsourcing platform, resulting in difficulties in implementing the method and inaccuracy of the estimation. How to automatically adjust these thresholds based on data available in the platform so as to improve the estimation accuracy and reduce the computation cost is the topic to study in the future.

References

1. Feng, J.H., Li, G.: A survey on crowdsourcing. Chinese J. Comput. (2015)
2. Kotenko, I., Saenko, I.: Improved genetic algorithms for solving the optimisation tasks for design of access control schemes in computer networks. Int. J. Bio Inspired Comput. 7(2), 98–110 (2015)
3. Shen, L.M., Yang, Y.L., Chen, Z.: Collaborative prediction of web service QOS considering similarity ratio. Comput. Integr. Manuf. Syst. 22(1), 144–154 (2016)
4. Umbarkar, A.J., Joshi, M.S., Hong, W.-C.: Comparative study of diversity based parallel dual population genetic algorithm for unconstrained function optimisations. Int. J. Bio Inspired Comput. 8(4), 248–263 (2016)
5. Vasileios, T., Choonhwa, L., Muhammad, H., Eunsam, K., Sumi, H.: VM capacity-aware scheduling within budget constraints in IaaS clouds. Plos One 11(8), e0160456 (2016)
6. Wazir, H., Jan, M.A., Mashwani, W.K., Shah, T.T.: A penalty function based differential evolution algorithm for constrained optimization. Nucleus 53(1), 155–166 (2016)

A Genetic Mission Planner for Solving Temporal Multi-agent Problems with Concurrent Tasks

Branko Miloradović[(✉)], Baran Çürüklü, and Mikael Ekström

School of Innovation, Design, and Engineering,
Mälardalen University, Västerås, Sweden
branko.miloradovic@mdh.se

Abstract. In this paper, a centralized mission planner is presented. The planner employs a genetic algorithm for the optimization of the temporal planning problem. With the knowledge of agents' specification and capabilities, as well as constraints and parameters for each task, the planner can produce plans that utilize multi-agent tasks, concurrency on agent level, and heterogeneous agents. Numerous optimization criteria that can be of use to the mission operator are tested on the same mission data set. Promising results and effectiveness of this approach are presented in the case study section.

Keywords: Genetic algorithms · Mission planning · Concurrent tasks · Multi-agent systems · Underwater robotics

1 Introduction

A multi-agent system (MAS) is a system where two or more agents interact to solve problems with a joint effort, i.e. to achieve a certain set of goals that may be well beyond individual capacities and knowledge of a single agent. A MAS can consist of a heterogeneous set of agents having different capabilities and roles. Agents can also take different forms, e.g. software and/or hardware agents. The former are computer programs, whereas the latter refer to physical robots. A MAS can be seen as a powerful tool with the ability to exchange information, making decisions and plans.

In the context of this paper planning is a centralized mechanism that produces a plan based on the capabilities and roles of the agents that are part of a MAS. The output of a plan is a set of tasks dedicated to the set of agents. The plan will advance towards to the final goal as the agents perform their tasks. The planning algorithm presented in this paper assumes that the global mission objective is met through dividing the mission into sub-goals. A search is done based on the constraints and available resources (agents). More concretely, the planner assigns an agent, or a set of agents a sub-goal. Thus, this scheme assumes that agents that are assigned a task will collaborate to address the objective of their sub-goal. These agents do not necessarily have to be identical since generally, it is plausible to assume that agents may have different capabilities and that sometimes this is even required for more complex tasks. The output of the planner is a global plan of concurrent activities for the set of selected agents with temporal constraints.

© Springer International Publishing AG 2017
Y. Tan et al. (Eds.): ICSI 2017, Part II, LNCS 10386, pp. 481–493, 2017.
DOI: 10.1007/978-3-319-61833-3_51

This type of problem is a mixed planning/scheduling problem which well belongs to the more general domain of combinatorial problems. The planner, in this work, is a genetic algorithm (GA) [1] adapted to this specific problem.

1.1 Multi-agent Planning

In the approach presented in this paper interaction between individual agents in a task is assumed, however, these interactions are not addressed directly since they are part of a low-level mechanism. A simple example of the need for multi-agent tasks (MATs) in an underwater application is when one or more agents monitor the execution of a certain task performed by other agents. Due to time constraints, a single agent task (e.g. scanning seabed) can become a multi-agent task where the desired area is divided amongst multiple agents. However, tasks running concurrently on a single agent could possibly have negative effects on a mission plan. In the case with seabed scanning, an acoustic modem mounted on an underwater vehicle sends data, and by doing this interferes with a task for mapping seabed by sonars. This is why this information is taken into account by the planner so that these tasks are not executed at the same time. A list of allowed concurrent tasks has to be created *a priori* for each task in order to avoid such problem. This list consists of tasks that can be performed concurrently with a given current task. Tasks that are not in the list are automatically assumed to be invalid for a concurrent run with this task.

In a general case, different agents have different capabilities, sensors, and performances. This gives the necessary heterogeneity to the MAS and creates a possibility to utilize the full strength of the swarms approach. These constraints can be handled through different approaches. In this paper, it is handled through integration into the planning algorithm, more specifically as part of the fitness function of the GA. Since the planner presented in this paper is a temporal planner, it is the planner's job to ensure that the execution of concurrent tasks is synchronized. Single agent tasks are considered to be atomic, i.e. cannot be divided into sub-tasks. The same is assumed for MATs.

Multi-agent tasks must have a simultaneous start time for all the agents involved. It is the planner's job to schedule this and sync starting times. The role of each agent in that task is not considered in this paper and it is assumed that an onboard low-level planning algorithm will handle this as depicted in Fig. 1. In this example, 3 agents are

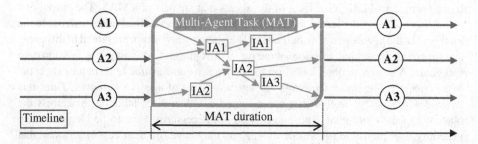

Fig. 1. Schematic view of an example of Multi-Agent Task. A1-3 are agents, while IA represents individual actions of those agents within MAT, and JA is the possible joint actions.

scheduled for the same task. Inside the task, each agent has individual action, which can also be joint actions. These individual actions are out of the scope of the paper, as the accent here is on the high-level planning.

1.2   Genetic Algorithms

A large number of conventional planning algorithms handle planning by mapping the search space into a graph or a tree, searching through the nodes using heuristic functions, cutting infeasible branches or backtracking from dead-ends. GA solves the planning problem with a different method, sometimes leading to sub-optimal solutions.

Genetic Algorithms are adaptive heuristic search methods for solving both constrained and unconstrained optimization problems that use concepts from evolutionary biology to search for an optimal solution. GA works by starting with an initial generation of candidate solutions that are tested against the objective function. Subsequent generations evolve from the previous generation through genetic operators. Reseeding of the population across the search space is done according to probabilistic selection, crossover, and mutation. Crossover operator allows big jumps in the search space by combining several candidate solutions. Mutation allows the genetic algorithm to avoid falling into local minima by maintaining genetic diversity in the population. The goal of retaining the best performing chromosomes from one generation to the next, along with choosing the parents is carried out by selection operator.

1.3   Related Work

A formal analysis of multi-robot task allocation (MRTA) is given in [2], where three different axes for use in describing MRTA problems are proposed. The genetic planner presented in this work covers all of these problems.

Plans that are to be executed in a distributed fashion can nonetheless be produced by a centralized planner. A planner breaks a mission plan into smaller pieces that are sent to the appropriate agent for execution. In one of the possible approaches, a partial order planner generates plans where there need not be a strict ordering between some of the actions, and in fact where those actions can be executed concurrently [3]. The authors there define concurrency on actions, specifying which actions can be performed simultaneously. A similar approach is used in the current paper, where concurrency constraints are defined for each task. The opposite approach was taken in [4], where the authors investigate how centralized, cooperative, multi-agent planning problems with concurrent action constraints and heterogeneous agents can be encoded to Planning Domain Definition Language (PDDL). They encode concurrency constraints on objects and determine conditions under which a certain object can be used concurrently. In both above-mentioned approaches, it is assumed that there cannot be concurrent actions on a single agent. Genetic Algorithm Inspired Descent (GAID) is used in solving multi-agent planning problem related to Traveling Salesman Problem (TSP) in [5]. However, the planner does not support durative actions but show promising results in the routing problem optimization. Different crossover operators are tested against

Vehicle Routing Problem (VRP) and TSP in [6], showing the different behavior of crossover operators between these two problems. A simple genetic algorithm with several enhancements and PDDL modeling language were implemented in [7]. The proposed genetic planner utilized the approach of variable chromosome length, resulting in efficient memory usage. On the other hand, this approach cannot be used for multi-agent planning or planning concurrent actions. A solution to the multi-UAV mission planning problem is proposed in [8]. A multi-objective GA is used for producing a solution to this Temporal Constraint Satisfaction Problem (TCSP). As a result, the planner can produce a plan utilizing different agents and tasks while avoiding forbidden areas and complying with the task ordering constraints. Another popular way for solving Resource Constrained Problems (RCP) is by using GA, where chromosome representation is based on random keys as presented in [9]. The schedule is constructed by using a heuristic priority rule in which the priorities and delay times of the activities are defined by the genetic algorithm. However, joint or concurrent actions are not possible in the latter two approaches. In conclusion, a review of the current state of multi-agent cooperative planning architectures is done in [10].

Since the research done here is a part of the Smart and Networking Underwater Robots in Cooperation Meshes (SWARMs) project, the focus is on the underwater mission planning, using different autonomous underwater vehicles (AUVs) as agents. However, the solution to the described planning problem shown here is generic, and it is not domain dependent.

2   The Genetic Planner

The presented genetic mission planner solves temporal planning problem that includes usage of multi-agent tasks and task concurrency on agent level. In addition, the knowledge from the input data is utilized for the creation of the initial population. However, it is still not enough to create a population of feasible solutions. That is left to the planner. Furthermore, it is important to indicate that none of the above-mentioned approaches allow concurrent action/tasks on agent level. It is important to note that some simplifications are assumed, as atomic time, deterministic effects, and omniscience – world state is completely known. Since this is high-level planning, only high-level tasks are considered, we refer to them as "tasks". Low-level tasks are out of the scope of this paper and are referred to as actions.

2.1   Input and Initial Population

In the proposed genetic planner the input that initiates the search is generated by the user, usually an operator/expert in the application domain, in the mission management tool (MMT). This input consists of: (1) a set of available agents/robots, and (2) a set of tasks to be done. When the latter are addressed by the former the mission is completed. It is assumed that agents' specification and capabilities, as well as constraints and parameters for each task, are known before the mission. Thus, the planner also has access to a set of tasks definitions that will be used in producing the plan. This information is also necessary for the calculation of task parameters such as duration and energy consumption. These two parameters depend on the type of agent used for their execution, e.g. if the agent A_i has higher velocity then the agent A_j it is only logical for the agent A_i to finish a certain task quicker than the agent A_j. While doing this A_i may or may not use more energy depending on the agent's efficiency. This data can be pre-processed and delivered to the planner, or a function can be provided to calculate these values ad-hoc. In this case the planner is supplied with the pre-processed data for each task and agent, respectively.

With the input delivered to the planner, initial population set is being created based on the knowledge that can be extracted from the input data. This is the important step in any optimization process since it has a direct influence on the efficiency of the algorithm. The planner can also work even with a completely random initial population set but it would be less efficient while producing worse solutions. What can be extracted from the input is the minimum number of agents and tasks required as well as tasks that require multiple agents for their execution. It is important to emphasize that the initial population consists of the infeasible candidate solutions.

Since the actual length of a mission plan is not known a priori, chromosomes have a variable length. The approach from our previous work [11] is kept. This means that all chromosomes will have the same length, but the length changes based on the longest one. Chromosomes are filled with the inactive "dummy" genes up to the given length. Initial chromosome length is not strictly defined and it can vary depending on the problem at hand. A minimum number of inactive genes per chromosome is one.

The population set consists of chromosomes. They can be represented in many different ways as it is presented in [12]. Bit strings are the most common way, but here integer identifiers are used, i.e. one gene represents either task, agent, or parameter. So far described population set is two-dimensional array, but there is a third dimension filled with parameter identifiers as is shown in Fig. 2a. For example, if the mission requires four areas to be scanned, there would have to be four tasks with the same integer identifier, but parameter identifiers would be different for each task.

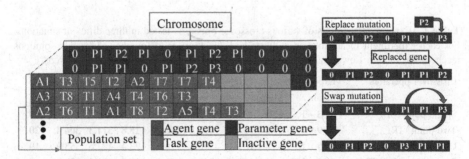

Fig. 2. a. An example of the population set (on the left), **b.** Scheme of work for parameter replace and swap mutation (on the right).

2.2   Genetic Operators

The genetic operators in the proposed mission planner correspond to the standard GA paradigm: selection, crossover, mutation, and elitism. For selection operator rank selection is chosen as it yielded the best results in this case. It might be slower than the other known selection operators, but it also has the advantage of being more resilient to the premature convergence i.e. has less chance to get stuck in the local minima. This method also ensures good diversity in the population set since all chromosomes have a chance to be selected for mating.

Once parents have been selected, crossover operator combines them in order to produce an offspring. For this type of problem, various crossover operators were tested in three different situations. In the first one (Experiment 1), only crossover operator was used, mutation rate was set to zero. The second test (Experiment 2) was done with the initial population already consisting of feasible solutions. Finally, in the last test (Experiment 3) initial population was not feasible in the beginning and mutation operators were included. Results of these tests are presented in Table 1. It can be noted that the two point (TP) crossover operator yield best results in experiment 1, while partially mapped crossover (PMX) had best results in experiments 2 and 3. It seems that TP works better with proposed mutation operators and has better exploration power, whereas PMX has better exploitation capabilities and slow but steady convergence rate. This is why it was the only crossover operator able to produce feasible solutions without mutation operator involved (Experiment 1). This crossover method also requires a fix to be applied to the resulting offspring. During the crossover it can happen, quite often, that offspring does not start with an agent gene, which is illegal. A fixing operation consists of searching for the agent gene in the chromosome and changing its location to be the first gene of the chromosome. If the chromosomes start with an agent gene, a fixing of this type is not needed.

By assuming that the samples used in these experiments are random and independent, a series of statistical hypotheses tests have been conducted. Since the sample data can be highly skewed and have extended tails, an average of that data can produce a value that behaves non-intuitively, thus median was used instead of the mean. This

Table 1. Experimental results of various crossover operators tested in three different situations. For each experiment mean, median and standard deviation or #f (number of feasible solutions reached) are shown. Lower values are better, except for #f. All experiments had 20 runs.

Crossover operators	Experiment 1			Experiment 2			Experiment 3		
	Mean	Median	#f	Mean	Median	Std.	Mean	Median	#f
Single point (SP)	$6.83 \cdot 10^4$	$6.94 \cdot 10^4$	0	271.1	271.3	22.4	369	315.3	19
Two point (TP)	$6.49 \cdot 10^4$	$6.45 \cdot 10^4$	0	270.8	272.1	**15.8**	**311.3**	**306.4**	**20**
Uniform (UN)	$6.35 \cdot 10^4$	$6.34 \cdot 10^4$	0	275.7	276.8	23.4	422.6	355.9	19
Partially mapped (PM)	**$1.56 \cdot 10^4$**	**$1.40 \cdot 10^4$**	**2**	**260**	**256.1**	19.3	881.4	329.6	16
Edge recombination (ER)	$8.48 \cdot 10^4$	$8.57 \cdot 10^4$	0	283	280	22.3	824.8	335	17

Table 2. Statistical tests performed for each experiment between the crossover operator with the best median value versus the rest.

PM vs.	Experiment 1		PM vs.	Experiment 2		TP vs.	Experiment 3	
	p-value	Critical value		p-value	Critical value		p-value	Critical value
ER	$0.84 \cdot 10^{-8}$	0.0250	ER	0.0023	0.0250	ER	0.0098	0.0250
SP	$1.97 \cdot 10^{-8}$	0.0500	UN	0.0256	0.0500	UN	0.0123	0.0500
TP	$3.26 \cdot 10^{-8}$	0.0750	TP	0.0439	0.0750	PM	0.0810	0.0750
UN	$3.55 \cdot 10^{-8}$	0.1000	SP	0.1108	0.1000	SP	0.2733	0.1000

means that the non-parametric test is used, specifically Mann-Whitney-Wilcoxon test. Since a multiple comparison is performed, Benjamini-Hochberg (BH) [13] procedure is applied in order to adjust false discovery rate. BH critical value is calculated as $(i/m) \cdot Q$, where i is the rank, m number of tests and Q false discovery rate set to 0.1. The crossover operator that had the best median value is compared to all other crossover operators in the experiments 1, 2, and 3. The difference between PM and other operators, in the experiment 1, is statistically significant at the significance level 0.05. In the experiment 2 there is no statistically significant difference between PM and SP, while in the experiment 3, there is not statistically significant difference between TP and PM, SP. Results are shown in Table 2.

The mutation operator is very important for this type of optimization problem that is very sensitive even to the small changes. Mutation can recover some of the lost genetic material. In the addition of the four mutation operators described in our previous work [11], two new parameter mutations were added, while agents and tasks kept their old mutation implementation. The new parameters mutation consists of "replace" mutation, where one parameter is replaced with a new one from the set, and a swap mutation where the location of two parameters is swapped in the chromosome. Graphical representation of the described process can be seen in Fig. 2b.

The mutation process can be guided or uninformed. Uninformed mutation is completely random. Guided mutation uses information from fitness function to better choose mutating genes. This information is acquired from the plan execution simulation part of the fitness function. In this part of the fitness function number of conflicting genes is counted and its location is stored in an array. In the mutation process, the location of the mutating gene in the chromosome is randomly selected from that array.

The uninformed uniform mutation works better in the early stages of the algorithm, i.e. in the exploration phase of the algorithm. However, in the later stages, guided mutation takes the lead. This is due to the lower number of conflicts thus the targeted mutation has more success in comparison with the uninformed uniform approach. A logical step was the combination of these two approaches. The algorithm starts with uninformed uniform mutation and then changes to the guided mutation. This switch is done when the convergence rate (change of gradient) slows down. This point of change is arbitrary.

As already mentioned, the mission plan is very sensitive to changes in the chromosome. This can lead to the divergence of the population. To solve this problem, the elitism operator is introduced. It allows a certain number of chromosomes from the current generation to be transferred to the next generation unaltered. If too many unaltered chromosomes are transferred from one generation to another it can lead to the loss of population diversity, and therefore should be used with caution.

2.3 Fitness Function

Fitness function combines different variables and various user defined weights into a solution's cost. It is not sufficient for a solution to be only feasible, but the goal is to reach a globally optimal solution in as many cases as possible. In this work only two parameters are being optimized: time and energy or the linear combination of these two. While mission makespan optimization for a single agent case is a quite straight forward process, in MAS that is not the case. Mission makespan is defined, in this paper, as the time needed for the agent with the longest plan to complete its mission. Time is used as the optimization criterion in three different cases.

Let T_j be the finish time of the last gene in an agent's plan vector $(T_1, T_2, \ldots, T_j)$ and n be the number of agents involved in a mission, then A_i is the makespan of i-th agent's plan, and is described in (1) as:

$$A_i = (T_1, T_2, \ldots, T_j) \tag{1}$$

The first optimization function is shown in (2) where the planner tries to minimize the sum of the plan makespan of every agent involved.

$$f_t = \min\left(\sum_{i=1}^{n} A_i\right) \tag{2}$$

This approach can produce a solution where one agent does 90% of the work, while the rest of the work is split amongst other agents. By our definition, the mission will then last as long as the longest agent's makespan. This approach doesn't fully utilize the power of MAS, but can be useful in some cases when few missions are run in parallel and agents are beings shared between missions. Based on the desire to minimize the mission duration, this second optimization function (3) is derived. This is done by minimizing the makespan of the agent that takes the longest time to complete its plan.

$$f_t = \min(\max(\{A_1, \ldots, A_n\})) \tag{3}$$

While this function tries to minimize mission time, there are still no constraints imposed for other agents used in a mission. One way of imposing constraints to all agents involved in a mission is to minimize the difference between each makespan of every agent in a mission, as it is shown in (4).

$$f_t = \min\left(\sum_{i=1}^{n-1} \sum_{j=i+1}^{n} |A_i - A_j|\right) \tag{4}$$

By doing this, the planner tries to force agents to complete the mission approximately at the same time. Of course, a linear combination of above-mentioned approaches is also possible or even some new criteria can be added. For example, the criterion of using the least number of agents or the completely opposite approach where the use of as many agents as possible is rewarded.

What is maybe even more important than the mission duration, especially in the case of autonomous agents, is the optimization of the energy used (5).

$$f_e = \min(\sum_{i=1}^{n} E_i) \tag{5}$$

In this case, the approach of optimizing the sum of the energy needed for the each task in the mission works well. However, a constraint is imposed that no agent can execute a plan that requires more energy than it currently has. The linear combination of time and energy is also commonly used to calculate the fitness of the solution, as it is shown in (6).

$$f_{te} = \min(W_i \cdot f_t + W_k \cdot f_e + P + D) \tag{6}$$

Where W_i and W_k are user defined weights, P is the sum of all penalties (feasible solutions have penalties equal to zero), and D stands for the sum of awards (different behaviors can be rewarded).

A check has been conducted to determine whether the task is assigned to an agent that has necessary sensors available to fulfill it. Let *sensors (a)* denote available sensors of the agent *a*, and *sensors (t)* denote the task *t* requiring the use of that specific sensor, then the penalty is calculated as shown in (7):

$$\forall t \in T, \quad \forall a \in A, \quad penalty = \left\{ \begin{array}{ll} 0, & if \ |sensors(a) \ \cap \ sensors(t)| \ = \ true \\ 1 & otherwise. \end{array} \right\} \tag{7}$$

Other conflicts are penalized too, if: there is a repeating agent in a plan; there are more agents than available; there are no agents at all; the necessary task is missing; if there is an agent without any task assigned to it; multi-agent task is missing an agent; pre/post conditions are not fulfilled; location condition is not met. The reward is awarded if a candidate solution has the sum of penalties equal to zero. This means that the plan is feasible.

Penalties for the MATs are a bit more complicated. First, it is checked if the agents necessary for the specific task are at the predefined location and at specific time step. If not, a penalty equal to the sum of the start time differences is assigned. Before every MAT a delay is added. The agent that arrives the last at the defined location has a delay equal to zero. Delays for the other agents involved is calculated based on the time difference between last agent's time and their own time as shown in (8):

$$delay_i = \sum_{i=1}^{n} T_n - T_i \tag{8}$$

where T_i is the time when an agent *i* reaches desired location, and *n* is the number of agents. The agent that takes the most time (T_n) to reach its location has zero delay. This delay is added to the agent's plan makespan and tasks coming afterward are shifted for that duration as well. On the agent's level, a task is tested against tasks scheduled before it. If both tasks can be run concurrently, the procedure is continued until there are no more tasks in front of that task, or a task is reached that cannot be run concurrently with the current task. This means that a task can only start, at zero time or when another task is starting or ending.

3 Case Study

The presented planner is an extension of our previous work [11] and is implemented in MATLAB as well. A maximum number of generations is chosen as the stopping criterion and it is set to 500. Population size is set to 10. Chosen crossover operator is TP and crossover rate is 85%, while mutation rate is kept low, at 15%. In this case study, 20% of the best chromosomes are transferred to the next generation. This ensures a steady convergence process. The mission in this case study consists of 2 MATs (Task 3) requiring 3 agents, 4 MATs (Task 8) requiring 2 agents, and 3 single agent tasks (Task 5). Task 5 also require Task 6 to be executed after it. Task 6 is not defined by the operator in the MMT, it is added to the plan through task constraints by the planner. In this underwater scenario Task 3 represents a task of scanning a seabed with three AUVs. For the inspection of the underwater oil pipes (Task 8) two AUVs are needed, one for the ultrasound crack detection, and the other one for the visual inspection. Finally, Task 5 and 6 represents measurements of the hydrogen sulfide (H_2S) at locations of interest and sending collected data back to the command center, respectively. To sum up, the mission consists of two areas to be scanned, four oil pipes to be examined and three measurements of H_2S to be taken at three different locations. The planner has 7 agents at its disposal. A number of agents used in a plan usually depends on the optimization criterion. All agents have different equipment, velocity, and energy consumption per task.

The output of the planner is a plan that is represented in a form of a Gantt chart as it can be seen in the Figs. 3 and 4. Both figures are showing the result of the same mission, but with different optimization criterion. In the Fig. 3a it can be seen a Gantt chart of a mission that is being optimized for the total sum of individual agent's makespan. It can be noted that tasks 4 and 8 are executed at the same time with 3 and 2 agents, respectively. Task 6 can run concurrently with task 8 as can be seen in the plan of agent number 2. Blue lines represent task 2 (go to the task), that is basically a duration for going from one task to another. Green lines represent a delay, i.e. idle regime in the sense of the movement. An agent may be in the idle regime, waiting for other agents and still run task 6 while waiting, as it can be seen in the Fig. 3b in the agent's 7 plan. Since the optimization criterion was to minimize the mission duration, a produced plan has makespan less than 400 s. Figure 4 shows two more outputs of a mission planner. Figure 4a presents a solution of a mission when the optimization criterion was that all agents are to finish the mission at approximately the same time. It can be seen that this criterion has been fulfilled, although that the overall mission duration is longer than the one in the Fig. 3b. Again it can be seen that the task 6 runs concurrently with the task 8 on agents 5 and 7. The last Gantt chart in the Fig. 4b shows the output from the planner for the energy optimizing criterion. The interesting

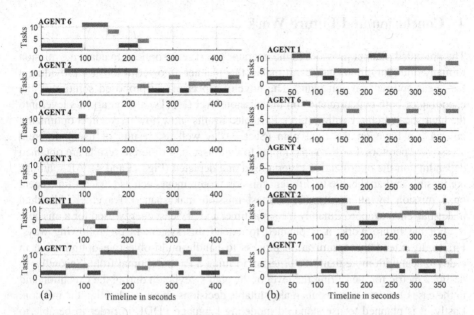

Fig. 3. (a) Least time spent on a mission by all agents, (b) shortest mission makespan (Color figure online)

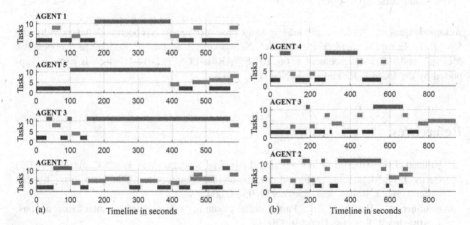

Fig. 4. (a) Approximately the same finishing time, (b) minimum energy used (Color figure online)

part is that the planner now uses only 3 agents, what is the bare minimum for this mission to be executed successfully. The mission duration is the longest of all so far, and the agents used are the slowest ones, but the most energy efficient. The mission lasts more than twice as long as the one in the Fig. 3b, but uses the least amount of energy.

4 Conclusion and Future Work

The presented planner provides to the operator a vast set of options to adapt the mission to many different circumstances. It makes this planner a powerful tool for providing a fast, near optimal mission plans. The novelty aspect of the proposed solution is the encoding of both concurrencies on the mission level (MATs) and on agent's level into the heuristic search algorithm (GA). Presented results show how large effect optimizing criteria can have on the final plan (Sect. 3), as well as, results of using different crossover operators. For the same inputs to the mission, 4 different results are obtained depending on the optimization criteria used and depicted (Figs. 3 and 4). All solutions are feasible and applicable to different mission requirement such as (1) least time spent on a mission by all agents, (2) shortest mission makespan, (3) agents' individual makespan ends at approximately the same time, and (4) least energy used for a mission.

For the future work, it is planned to extend this system in the following ways. Firstly, to extend the optimization process to handle multi-objective optimization in order to deal with more than two criterion forming a Pareto optimal front. Secondly, to design a metric for measuring the level of the plan's deviation during the execution and in the case of unforeseen events make suitable decisions about re-planning the mission. Lastly, it is planned to use standard modeling language PDDL in order to be able to benchmark proposed planner with other available solutions, although there are not many benchmarking data sets for the planning problem with task concurrency on both agent's and mission's level.

Acknowledgment. The research leading to the presented results has been undertaken within the SWARMs European project (Smart and Networking Underwater Robots in Cooperation Meshes), under Grant Agreement n. 662107-SWARMs-ECSEL-2014-1, which is partially supported by the ECSEL JU and the VINNOVA.

References

1. Holland, J.H.: Adaptation in Natural and Artificial Systems. MIT Press, Cambridge (1975)
2. Gerky, B.P., Maja, J.M.: A formal analysis and taxonomy of task allocation in multi-robot systems. Int. J. Robot. Res. **23**(9), 939–954 (2004)
3. Boutilier, C., Brafman, R.I.: Partial-order planning with concurrent interacting actions. J. Artif. Intell. Res. **14**, 105–136 (2001)
4. Crosby, M., Patrick, R.P.A.: Temporal multiagent planning with concurrent action constraints. In: Association for the Advancement of Artificial Intelligence, Quebec (2014)
5. Kalmár-Nagy, T., Giardini, G., Bak, B.D.: The multiagent planning problem. Complexity **2017**, 12 (2017)
6. Puljic, K., Manger, R.: Comparison of eight evolutionary crossover operators for the vehicle routing problem. Math. Commun. **18**(2), 350–375 (2013)
7. Brie, A.H., Morignot, P.: Genetic planning using variable length chromosomes. In: International Conference on Automated Planning and Scheduling, Monterey (2005)

8. Ramirez-Atencia, C., Bello-Orgaz, G., R-Moreno, M.D., Camacho, D.: Solving complex multi-UAV mission planning problems using multi-objective genetic algorithms. Soft. Comput. **20**, 1–19 (2016)
9. Magalhaes-Mendes, J.: Project scheduling using a competitive genetic algorithm. In: International Conference on Simulation, Modelling and Optimization, Santander (2008)
10. Evans, J., Sotzing, C., Patron, P., Lane, D.: Cooperative planning architectures for multi-vehicle autonomous operations. In: 1st SEAS DTC Technical Conference, Edinburgh (2006)
11. Miloradović, B., Çürüklü, B., Ekström, M.: A genetic planner for mission planning of cooperative agents in an underwater environment. In: Symposium Series on Computational Intelligence, Athens (2016)
12. Michalewicz, Z.: Genetic Algorithms + Data Structures = Evolution Programs. Springer, Heidelberg (1996)
13. Benjamini, Y., Hochberg, Y.: Controlling the false discovery rate: a practical and powerful approach to multiple testing. J. Roy. Stat. Soc. **57**(1), 289–300 (1995)

Reformulation and Metaheuristic for the Team Orienteering Arc Routing Problem

Liangjun Ke[✉] and Weibo Yang

The State Key Laboratory for Manufacturing Systems Engineering,
Xian Jiaotong University, Xi'an 710049, China
keljxjtu@xjtu.edu.cn

Abstract. The team orienteering arc routing problem (TOARP) is a relatively new vehicle routing problem. In this problem, a fleet of vehicles are available to serve two sets of customers, each of which is associated with an arc of a directed graph. The customers of the first set are required to be served whereas the ones of the second set are potential and may be not served. Each potential customer is associated with a profit, and the profit can be gained at most once when it is served. The TOARP aims to maximize the total profit gained by serving the customers while each vehicle must start from and end at a depot within a permitted maximum traveling time. This paper shows that the TOARP can be transformed into a team orienteering problem defined on a directed graph. To solve the TOARP, an iterated local search based algorithm is presented. The effectiveness of the proposed algorithm is studied on the benchmark instances.

Keywords: Vehicle routing problem · Metaheuristic · Team orienteering problem · Arc routing problem · Iterated local search

1 Introduction

Routing problem is an essential problem in real world [7]. In this problem, a fleet of vehicles are used to serve a set of customers distributed in a graph. It aims to find optimal routes for the vehicles under some constraints, e.g., costs, demands, and time windows. Under various objectives and constraints, a large number of routing problems have been considered in the literature [11].

The team orienteering arc routing problem (TOARP) is a routing problem which was first formulated in [3]. The TOARP is defined on a directed graph $G = (V, A)$. In this problem, there are two sets of customers. The first set A_R consists of the customers required to be served. The second one A_P consists of the potential customers which may be not served. Each customer is located at an arc of the directed graph. For each arc, there is a time cost t_{ij}. For each potential customer, a profit $p_{ij} \geq 0$ is assigned to it. There are K vehicles. Each vehicle must start from and end at the depot within the time limit T_{max}. The profit gained by serving each potential customer can be accumulated at most once. The goal of the TOARP is to maximize the total profit collected.

© Springer International Publishing AG 2017
Y. Tan et al. (Eds.): ICSI 2017, Part II, LNCS 10386, pp. 494–501, 2017.
DOI: 10.1007/978-3-319-61833-3_52

The TOARP was first solved by a branch-and-cut [3]. Later, a matheuristic was used to deal with the TOARP [1]. This algorithm uses tabu search to explore the neighborhood of the current solution. The solution found during the tabu search phase will be improved by an intensification phase according to the integer linear programming models. The final step of the tabu search phase uses a diversification mechanism with the aim of searching in a totally new sub-region of the solution space.

In this paper, we show that the TOARP can be transformed into a routing problem defined on a new graph, each node of which corresponds to a potential or required arc of the original problem. We notice that some efforts have been devoted to transforming a (capacitated) arc routing problem into a node routing problem [8]. Here we deduce the transformed problem according to the feature of the TOARP. The transformed problem is defined on a new graph where every node corresponds to every potential arc of primal problem. The distance between two arcs is the minimal shortest path distance calculated by Bellman-Ford algorithm. In addition, we propose an iterated local search based algorithm to deal with the transformed problem. The comparison results on the benchmark instances support the performance of the proposed algorithm is competitive.

The remainder of this paper is organized as follows. Section 2 presents the problem transformed from the TOARP. Our proposed algorithm is described in Sect. 3. Section 4 presents the Simulation study. Section 5 concludes our work.

2   A New Formulation of the TOARP

As for the TOARP, we can define a new directed graph $G' = (V', E')$ as follows:

(1) A dummy node 0: It corresponds to arc $(0,0) \in A$.

(2) The set of nodes V': $V' = \{0, 1, \cdots, m\}$ where m is the number of the required and potential arcs. Each node except node 0 corresponds to a required or potential arc of G. Suppose that node $i \in V' \backslash \{0\}$ corresponds to arc $(n_{is}, n_{ie}) \in A_P \cup A_R$. We can assign a service time s_i and a profit q_i to node i. The service time s_i is defined as

$$s_i = t_{n_{is}, n_{ie}}. \tag{1}$$

The profit q_i is defined as

$$q_i = \begin{cases} p_{n_{is}, n_{ie}} & \text{if } (n_{is}, n_{ie}) \in A_P \\ M & \text{if } (n_{is}, n_{ie}) \in A_R, \end{cases} \tag{2}$$

where M is a sufficient large positive number, say $M = \sum_{b \in A_P} p_b$ (that is, the total profit of all potential arcs in A_P). Accordingly, the nodes in G' corresponding to the required arcs are called the required nodes. Let A'_R be the set of the required nodes.

(3) The set of arcs E': Each arc connects every two nodes in V'. Let us consider two nodes i and $j \in V'$. They correspond to (n_{is}, n_{ie}) and $(n_{js}, n_{je}) \in A_P \cup A_R$

respectively. Let $\mathfrak{P}$ is the shortest path through a subset of arcs in A starting from n_{ie} and ending at n_{js}. The traveling time of arc $(i, j) \in E'$, denoted by d_{ij}, is defined as the distance of the shortest path $\mathfrak{P}$. To calculate the distance of path $\mathfrak{P}$, we implement a label based Bellman-Ford algorithm [5] which is a dynamic programming approach.

In this way, we can define a new routing problem: Given a directed graph G' mentioned above, the goal is to determine K routes, each of which is limited by the time budget T_{max} and starts from and ends at node 0 and visits nodes in V', such that the total collected profit is maximized. It is assumed that the profit of a visited node can be collected at most once. According to the definition of the TOP, one can notice that the routing problem is a TOP defined on a directed graph. In the following, the transformed problem is denoted as DG-TOP.

3   The Proposed Algorithm

In this paper, we present an iterated local search [9] based algorithm to solve the DG-TOP. Iterated local search (ILS) is a singleton search technique, which evolves an incumbent solution over time. Let $\mathbf{s}_b$ and $\mathbf{s}_c$ be the best-so-far and incumbent solution respectively. ILS works as follows: At first, $\mathbf{s}_c$ is initialized and improved by local search, and let $\mathbf{s}_b = \mathbf{s}_c$. After that, four steps are repeated until a termination condition is satisfied. At the first step, a new incumbent solution $\mathbf{s}_c$ is obtained by perturbation. At the second step, $\mathbf{s}_c$ is improved by local search. At the third step, the best-so-far solution $\mathbf{s}_b$ will be replaced by $\mathbf{s}_c$ if $F(\mathbf{s}_c) \geq F(\mathbf{s}_b)$ (i.e., $\mathbf{s}_c$ is better than $\mathbf{s}_b$) where F is the objective value. At the fourth step, the incumbent solution $\mathbf{s}_c$ will be updated according to the solution acceptance criterion. The main procedure of ILS is described in Algorithm 1.

Algorithm 1. The main procedure of ILS

Input: a DG-TOP instance to be solved
Output: the best-so-far solution $\mathbf{s}_b$
 1: initialize the incumbent solution $\mathbf{s}_c$ and improve it by local search
 2: set $\mathbf{s}_c$ to $\mathbf{s}_b$
 3: **while** the stopping condition is not reached **do**
 4:    obtain a new solution $\mathbf{s}$ by performing a perturbation operator on $\mathbf{s}_c$
 5:    perform local search on $\mathbf{s}$
 6:    replace $\mathbf{s}_b$ by $\mathbf{s}$ if $F(\mathbf{s}) \geq F(\mathbf{s}_b)$
 7:    update $\mathbf{s}_c$ according to the solution acceptance criterion
 8: **end while**

3.1   Solution Initialization

The incumbent solution is initialized by two steps. The first step only considers the required nodes. The second step tries to insert some non-required nodes.

The first step constructs a solution as follows: Each route starts from node 0. Afterwards, they are probed in ascending order of their travel time. For a route, suppose its last node is i, an unvisited node is said to be feasible if it obeys the inequality $T(i) + d_{ij} + d_{j0} \leq T_{max}$ where $T(i)$ is the total travel time of the partial route from node 0 to i. Let C be the set of the feasible required nodes. If C is empty, then this route is finished and the next route is tried. Otherwise, a node is chosen from C according to the following probability

$$P(j) = \frac{\frac{1}{d_{ij}}}{\sum_{k \in C} \frac{1}{d_{ik}}}, \quad \forall j \in C \tag{3}$$

If there are still some unvisited required nodes after using the above procedure, the following procedure will be adopted. Firstly, γ required nodes are removed from the routes generated by the above procedure. Then, all unvisited required nodes are inserted based on regret value (that is, the nodes are inserted one by one. At each step, the node with the largest regret value is inserted). The concept of regret value has been adopted in [10]. This procedure is repeated until all required nodes are visited.

The second step repeats the following procedure until no non-required node can be inserted: Let C_{nr} be the set of the remaining non-required nodes which is feasible to be inserted. For each node $k \in C_{nr}$, we define a preference value as $\varphi_k = q_k \delta_k$ where δ_k is the regret value of node k. That is, a node with larger profit and regret value is more desirable. A node $j \in C_{nr}$ is selected according to the probability

$$P(j) = \frac{\varphi_k}{\sum_{k \in C_{nr}} \varphi_k}, \quad \forall j \in C_{nr} \tag{4}$$

3.2 Perturbation

Perturbation aims to make the algorithm search in a new area of the solution space. A remove-insert-based and exchange-based operators are used to disturb the incumbent solution. The remove-insert-based removes γ_1 nodes and inserts the removed required back one by one based on regret value. The exchange-based operator only works on the visited nodes. It repeats the following procedure γ_2 times: randomly choose two nodes from different routes and exchange their positions. If the resulting solution is infeasible, some visited non-required nodes will be removed based on preference value until the solution becomes feasible.

3.3 Local Search

Local search tries to find a better solution in the neighborhood of a starting solution. However, searching in a larger neighborhood is usually time-consuming. To accelerate local search, the mechanism of "do not look bits" [4] is adopted.

We use a so-called *active list*, denoted by AL, to record those active nodes. The active list initially consists of all visited nodes. The active nodes are checked in a random order. Given an active node i, suppose its candidate list is CL. For a node j in CL, three possible cases may occur:

(1) both i and j belong to the same route: a 2-opt move will be used. The resulting solution will be accepted once the total travel time is shortened.
(2) i and j belongs to two different routes: the best move between an exchange move and a relocation move is used. The resulting solution will be accepted once the total travel time is shortened.
(3) j is an unvisited node: at first, a relocation move is used, the resulting solution will be accepted if it is feasible. In this case, the total profit will be increased. Otherwise, i and j are tried to exchange. Let the forward node and backward node of i be i_f and i_b respectively, then the resulting solution will be accepted if it is feasible and if $\frac{p_j}{d_{i_f,j}+d_{j,i_b}-d_{i_f,i_b}} \geq \frac{p_i}{d_{i_f,i}+d_{i,i_b}-d_{i_f,i_b}}$.

If a better solution is found, node i, node j, and their forward and backward nodes in their corresponding routes are added into an auxiliary list. If all active nodes have been tried, the active node list is replaced by the auxiliary list. The procedure is repeated until the auxiliary list is empty.

3.4   Solution Acceptance Criterion

A simulated annealing like criterion [9] is used. It permits some inferior solutions to replace the old incumbent solution with a small probability, which is beneficial to preserve diversity. Formally, let the new solution be $\mathbf{s}$, then

$$\mathbf{s}_c = \begin{cases} \mathbf{s} & \text{if } F(\mathbf{s}) \geq F(\mathbf{s}_b) \\ \mathbf{s} & \text{if } \exp(\frac{F(\mathbf{s})-F(\mathbf{s}_b)}{\theta F(\mathbf{s}_b)}) \geq \varepsilon \\ \mathbf{s}_b & \text{otherwise} \end{cases} \tag{5}$$

where ε is a random number generated from $[0, 1]$. θ is a parameter.

4   Simulation Study

To study the performance of the proposed algorithm (ILS), we implemented it in C++ and tested it on a PC equipped with Pentium 4, 2.4 GHz CPU, and 4 GB RAM. For each instance, ILS was stopped when one of the following conditions is satisfied: (1) the number of iterations reaches 40000, (2) the running time reaches 200 s. Parameters γ, γ_1, and γ_2 are an integer randomly sampled from interval $[5, 15]$ at each time. Parameter $\theta = 0.1$. These parameters are determined according to extensive test. We compare ILS with the matheuristic algorithm proposed in [1], denoted by MAT. MAT was performed on a PC with Athlon 64 X2 Dual Core Processor 5600 + 2.89 GHz CPU, and 3.37 GB RAM. The stopping criterion of MAT was set to 30 min.

4.1   Test Instances

We used $D36$, $D64$, and $D100$ in [1] to test the algorithm. Parameter p determines the probability of an arc is declared required when generating instances of the TOARP. For each instance of RPP in [6], nine TOARP instances were generated by taking p and K from $\{0, 0.25, 0.5\}$ and $\{2, 3, 4\}$, respectively. In Table 1, the fist four columns present the name of a set, the number of instances, the minimum and maximum number of vertices and arcs, respectively. From columns

Table 1. The information of class D

set	#Inst	$|V|$	$|A|$	$p=0$		$p=0.25$		$p=0.5$	
				$|A_R|$	$|A_P|$	$|A_R|$	$|A_P|$	$|A_R|$	$|A_P|$
D36	9	17–36	96–270	0	10–38	2–10	6–30	6–20	4–23
D64	9	37–62	264–482	0	27–75	4–21	22–54	11–38	15–37
D100	9	68–100	544–846	0	50–121	9–28	37–95	26–64	20–70

Table 2. Computational results obtained by MAT and ILS for class D

Instance				MAT			ILS			
K	p	Set	#Solved	Opt	Av.Gap	Max.Gap	Opt	Av.Gap	Max.Gap	Time(s)
2	0	D36	9	9	0.00	0.00	9	0.00	0.00	5.38
		D64	9	5	0.48	1.79	**7**	**0.10**	**0.54**	42.28
		D100	9	0	2.58	5.40	**3**	**0.76**	**2.02**	168.37
		D36	9	9	0.00	0.00	9	0.00	0.00	4.88
	0.25	D64	9	5	0.26	1.20	**6**	**0.09**	**0.44**	28.4
		D100	9	1	2.92	10.68	**4**	**0.57**	**2.08**	1.58
		D36	9	9	0.00	0.00	9	0.00	0.00	4.88
	0.5	D64	9	4	1.10	5.12	**6**	**0.25**	**1.20**	34.8
		D100	9	1	4.71	12.42	**3**	**1.22**	**2.98**	153.71
3	0	D36	9	**9**	**0.00**	**0.00**	8	0.02	0.14	2.57
		D64	9	**7**	**0.10**	**0.75**	6	0.26	1.63	17.14
		D100	4	**2**	3.31	12.50	1	**1.97**	**5.89**	92.44
		D36	9	8	**0.08**	**0.74**	8	0.17	1.05	5.40
	0.25	D64	9	6	**0.42**	**1.64**	6	0.48	2.89	41.04
		D100	5	**3**	4.14	9.34	2	**1.57**	**5.29**	48.8
		D36	9	9	0.00	0.00	9	0.00	0.00	3.04
	0.5	D64	9	5	2.05	5.78	**6**	**0.99**	**4.83**	15.89
		D100	7	1	20.45	20.07	**4**	**1.87**	**8.02**	70.75
4	0	D36	9	9	0.00	0.00	9	0.00	0.00	2.02
		D64	5	4	1.42	4.05	4	**1.30**	**3.89**	13.4
		D100	2	2	4.58	10.98	2	**4.15**	**8.91**	56.66
		D36	9	9	0.00	0.00	9	0.00	0.00	2.14
	0.25	D64	7	6	0.68	4.57	6	**0.65**	**4.24**	13.28
		D100	4	3	3.02	9.27	**4**	**2.32**	**6.47**	69.13
		D36	9	9	0.00	0.00	9	0.00	0.00	2.16
	0.5	D64	7	**6**	**1.21**	5.36	5	1.25	**5.22**	12.68
		D100	5	3	7.53	22.07	**5**	**3.02**	**9.73**	58.32

5 to 10, the minimum and maximum number of required arcs and potential arcs are shown for different values of p.

4.2 Results

Table 2 reports the results obtained on class D. By varying K, p, and *set*, 27 combinations are obtained. The column '#*solved*' reports the number of instances which can be solved to optimality by the branch-and-cut algorithm in [2]. For each algorithm, we report the following results:

– #*opt*: the number of instances of which an optimal solution can be found,
– *Av.Gap* and *Max.Gap*: the average and maximum percentage gap of the solution found with respect to the upper bound found by the branch-and-cut algorithm in [2].

Table 2 presents the results obtained on class D. When $K = 2$, ILS can find better solutions for 6 out of 9 combinations. For the other three combinations, ILS and MAT are even. When $K = 3$, in terms of #*opt*, MAT works better on 4 combinations and ILS works better on 2 other combinations. In terms of *Av.Gap* and *Max.Gap*, both MAT and ILS work better on 4 different combinations. When $K = 4$, in terms of #*opt*, MAT works better on 1 combination and ILS works better on 2 other combinations. In terms of *Av.Gap*, ILS works better on 5 combinations and MAT works better on only one combination. In terms of *Max.Gap*, ILS works better on 6 combinations. Therefore, ILS works slightly better than MAT on class D. The maximum time spent by ILS is less than 169 s.

5 Conclusion

The team orienteering arc routing problem aggregates some features of the team orienteering problem and arc routing problem, which incurs extra requirements to the current solution techniques. This paper first transforms the TOARP to a node routing problem, called DG-TOP. The difference between DG-TOP and the traditional TOP are analyzed. According to the characteristics of the DG-TOP, an ILS-based algorithm is proposed to solve the DG-TOP. Special considerations have been devoted to the required nodes. Moreover, the algorithm uses two operators to perturb the incumbent solution. In addition, a fast local search is presented. Based on the experimental results, the proposed algorithm can find promising solutions for the tested instances within short time.

Acknowledgements. The authors would like to thank the anonymous reviewers for their insightful comments. This work was supported by National Natural Science Foundation of China (No. 61573277, 71471158), the Research Grants Council of the Hong Kong Special Administrative Region, China (Project No. PolyU 15201414), the Fundamental Research Funds for the Central Universities, the Open Research Fund of the State Key Laboratory of Astronautic Dynamics under Grant 2015ADL-DW403, and the Scientific Research Foundation for the Returned Overseas Chinese Scholars, State

Education Ministry, Natural Science Basic Research Plan in Shaanxi Province of China (No. 2015JM6316). The authors also would like to thank The Hong Kong Polytechnic University Research Committee for financial and technical support.

References

1. Archetti, C., Corberán, Á., Plana, I., Sanchis, J.M., Speranza, M.G.: A matheuristic for the team orienteering arc routing problem. Eur. J. Oper. Res. **245**(2), 392–401 (2015)
2. Archetti, C., Speranza, M.G.: Arc routing problems with profits. In: Arc Routing: Problems, Methods, and Applications. MOS-SIAM Series on Optimization, pp. 257–284 (2013)
3. Archetti, C., Speranza, M.G., Corberán, Á., Sanchis, J.M., Plana, I.: The team orienteering arc routing problem. Transp. Sci. **48**(3), 442–457 (2013)
4. Bentley, J.J.: Fast algorithms for geometric traveling salesman problems. ORSA J. Comput. **4**(4), 387–411 (1992)
5. Goldberg, A.V., Radzik, T.: A heuristic improvement of the Bellman-Ford algorithm. Appl. Math. Lett. **6**(3), 3–6 (1993)
6. Hertz, A., Laporte, G., Hugo, P.N.: Improvement procedures for the undirected rural postman problem. INFORMS J. Comput. **11**(1), 53–62 (1999)
7. Laporte, G.: Fifty years of vehicle routing. Transp. Sci. **43**(4), 408–416 (2009)
8. Longo, H., De Aragaao, M.P., Uchoa, E.: Solving capacitated arc routing problems using a transformation to the CVRP. Comput. Oper. Res. **33**(6), 1823–1837 (2006)
9. Lourenço, H.R., Martin, O.C., Stützle, T.: Iterated local search: framework and applications. In: Gendreau, M., Potvin, J.-Y. (eds.) Handbook of Metaheuristics, pp. 363–397. Springer, New York (2010)
10. Ropke, S., Pisinger, D.: An adaptive large neighborhood search heuristic for the pickup and delivery problem with time windows. Transp. Sci. **40**(4), 455–472 (2006)
11. Toth, P., Vigo, D.: Vehicle Routing: Problems, Methods, and Applications, vol. 18. SIAM, Philadelphia (2014)

Application of Smell Detection Agent Based Algorithm for Optimal Path Identification by SDN Controllers

R. Ananthalakshmi Ammal[1]([⊠]), P.C. Sajimon[1],
and S.S. Vinodchandra[2,3]

[1] Centre for Development of Advanced Computing, Thiruvananthapuram, India
{lakshmi,pcsaji}@cdac.in
[2] Computer Centre, University of Kerala, Thiruvananthapuram, India
[3] Department of Computational Biology, University of Kerala,
Thiruvananthapuram, India
vinod@keralauniversity.ac.in

Abstract. Software Defined Networking separates the control plane and data plane with which the switches and routers become simply packet forwarding devices. The decision related to the path to be taken by the packet from the source to the destination is taken at the control plane. Thus the SDN controller has to identify the optimal path for the packets. Many of the SDN controllers use Dijkstra's algorithm for computing the shortest path and subsequently update the data plane devices. Many path computation algorithms including bio-inspired algorithms are published and are in use today in computer networks. In this paper, a novel bio inspired algorithm namely Smell Detection Agent based path computation algorithm is applied and studied for its performance in comparison with Dijkstra's algorithm, Extended Dijkstra's algorithm and the most commonly used bio inspired algorithm based on Ant Colony Optimisation. The Smell Detection Agent based algorithm inspired from the dog's smell detection capability for tracing and reaching a destination is found to be very useful and providing better results compared to the other algorithms.

Keywords: SDN · OpenFlow · Shortest path · Bio-inspired · Smell detection algorithm

1 Introduction

Software Defined Networking (SDN) is becoming popular in computer networking over the last few years and is presently a hot topic of research. In traditional networks, both the control plane and data plane reside in the routers and switches. The control plane in each device decides on the way the packets have to be forwarded. Whenever changes happen in network state or traffic patterns, there is lack of flexibility in accommodating those changes dynamically in the forwarding paths. With SDN, the data plane and control plane are separated [1]. There is a logically centralised SDN controller that configures the forwarding table in each data plane device, based on which the packets are forwarded in a network. The SDN controller is also responsible

© Springer International Publishing AG 2017
Y. Tan et al. (Eds.): ICSI 2017, Part II, LNCS 10386, pp. 502–510, 2017.
DOI: 10.1007/978-3-319-61833-3_53

for network resource discovery and topology discovery. The controller gathers information about the state of the network including availability status, performance status including utilisation in a periodic manner. This enables the controller to identify the optimal path for the packets from source to destination and update the forwarding tables in the data plane devices in a dynamic manner. Thus SDN provides the required flexibility for efficiently managing traffic.

OpenFlow is one of the widely used protocols for communication between the controller and the data plane devices [2]. SDN controller uses OpenFlow Protocol for setting up flows for the switches, i.e. building up flow tables in switches. Whenever a new packet arrives at the switch port, resulting in a flow table miss, the switch can be programmed to send the packet to the controller which in turn receives the packet-in messages. The Controller intelligently sets the path to be traversed by the packet flows in the network. A controller has to be aware of the topology of the switch network and compute the packet traversal path. This gives rise to an optimisation challenge where the controller has to compute the optimal packet flow traversal path. Currently many of the SDN controllers use Dijkstra's algorithm for computing the shortest path and does not take into account the network link congestion, under-utilized state and network latency.

The problem is to compute optimal path from source to destination by the SDN Controller, given the network resource & topology discovery of the network are completed and the current status of the network resources such as device and link availability, bandwidth utilization are known to the Controller. To create path between source and destination within a network, some of the operational challenges include network link/resource failure condition, network link congestion condition, and network link under utilised condition. Dynamically finding the optimal path in shortest possible time based on network state changes, to reduce network latency is a major challenge.

In this paper, the application of bio inspired Smell Detection Agent based path computation algorithm [3] is applied and studied for optimal network path finding. The performance is compared with that of Dijkstra's [4] and Extended Dijkstra algorithms [5] and a variant of the most commonly used bio inspired algorithm based on Ant Colony Optimisation (ACO) [6].

2 Related Works

A few studies regarding the shortest or optimal path to be taken by the packet flows in Software Defined Networks have been done. The implementation issues in the modified Dijkstra's algorithm and the modified Floyd-Warshall shortest path algorithm in OpenFlow have been studied by Rus et al. [7]. Jehn-Ruey Jiang et al. have simulated the Extended Dijkstra's algorithm for SDN in a Mininet environment under the Abilene network topology [5]. Adnan Shahid et al. [8] have modified the SDN Controller to find out the highest bandwidth path instead of the shortest path.

Bio inspired algorithms for optimal path computations are in use for quite some time. The famous AntNet algorithm based on Ant Colony Optimisation [9] is used in many telecommunication Networks for routing purposes. Other bio inspired algorithms used for path computation include Bee Colony Optimisation, Genetic Algorithms etc. Meta heuristics algorithms such as Particle Swarm Optimisation and its variants [10–12] are also used to solve hard optimisation problems. In SDN, study of application of bio-inspired algorithms is an emerging area. Ant Colony Optimisation approach to Quality of Experience (QoE) based flow routing [13] in SDN has been studied by Ognjen Dobrijevic et al. To the best of our knowledge, the application of Smell Detection Agent based model for optimal path finding by SDN Controllers is the first of its kind.

3 Smell Detection Agent (SDA) Based Algorithm

SDA based algorithm [3] is a novel bio inspired optimization algorithm based on the trained behavior of dogs in detecting smell trails. The olfactory mechanism of dogs can detect as well as memorize different smell signatures. Dogs also urinate in different spots to mark their territory as occupied. These two properties of dogs have been used to create the SDA to mark the path they undertake to reach the destination.

It is a known factor that different SDAs have varying olfactory capabilities and all the points within a territory cannot be traversed. The selected points within a territory are the smell spots which are visited by SDA. Each smell spot is characterized by two values, one is the signature related to the visit of an SDA to the smell spot and the other is the smell trail, which is value from destination. Each SDA is also characterized by two values, one is their signature value to be marked in smell spots and the other is the radius value indicating their olfactory capability. Thus the whole algorithm is based on a source, destination, smell spots and the smell radius for SDA to traverse. At any instance during the execution of the algorithm, the parameters related to the SDA include the signature value of SDA, total length traversed by the SDA and the current smell spot location of the SDA. Similarly for any smell spot, the parameters include the smell value and the signature of the visited SDA.

Application of SDA Based Algorithm in SDN Environment for Optimal Path Problem

The network domain under the control of the SDN Controller can be considered as the territory, a surface with smell trails and the agents inspired from dogs can be used to detect the optimal path. Each data plane device through which the packet flow has to traverse is considered as a smell spot. The smell value of the smell spot is assigned proportional to the number of interfaces available on the node. Thus the node having the highest number of interfaces gets maximum smell value and least one gets the minimal smell value. The SDAs start from the source node to the adjacent nodes and the visited nodes are marked with the signature of the SDA. Each SDA thus seek the path to the destination by traversing through the 'not visited' nodes. Since the visited nodes are marked, each SDA seeks disjoint path. Thus the execution the algorithm result in multiple possible paths to destination.

Algorithm: SDA based Algorithm for optimal path finding

```
Let
N_sda: Number of agents
N_ss: Number of smell spots
1. Initialise N_sda=Number of interfaces of the source
   node
2. Initialise N_ss=Number of data plane devices in the
   network
3. For i = 1 to N_sda
      Signature Value = i;
4. For i = 1 to N_ss
      Smell value = number of node interfaces *(1 /
      highest interface count of a node in network)
5. For every agent 1 to N_sda,
      Update each agent to the next unvisited node with
      highest smell value
      Update the path including current node
      Update the total link weight value
6. Repeat step 5 until the SDAs reach the destination
```

The smell value of the nodes are decremented by the SDN Controller whenever there is a link failure, interface failure or high bandwidth utilization of the link indicating a possible congestion. The smell values are incremented when the interface become available or bandwidth utilization is below the threshold value. The optimal path has to be selected by comparing all the paths. To find the optimal path among the multiple identified paths, link weight is taken into consideration, apart from smell value of nodes. Each link in the network is assigned a link weight proportional to the interface bandwidth capacity, highest bandwidth is assigned the minimum link weight and the least bandwidth link is assigned the maximum link weight. The path with lowest link weight is chosen as the optimal path.

4 Test Bed Simulation and Results

Mininet
Mininet is an open source network emulation orchestration system for prototyping a large network on a single machine [14, 15]. It can create a network of virtual hosts, switches, links and support SDN Controller and OpenFlow protocol. Mininet runs on standard Linux system and uses virtualisation to emulate a complete network. The Mininet virtual hosts, switches and routers behave like real hardware, though they are created using software and we can send packets through Ethernet interfaces with specified link speed and delay. The advantage is that the same binary code and applications which we run on an actual network can be run in a Mininet network also.

As part of our study, Mininet has been used to create a network test bed for testing the different algorithms with a basic set of parametrised topology, where a set of parameters are passed for a flexible topology.

Floodlight SDN Controller

Floodlight which is an Open Source Apache Licensed Java based SDN controller, is used for the simulations [16]. Being a module loading system, Floodlight SDN controller is easy to extend and enhance. We can consider Floodlight as a set of applications built over a Controller. Floodlight provides the core network services such as Device Manager, Link discovery, Topology Manager and application services such as forwarding, access control and firewall. By default, Floodlight SDN controller uses Dijkstra's algorithm to compute the path from source node to destination node. This algorithm has been replaced by different algorithms and the performance is compared to study the effectiveness of the bio inspired Smell Detection Agent based algorithm.

Test Bed Setup

Mininet is used to create a network topology for evaluation as shown in Fig. 1. The network consists of one SDN controller **C0** and 12 data plane devices from **s1** to **s12**. The source and destination nodes are two hosts **h1** and **h2** connected to data plane devices **s1** and **s12** respectively. Each data plane device is connected to the Controller. In the Floodlight SDN Controller the files related to topology such as *floodlightcontroller/topology/TopologyInstance.java, Topology Manager.java* were modified to incorporate the SDA based algorithm. The smell value of a node is assigned based on the number of interfaces in the node and the highest interface count of a node in network. The smell value ranges between 0 and 1. In the network topology shown in Fig. 1, the node **s4** has the highest number of interfaces on network, i.e. 5. Then **s4** has a smell value of $5/(1/5)$, equals to 1. Smell value of **s1** is $3/(1/5)$, equals to 0.6. The agents pass through the "not visited" nodes and the number of hops to reach the destination is an indicator of the distance to be traversed. In the present example shown in Fig. 1, the agents start from the node s1 and seek the unmarked next hops with the highest smell value and mark the smell spot as visited. This is iteratively done till the destination node of s12 is reached. Thus from s1, three paths are chosen to the destination:

Path 1: **s1** $\leftrightarrow$ **s4** $\leftrightarrow$ **s6** $\leftrightarrow$ **s12** – Distance = 3 hops
Path 2: **s1** $\leftrightarrow$ **s3** $\leftrightarrow$ **s5** $\leftrightarrow$ **s9** $\leftrightarrow$ **s12** – Distance = 4 hops
Path 3: **s1** $\leftrightarrow$ **s2** $\leftrightarrow$ **s8** $\leftrightarrow$ **s11** $\leftrightarrow$ **s12** – Distance = 4 hops

To find the optimal path among the three identified paths, link weight is considered. The assigned link weights in the test bed topology for each link are marked in Fig. 1. Accordingly from the multiple paths that reach the destination, the one with the least weight is selected as the optimal path. The added link weight for the identified paths are 21, 48 and 18 for Path1, Path2 and Path3 respectively and the optimal path selected in this test bed is Path 3. This was verified by looking at the data plane devices flow table updates by the Controller with the help of modules for routing such as floodlightcontroller/ routing/ForwardingBase.java. The smell value updates were done after getting the network monitoring statistics with the help of modules *floodlightcontroller/statistics/*

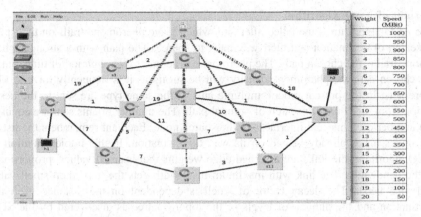

Fig. 1. Network topology for evaluation

StatisticsCollector.java. These modules provide the performance statistics related to bandwidth utilisation and port status.

5 Evaluation and Discussion

Benchmarks allow the evaluation of different algorithms for their performance. Three algorithms were executed on the same test bed after modifying the Floodlight SDN Controller. The algorithms include Dijkstra's algorithm,.Extended Dijkstra's algorithm and a variant of ACO algorithm, which is a bio inspired one.

Dijkstra's Algorithm

The default algorithm for path computation in the Floodlight Controller is classical Dijkstra's Algorithm. In the original Dijkstra's algorithm, neither the nodes nor the links are associated with any weights. In other words, the algorithm does not take into consideration, either the bandwidth capacity of the links or the number of interfaces in the data plane node. Floodlight in SDN controller uses classical Dijkstra algorithm and it considers unit cost on all edges. Dijkstra's algorithm returns the shortest path from host **h1** to host **h2**, and is found to be **s1** ↔ **s4** ↔ **s6** ↔ **s12**. The distance is 3 hops.

Extended Dijkstra's Algorithm

In the Extended Dijkstra's algorithm, both link weight and node weight are taken into consideration. Here each link is assigned a weight proportional to the bandwidth capacity, as shown in Fig. 1. Each node is assigned a node weight based on the number of active interfaces among the total number of available interfaces. For the optimal path identification, the sum of link weights from source to destination is also considered. The link with the lowest weight is considered as the optimal path. Thus in the present example, the path chosen is **s1** ↔ **s4** ↔ **s3** ↔ **s5** ↔ **s9** ↔ **s6** ↔ **s12** with distance as 6 hops and the link weight as 17. The path chosen comprises the high bandwidth links though the number of hops are high.

Variant of ACO Algorithm

The ACO algorithm is modelled after ants which leave pheromone trails on their path in search of destination and their way back to source. The path with a stronger pheromone trail is the chosen path. The pheromone evaporates over a period of time, but in the chosen path the pheromone density will be strong as it is frequently traversed by more ants. In the present model, multiple ants of the same type are sent to the destination from the source in search of optimal path. The number of ants sent are equal to the number of connected interfaces in the source node. Each ant remembers the list of the nodes it has already visited on its way to destination. In this implementation of ACO algorithm, the links are assigned edge weights that is smell values, proportional to the bandwidth. The link with maximum bandwidth gets the maximum smell value and vice versa. The decay factor of smell is dependent on the distance from the destination and the number of levels of the topology tree as discovered by the SDN Controller. Thus as part of ACO algorithm implementation in SDN,

$$\text{Initial smell of link} = (\text{Weight of link/Maximum link weight on network}) * 100 \quad (1)$$

$$\text{Decay factor of smell} = \text{node distance} * (1/(\text{maximum level of tree})) \quad (2)$$

$$\text{Updated smell of link} = \text{link smell} + (\text{link smell} * \text{decay factor}) \quad (3)$$

In the present example, smell values of edges are assigned between *0 and 100* based on bandwidth of links. The smell is updated by considering decay factor on path. The highest smell path selected as the optimal path. Thus from the source host **h1** to destination host **h2**, the multiple paths, their smell values and decay values in the present example are given in Table 1. The path with highest smell value that is Path 2 is selected.

Table 1. Path computation using variant of ACO algorithm in SDN Controller

Path no:	Path nodes	Smell value	Decay factor	Total
Path 1	s1 ↔ s3 ↔ s5 ↔ s9 ↔ s12	150	72	78
Path 2	s1 ↔ s4 ↔ s2 ↔ s8 ↔ s11 ↔ s12	415	251	164
Path 3	s1 ↔ s4 ↔ s6 ↔ s12	195	78	117
Path 4	s1 ↔ s2 ↔ s8 ↔ s11 ↔ s12	310	166	144

We used *Iperf* network bandwidth measurement tool to test the TCP bandwidth performance. The testing time was set to 120 s. The four path finding algorithms were executed one after the other in a Floodlight Controller with the hosts **h1** and **h2** switching as client and server. The *Iperf* tool provided the bandwidth usage in all the four cases where the algorithms were executed by the Floodlight Controller. The results are shown in Fig. 2.

The maximum throughput of 681 Mbps was found to be for SDA based algorithm, followed by variant of ACO algorithm which was equal to 676 Mbps. The least

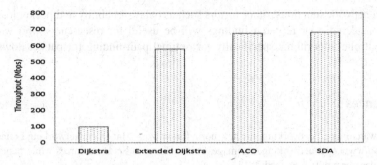

Fig. 2. Iperf bandwidth usage results

throughput of 99.4 Mbps was obtained for the unit weight classical Dijkstra's algorithm. The path selected by each of the algorithm for packet flows from **h1** to **h2** and back during the Iperf execution in the test bed is as given in Table 2. It is seen that the shortest path is selected by the unit weight Dijkstra's algorithm. The SDA based algorithm is the next shorter path with lesser number of hops compared to the other algorithms. When we compare both, the path taken and throughput, it is evident that SDA based algorithm performs better than all the other chosen algorithms.

Table 2. Details of selected paths

Sl No.	Algorithm	Path	Flows
1	Dijkstra	**h1 − > h2**	s1, s4, s6, s12
		h2 − > h1	s12, s6, s4, s1
2	Extended Dijkstra	**h1 − > h2**	s1, s4, s3, s5, s9, s6, s12
		h2 − > h1	s12, s6, s9, s5, s3, s4, s1
3	ACO variant	**h1 − > h2**	s1, s4, s2, s8, s11, s12
		h2 − > h1	s12, s11, s8, s2, s4, s1
4	SDA	**h1 − > h2**	s1, s2, s8, s11, s12
		h2 − > h1	s12, s11, s8, s2, s1

6   Conclusions

The simulation results clearly demonstrate the effectiveness of SDA based algorithm in finding the optimal path. Another point to be noted is that the bio inspired algorithms have shown better performance compared to the conventional algorithms. One of the limitations we faced in Mininet is that the experiments have to be conducted with slower links of 10–100 Mb/s. Here the packets were forwarded through OpenVSwitch rather than dedicated switching hardware. So the experiments have to be conducted with SDN Enabled switching platforms with high speed links of Gb/sec. Moreover realistic traffic load has also to be taken into consideration in the physical SDN test bed. In the present example only two attributes were considered in computing the optimal path which include the distance and the cost of edge in terms of link bandwidth. But in practical network situations, other factors such as delay, reliability and packet loss have to be considered. Our research findings will be useful for researchers who work on SDN controller algorithms specifically for optimal path finding for packet flows.

References

1. Software-defined networking: the new norm for networks. In: Open Networking Foundation. https://www.opennetworking.org/images/stories/downloads/sdn-resources/white-papers/wp-sdn-newnorm.pdf. Accessed 2012
2. McKeown, N., Anderson, T., Balakrishnan, H., Parulkar, G., Peterson, L., Rexford, J., Shenker, S., Turner, J.: OpenFlow: enabling innovation in campus networks. ACM SIGCOMM Comput. Commun. Rev. **38**, 69–74 (2008)
3. Vinod-Chandra, S.S.: Smell detection agent based optimization algorithm. J. Inst. Eng. India Ser. B **97**, 431–436 (2016)
4. Donald, J.: A note on Dijkstra's shortest path algorithm. J. ACM **20**(3), 385–388 (1973)
5. Jiang, J.R., Huang, H.-W., Liao, J.H., Chen, S.Y.: Extending Dijkstra's shortest path algorithm for software defined networking. In: Proceedings of the of IECiE APNOMS (2014)
6. Dorigo, M., Blum, C..: Ant colony optimization theory: a survey (2007)
7. Furculita, M.: Implementation issues for modified Dijkstra's and Floyd-Warshall algorithms in OpenFlow. In: Proceedings of RoEduNet International Conference on Networking in Education and Research (2013)
8. Shahid, A., Fiaidhi, J., Mohammed, S.: Implementing Innovative Routing Using Software Defined Networking (SDN). Int. J. Multimedia Ubiquit. Eng. **11**, 159–172 (2016)
9. Ducatelle, F., Di Caro, G.A., Gambardella, L.M.: Principles and applications of swarm intelligence for adaptive routing in telecommunications networks. Swarm Intell. **4**, 173–198 (2010)
10. Qin, Q., Cheng, S., Zhang, Q., Li, L., Shi, Y.: Particle swarm optimization with interswarm interactive learning strategy. IEEE Trans. Cybern. **46**, 2238–2251 (2016)
11. Zhang, J., Lina, N.I., Chen, X.I.E., Ying, T.A.N., Zheng, T.A.N.G.: AMT-PSO: an adaptive magnification transformation based particle swarm optimizer. IEICE Trans. Inf. Syst. **94**, 786–797 (2011)
12. Solos, I.P., Tassopoulos, I.X., Beligiannis, G.N.: Optimizing shift scheduling for tank trucks using an effective stochastic variable neighbourhood approach. Int. J. Artif. Intell. **14**, 1–26 (2016)
13. Dobrijevic, O., Santl, M., Matijasevic, M.: Ant colony optimization for QoE-centric flow routing in software-defined networks. In: IFIP CNSM (2015)
14. Lantz, B., Heller, B., McKeown, N.: A network in a laptop: rapid prototyping for software-defined networks. In: Proceedings of ACM HotNets 2010 (2010)
15. Mininet. http://mininet.org
16. Floodlight controller. http://www.projectfloodlight.org/floodlight/

A Comparison of Heuristic Algorithms for Bus Dispatch

Hong Wang[1,2], Lulu Zuo[1], Jia Liu[1], Chen Yang[1(✉)],
Ya Li[3(✉)], and Jaejong Baek[4(✉)]

[1] College of Management, Shenzhen University, Shenzhen 518060, China
yangc0201@gmail.com
[2] Department of Mechanical Engineering, Hong Kong Polytechnic University,
Hong Kong, China
[3] School of Computer and Information Science,
Southwest University, Chongqing 400715, China
liyaswu@163.com
[4] School of Computing, Informatics and Decision Systems Engineering,
Arizona State University, Tempe, AZ 85281, USA
jbaek7@asu.edu

Abstract. Bus dispatch (BD) system plays an essential role to ensure the efficiency of public transportation, which has been frequently addressed by the heuristic algorithms. In this paper, five well-exploited heuristic algorithms, i.e. Genetic algorithm (GA), Particle Swarm Optimization (PSO), Artificial Bee Colony algorithm (ABC), Bacterial Foraging Optimization (BFO) and Differential Evolution algorithm (DE), are employed and compared for solving the problem of BD. The comparison results indicate that DE is the best method in dealing with the problem of BD in terms of mean, minimum, and maximum, while BFO obtains the minor lower value of standard deviation and achieves the similar convergence speed in comparison to DE. The performance of PSO seems to outperform the remaining two algorithms (i.e. ABC and GA) in most cases. However, among five algorithms, GA achieves the worst results in terms of the weight estimated objective (i.e. number of departures and average waiting time).

Keywords: Heuristic algorithm · Multiple objective optimization · Dispatch time interval · Bus dispatch

1 Introduction

In recent years, bus dispatch (BD) system has gained the great popularity in the planning of public transportation. The operational cost of Bus Company and the satisfaction of passengers are two main objectives considered [1]. However, those two main objectives involved in BD system are contradictory. The first objective is to reduce the capital expense. However, the decrease of the operational cost of Bus Company would bring about the reduction of the number of departures, which contributes to the longer waiting time of passengers. Therefore, it is essentially important to arrange a reasonable bus dispatching interval for BD.

© Springer International Publishing AG 2017
Y. Tan et al. (Eds.): ICSI 2017, Part II, LNCS 10386, pp. 511–518, 2017.
DOI: 10.1007/978-3-319-61833-3_54

Some researchers have applied the heuristic algorithms to the BD system such as GA, PSO, and BFO. A multi-objective optimization model of BD system was established in [2] to maximize bus company's interests and passengers' satisfaction, and GA was applied to solve this multiple objective problem. Wang et al. [3] presented an improvement of Particle Swarm Optimization by adopting a dispersing strategy to converge to the better solution. In [4], the BD problem that was minimizing the operation cost of bus company and the mean waiting time of all passengers was be solved by an adaptive Bacterial Foraging Optimization. Additionally, an improvement of BFO using differential evolution was proposed in [5], which adopted an adaptive strategy to update the position of the bacteria in chemotaxis process and was demonstrated to be effective in dealing with the BD problem.

Earlier studies mainly focus on the development of the improvements of the standard heuristic algorithms. To verify the performance of those earlier contributions in solving the problem of BD, this paper compares five well-exploited algorithms: BFO, PSO, ABC, GA and DE according to their efficiencies and the convergence speed. This paper is to make the comparison of the five well-exploited heuristic algorithms and discuss the advantages and disadvantages of them. Those five algorithms are all initializing the population, and evaluating the fitness of initialized individuals. The new individual is obtained through the iteration process until the optimal value is found or the termination of iterative numbers is reached.

The rest of the paper is organized as follows: Sect. 2 outlines a briefly introduction of the BD system. In Sect. 3, five heuristic algorithms are described, separately. Section 4 provides the comparison results and discussions. Finally, the summary is presented in Sect. 5.

2 Description of Problems

The BD problem [6, 7] is described as follows. The fitness function (see Eq. (1)) of BD contains two main factors: number of departures and average waiting time. To transform the two objectives as a single objective optimization problem, two weight coefficients (i.e. α and β, where $\alpha + \beta = 1$) are adopted. The mathematic formulation of BD is shown as follows:

$$fit = \alpha \times \frac{\sum_{m=1}^{M} (T_m/\Delta t_m)}{T_s/\Delta t_{\min}} + \beta \times \frac{(\sum_{m=1}^{M} \sum_{n=1}^{N} I_m \times \rho_{mn} \times \frac{\Delta t_m^2}{2})/ \sum_{m=1}^{M} \sum_{n=1}^{N} \lambda_{mn}}{\Delta t_{\max}} \quad (1)$$

s.t.

$$h_{m\min} \leq \Delta t_m \leq h_{m\max} \quad (2)$$

$$\sum_{m=1}^{M} \sum_{n=1}^{N} \lambda_{mn} / Q \sum_{m=1}^{M} (T_m/\Delta t_m) \geq 75\% \quad (3)$$

$$\sum_{m=1}^{M}\sum_{n=1}^{N}\lambda_{mn} > 2.5 \times L \times \sum_{m=1}^{M}(T_m/\Delta t_m) \qquad (4)$$

where $m \in (1\ldots m \ldots M)$, $n \in (1\ldots n \ldots N)$ are the mth period and the nth station. T_s is the total operational period. The dispatching interval in the mth period is Δt_m. The time duration in the mth period is T_m. The total number of departures in the mth period is I_m. The arriving passenger's number and passenger's arrival rate in the mth period at the nth station are respectively λ_{mn} and ρ_{mn}. L represents the length of bus line and Q is the passenger capacity. While the range of Δt_m is h_{mmin} to h_{mmax} in the mth period. The penalty function factor is χ, and $\chi = 1000$.

3 Heuristic Algorithms

3.1 GA

In the simulation experiment of the GA [8], the first stage is to randomly generate candidate solutions, and sort the fitness of the candidate solutions. Secondly, select the better individuals as parent solutions according to the fitness values and update new individuals through crossover and mutation. Finally, the new individuals are evaluated according to the fitness value.

3.2 PSO

In PSO [9], it is illustrated that each particle can communicate and share information with others. The particles estimate the fitness of current location and make the record of the best position p_{id} by comparing with other locations as well as the global optimal position p_{gd} by comparing with other particles. Then the new position of the particle x_{id} is updated according to the shared information and the velocity v_{id} using Eqs. (5) and (6).

$$v_{id}^{t+1} = \omega v_{id}^{t} + c_1 rand(0,1)(p_{id}^{t} - x_{id}^{t}) + c_2 rand(0,1)(p_{gd}^{t} - x_{id}^{t}) \qquad (5)$$

$$x_{id}^{t+1} = x_{id}^{t} + v_{id}^{t+1}, \, i = 1,\ldots, N, \, d = 1,\ldots, D \qquad (6)$$

where N is particles' number, D is the dimensions of search space, and t represents the iterative number. ω indicates the inertia weight. Additionally, parameters c_1 and c_2 are acceleration factors.

3.3 ABC

The employed bee in ABC method [10, 11] represents the information of the food source. We should initialize the food source x_i firstly. Equation (7) shows the process of new food source v_i searching which will be conveyed to the onlookers. The new food source is chosen randomly by each onlooker bee according to a probability p_i provided by Eq. (8). Additionally, a scout will be employed for moving to new food sources using Eq. (9) when the performance of the employed bee cannot be improved after several evolutions.

$$v_{ij} = x_{ij} + \phi_{ij}(x_{ij} - x_{kj}) \tag{7}$$

$$p_i = fit_i / \sum_{n=1}^{N} fit_n \tag{8}$$

$$x_{ij} = lb_j + rand(0,1)(ub_j - lb_j), \ i = 1, \ldots N, j = 1, \ldots, D \tag{9}$$

where N is the food sources' number, while the D is the variables' number. ϕ_{ij} is a random number between 1 and -1, k is the index of a randomly selected solution, fit_i is the fitness of the ith food source, lb_j and ub_j are the lower and upper limits of problem variable j.

3.4   BFO

In the standard BFO [12], chemotaxis process of bacteria containing tumbling and swimming is the major behavior for the optimal solution. The chemotaxis loop is given in Eq. (10). After that, the bacteria of weak foraging ability are waived, and the individuals with better performance are reproduced. Additionally, the elimination & disperse process is employed after the chemotactic progress to avoid getting caught in the local convergence.

$$\theta^i(j+1, k, l) = \theta^i(j, k, l) + C(i) \times \Delta(i) / \sqrt{\Delta^T(i)\Delta(i)} \tag{10}$$

where $\theta^i(j, k, l)$ indicates the ith bacterium in the jth chemotactic, the kth reproductive, the lth elimination & dispersal step, C is the step size, and Δ is given a vector in the random direction between -1 and 1.

3.5   DE

DE [13] is a typical evolutionary method to improve the candidate solution iteratively. Equations (11) and (12) shows the process of mutation and crossing to generate new generation. If the current fitness is superior to the previous best, the best solution need to be updated.

$$v_i = x_{r1} + F(x_{r2} - x_{r3}) \tag{11}$$

$$u_{ij} = \begin{cases} v_{ij} & if(rand_j(0,1) \leq CR)or\ j = j_{rand} \\ x_{ij} & otherwise \end{cases}, \ i = 1, \ldots N, j = 1, \ldots, D \tag{12}$$

where N is the population size, D is the dimensionality of variables, while F is constant mutation factor. $r1$, $r2$ and $r3$ are randomly selected indexes ranging in $[0, N]$. CR $\in$ $[0, 1]$ is a crossover constant and $j_{rand} \in [0, D]$ is a randomly selected index.

3.6 Pseudo-code

Table 1 shows the pseudo-code of heuristic algorithms for BD model.

Table 1. The pseudo-code of algorithms

01	Parameters initialization
02	Generate initial population X;
03	Evaluate the fitness of the population
04	$X_{best} = X$;
05	$P_{best} = $ fitness(X) ;
06	$CheIter = 1$;
07	While $CheIter \leq MaxEFs$
08	Update new individuals X
09	Evaluate each individual fitness;
10	If fitness(X)$<P_{best}$
11	$X_{best} = X$;
12	$P_{best} = $ fitness(X) ;
13	End
14	$CheIter = 1 + CheIter$;
15	End
16	Output X_{best} , P_{best};

4 Simulation Test and Discussion

4.1 Parameter Settings and Encoding Fitness

Each object represents a potential solution in solving BD problem. That is $\theta = [\Delta t_1, \Delta t_2 \ldots \Delta t_D]$. If Constraint 1 (i.e. Eq. (2)) is satisfied, the solution would be a feasible for the optimization problem. Otherwise, the solution should be removed. Constraints 2 and 3 (i.e. Eqs. (3) and (4)) are the penalty functions. Thus, the fitness function of the real encoding process is formulated as follows:

$$fit = (\alpha \times \frac{\sum\limits_{m=1}^{M} \frac{T_m}{\Delta t_m}}{T_s / \Delta t_{\min}} + \beta \times \frac{(\sum\limits_{m=1}^{M} \sum\limits_{n=1}^{N} I_m \times \rho_{mn} \times \frac{\Delta t_m^2}{2}) / \sum\limits_{m=1}^{M} \sum\limits_{n=1}^{N} \lambda_{mn}}{\Delta t_{\max}} +$$
$$\chi \times (\left| \sum\limits_{m=1}^{M} \sum\limits_{n=1}^{N} \lambda_{mn} / (Q \sum\limits_{m=1}^{M} \frac{T_m}{\Delta t_m}) - 75\% \right| + \left| \sum\limits_{m=1}^{M} \sum\limits_{n=1}^{N} \lambda_{mn} - 2.5 \times L \times \sum\limits_{m=1}^{M} \frac{T_m}{\Delta t_m} \right|)) \tag{13}$$

The parameters settings are referred to literature [14, 15]. The swarm size is 40, 1000 is the maximum iterations, and the operation times is set to 10 with the search space dimension 16. The crossover probability of GA is 0.65, and the mutation probability is 0.1. In BFO method, the number of swims, chemotaxis, reproduction and elimination & dispersal are separately $N_s = 4$, $N_e = 100$, $N_{re} = 5$ and $N_{ed} = 2$. $P_{ed} = 0.25$ is the probability of elimination and dispersal while the step size C is 0.1. More parameters settings of PSO and DE are displayed as follows: In PSO, $c_1 = c_2 = 1.193$, and $\omega = 0.721$; In DE, $F = CR = 0.9$.

The basic data and value of variables from the BD system are referred to [6]. The operation period is 6:00–22:00 and divides into $m = 16$ h. $n = 13$, $L = 15$ km, $Q = 60$, and $h_{m\,min} = 2$. The upper bounds of bus dispatching intervals $h_{m\,max}$ are 16, 8, 8, 16, 16, 8, 8, 16, 16, 16, 16, 8, 8, 16, 16, 16 min, separately. The volume of passenger in each time interval at each station is also in [6].

4.2   Experiment Results and Discussion

The comparison results are presented in Table 2, and the best solutions are highlighted in bold. Figures 1 and 2 shows the different average convergence curves. As shown in Table 2, the running time obtained by BFO is the longest, while GA consumes the shortest running time. In addition, DE can provide the best results of mean, minimum and maximum, and the results obtained by GA are worst. BFO and ABC almost reach to the lower standard deviation, while PSO achieves to the maximum.

The iteration processes of algorithms are drawn in terms of iterations and the corresponding optimal fitness value. From Figs. 1 and 2, DE is obviously more conductive in optimal search and converges to the better solution at earlier stage. The reason might rely on the mutation equation for DE which generates new variable at a time from previous multiple variables. Except for the DE, the BFO method is superior to other three algorithms in convergence speed. The GA and ABC methods are less conductive in comparison to other algorithms.

Table 2. The fitness values and computational cost of the algorithms

α		PSO	BFO	ABC	DE	GA
0.2	Values	0.059 ± 0.001	0.057 ± 0.001	0.061 ± 0.001	**0.053 ± 0.001**	0.062 ± 0.002
	Time	8.965	30.259	16.792	8.111	**5.604**
0.4	Values	0.115 ± 0.003	0.109 ± 0.002	0.117 ± 0.002	**0.100 ± 0.002**	0.118 ± 0.002
	Time	8.074	30.937	16.852	8.223	**5.334**
0.6	Values	0.168 ± 0.005	0.161 ± 0.002	0.171 ± 0.002	**0.143 ± 0.003**	0.172 ± 0.004
	Time	8.185	31.219	21.250	8.587	**6.224**
0.8	Values	0.224 ± 0.008	0.214 ± 0.004	0.225 ± 0.005	**0.189 ± 0.005**	0.226 ± 0.006
	Time	8.095	37.627	18.290	9.470	**6.137**

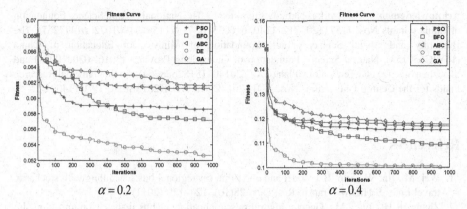

$\alpha = 0.2$ $\alpha = 0.4$

Fig. 1. The iteration process of the algorithms when $\alpha = 0.2$ and $\alpha = 0.4$

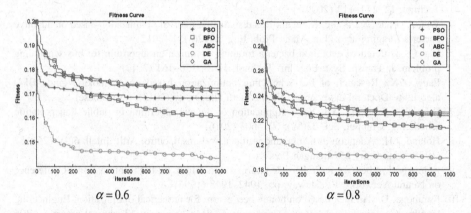

$\alpha = 0.6$ $\alpha = 0.8$

Fig. 2. The iteration process of the algorithms when $\alpha = 0.6$ and $\alpha = 0.8$

5 Conclusion

In this study, the typical heuristic algorithms have been compared in solving the problem of BD. The comparison results of the mean, minimum and maximum indicate that DE can find the best solution. In terms of the efficiency, GA can achieve the shortest running time, but the weight fitness is the worst. Though the BFO consumes larger computational complexity, the weight fitness value is also slightly worse than the DE. For the standard deviation, BFO and ABC reach to the lower standard deviation with higher stability, while PSO reaches to the maximum with larger randomness. Therefore, it is appropriate to choose the suitable algorithms according to the specific requirement in the real-applications.

Acknowledgment. This work is partially supported by The National Natural Science Foundation of China (Grants Nos. 71571120, 71271140, 61603310, 71471158, 71001072, 61472257), The Humanity and Social Science Youth Foundation of Ministry of Education of China (16YJC630153), Natural Science Foundation of Guangdong Province (2016A030310074) and Shenzhen Science and Technology Plan (CXZZ20140418182638764), the Fundamental Research Funds for the Central Universities Nos. XDJK2014C082, XDJK2013B029, SWU114091.

References

1. Wei, M., Jin, W., Sun, B.: Model and algorithm for regional bus scheduling with stochastic travel time. J. Highw. Transp. Res. Dev. **28**(10), 124–129 (2011)
2. Zhang, R.H., Jia, J.M.: Genetic algorithm's application in bus dispatch optimization. In: International Conference of Chinese Transportation Professionals, pp. 137–146 (2011)
3. Wang, M., Wang, K.: Study on bus scheduling based on particle swarm optimization. Inf. Technol. **12**, 111–113 (2009)
4. Wei, Z., Zhao, X., Wang, K., et al.: Bus dispatching interval optimization based on adaptive bacteria foraging algorithm. Math. Prob. Eng. **2012**(3), 1 (2012)
5. Liu, Q.: Differential evolution bacteria foraging optimization algorithm for bus scheduling problem. J. Transp. Syst. Eng. Inf. Technol. **12**(2), 156–161 (2012)
6. Fang, Z.X.: Research of bus scheduling optimization based on chemokine guide BFO algorithm. Doctoral dissertation, Northeastern University (2013). (in Chinese)
7. Ding, Y., Jiang, F., Wu, Y.Y.: Application of genetic algorithm in public transportation scheduling. Comput. Sci. **43**(S2), 601–603 (2016)
8. Holand, J.H.: Adaption in natural and artificial systems. Control Artif. Intell. **6**(2), 126–137 (1975). University of Michigan Press
9. Kennedy, J., Eberhart, R.C.: Particle swarm optimization. In: IEEE International Conference on Neural Networks, Piscataway, pp. 1942–1948 (1995)
10. Karaboga, D.: An idea based on honey bee swarm for numerical optimization. Engineering Faculty, Computer Engineering Department, Erciyes University, Technical report - TR06 (2005)
11. Karaboga, D., Basturk, B.: A powerful and efficient algorithm for numerical function optimization: artificial bee colony (ABC) algorithm. J. Glob. Optim. **39**(3), 459–471 (2007)
12. Passino, K.M.: Biomimicry of bacterial foraging for distributed optimization and control. IEEE Control Syst. **22**(3), 52–67 (2002)
13. Storn, R., Price, K.: Differential evolution – a simple and efficient heuristic for global optimization over continuous spaces. J. Glob. Optim. **11**(4), 341–359 (1997)
14. Niu, B., Wang, J., Wang, H.: Bacterial-inspired algorithms for solving constrained optimization problems. Neurocomputing **148**, 54–62 (2015)
15. El-Abd, M.: Performance assessment of foraging algorithms vs evolutionary algorithms. Inf. Sci. **182**(1), 243–263 (2012)

Simulation and Application of Algorithms CVRP to Optimize the Transport of Minerals Metallic and Nonmetallic by Rail for Export

Lourdes Margain[1(✉)], Edna Cruz[2], Alberto Ochoa[3(✉)],
Alberto Hernández[4], and Jacqueline Ramos Landeros[1]

[1] Universidad Politécnica de Aguascalientes, Aguascalientes, Mexico
lourdes.margain@upa.edu.mx
[2] Instituto Tecnológico Superior de Naranjos, Naranjos, Mexico
[3] Maestría en Cómputo Aplicado, UACJ, Ciudad Juárez, Mexico
alberto.ochoa@uacj.mx
[4] FCAeI, Universidad Autónoma del Estado de Morelos, Cuernavaca, Mexico

Abstract. Metallic and nonmetallic minerals produced by the State of Puebla, for convenience, can be transported in containers by train to a seaport cargo exported to other countries; for transporting minerals by railways, it must analyze what the most optimal route to bring the product, taking into account various factors involved through each route in order to get a greater benefit for the transportation of minerals by train. This article comparing two metaheuristic algorithms applied to Capacitated Vehicle Routing Problem (CVRP), in order to determine which algorithm gives better optimization solutions that help you make the best route for the transfer. As a final result the implementation of an optimization algorithm Ant Colony was more successful in the runtime that on Genetic Algorithm; because it is slow to find an optimal solution among all generations.

Keywords: Metallic and nonmetallic · Export · Container train · Sea port · Metaheuristics algorithms · Capacitated Vehicle Routing Problem (CVRP)

1 Introduction

The mining sector is one of the main sources of economy in Mexico, where the industrial sector, which provides social benefits that put Mexico in the main destination for investment in mineral exports in developed in 24 of the 32 states of the Republic Latin America and the fourth worldwide [1]. The state of Puebla is located in the east central area of Mexico and is strategically located within four physiographic regions; Neo Volcanic Axis, the Sierra Madre del Sur, the Sierra Madre Oriental and the Gulf Coastal Plain. Which makes the state a major producer mostly non-metallic minerals [2] (Table 1).

© Springer International Publishing AG 2017
Y. Tan et al. (Eds.): ICSI 2017, Part II, LNCS 10386, pp. 519–525, 2017.
DOI: 10.1007/978-3-319-61833-3_55

Table 1. Volume of mining production 2012–2013 in the state of Puebla (tonnes).

Minerals	2012	2013
Metallic		
Fierro	6, 925 t	6, 925 t
Non-metallic		
Caliza	5,220,082 t	4,437,070 t
Arena	4,292,000 t	3,651,563 t
Grava	2,007,008 t	1,705,957 t
Calcita	1,567,678 t	1,506,767 t
Arcilla	555,000 t	471,750 t
Yeso	258,980 t	220,133 t
Feldespato	352,308 t	138,279 t

Documentary Management Control and Strategic Indicators, Ministry of Economy, National Institute of Statistics and Geography, S.H.C.P.

2   Problematic

Minerals like Iron, feldspar and gypsum exploited by the state of Puebla, are on the list of major minerals exported our country [2]; so as the need to exploit the mining sector of the State of Puebla to transport their minerals by railways to a nearby seaport where they can be exported. To do this you need to consider various factors that influence the transfer from Puebla to the seaport in order to find the best route for transporting minerals.

3   Justification

In the railway system in Mexico we can find railways intermodal corridors for maritime type where each route has its weight restrictions (Fig. 1).

In this investigation will be comparing the routes the train station in Puebla to the station near the seaport in Veracruz Port by train railways and the second route to the train station at the seaport of Altamira, Tamaulipas; by metaheuristics techniques into account the characteristics of minerals to be transported, in this case to the Fierro, Feldspar and Yeso in this case to choose the right type container train to use. In the course of the two routes, the train passes through different stations to which call nodes; with the help of algorithms routes will be analyzed to compare the benefits of both. According to the characteristics of the mineral, we implemented the wagon called nacelle, specialized to transport goods as ore, coal, metals, scrap, etc. [4] (Table 2).

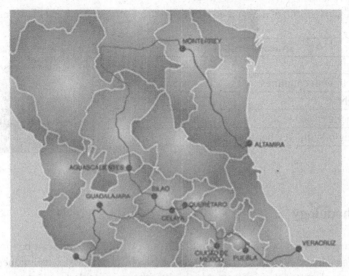

a)   Intermodal maritime corridors

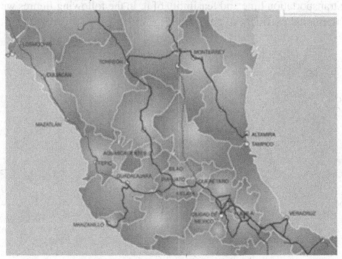

b)   Railways classified by weight capacity.

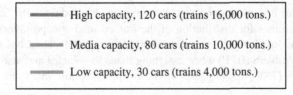

High capacity, 120 cars (trains 16,000 tons.)

Media capacity, 80 cars (trains 10,000 tons.)

Low capacity, 30 cars (trains 4,000 tons.)

Fig. 1. Railways with maritime intermodal corridors and symbology weight capacity on lines [4].

Table 2. Characteristics of the car used for transporting Fierro, Feldspar and Plaster mineral [4].

Nacelle bulk	
Length	15.8 mts.
Lenhth with couplings	17.4 mts.
Height	2.87 mts.
Capacity	90 a 100 tons
Net Weight Without Charge	29.7 tons

4 Methodology

The logistics involved in the transport of minerals by rail for export can be modeled by CVRP (Capacitated Vehicle Routing Problem) with metaheuristic techniques to analyze a series of routes that indicate which algorithm is suitable to adapt to each route optimizing transportation time and feasibility of it. In the following figures we show the maritime intermodal corridors to compare (Fig. 2).

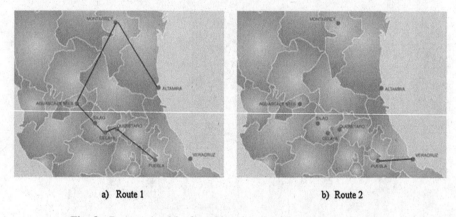

a) Route 1   b) Route 2

Fig. 2. Routes a1 y b2 of maritime intermodal corridors to sea port.

In the literature it is mentioned that the optimization model CVRP, it is to determine the optimal routes for distribution of the limited load transportation to different geographical locations; It is an optimization problem type NP- Hard has to do with
Bin Packing Problem (BPP) when assigning loads to vehicles and also a Traveling Salesman Problem (TSP) that suggests the best route to follow [5].

4.1   Genetic Algorithms (GA)

The function of algorithm executed is to cross of individual's greater ability to improve the population and eliminate individuals of lesser capacity, to find the best chromosome that is the solution of the problem [6, 7]. In the literature there are some basic concepts to better understanding of genetic algorithms [6]:

- Mutation: operation where selected at random and change one or more genes on chromosome; it occurs with low probability.
- An example taken from literature Sait and Youssef (1999), a genetic algorithm follows defined [6]:

Genetic Algorithm for TSP. These algorithms applied to the TSP, in execution represent a solution as a sequence of nodes through permutations (1, ..., n) while operators crossover and mutation may generate n-úplas that not be permutations

4.2   Ant Colony Optimization (ACO)

The ant colony optimization gives various solutions by iterative procedures where each artificial colony casts a solution at each iteration, respectively until an end condition. The first such algorithm was applied to the traveling salesman problem (TSP) to obtain optimum results (Guirao et al. 1996) [8, 9].

Travelling Salesman Problem (TSP) Applied to Ant Colony Optimization (ACO).
Ants provide solutions probabilistically with pheromones that are track; for the case of TSP the following formulas represent one rule to follow:

$$p_{ij}^k(t) = \frac{[\tau_{ij}(t)]^\alpha \cdot [\eta_{ij}]^\beta}{\sum_{l \in N_i^k} [\tau_{ij}(t)]^\alpha \cdot [\eta_{ij}]^\beta} \mid \text{With} \, j \in N_i^k \tag{1}$$

Where $p_{ij}^k(t)$ is the probability that an iteration t algorithm, the ant k, currently located in the city i, choose the city j as next stop. N_i^k it is the set of cities not yet visited by the ant k. $\tau_{ij}(t)$ is the cumulative amount of pheromone on the arc (i, j) of the network in iteration t. η_{ij} is the heuristic information for which in the case of TSP, the inverse of the distance between the cities is used i and j. α and β are two parameters of the algorithm that can be edited [9].

4.3   Design of Experiment

The optimization algorithms based on Ant Colony are intended to represent various agents using random methods, while sharing information routes that are traveled. In genetic algorithms it is simulated genetic evolution where in each generation are selected the fittest and are crossed to generate better adapted descendants; considering that mutations are also generated in the process.

Testing by Genetic Algorithm. With a genetic algorithm executed by a Java applet, testing for route optimization are performed in the first test was conducted between 10 nodes that simulate the cities featured in the journey by train from Puebla to the port of Altamira. In the second test simulate 7 nodes representing the number of stations that are in the path of the state of Puebla to the port of Altamira. **Test 1 by genetic algorithm for TSP:** In the test parameters 1, the simulation of ten nodes are with mutation probability of 10%, the probability of crossing 60% and a maximum number generations of 10. Six tests which showed different routes to reach the optimum were made.

Testing by Algorithm Ant Colony Optimization. With an ACO algorithm for TSP executed on a Java applet, two tests where the first one route with 10 nodes representing cities that appear in the journey by train from Puebla to the port of Altamira analyzes them were performed. In the second test in 7 nodes representing the number of stations that are in the path of the state of Puebla to the port of Altamira they are simulated. **Test 1 by ant colony optimization algorithm for TSP:** Tests conducted with 10 nodes representing cities that are in the path of the train from Puebla to the port of Altamira. Pheromone initial value on each line of 2.00 and ants value of 2.00 to select the next node.

5   Results

To compare the efficiency of two algorithms two tests were conducted, the first was performed with a map of 10 nodes of which the location is the same for the different executions. In the second test it was performed with a map 7 nodes where they remained in the same location for each run (Table 3).

Table 3. Results shown in the execution of the AG and ACO algorithms for Travelling Salesman Problem.

Algorithm	E1	E2	E3	E4	E5	E6
	Tests with 10 nodes					
AG	Route 1	Route 2	Route 3	Route 4	Route 5	Route optimal
ACO	Route 1	Route 2	Route 3	Route optimal		
	Tests with 7 nodes					
AG	Route optimal	Route optimal	Route optimal			
ACO	Route 1	Route 2	Route 3	Route optimal		

Own source made with data obtained with the implementation of algorithms in java applets [10, 11]

6   Analysis and Discussion

In the algorithm Ant Colony Optimization by Ant cycle model; the ants represent the train carrying minerals are dispersed among the nodes representing the cities that exist between the state of Puebla and Altamira. With the application of an algorithm of this kind, it is favored by the local update of pheromones used to detect the busiest and most convenient routes for transport.

7   Conclusion and Future Work

Each optimization problem contains certain parameters specific; to implement an optimization algorithm to some problems need to first identify the components that carry the same, for choose which metaheuristic technique is closer to solution. In conclusion an algorithm of Ant Colony Optimization it is more adaptable to providing solutions to Travelling Salesman Problem, because approaching showing for the routes closest to go, in less runtime. For the genetic algorithm, when exist fewer nodes on the route, yields better results; however, as more nodes or cities are added and increases the complexity, the ACO algorithms have better runtime.

References

1. Web Reference: The importance of mining in Mexico. http://www.industriamineramexicana.com/2013/02/
2. Mexico, D.F., Salazar, D.Q.: National Institute of Statistics and Geography
3. Bravo Diaz, M.A.R.U.O., Maritima, I.C.E.P.V.: How to export seaborne
4. Web Reference: Ferromex. http://www.ferromex.com.mx/
5. Cardozo, J.P.O.: Solution to the problem of vehicle routing with limited capacity "PTRC" through heuristic scanning and the implementation of the genetic algorithm Chu-Beasley (2013)
6. Vargas, G.G., Aristizabal, F.G.: Metaheuristics applied to vehicle routing. A case study: Part 1: formulation of the problem. Eng. Res. **26**(3), 149–156 (2006)
7. Valencia, E.: Optimization through genetic algorithms. An. Inst. Eng. Chile **109**(2), 83–92 (1997)
8. Guirao, D.A.: Application of ant algorithms for solving a problem balancing robotic assembly lines (2012)
9. Boats, L., Rodriguez, V., Alvarez, M.J., Robusté, F.: Algorithm based on ant colony optimization by solving the problem of freight from multiple sources to multiple destinos. Santander, Spain (2002)
10. Java Applet: Alberto Arceti. Traveling Salesman Problem (Genetic Algorithm). https://soporte900.wordpress.com/2011/11/13/proproblema-del-viajante-tsp/
11. Java Applet: Марк Биргер. Ant Colony Optimization Demo. https://github.com/kusha/ant-colony

Dialog System Applications

User Intention Classification in an Entities Missed In-vehicle Dialog System

Ke Zhang[1,2(✉)], Qingjie Zhu[3,4], Naiqian Zhang[1,5], Zhixin Shi[3],
and Yongsong Zhan[6]

[1] Guangxi Key Laboratory of Trusted Software,
Guilin University of Electronic Technology, Guilin 541004, Guangxi, China
Zhuatou2014@163.com
[2] Guangxi Colleges and Universities Key Laboratory of Intelligent Processing
of Computer Image and Graphics, Guilin University of Electronic Technology,
Guilin 541004, Guangxi, China
[3] Guangxi Key Laboratory of Cryptography and Information Security,
Guilin University of Electronic Technology, Guilin 541004, Guangxi, China
[4] Guangxi Cooperative Innovation Center of Cloud Computing and Big Data,
Guilin University of Electronic Technology, Guilin 541004, Guangxi, China
[5] Key Laboratory of Cloud Computing and Complex System,
Guilin University of Electronic Technology, Guilin 541004, Guangxi, China
[6] Guangxi Experiment Center of Information Science,
Guilin University of Electronic Technology, Guilin 541004, Guangxi, China

Abstract. In the human computer dialog system in vehicle environment, some dialogue entities are usually left out by human after several dialog turns. This causes troubles to classify user's intention in a period of chat history. A usual solution for this problem is adding context information to expand the current question. This method causes a trend to generate multiple entities in the expanded question and decreases the classification accuracy of users' intention. In this paper, an RNN based entity recognition model is built to recognize entities in the current problem. If the topic related entities are recognized, the intention and property are classified respectively using LDA and word2vec models; otherwise entities in context information are added to complete the question before intention classification. Experiments show that the proposed method has about 9.4% improvement in precision and 2.3% improvement in recall compared with the traditional context expansion method.

Keywords: Intention classification · Entity identification · Question expansion

1 Introduction

Along with the rapid development of speech recognition, speech synthesis and smart mobiles and other devices, natural dialogue systems play more important roles in human-machine interaction area, such as the BML Realizer "Elckerlyc" [1], Digital intelligent life forms at the University of Southern California (Creative Agent) [2]. These systems can answer the user's daily questions or provide ticket information query service, even with inaccurate speech recognition results. In some special

© Springer International Publishing AG 2017
Y. Tan et al. (Eds.): ICSI 2017, Part II, LNCS 10386, pp. 529–537, 2017.
DOI: 10.1007/978-3-319-61833-3_56

situations such as operating the air conditioner or rearview mirror when driving a car, the natural dialogue system can simplify the car operation and improve the safety of driving [3]. Recent researches demonstrate that supporting human-computer interaction becomes a creative trend of vehicle intelligent systems [4, 5]. Currently, the dominate vehicle natural dialog system is based on keyword spotting (KWS), including entity names, voice control commands and so on. However, not all the necessary information is provided at one user query, and then the intention of an incomplete question cannot be accurately classified. In such cases, the KWS-based method fails to reply. An example of this problem is shown below:

U1:   What is automatic mode of air conditioning?
S1:   Automatic operation mode means the air conditioning automatically adjust air flow, air distribution and temperature
U2:   How can I turn it on?
S2:   Please press the "air-conditioned 3" button to turn on the automatic operation mode.

In the above example, U2 is difficult to understand if we do not look back to U1 and S1. To solve this problem, Bhargava et al. [6] added context information to expand input questions and used the SVM-HMM model for intent prediction. Xu et al. [7] proposed a recurrent neural network (RNN) based approach that directly used the previous model prediction as additional features for the current round and the resulting model significantly outperforms SVM with contextual features. Shi et al. [8] compared the models of single/multi SVM plus trigram and joint training RNN models, and demonstrated the performance of the joint training multi-RNN model with context information in each query. However, after expanding every input question with context information, there may be multiple entities and other interference information, which conversely causes a decrease of intention classification accuracy.

In this paper, a novel question intention classification method is proposed based on entity identification and context-sensitive information. At first, an RNN based entity recognition model is built to recognize entities in the current question. If any entities are recognized, the question intention and property will be classified respectively using the LDA and word vector models; otherwise entities in context information are added to complete the question before intention classification. According to the analysis of large numbers of in-vehicle dialogue statements, a dialogue with the same intention is generally no more than three rounds. By distinctively using context information, the proposed method has improved the intention classification accuracy of incomplete questions while ensuring the accuracy of complete questions.

This paper is organized as follows: Sect. 2 introduces the system framework. Section 3 describes the dialogue intention classification with contextual information, and question property classification. Experiments are given in Sect. 4, which compares the performance of the proposed method and the traditional context extension method. Section 5 is our summary.

2 System Framework

The framework of in-vehicle dialog system is shown in Fig. 1. It consists of two components: control order module and query dialog module. Control order (CO) module is based on slot-filling method [6]. The query dialog module, dialog management module are based on the finite state transition diagram method. In this work, we classify the entire conversation into nine states according to the theme and actions, which are air conditioning, rear view mirror, headlight, seat, seat belt, headrest, steering wheel, child seat and door lock. Dialogue will switch back and forth in these nine states.

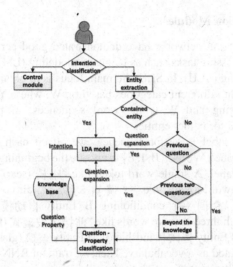

Fig. 1. The framework of system

The workflow of this system is that user speaks to the dialog system, then the system will judge the user's request is vehicle control order or car manual query. If it is a control order, the control order module will find the most possible control order and check whether the parameters are given completely and accurately. If all the parameters are given, the control order will be sent to the operation system. If lacking some parameters, the dialogue module will prompt missing parameters to users until all the parameters are ready. If it is about vehicles manual, the query dialogue module will be activated, and the processing flow is as follows:

- Extract entities from sentence using entity extraction tools based on RNN(recurrent neural network) model [9].
- Judge whether the current question contains entities, if not, last two dialogues will be used to expand current question.
- Generate answers to user according to semantic based search results.

The control module is based on slot-filling method, which is a typical method used in dialogue systems, and usually has high accuracy; therefore this article will not detail the implementation process of control module.

3   Contextual Intention Classification

This part introduces the proposed method in detail. As a result of the system is Chinese based, some Chinese words will be used in part 3 and part 4.

3.1   Entity Extraction Module

Recently, recurrent neural networks have demonstrated good performance in various natural language processing tasks such as language modeling (LM) [9, 10], and spoken language understanding (SLU). In SLU, the main focus is to obtain a optimal model P (Y|W) to classify Y into different categories based on W, where Y can be a sequential variable in slot tagging and W is the word sequences. In this article, we use Yik-Cheung's method to extract entities.

Since the output label is a slot sequence, we attach each word token with a position-specific slot label. We use {B, I, E} to encode the beginning, middle, and ending positions of the slot label. A single word token like "座椅 (seat)", it becomes "座椅 (seat):Entity". A two-word tokens like "空调自动模式 (air conditioning automatic mode)", it becomes "空调 (air conditioning):B Entity 自动模 (automatic mode): E Entity". A token with three or more words like "前挡风玻璃式 (front windshield)", it becomes "前 (front):B Entity 挡风 (windshield):M Entity 玻璃 (glass):E Entity". A joint word slot token is treated as a vocabulary. Then we train an RNN language model per domain to model by $P(W, Y, \Delta d) = \prod_{i=1}^{N} p((w : y)|(w : y)_{i-1}, h_{i-1}, \Delta d)$, where Δd is the parameter of domain d. Figure 2 shows the RNN architecture.

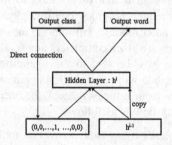

Fig. 2.   RNN data generation for slot tagging

In the training phase, we use four categories (conditioning, rear view mirror, headlight and seat) of real user questions which have marked entities, and we used RNN training set to train a tri-gram language model with word tokens attached with slot labels.

3.2   User Intention Classification

In the vehicle interactive system, user intention classification controls the transition of the entire conversation state. The intention module of the system is divided into two layers. The first layer of classification decides the theme of dialogue intention, and the second layer of intention classification determines the attributes of the current state of the dialogue statement. In the word meaning of a text-based classification task, the widely used methods are DBN, LDA model and vector classification method which based on words. DBN achieved good results in the long text classification, however, dialogue systems are often dozens of phrase or word level of intention judgment. Therefore, we need to find a short-text oriented classification method shows that LDA model can achieve well in short text classification. Thus, we use LDA model in the first layer of intention classification.

In order to solve the problem of the low accuracy of intention classification from information-incomplete questions, RNN model are used to recognize entities, analyze whether the question is lack of information or not. Context entity which be added in information-incomplete question then conducts intention after expansion.

The main process of LDA used in text classification is shown as follows: Assume that the text is expressed as a probability distribution on several linear latent topics/themes. The probability distribution of the text on the whole topic set, i.e., $d = \{K_1, K_2, \ldots\ldots, K_T\}$. (T is the number of themes). Notation conventions of LDA model is applied to the document set topic modeling are as follows:

- Word is the basic unit of text data, with $\{1, 2, \ldots V\}$ sub-index vocabularies. The first V vocabulary word was represented by an N-dimensional vector, wherein for any $u \neq v, W_v = 1, W_u = 0$.
- The document is a sequence of N-word, represented by $d = \{W_1, W_2, \ldots\ldots, W_n\}$, W_n is the sequence of the n-th word.
- The documentation set is a collection of M documents, represented by $D = \{d_1, d_2, \ldots\ldots, d_m\}$. Considering that there are k topics, the probability of document d in the first word W_n can be expressed as follows:

$$P(w_i) = \sum_{j=1}^{T} P(w_i \mid z_i = j) P(z_i = j). \tag{1}$$

wherein, z_i is latent variable, which represents the ith vocabulary w_i taken from the subject; $P(w_i|z_i = j)$ is the probability of vocabulary w_i belonging to theme j; $P(z_i = j)$ is the probability of document d belongs to the theme of j. The jth theme is expressed as a polynomial distribution of V vocabulary word:

$$\varphi_{W_i}^{j} = P(w_i \mid z_i = j). \tag{2}$$

Text is expressed as the K randomly mixed theme:

$$\theta_j^d = P(z_i = j). \tag{3}$$

Therefore the probability of text d "have" vocabulary for w:

$$P(w \mid d = \sum_{j=1}^{T} \varphi_{w_i}^j \cdot \theta_j^d). \tag{4}$$

Calculating the maximum likelihood function by EM (Expectation Maximization):

$$l(\alpha, \beta) = \sum_{i=1}^{M} \log p(d_i \mid \alpha, \beta). \tag{5}$$

The maximum likelihood estimator α, β estimated parameter values of α, β, and then determined LDA Model. Wherein the conditional probability of "happen" text d is:

$$P(d \mid \alpha, \beta) = \frac{\tau(\sum_i a_i)}{\prod_i \tau(a_i)} \int (\prod_{i=1}^{k} \theta_i^{\alpha_i - 1})(\sum_{n=1}^{N} \sum_{i=1}^{k} \prod_{j=1}^{V} (\theta_i \beta_{ij})^{W_n^j}) d\theta. \tag{6}$$

In the LDA model, approximate reasoning and other parameters values can be calculated through the Laplace approximation, vibration inference (Vibration Inference). Gibbs sampling and expects a diffusion (expectation propagation) algorithm.

4 Experiments

In this part, we compare the performance of the systems with the following three methods:

- Method 1: Without context information, only use the current dialogue to classify intention with LDA model.
- Method 2: With context information, don't judge whether the question is information incomplete and the context information is added directly to expand the current question.
- Method 3: With context information, judge and join entity recognition and integration of context information for extending problem.

4.1 Text Data Set

The proposed system could cover different topics (air conditioning, rear view mirror, headlight and seat, seat belt, headrest, steering wheel, child seat and door lock) and intentions (find location, confirm function, how-to and so on) in a common vehicle dialog system. We select 80 scenarios for testing, including all topics and intentions. 23 volunteers of native Chinese participate in this test and are divided into 4 groups (Table 1).

Table 1. Question category

Category	Instances
Reasonable	发动机故障灯亮了该怎么处理 (When enginemal function indicator is on, how to deal with it)
Unreasonable	什么故障？怎么解决？ (What' wrong? how to deal with it?)
Unreasonable	这个是什么灯？ (What light is it?)
Reasonable	怎么处理发动机故障灯亮了？ (how to deal with the engine malfunction indicator is on?)

Every group has 20 scenarios, in group A, a total of 148 questions are collected, average 7.4 questions for each scenario. In group B, a total of 220 questions are collected, and in average 11 questions for each scenario. In group C, a total of 117 questions are collected, and in average 5.85 questions for each scenario. In group D, a total of 147 questions are collected, and in average 7.35 questions for each scenario (Table 2).

Table 2. Correct answer and incorrect answer

	Reasonable question	Unreasonable question
Correct answer	TP: System answer matches with reference	TN: "sorry, I am growing!"
Incorrect answer	FN: Any other answers	FP: Any other answers

TP: True Positives, TN: True Negatives, FP: False Positives, FN: False Negatives

To evaluate the performance, all questions are divided into reasonable and unreasonable parts:

By comparing the system's reply with reference answers, we can get the precision and F-1 measure of the proposed system.

4.2 System Performance Comparison

Table 3 shows the overall performance with the three different methods.

Table 3. System results with the three methods

Method	TP	FN	TN	FP	Precision	Recall	F1
I: current dialogue	396	114	25	68	0.853	0.776	0.813
II: context information	452	80	10	61	0.881	0.850	0.865
III: context information + entity recognition	503	74	16	13	0.974	0.872	0.920

Experimental results show that after integrating entity recognition method and context information expansion, the precision and recall rate are significantly higher than those of other methods.

5   Conclusion

Improving the accuracy of the intention of the questions which have incomplete information is the main goal of this paper. An RNN-based entity recognition method is employed to determine whether the current question has incomplete information, then the information-incomplete question will be expanded by combining one or two rounds of dialogue. Our method not only helps a lot in the case of information-inadequate questions but also ensures the accuracy of information-complete questions. Experiments show that the proposed method has a 9.4% improvement in precision and a 2.3% improvement in recall compared with the traditional context expansion method.

Acknowledgments. This research work is supported by the grant of Guangxi science and technology development project (No: AC16380124, 1598018-6), the grant of Guangxi Key Laboratory of Cryptography & Information Security of Guilin University of Electronic Technology (No: GCIS201601), the grant of Guangxi Colleges and Universities Key Laboratory of Intelligent Processing of Computer Images and Graphics of Guilin University of Electronic Technology (No: GIIP201602), the grant of Guangxi Cooperative Innovation Center of Cloud Computing and Big Data of Guilin University of Electronic Technology (No: YD16E11), the grant of Guangxi Key Laboratory of Trusted Software of Guilin University of Electronic Technology (No: KX201514), the grant of Guangxi Experiment Center of Information Science of Guilin University of Electronic Technology (No: 20140208), the grant of Key Laboratory of Cloud Computing & Complex System of Guilin University of Electronic Technology (No: 15210).

References

1. Wikipedia, http://en.wikipedia.org/wiki/Ananova
2. Morbini, F., DeVault, D., Sagae, K., Gerten, J., Nazarian, A., Traum, D.: FLoReS: A Forward Looking, Reward Seeking, Dialogue Manager. In: Mariani, J., Rosset, S., Garnier-Rizet, M., Devillers, L. (eds.) Natural Interaction with Robots, Knowbots and Smartphones, pp. 313–325. Springer, New York (2014)
3. Courgeon, M., et al.: Life-sized audiovisual spatial social scenes with multiple characters: MARC & SMART-I. In: Proceedings of Theèmes Journées De Lafrv (2010)
4. Design and Implementation of Human-machine Speech Interaction in Vehicle Navigation. Electronic Engineering & Product World (2007)
5. Reichardt, D., et al.: CarTALK 2000: safe and comfortable driving based upon inter-vehicle-communication. In: Intelligent Vehicle Symposium IEEE Xplore, 2002, vol. 2, pp. 545–550 (2002)
6. Bhargava, A., et al.: Easy contextual intent prediction and slot detection. In: IEEE International Conference on Acoustics, Speech and Signal Processing, pp. 8337–8341. IEEE (2013)

7. Xu, P., Sarikaya, R.: Contextual domain classification in spoken language understanding systems using recurrent neural network. In: IEEE International Conference on Acoustics, Speech and Signal Processing, pp. 136–140. IEEE (2014)
8. Shi, Y., et al.: Contextual spoken language understanding using recurrent neural networks. In: IEEE International Conference on Acoustics, pp. 5271–5275 (2015)
9. Mikolov, T., Kombrink, S., Deoras, A., Burget, L., Cernocky, J.: RNNLM – recurrent neural network language modeling toolkit. In: ASRU (2011)
10. Deoras, A., et al.: Variational approximation of long-span language models for LVCSR, vol. 125(3), pp. 5532–5535 (2011)

An Exploratory Study of Factors Affecting Number of Fans on Facebook Based on Dialogic Theory

Hui Chi Chen[1], Ping Yu Hsu[1], Ming Shien Cheng[2(✉)],
Hong Tsuen Lei[1], and Ching Fen Wu[1]

[1] Department of Business Administration, National Central University,
No.300, Jhongda Road, Jhongli City, Taoyuan County 32001, Taiwan (R.O.C.)
984401019@cc.ncu.edu.tw
[2] Department of Industrial Engineering and Management,
Ming Chi University of Technology, No.84, Gongzhuan Road,
Taishan District, New Taipei City 24301, Taiwan (R.O.C.)
mscheng@mail.mcut.edu.tw

Abstract. The invention of Social Networking Site not only changes the way people searching information and communicating, but also business marketing strategy. Before Social Networking Site came out, the usual way business marketing strategies are TV commercials, print advertisements and activities. After Social Networking Site got popular, more and more businesses turned to use the Internet as a strategy of marketing because the cost is inexpensive and it is easier to gain potential clients through the whole world. One of the Internet's characteristics is virtually, which is a difficulty for business to run their website because they cannot actually observe their customers' behavior. In this study, we try to use the data that we collect to analysis the behavior of iCook fan Page's fans based on Dialogic Theory. Dialogic Theory has five facets, Dialogic Loop, Usefulness of Information, Generation of Return Visits, Ease of the Interface and Conservation of Visitors. Most of the studies used content analysis to analysis the criteria they defined, we define the criteria form user's perspective and collect the data in a more convenient way. Moreover, we attempt to find out the relationship between Dialogic Theory and the change of fan number. The study found out that some criteria have relationship with fan's number, which means if a business wants to improve their public strategy on fanpage, they can start from looking for these criteria. These criteria can help a business to run their fanpage more effectively; it also helps to gain fans' value on the Internet.

Keywords: Facebook · Fanpage · Dialogic theory · Social networking site

1 Introduction

Both quantity and quality of fans are the important goals of fan page. For a Facebook fan page created by an enterprise, the primary aim is to increase the number of fans; because of the increased number is an indicator very intuitive and easy to measure. Operators can spread messages of events as media, to attract users interested in the

© Springer International Publishing AG 2017
Y. Tan et al. (Eds.): ICSI 2017, Part II, LNCS 10386, pp. 538–546, 2017.
DOI: 10.1007/978-3-319-61833-3_57

events, starting with the relationship between friends, who then become fans, and then followers. The most valuable customers tend to be followers, who not only have a high degree of brand loyalty but also even help spread the brand awareness by sharing relevant information about the brand. An enterprise may attract fans when it starts running its fan page. However, when the cumulative number of fans reached a certain amount, it is important to make the fans remain in the fans club for interaction with each other or between fans and the operator, because an active fan page can increase the fan page or brand awareness, or attract more fans to participate or expand the customer base to boost sales of products, so as to achieve the final target of enterprise's creating the fan page.

Researches on fan pages in the past were mainly aimed to study whether enterprises' running fan pages affected the user's willingness to buy, etc., so when creating fan pages, they usually put photos of physical goods on the pages. However, sometimes over-marketing may cause consumers to resist the information. Therefore, the characteristics fan pages explored in this study are different from those researched in the past. In order to simply explore the operating results of fan pages, the subjects of this study are those fan pages not aimed to market goods. So we can further observe whether operators' practices used for running their fan page can really increase the number of fans.

This study is mainly aimed to collect data, use the five dimensions of Dialogic Theory as principles, assume the impact factors in the five dimensions, and explore the whether the factors have a direct impact on the growth of the numbers of fans. Based on literature review, the features on which the specific effects of dialogic dimensions have defined from the standpoint of fans. The measurement indicators are assumed from the view of a manager of the fan page. The collected data is used for verifying whether the measurement indicators can be used to evaluate the operating effectiveness of the fan page.

This paper is organized as follow: (1) Literature review: This part contains a review of scholars' researches on social networking sites, five dimensions of dialogic public relation theory; (2) Research Methodology: This part describes the content of research hypothesis in this study; (3) Result Discussion and Management Implications: this part describes the result of research hypothesis test and discusses management implication; (4) Conclusion and future research: The contribution of the hypothesis developed in this study is proposed, and future research is discussed.

2 Literature Review

2.1 Social Networking Sites

The rise of social networking sites further stimulates the construction of new marketing models, e.g. the so-called social media marketing, which can usually achieve the following objectives:

(1) Increase website traffic, and promote shopping;
(2) Establish brand image, brand awareness;
(3) Long-term information delivery, short-term preferential exposure;
(4) Create word of mouth;

(5) Observe first-hand market trends;
(6) Establish relations with surfers, to create "loyal fans".

According to the forecast made by eMarketer in 2015, compared to other social networking sites, Facebook will still have the largest number of users by 2016, and its number of users will reach 160 million. Time has proven that is true. In addition, it is worth noting that the ratio of users aged over 65 is 7.6% in 2016, up from 6.8% in 2015, growing faster than all other age groups. The ratio of users aged 18-24 is 16.4% in 2016, slightly dropped from 16.7% in 2015. This is different from what we perceived in the past, so social media should not simply focus on young people.

2.2 Five Dimensions of Dialogic Public Relation Theory

Kent and Taylor in 1998 [2] listed the five dimensions of public relations established by companies through websites. They extended Grunig's dialogic communication theory, to establish the specific analysis indicators for website communication strategies. The five dimensions are described as follows:

1. The Dialogic Loop: The dialogic loop can not only allow consumers to directly ask companies questions through the Internet, but also allow companies to directly answer consumers' questions, things consumers care about or related issues, etc. through the Internet. This is the starting point for companies to establish relations with the public.
2. The Usefulness of Information: A corporate website should be dedicated to providing information to the public. With the development of Internet, many scholars have also found that the content of information was the key factor that influenced the effectiveness of the website, rather than just to provide some flashy pictures or text. A good website will be constantly visited by users because the company can continue to provide valuable information to the public. This feature is one of important dimensions of dialogic theory.
3. The Generation of Return Visits (RV): The term "return visit" is also known as stickiness. On a website updating information, changing or creating new topics, answering users' questions, encouraging users to add the website to their "Favorites", providing users forums, providing platforms for users to communicate with each other, and allowing users to subscribe to the latest news will be able to increase return visits. In terms of regular updates, the information content must be practical and can bring value to users. Regularly updating information is also a simple strategy for creating relationship of communication.
4. The Intuitiveness/Ease of the Interface: Users visit a website usually owing to special purpose or curiosity, so the interface of the website should be designed from the standpoint of users, to allow users to be able to understand in a short time and clearly knows how to find the target they want. Based on the second dimension - information usability, the presentation of field on the home page of a website must be organized and hierarchical, rather than randomly classified. A good home page should be text-based instead of images, because loading text is faster than loading a picture, and can help the more urgent user search the target in a short time.

5. The Rule of Conservation of Visitors: When a website is designed, it should be carefully considered to provide external users with hyperlinks, which a company provides may allow users to quickly determine whether they will visit the website next time. If the link gives users little help, or even leads users to a page they do not need, this process would give potential consumers a bad image of the company and they will not visit the website again. In addition, the number of links should be appropriate, and too many advertisements would annoy or disgust users.

3 Research Methodology

In this study, the observation method and secondary data analysis are used for the usefulness of information, conservation of visitors, and revisit rate to define the affecting factors through result orientation. Then, whether the three dimensions really affect the number of fans is further explored based on the definitions of this study. The subject of this study is the single fan page of iCook, without other fan pages for comparison. Based on observation, iCook fan page has been aimed to let more people know iCook's recipes, and the company has not had too many interactive questions and answers, so the dialogue loop will not be discussed in this study.

In this study, the software fan-page Karma provided through the Internet has been used to collect the relevant information necessary for this study. A user can further observe the fan page he/she wants to, even if the user is not an administrator. Unlike other software, Fan page Karma provides a broader range of analyses and more analytical indicators, and users also can make an observation for a longer time free of charge. It is the best choice for a student to use it.

3.1 Develop Analytic Indicators

After the literature review, given the differences of marketing characteristics between Facebook fan page and iCook, the result orientation will be used to define the five indicators, so as to explore the correlation between business practices of iCook fan page and its number of fans. The definitions of the analytic dimensions in this study are as follows:

1. Usefulness of Information: In this study, the useful value of the information will be defined based on the behavior of the fans in the fan page. The analytic indicators are: (1) Comments: When a user presses "like" on Facebook, he/she usually agrees to the ideas in the information, and even wants to know more about the information. (2) Shares: When a user presses "like" on Facebook, he/she usually agrees to the ideas in the information, wants to know more about the information, and even loves to share the information with his/her relatives and friends.
2. Conservation of Visitors: According to Elizabeth Muckensturm [3], the indicators for measuring the conservation of visitors include posting frequency, whether there is a link with the company's website and blog. Based on this, in this study the indicators used to analyze the conservation of visitors are: 1) Posting Frequency:

A Facebook fan page providing certain amount of information daily helps make fans stay on the website for a longer time [4]. 2) Websites contain links: If a Facebook fan page contains a link to the company's official website, fans can read more information via a link to the official website [3].

3. Generation of Returns Visits: According to the literature review, it is found that many definitions of the indicators of dimensions have been established from the corporate perspective. In this study the generation of return visits is calculated directly based on the collected data. It is defined as follows: Returns: Fans who give comments or write posts on the Facebook page are called active fans.

The number of fans is often used as an indicator to measure operating effectiveness. This study has observed fan actions that often appear on fan page, including "pressing like", "leaving comments", "sharing the article," and "revisiting". For enterprises, they often run their fan pages by "posting articles" and "adding links to external websites to articles". This study suggests that either the operator or a fan their actions mentioned above possibly affect the daily number of fans, and the related analysis is as follows: 1) Number of comments: A large number of comments mean not only the information is much helpful for fans but also the fan page has attractive many fans. 2) Number of shares: a larger number of shares represent much more fans' participation. When a fan shares an article to individual wall, the fan's friends are likely to become a new fan of the corporate fan page because he/she reads the shared article. 3) Number of posts: when a company posts a certain number of articles per day, fans tagging the fan page will receive the messages from the fan page daily, and then log into the fan page daily. 4) The posted article contains a link to an external website links: in addition to posting certain number of articles, a company can create a link in an article to an external website, typically the company's official website. 5) Number of return visits: a fan is willing to revisit fan-page means that the fan is interested in the company's products or value, and used to regularly visiting the fan page.

In this study, referring to the research framework of Chang Ting-Han [1], based on the previous related researches and the dialogic public relation theory proposed by Kent and Taylor [2], the number of fans is dependent variable, the relevant indicators are defined and the research issues are listed as follows:

H1:   the number of comments is positively correlated with the number of fans per day

H2:   the number of shares is positively correlated with the number of fans per day

H3:   the number of posts by manager is positively correlated with the number of fans per day

H4:   the number of posts with links to external websites is positively correlated with the number of fans per day

H5:   the number of return visits positively correlated with the number of fans per day.

According to the dialogic public relation theory, the assumed construction defined is shown as the Fig. 1 below:

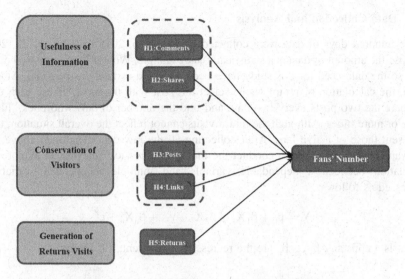

Fig. 1. Dimension indicator of research definition

3.2 Research Subject and Samples

The subject of this study is iCook (Fig. 2), which is Taiwan's largest recipe brand with more than seventy thousand happy dishes shared by surfers. It has its own official website, on which surfers can share dishes in a very simple way. In addition, iCook has built its official blog and Facebook fan page. The former contains knowledge related to cuisine, and interacts with surfers on behalf of iCook. The latter allows more users to share the surfers' recipes through Facebook's channel. These three websites constitute a solid social network. Less than two months after the website was put online, more than 20,000 people have logged into the Facebook fan page. Until now nearly 1.6 million users have logged into the fan page. The monthly views of iCook have reached 160,000, a total of more than 3,000 members. The fans of iCook are active, so iCook ranks high among the popular brands in Taiwan.

Fig. 2. iCook official website

3.3 Data Collection and Analysis

Three hundred days of data were collected from May 6, 2015 to February 29, 2016 because the amount of data is a statistically large sample. Websites are improved every day, so the data used for this study have been collected as lately as possible. Among them, the calculation of return visits: select the post with the most "likes" each day, compare the two posts every two day, and find the number of fans who click "likes" twice or more times. Although the return visits cannot reflect the overall situation, they represent those of active fans. After collecting all the raw data mentioned above, the software Statistica is used to conduct the multiple regression analysis, to explore the correlation between a dependable variable and multiple independent variables, expressed as follows:

$$Y = \beta_0 + \beta_1 X_1 + \beta_2 X_2 + \ldots + \beta_n X_n + \varepsilon \qquad (1)$$

β_0 is a constant, $\beta_1 \ldots \beta_n$ are the regression coefficient, ε is the error.

4 Research Results Discussion and Management Implication

According to the results of this study, there is no significant difference between "daily number of posts" and "the number of posts containing link to external websites". The further discussions are described as follows:

According to Devon Lai, who conducted a research on the life cycle of fan page based on "product life cycle" and "consumer decision-making model", this period is at the stable development stage for the number of fans. During the stable development stage of fan page life cycle, the main development objective is to be able to develop good and long-term relationship with the existing fans. For a company running its fan page, regular "number of posts" allows fans to develop a habit of visiting the fan page, so that they will become followers. According to the results of this study, the indicator of "number of posts" may have an influence the number of fans at the last stage of the life cycle, with no direct correlation with the number of fans at the early stage (Table 1).

Table 1. Research result

Hypothesis	R^2	P value	Significant
H1	0.55451458	0.013437	Yes
H2	0.55451458	0.000192	Yes
H3	0.55451458	0.198068	Yes
H4	0.55451458	0.946479	Yes
H5	0.55451458	0.00000	Yes

The stable period in fan page's life cycle, increasing the number of fans is less important than emphasizing the values of the existing fans, such as interaction and conversion rates. The data of links to external website are used to measure conversion rates, so it is reasonable to infer that "the number of posts containing links to external websites" is more correlated with conversion rate than with the number of fans.

According to the results of this study, "the number of comments", "the number of shares", and "the generation of return visits" are correlated with "the number of fans". This means that these redefined indicators fit the dialogic public relation theory proposed by Kent and Taylor [2]. In other words, it is feasible to use these defined indicators to carry out measurement based on the dialogic public relation theory.

The dialogic public relation theory can be used to measure the dialogic capacity of a company's fan page. The measurement indicators proposed by this study can be used to measure the dialogic effectiveness by directly collecting the data from the fan page. In addition, the defined indicators can help us further understand the impact of factors on the number of fans.

5 Conclusion and Future Research

In the past the content analysis method was mainly adopted and human judgment and coding method were used to conduct researched, not only time-consuming but also labor-intensive. On the other hand, the data collected for this study were those provided freely through the Internet, so that the time and costs of collecting data can be reduced. The results confirmed that four of the indicators defined for this study have significant differences. The number of likes, the number of shares, the number of posts, and the generation of return visits are correlated with the number of fans. And the number of comments and the number of posts containing links to external websites have no significant difference.

The research limitations and future research recommendations will discuss below:

1. Limitations of Data: the subject of this study is the only one type of fan page, and the scope of subject can be expanded in the future.
2. Collected data is insufficient: Perhaps exploring whether the five dimensions are correlated to network reach is a better issue. If the data analysis of administrator's dynamic report can be obtained, there will be more data can be used.
3. Recommendation: the application of other analytical data: The analytic data used for this study are the number of likes, the number of comments, the number of shares, the number of posts, and the calculated number of return visits. Fan page Karma offers many other analytic data such as the number of films, the degree of fan participation, the engagement rate of interaction. It is recommended the data that has not adopted yet can be used for further research.

References

1. Chang, T.-H.: How to manage fan-page of the Facebook? (2014)
2. Kent, M.L., Taylor, M.: Building dialogic relationships through the World Wide Web. Public Relat. Rev. **24**(3), 321–334 (1998)
3. Muckensturm, E.: Using dialogic principles of Facebook: how the accommodation sectors communicating with its consumers (2013)
4. Rybalko, S., Seltzer, T.: Dialogic communication in 140 characters or less: how fortune 500 companies engage stakeholders using Twitter. Public Relat. Rev. **36**(4), 336–341 (2010)
5. Wellman, B., Berkowitz, S.D.: Social structures: a network approach, vol. 2. CUP Archive (1988)

Assembling Chinese-Mongolian Speech Corpus via Crowdsourcing

Rihai Su[1], Shumin Shi[1,2(✉)], Meng Zhao[1], and Heyan Huang[1,2]

[1] School of Computer Sciences and Technology,
Beijing Institute of Technology, Beijing, China
bjssm@bit.edu.cn
[2] Beijing Engineering Research Centre of High Volume Language Information
Processing and Cloud Computing Applications, Beijing, China

Abstract. Chinese-Mongolian Speech Corpus (CMSC) is utilized in many practical applications in recent years, and it is a kind of low-resource corpus due to its high-cost construction. We describe a crowdsourcing method to build a collection of bilingual speech corpus through the use of a messaging app called WeChat, in which followers can send voice and text message to our Official Account Platform freely. Owing to most followers are fluent in Chinese and Mongolian, we gathered natural speech recordings in our daily life, and constructed a parallel speech corpus of 20547 utterances from 296 speakers, totalling 21.43 h of speech, during the first 25 days that collecting notification was pushed. Moreover, we present a quality control measure in the evaluation part that independent subscribers voted on the translations of each source sentence and it improves the quality of corpus markedly. We show that WeChat Official Account Platform can be used to assemble speech corpus quickly and cheaply, with near-expert accuracy. As the basic research content of natural language processing (NLP), the construction of bilingual speech corpus via crowdsourcing has a reference value for the similar studies.

Keywords: Crowdsourcing · Speech corpus · WeChat · Mongolian

1 Introduction

Collecting data from the web for commercial and research purposes has become a popular task, used for a wide variety of purposes in text and speech processing. However, to date, most of this data collection has been done for English and other High Resource Languages (HRLs). These languages are characterized by having extensive computational tools and large amounts of readily available web data and include languages such as French, Spanish, German and Japanese. Low Resource Languages (LRLs), although millions of people are much less likely and much more difficult to collect, due speak many largely to the smaller presence these languages have on the web. These include languages such as Igbo, Amharic, Pashto and what we are going to talk-Mongolian. The constructing process of traditional speech corpus is time-consuming and labor-intensive, which is recorded by speakers in professional studio [1–3]. Thus, the approach utilizing

© Springer International Publishing AG 2017
Y. Tan et al. (Eds.): ICSI 2017, Part II, LNCS 10386, pp. 547–555, 2017.
DOI: 10.1007/978-3-319-61833-3_58

crowdsourcing technologies for corpus generation has gradually become a research hot topic [4].

In this paper, we describe a new approach that addresses the problem of collecting large amounts of CMSC using WeChat Official Account Platform (WOAP), which can be served as a crowdsourcing system. Unlike current laboratory-generated corpus, WOAP provides a targeted collection pipeline for social networks and conversational style text. The purpose of this corpus collection is to augment the training data used by Automatic Speech Recognition (ASR) and for corpus resources for Mongolian NLP research. The more specific goal is to reduce the translation cost and improve the efficiency, and thus the corpus quality may improve sensibly. As we are aiming at spoken language, we try to collect audio recording from the real world environment. We performed an experiment of constructing CMSC via crowdsourcing technologies, all of them somewhat novel to the Mongolian NLP community but with potential for future research in other ethnic minority computational linguistics. Additionally, we discuss methods for validating corpus quality, finding the crowdsourced results are relatively better compared with controlled laboratory experiments.

The rest of the paper is organized as follows: Sect. 2 reviews the existing work; Sect. 3 describes the corpus constructing method in detail; Sect. 4 is the corpus evaluation part; we draw conclusions and discuss future work in Sect. 5.

2 Previous Research

The corpus research began in the 1970s. Along with the development of corpus linguistics, a large number of corpus such as Peking University Corpus, LDC Chinese Tree Bank, LOB and BROWN have been built up at home and abroad [5].

Although the construction of ethnic minority language corpus in our country started relatively late compared with the existing large-scale Chinese corpus, the ethnic minority speech corpus construction and speech recognition headed by Mongolian and Uygur have also been carried out fundamental research [6–11]. Dawa Yi et al. have described their work on the Mongolian multilingual speech corpus and its application in Japan [12]. RongYu et al. have constructed "Mongolian speech corpus" towards Mongolian speech grammar research, which is made up of 10 million words recorded in various genres such as movies, jokes, stories, daily conversations and composition reading. Academy of Social Sciences of Inner Mongolia in china built the "Mongolian Speech Corpus". Recorders were selected from the eight provinces of Chinese 53 collection points, 10 points in Mongolia, the Russian Federation 7 points, recording the contents of free dialogue between native locals [13]. RongYu et al. introduced the general situation of Mongolian standard speech-language dialogue corpus, and expounded some problems encountered in recording annotation and the way of solutions [14]. Yating Yang et al. proposed a reasonable specification, a method of speech acquisition and annotation, and established a 300-hour recording of speech Uighur language [15].

In the field of global NLP research, crowdsourcing has become a popular collaborative approach as well, which is utilized for the acquisition of annotated corpora and a wide range of other linguistic resources. The results of these studies have provided a good reference for the construction of CMSC [16–18]. It is possible to achieve ideal results by using crowdsourcing technology in corpus processing and annotation [19]. Posting on the platform AMT (Amazon's Mechanical Turk), a collection of parallel corpora between English and six languages from the Indian subcontinent: Bengali, Hindi, Malayalam, Tamil, Telugu, and Urdu have been built [20]. In the study of domestic crowdsourcing, a public opinion corpus of Uygur, Kazak and Kirgiz language based on crowdsourcing is proposed, which provides an essential resource support for the study of ethnic minority corpus [21].

3 Chinese-Mongolian Speech Corpus (CMSC)

In this paper, we utilize the crowdsourcing technology in the process of collection and evaluation. We exploit an interface-based WeChat Official Account Platform (WOAP) that allows us to collect audio recordings send from subscribers' mobile phone.

3.1 Basic Idea

WOAP is an open mobile Internet environment, which can be exploited for the collection and evaluation of corpus in a reasonable way. "Mother Tongue" is a WOAP oriented towards Mongolian users, on which we can interact with our followers unrestrictedly. And most of the subscribers are fluent in both Chinese and Mongolian, making it easy to meet the translation task requirements. Crowdsourcing enables real-world, large-scale corpus studies to be more affordable and convenient than expensive, time-consuming lab-based studies.

WOAP is flexible, relatively easy to use, and capable of collecting noisy, real-world audio data effectively. To sum up, this paper proposes a straightforward method for gathering Chinese-Mongolian speech data, which can be utilized to form a large-scale, low-cost CMSC in a short time. Figure 1 contains an example of audio recording transcribed into text by Mongolian and English.

e.g.1: 我想要这两本书。

Mon: ᠪᠢ ᠡᠨᠡ ᠬᠣᠶᠠᠷ ᠨᠣᠮ ᠢ ᠠᠪᠬᠤ ᠰᠠᠨᠠᠲᠠᠢ ᠪᠠᠢᠨᠠ᠃

En: I want these two books.

Fig. 1. An example of audio recording transcribed into text by Mongolian and English.

3.2 Experiment Design

3.2.1 Setup

Different from the traditional lab-based corpus, our current experiments have relatively low requirement for equipment. And just a computer with Internet access is needed. We perform the experiment on the webpage through Mother Tongue (WOAP).

3.2.2   Speaker

A key component of managing WeChat subscribers is to ensure them competently and conscientiously undertaking the tasks. As a quality-control measure, we release an Mongolian proficiency test, consists of the basic subscriber attributes (WeChat nickname, user age, education level and location) and 15 test questions. The test comprises questions of short phrases, sentences in Chinese and answers in Mongolian. Candidates are asked to vote on one suitable translated answer among 4 choices, and candidates who pass the test can be considered as native speaker or proficient in Mongolian. The statistical results of the test show that 296 out of the 624 participants have passed the test, and these qualified subscribers are the main contributors in the process of corpus gathering.

3.2.3   Data Preparation

For the purpose of this study, the final corpus paradigm is composed of Chinese text and Mongolian recordings (partially transcribed). The main sources of data are Chinese conversations that embody the daily-life features in content. So the original data consists of following two parts: (1) Digitized forms of books series including Chinese-Mongolian daily conversations. (2) Texts downloaded from websites.

According to the distribution of corpus and the frequency of use, we gather the original Chinese texts considering the categories and corpus proportion. Concerning about the robustness of ultimate corpus, it is necessary to ensure the collection of original data is extensive and diversiform. Table 1 provides a general quantitative description of the corpus.

Table 1. A quantitative description of the corpus.

Categories	#sens	#words
Education	3904	2489
Recreation	5137	3954
Tourism	4726	2782
Diet	4520	3368
Baidu Tieba	2260	2050
Total	20547	14643

3.2.4   Experiment

According to the management regulations, each WOAP can only push one article to their subscribers per day. After taking into account the data collection process and experimental cycle, we push an article contains 15 Chinese sentences by the Mother Tongue (WOAP) everyday. The article includes the translating specifications as follows: (1) each subscriber selects one or more of the 15 Chinese sentences in the article to translate, and response us with Mongolian audio recording. (2) Those who provide high-quality audio can receive monetary rewards. (3) Empty or very noisy responses won't be included in the final set. Figure 2 shows an article contains translating norms and daily sentences.

Fig. 2. An article contains translating norms and daily sentences.

3.2.5 Incentive

There are many ways to motivate participants, and this paper exploits a traditional and common monetary incentive mechanism. In such crowdsourcing systems that base themselves on monetary payment, the incentive mechanism has to be effectively designed to be economical and fair. In general, each feedback audio recording of subscribers took on an average 20–40 s to complete and a subscriber was paid 5 cents per task including a bonus that was paid on completion of five tasks. It should be noted that subscribers who pass qualification review could receive the corresponding reward.

3.2.6 Quality Control

Obtaining translations of high quality is one of the primary concerns in a translation crowdsourcing system. In this work, not having any, external mechanism can improve the quality of the result; we had to turn to the task design process.

Firstly, different subscribers complete each Chinese sentence in the daily article. For every translation task, we ensure the audio recordings contributed by subscribers are more than one in this way, and select the most reasonable result from candidates with a feasible strategies detailed in Sect. 4. Secondly, the design of the Mongolian proficiency test above is another quality control mechanism. The difficulty of level test should be moderate; if the test is difficult, then the subscribers feel stressed and lost patience, resulting in the loss of the subscribers; if the test is too simple, then the filter-out process will become invalid and we cannot achieve high-quality results. Figure 3 shows the test cases of the Mongolian speaker's proficiency.

Fig. 3. Test cases of the Mongolian speaker's proficiency.

4 Evaluation

Although a testing process is introduced before the translation task, there are inevitably some subscribers who misunderstand the task or intentionally provide some erroneous or random results. In order to achieve better results, we take full advantage of the crowdsourcing approach. Since speech corpus could be contributed by the public, the same can also be done to validate the results of the corpus. In the process of the evaluation, we use the public voting strategy to screen out further qualified results.

Motivated by desire to have some measure of the relative quality and variance of the translations, we designed another task in which we presented an independent set of subscribers with an original sentence and its four transcribed translations, and asked them to vote on which was best. Five independent subscribers voted on the translations of each source sentence. Tallying the resulting votes, we found that roughly 65% of the sentences had five votes cast on just one or two of the translations, and about 95% of the sentences had all the votes cast on one, two, or three sentences. This suggests both (1) that there is a difference in the quality of the translations, and (2) the voters are able to discern these differences, and took their task seriously enough to report them. Figure 4 is an example of all the 5 votes cast on one translation.

In total, the corpus contains 20547 sentences in 14643 distinct words, in a recording time of about 21.43 h. The total cost of this work is 42.14 $, the average cost of each recording is 0.01 $. Our results are taken from 296 subscribers, ranging from age 14–65, with the majority of which belong to Inner Mongolia and Beijing. They represent a range of education levels, the majority had been to college: about 64.42% had bachelor's degree, and 16.51% had master's degree, proving that most of the contributors have a high degree of education. Most subscribers responded an average of three audio recordings, with a few qualified chosen subscribers feedbacks plenty of recordings ranging from 20–50. It should be noted that few subscribers who contributed sufficient high-quality audio recordings are the candidates who achieved

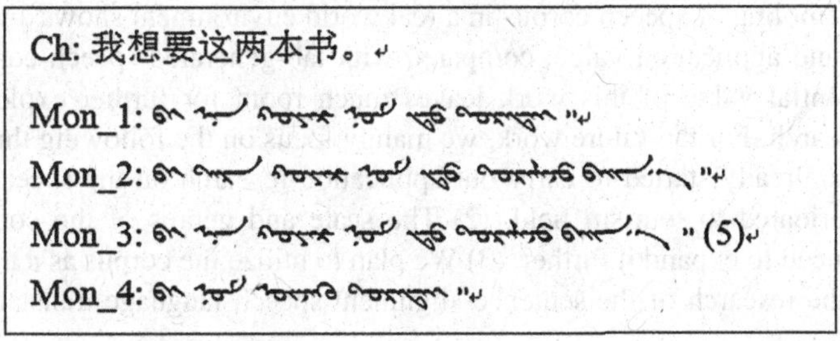

Ch: 我想要这两本书。

Mon_1: [Mongolian script]

Mon_2: [Mongolian script]

Mon_3: [Mongolian script] (5)

Mon_4: [Mongolian script]

Fig. 4. An example of all the 5 votes cast on one translation; the parenthesized number indicates the number of votes. Translation: I want these two books.

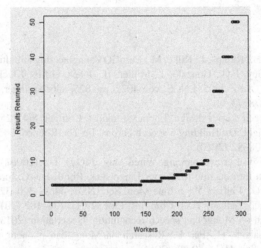

Fig. 5. About half of the subscribers returned more than 3 audio recordings and a few chosen subscribers accounted for most of our data.

highest score in the proficiency test above. We did not strictly balance the speakers by their gender due to the difficulties in finding the qualified subscribers. Figure 5 shows the details.

5 Conclusion and Future Work

In this work, we have shown that crowdsourcing platform such as WeChat can be significantly exploited for the resource construction in NLP area. We have described the design and construction of Chinese-Mongolian speech corpus (CMSC) based on WeChat official account platform (WOAP). A parallel speech corpus of 296 speakers and 21.43 h recordings is finally constructed. We present a novel idea for the construction of other languages that are low-resource and understudied, and it is feasible and valuable to bilingual resources building and Automatic Speech Recognition (ASR).

And our assembling of speech corpus in a real world environment shows a wider range of research and application value, compared with lab-generated speech corpus.

The potential value of this work leaves much room for further exploratory and practical research. For the future work, we mainly focus on the following three aspects. (1) We have already started to carry out application research on the speech language translation oriented to tourism field. (2) The scale and genres of the corpus is still limited, we need to expand it further. (3) We plan to utilize the corpus as a training data to develop the research of the sentence-alignment speech language translation.

Acknowledgments. We thank reviewers for their constructive comments, and gratefully acknowledge the support of Natural Science Foundation of China (61671064) and BIT Basic Research Fund (20160742017).

References

1. Sigurbjörnsson, B., Kamps, J., Rijke, M.: EuroGOV: engineering a multilingual web corpus. In: Peters, C., Gey, F.C., Gonzalo, J., Müller, H., Jones, G.J.F., Kluck, M., Magnini, B., Rijke, M. (eds.) CLEF 2005. LNCS, vol. 4022, pp. 825–836. Springer, Heidelberg (2006). doi:10.1007/11878773_90
2. Crowdy, S.: Speech corpus design. Literary Linguist. Comput. **8**(4), 259–265 (1993)
3. Adolphs, S., Knight, D.: Building a speech corpus. In: The Routledge Handbook of Corpus Linguistics, pp. 38–52 (2010)
4. Howe, J.: The rise of crowdsourcing. Wired Mag. **14**(14), 1–5 (2006)
5. Kennedy, G.: An Introduction to Corpus Linguistics. Routledge, Oxford (2014)
6. Fei, L., Laigao, G., Laibao, Y.: J. Inne Mon. Sci. (NSE) **44**(3), 320–323 (2013)
7. Fei, L., Laigao, G., Laibao, Y.: J. Chin. Inf. Proc. **29**(1), 178–182 (2015)
8. Mu, R.: Research on Mongolian speech recognition. Dissertation (2013)
9. Dongzhao, J., Laigao, G., Fei, L.: Research on Mongolian phonetic synthesis based on HMM. Comput. Sci. **41**(1), 80–82 (2014)
10. Reyiman, T., Yipitihaer, M., Wushouer, S.: J. XJ Sci. (NSE), **30**(2), 199–203 (2013)
11. Jiang, D.: J. Chin. Inf. Proc. **29**(1), 178–182 (2015)
12. Dawa, I., Zhang, Y., Uezono, K., Zhang, S.: Processing of Mongolian by computer. J. Chin. Inf. Proc. **20**(4), 56–62 (2006)
13. Xingwu, J.: Lexical tagging of Mongolian corpus. S.C. Inne Mon., 59–63 (2013)
14. Yu, R., et al.: Problems of recording tagging and solutions in "Mongolian Speech Corpus". In: PCC (2012)
15. Tingyang, Y., Huadong, X., Wang, L.: Research on Uyghur speech language speech corpus of telephone channel. Comput. Eng. Appl. **47**(23), 150–153 (2011)
16. Finin, T., Murnane, W., Karandikar, V., Keller, N., Martineau, J., Dredze, M.: Annotating named entities in Twitter data with crowdsourcing. In: ACL, pp. 80–88 (2010)
17. Sabou, M., Bontcheva, K., Derczynski, L., Scharl, A.: Corpus annotation through crowdsourcing: towards best practice guidelines. In: LREC, pp. 859–866 (2014)
18. Filatova, E.: Irony and sarcasm: corpus generation and analysis using crowdsourcing. In: LREC, pp. 392–398 (2012)

19. Munro, R., Bethard, S., Kuperman, V., Lai, V.T., Melnick, R., Potts, C., Schnoebelen, T., Tily, H.: Crowdsourcing and language studies: the new generation of linguistic data. In: ACL, pp. 122–130 (2010)
20. Post, M., Callison-Burch, C., Osborne, M.: Constructing parallel corpora for six indian languages via crowdsourcing. In: ACL, pp. 401–409 (2012)
21. Chen, H.: Research on the construction of Uygur, Kazak and Kirgiz public opinion tagging corpus based on crowdsourcing. MS thesis (2015)

Robotic Control

Developing Robot Drumming Skill
with Listening-Playing Loop

Xingfang Wu, Tianlin Liu, Yian Deng, Xihong Wu, and Dingsheng Luo[✉]

Key Laboratory of Machine Perception (MOE),
Department of Machine Intelligence,
School of Electronics Engineering and Computer Science,
Speech and Hearing Research Center, Peking University,
Beijing 100871, China
{wu.xf,liutl,yiandeng,xhwu,dsluo}@pku.edu.cn

Abstract. Reacting according to external sounds is an important ability in multi-robot and human-robot collaboration. Although network might be the first choice to connect multi-agents in the robot world, the unexpected connection snap would be a disaster to the whole system. Utilizing sounds is a feasible and supplementary way to transfer information between agents, which is also a smart and robust way to support swarm intelligence. In this paper, under the scenario of a robot band, the issue how each robot member achieves its performance ability is focused. Unlike most of the previous researches, we emphasize that robot's performance ability is achieved all by itself in an autonomous way. And an approach of Listening-Playing Loop (LPL) is proposed, where the developmental learning is involved. With a simple drumming robot, the proposed approach is evaluated. Experimental results show the proposed approach is effective, and via transferring raw audio data to the motion control, the robot successfully develops the drumming ability.

Keywords: Multi-robot collaboration · Listening-playing loop · Drumming skill acquisition · Cognitive robots · Developmental learning

1 Introduction

Robot drumming is a sub-domain of robotic musicianship. Unlike producing music by speakers, musical robots can directly play musical instruments, which can give visual and physical cues to human music performers. Further more, different musical robots can form a band and thus enhancing the richness of performances. Under this circumstance, swarm intelligence algorithms can be adopted and new finding may stimulate the invention of new algorithms conversely. Research of robotic musicianship is an interdisciplinary area that has been widely studied. Bretan et al. [1] enumerated the two primary research areas of robotic musicianship: *musical mechatronics* and *machine musicianship*. The former is the study and construction of physical systems that generate sound through mechanical means, while the latter focuses on developing algorithms

© Springer International Publishing AG 2017
Y. Tan et al. (Eds.): ICSI 2017, Part II, LNCS 10386, pp. 559–566, 2017.
DOI: 10.1007/978-3-319-61833-3_59

and cognitive models with respect to various aspects of music perception, composition, performance, and theory. Different from them, we focus this research on learning aspect of the task. We want to build a model for robots to acquire performance ability autonomously (i.e. robots keep practicing and gradually improving its ability). Therefore, a Listening-Playing Loop(LPL) has been suggested. To learn control parameters from the raw audio signal, a full-connected neural network has been adopted. Data acquired from robot arbitrary trials of shaking drumstick is used to training the neural network, which is accordant to the idea of developmental learning.

Developmental learning which is firstly put forward in developmental psychology area has the key idea of learning through continuously interacting with the outside world. Human infants tend to establish their proprioception through random actions. In robotics, the realizations tend to be the imitation of human infants' behavior: random moving to establish or improve the proprioception of doing a certain job.

To show our idea unequivocally and elegantly, we choose robot drumming as our concrete task, for it is relatively simple and can be done with little domain knowledge of music. In this task, we only consider the rhythm and magnitude of the drum beats. And the learning task mostly focuses on the mapping of magnitudes to control parameters. As for rhythm, we only extract onsets of a piece of drumbeats by the 'OnsetsDS' tool [2].

The remainder of the paper is organized as follows. We firstly give some introduction to related works about the drumming robots and developmental learning. Then, our method based on LPL is elaborated and the experiment comes and verifies the feasibility of the method proposed. Furthermore, the conclusions are drawn and future issues are argued.

2   Related Works

Music allows us to leap over cultural barriers, and it provides us a good research platform as it involved a lot of fields. Developing computer models that support robotic musicianship can have broader impacts outside of artistic applications. Issues regarding timing, anticipation, expression, mechanical dexterity and social interaction are pivotal to music and have numerous other functions in science as well [1]. So, researchers in many fields design instruments and algorithms to realize their own idea of robotic musicianship.

On the other side, playing musical instruments is such a challenge which involves many realms of robotics, such as perception, motion control and etc. Arturo et al. [3] proposed an active learning approach for robots to perform a virtual musical instrument. In their work, a system architecture which enables a robot to actively explore an object and obtain a playing model is proposed.

As for the specific drumming task, researchers of computer music take the lead. They want to combine art with computer technology for creating richer expressions of music. For example, Weinberg et al. [4] developed an interactive robotic percussionist, named 'Haile', which can analyze live musical input in real-time and react in an expressive manner by generating responsive acoustic responses that would inspire humans to interact with it. Kotosaka et al. [5] proposed a way using neural oscillators to achieve rhythmic movements to do drumming task. Their method can handle several kinds of periodic input signals but would be a little slow to catch up when the outside rhythm suddenly changes. Crick et al. [6] demonstrated an effective method for fusing diverse sources of oscillatory input of varying accuracy and phase shift in order to produce a reliable prediction of beats.

In general, actions by embodied agents automatically generate training data for the learning mechanisms and enhance its ability to do a certain job is the common process of developmental learning. As for developmental robots, self-exploration of the world automatically generates new training data to themselves and thus improving their accuracies of judgments or actions. Furthermore, the developmental learning of robots is a life-long process that imitating human beings' learning processes. When learning something new, humans would combine their existing experience with it and improve the learning efficiency. That is why humans are good at understanding visual properties of objects, even without ever acting on them.

Many researchers make their effort to model the learning process: Arsenio [7] proposed an embodied approach for learning task models while simultaneously extracting information about an object. The action of the robot on an object creates an event which will trigger the acquisition of further training data. Sigaud et al. [8] made a survey of deep learning techniques that may be used in developmental robotics and outlined the importance of unsupervised learning and hierarchical predictive processes.

In this research, we propose an approach to lead the developmental learning into the acquisition and enhancement of robot's drumming ability by a means of constructing the Listening-Playing Loop. In the loop, the developmental learning is taken advantage of to combine the past experience with current knowledge acquired by interacting with the drum.

3   Listening-Playing Loop

In the introduction part, we briefly introduce the platform on which our research based. It is a relatively simple model that we only need to concern a few parameters to control. The model greatly reduces the task into a parameters to parameters mapping problem. We continue to break down the drumming task into the auditory part and the motion part. The mapping between these two part would be gradually established and refined through developmental learning.

Firstly, we need to briefly introduce the learning process of the drumming task. The robot learns the mapping between volume and its correlate motion of a single drumbeat, and this process establishes the mapping which is similar to the proprioception of human beings. Then comes the listening process. The robot listens to a piece of drumbeats and tries to repeat it.

Due to the rhythm is relatively a complex concept to learning, the robot only extracts the onsets in a piece of drumbeats. According to the volume of each drumbeat, the robot would use the mapping built in the first stage to control its actuator. During this process, the robot listens to the drumbeats produced by itself and repeatedly adjusts the mapping to achieve a better mimic result. This is how the loop works.

3.1 Listening Module

As for auditory part, we only focus on two attributions of a piece of drumbeats: *Rhythm* and *Volume*. Rhythm depends on the initial moments of each strike, which is called *Onset*. Bello et al. [9] defined the term and make a clear distinction between the related concepts of *transients, onsets* and *attacks*. Stowell et al. [10] proposed a STFT-based onset detection method and was used to the implement of an onset detection library called *'OnsetsDS'*. Also, some other researches use neural networks to conduct onset detection [10].

Due to the limitation of the servo actuator's torque precision, we ignore the detail features of a single strike. Each strike is reduced to magnitude attribution. We simply extract the maximum magnitude of the wave created by a single beat. Maximum magnitude decides the strength of the strike which depends on the parameters given to the servo actuator.

3.2 Playing Module

Motion part of the closed-loop is also simplified. We only select two decisive control parameters of the actuator from all that we can alter. They are *speed* and *torque*. The speed is a float type data that controls the angular velocity of actuators. The torque of actuator can be set to a certain value according to the actuators' model. This two parameters can decide the striking strength of the drumstick. Every strike will trigger an analysis of the drumbeat and thus give the feedback of the strike action.

3.3 Developing Proprioception Based Drumming Skill

As mentioned above, we want to map the auditory parameters to motional parameters, and thus we can directly control the servo actuator according to the incoming acoustical signals. The neural networks offer a promising way of non-linear mapping and are good at making a representation of original raw data. So we employ a neural network to learn the mapping between the parameters mentioned above. The input of the network is the maximum magnitude and the output is the speed and torque of the actuator. From another point of view, this neural network can be thought as the proprioception of the robot on the special drumming task. Luo et al. [11] proposed using *autoencoder* to modeling the sense of robot's arm position and orientation. We borrow the key idea and applied it to our drumming task.

At present, the robot is ignorant of how to conduct the striking motion. So, randomly generated parameters are given to the actuator. At the same time, the motion triggers a recording process to take down the action and its consequences. When the data generated in this process accumulating to a certain amount, the training process of the neural network will be triggered. Therefore, the performance is improved and the actions become more accurate.

4   Experiments

This section introduces the drumming robot we design and reports the experiments conducted to evaluate the proposed approach for a robot to achieve the drumming skill, experiments are conducted in our specially designed robot platform and the results prove the proposed Listening-Playing Loop is effective.

4.1   Robot Platform

We make a simple drumming robot to simplify the task. The experimental platform is connected to a PC that works on Ubuntu 14.04 LTS and Robot Operating System (ROS) Indigo. The robot platform is made up of a Dynamixel RX-28 servo actuator and its connector, a hyper cardioid microphone, a panel, a cylinder drum and a drumstick. The hyper cardioid microphone which has a frequency response of 100 Hz to 16 kHz is adopted to attach a better hearing to the robot. The drumming robot has only one degree of freedom (DOF) that control the movement of the drumstick and one manual degree of freedom on the pillar of the panel which can be adjusted manually to control the height of the servo actuator. This robot platform greatly reduces the complexity of the task and eliminate the unconcerned factors, and thus make the idea of this paper concise. The robot platform is shown in Fig. 1.

Fig. 1. Robot platform developed for the experiment. The robot platform contains a cylinder drum, a 100 Hz to 16 kHz frequency response hyper cardioid microphone, a panel, a Dynamixel RX-28 servo actuator and its RS485 connector.

4.2   Experiments Process and Results

In the light of developmental learning, the robot would randomly generate sets of parameters in a rational range so that the speed and torque would not be too slow to drive the drumstick. Once the parameters are set, the robot would strike five times. Rather than only striking once, the robot would strike five times to decrease the error of the actuator's performances. At each strike, the robot would remember the maximum magnitude of the beat. After five trials, the robot would ignore the maximum and the minimum magnitudes and calculate the average magnitude of the left three. Then, the average magnitude will be stored with the parameters as one training sample. There are 1300 samples in Fig. 2. Intuitively, we could get some information from it. When speed is set to mid-range (e.g. Speed=6), the magnitude is approximately linear to the torque.

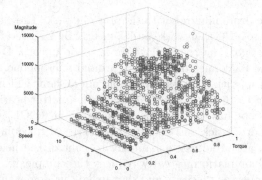

Fig. 2. The data generated through randomly striking. The three dimensions are speed, torque and the magnitude acquired by the method mentioned above.

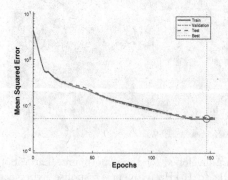

Fig. 3. The Mean Squared Error (MSE) of one training process. We can see that after 20 epochs, the MSE swoops to about 0.5, and decline slowly after 20 epochs. At 147 epochs, the MSE drops to the lowest and the best performance.

To confirm our intuition by the mapping constructed, we train the neural network. Due to the small scale of inputs and outputs, we choose a simple network structure that contains only one hidden layer with 50 neurons, whose activation function is sigmoid. The network is a nonlinear regression model which models the sensorimotor of robots (i.e. proprioception of robots).

There are two training methods we can choose: One is to incorporate all the data available into the training process of the neural network, the other is to divide the data into bunches and thus train the neural network several times. To compare two training methods, we firstly use all the 1300 pieces of data sampled to train the neural network. The result of training is shown in Fig. 3.

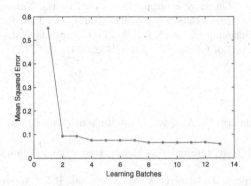

Fig. 4. The line chart of Mean Squared Error (MSE) in the developmental learning process. Each point represents one training process of the neural network. We see that the MSE tends to stable after 8 times.

After 6 epochs, the Mean Squared Error (MSE) swoops, and decline slowly after 10 epochs. At 94 epochs, the MSE drops to the lowest and the best performance in the validation set is achieved. Although the MSE still large, the decline proves that the proprioception is gradually built up.

Then, we train the network by data bunches, which is more aligned with the idea of developmental learning. We randomly divided the sampled data into 13 parts, and use each part to train the neural network and each training process meets its terminal condition. The result is shown in Fig. 4. Every point in this chart represents an outcome of a single training process which corresponds to the circled point in Fig. 3.

Comparing Figs. 4 and 3, we find that the two training methods mentioned above achieve similar results, which illustrates the second method is almost as effective as the first training method. However, the first method requires all the training data before training. This is unsuitable for developmental robots' common scenario.

5　Conclusions and Future Work

In this paper, we propose using a Listening-Playing Loop for a robot to acquire its drumming skill. Under the developmental learning paradigm, the iterative feedback improves the accuracy of striking action. The Listening-Playing Loop based on a neural network model facilitates the nonlinear mapping between audition and motion. With experiments on a simple drumming robot, the idea of this research is verified to be effective. In the future, with the proposed approach, a real robot case (e.g., PKU-HR6.0 which is developed by our lab and has 24 DOFs) will be further studied.

Acknowledgement. The work is supported in part by the National Natural Science Foundation of China (No. 11590773, No. 61421062), the Key Program of National Social Science Foundation of China (No. 12 & ZD119) and the National Basic Research Program (973 Program) of China (No. 2013CB329304).

References

1. Bretan, M., Weinberg, G.: A survey of robotic musicianship. Commun. ACM **59**(5), 100–109 (2016)
2. Onsetsds - real time musical onset detection c/c++ library. http://onsetsds.sourceforge.net
3. Ribes, A., Cerquides, J., Demiris, Y., de Mántaras, R.L.: Active learning of object and body models with time constraints on a humanoid robot. IEEE Trans. Cogn. Dev. Syst. **8**(1), 26–41 (2016)
4. Weinberg, G., Driscoll, S., Parry, M.: Haile-an interactive robotic percussionist. In: ICMC (2005)
5. Kotosaka, S., Schaal, S.: Synchronized robot drumming by neural oscillator. J. Rob. Soc. Japan **19**(1), 116–123 (2001)
6. Crick, C., Munz, M., Scassellati, B.: Synchronization in social tasks: robotic drumming. In: The 15th IEEE International Symposium on Robot and Human Interactive Communication, ROMAN 2006, pp. 97–102. IEEE (2006)
7. Arsenic, A.: Developmental learning on a humanoid robot. In: Proceedings of the 2004 IEEE International Joint Conference on Neural Networks, vol. 4, pp. 3167–3172. IEEE (2004)
8. Sigaud, O., Droniou, A.: Towards deep developmental learning. IEEE Trans. Cogn. Dev. Syst. **8**(2), 99–114 (2016)
9. Bello, J.P., Daudet, L., Abdallah, S., Duxbury, C., Davies, M., Sandler, M.B.: A tutorial on onset detection in music signals. IEEE Trans. Speech Audio Process. **13**(5), 1035–1047 (2005)
10. Stowell, D., Plumbley, M.: Adaptive whitening for improved real-time audio onset detection. In: Proceedings of the 2007 International Computer Music Conference, ICMC 2007, pp. 312–319 (2007)
11. Luo, D., Hu, F., Deng, Y., Liu, W., Wu, X.: An infant-inspired model for robot developing its reaching ability. In: IEEE International Conference on Developmental Learning and Epigenetic Robotics, pp. 1–8 (2016)

Evaluation of Parameters of Transactions When Remote Robot Control

Eugene Larkin[1]([✉]), Vladislav Kotov[1], Alexander Privalov[2], and Alexey Ivutin[1]

[1] Tula State University, Tula 300012, Russia
elarkin@mail.ru , vkotov@list.ru , alexey.ivutin@gmail.com
[2] Tula State Pedagogical University, Tula 300026, Russia
privalov.61@mail.ru

Abstract. It is shown that control of mobile robots is implemented on strategic, tactic and functional-logic levels. A strategy of mobile robot behavior is defined by human operator, being functioning in dialogue with dialogue computer. Tactic and functional robotic levels are realized by onboard computer. Human operator, dialogue and onboard computers are the subjects of control, who exchange data, volumes and content of which define characteristics of control. Data are transmitted, when a transaction between subjects occurs. So, working out the model of transactions and evaluation its parameters is the actual problem. The model of generator of transactions from human operator and mobile robot onboard computer to dialogue computer, and vice versa from dialogue computer to the human operator and onboard computer is worked out. It is shown that due to transactions in operators of algorithms competition process is developed. Such a process defines a value of parameters of flows of transactions. Formulae for the primary evaluation of parameters are obtained. Iteration procedure for elaboration of parameters of flows of transactions is worked our.

Keywords: Mobile robot · Control · Transaction · Flow · Function-logic level · Dialogue · Semi-markov process · "Competition" · Iteration

1 Introduction

Mobile robots (PR) at present are rather widely used at monitoring of environment [1], in industry [2,3] and other spheres of mankind activity [4–6]. Main feature of contemporary mobile robotics consists in a lack of hard/software intelligence. Due to the fact strategic functions of control, human operator pared with dialogue computer is executed. Tactic and functional-logic levels are realized in the robot control system itself. On such levels onboard computer receives from dialogue computer flow of commands, interpret them, and actuates onboard equipment control loops. So, main feature of tasks of such a level are rigid requirements to a lag of reactions of control system onto both external commands and sensors state. Second feature is the necessity of time coordination of

© Springer International Publishing AG 2017
Y. Tan et al. (Eds.): ICSI 2017, Part II, LNCS 10386, pp. 567–577, 2017.
DOI: 10.1007/978-3-319-61833-3_60

operation of onboard equipment, receiving commands, forming and dispatching messages to human, etc.

So evaluation time factor of dialogue regimes of control is the actual and at present non-solved problem.

2   Common Model of Control

Principle of mobile robot control is shown on Fig. 1 [7–9].

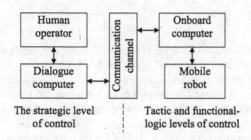

Fig. 1. Principle of mobile robot remote control

MR on the strategic level is managed by human operator, being situated at remote point of control, and maintaining an interactive dialogue with dialogue computer. Dialogue computer through communication channel is linked with onboard computer. In onboard computer external commands are interpreted, and both tactic and functional-logic levels of control are realized. Realization of command causes changes of onboard equipment states. Feedback information through onboard computer, communication channel and dialogue computer is transmitted to human opera-tor for making decision on the further continuation of control process.

In the functional diagram one can distinguish three active subjects, which operate, each on its own algorithm: human operator, dialogue computer and onboard computer. Mobile robot itself is the passive control object. As a result of control process, every subject generates transactions to the adjacent subject.

For evaluation of parameters of flows of transactions analytical model of such a system should be worked out. With taking into account features of algorithms under consideration (cyclic recurrence, quasi-stochastic nature of switches to adjacent operators, quasi-accidental time of interpretation of operators) [10] model of it is the ergodic semi-Markov process [11]. Operators of algorithm may be considered ad states of the process. Interpretation of algorithm may be considered as random wandering through the states of semi-Markov process.

In common semi-Markov process, which describe generators of transactions are as follows [10,11]:

$$^i\mu = \left\{ {}^iA, {}^i\boldsymbol{h}(t) \right\}, i = 1, 2, 3, \tag{1}$$

where $^1\mu$ – is the process, describing the human-operator; $^2\mu$ and $^3\mu$ – are processes, describing dialogue and onboard computers correspondingly; ^iA – is the set of states; $^i\boldsymbol{h}(t) = [h_{m_i n_i}(t)]$ – is the semi-Markov matrix of size $J_i \times J_i$; t – is the time;

$$^iA = \begin{cases} \{a_{1_i}, \ldots, a_{s_i}, \ldots, a_{S_i}, a_{S_i+1}, \ldots, a_{j_i}, \ldots, a_{J_i}\}, & \text{when } i = 1, 2; \\ \{a_{1_i}, \ldots, a_{s_i}, \ldots, a_{U_i}, a_{U_i+1}, \ldots, a_{S_i}, a_{S_i+1}, \ldots, a_{j_i}, \ldots, a_{J_i}\}, & \text{when } i = 3. \end{cases}$$
(2)

$$h_{m_i n_i}(t) = p_{m_i n_i} f_{m_i n_i}(t);$$
(3)

$p_{m_i n_i}$ –is the probability of switching from state $a_{m_i} \in^i A$ to state $a_{n_i} \in^i A$; $f_{m_i n_i}(t)$ – is density of time of residence in state $a_{m_i} \in^i A$ on condition of further switching to $a_{n_i} \in^i A$;

$$\sum_{n_i=1}^{J_i} p_{m_i n_i} = 1.$$
(4)

Nodes of graph with numbers from 1_i till S_i are analogues of states of transactions generation. In semi-Markov process, describing dialogue computer, nodes $A_{U_3} = \{a_{1_3}, \ldots, a_{u_3}, \ldots, a_{U_3}\}$ are analogue of states of generation of transactions from dialogue computer to human operator. Nodes $A_{S_3} = \{a_{U_3+1}, \ldots, a_{s_3}, \ldots, a_{S_3}\}$ are analogue of states of generation of transactions from dialogue to onboard computers. Nodes $A_{J_i} = \{a_{S_i+1}, \ldots, a_{j_i}, \ldots, a_{J_i}\}$ are analogue of other states of semi-Markov processes.

Transactions are generated in one of two cases:

1. When direct switching from states with numbers from 1_i till S_i to states with numbers from 1_i till S_i occurs;
2. When switching from states with numbers from 1_i till S_i to states with numbers from $S_i + 1_i$ till J_i with further wandering till states from 1_i till S_i occurs.

With use of methods described in [12,13], Semi-Markov processes may be reduced to processes, included generation transaction state only:

$$^i\mu \rightarrow^i \mu' = \left\{^iA', ^i\boldsymbol{h}'(t)\right\}, i = 1, 2, 3,$$
(5)

where $^iA'$ – is reduced set of states; $^i\boldsymbol{h}'(t)$ – is semi-Markov matrix of size $S_i \times S_i$;

$$^iA' = \begin{cases} \{a_{1_i}, \ldots, a'_{1_i}, \ldots, a'_{s_i}, \ldots, a'_{S_i}\}, & \text{when } i = 1, 2; \\ \{a_{1_i}, \ldots, a'_{1_i}, \ldots, a'_{u_i}, \ldots, a'_{U_i}, a'_{U_i+1}, \ldots, a'_{s_i}, \ldots, a'_{S_i}\}, & \text{when } i = 3. \end{cases}$$
(6)

$$^i\boldsymbol{h}'(t) = \left[h'_{m_i n_i}(t)\right].$$
(7)

At each switching of semi-Markov process (5) one transaction is generated. Due to the fact, that transformations applied are equivalent ones, processes (5) are ergodic too. For external observer probabilities of residence in states of ergodic semi-Markov process in steady regime of switching, are defined as follows:

$$\pi_{m_i} = \frac{T_{m_i}}{\tau_{m_i}} \tag{8}$$

where T_{m_i} – is the expectation of time of residence of ergodic semi-Markov process (1) in the state $a'_{m_i} \in^i A'$; τ_{m_i} – is the tine of return into state $a'_{m_i} \in^i A'$. Time of residence in state $a'_{m_i} \in^i A'$ is as follows

$$T_{m_i} = \int_0^\infty t \sum_{m_i=1_i}^{S_i} h'_{m_i n_i}(t) dt. \tag{9}$$

For evaluation of time τ_{m_i} one should to split the state a'_{m_i} of semi-Markov process (1) to $^b a'_{m_i}$ and $^e a'_{m_i}$. This leads to the transformation of matrix $h'_i(t)$ as follows:

- column with number m_i should be transmitted to column with number S_i+1;
- column with number m_i and row with number $S_i + 1$ should be fulfilled with zeros.

Matrix $\tilde{h}^{'}(t)$ after transformation is of size $(S_i + 1) \times (S_i + 1)$. Expectation of time of return is as follows:

$$\tau_{m_i} = \int_0^\infty t L^{-1r} I_{S_i+1} \sum_{k=1}^\infty \left\{ L\left[\tilde{h}'(t) \right] \right\}^k {}^c I_{m_i} dt, \tag{10}$$

where $^c I_{m_i}$ – is the column vector of size $S_i + 1$, m_i-th element of which is equal to one, and other elements are zeros; $^r I_{m_i}$ – is row vector of size S_i+1, S_i+1-th element of which is equal to one, and other elements are zeros; $L[\ldots], L^{-1}[\ldots]$ - are direct and inverse Laplace transforms correspondingly.

Due to (8) and property of ergodics of semi-Markov process under investigation, densities of time between neighboring transactions are as follows:

$$g_i(t) = \sum_{m_i=1_i}^{S_i} \pi_{m_i} \sum_{n_i=1_i}^{S_i} h'_{m_i n_i}(t), \qquad T_i = \int_0^T t g_i(t) dt,$$

$$D_i = \int_0^T (t - T_i)^2 g_i(t) dt, \qquad i = 1, 2, \tag{11}$$

where $g_i(t), T_i, D_i$ – are density, expectation and dispersion of time between transactions, correspondingly.

In such a way, processes $^1\mu'$ and $^2\mu'$ generate one stream each to adjacent process $^3\mu'$. Semi-Markov processes of generators after reduction are as follows:

$$^i\gamma = \{\{^i\alpha\}, [g_i(t)]\}, i = 1, 2. \tag{12}$$

Semi-Markov process $^3\mu'$ born two flows of transactions: to the process $^1\mu'$ and to the process $^2\mu'$. Probabilities of residence $^3\mu'$ in the state of generation transactions to $^1\mu'$ and $^2\mu'$ for the external observer are as follows:

$$\pi_{31} = \sum_{m_3=1_3}^{U_3} \pi_{m_3}; \qquad \pi_{32} = \sum_{m_3=U_3+1}^{S_3} \pi_{m_3}, \tag{13}$$

where π_{31} – is the probability of residence of $^3\mu'$ in the state of generation of trans-actions from $^3\mu'$ to $^1\mu'$; π_{32} – is the probability of residence of in the state of generation of transactions from $^3\mu'$ to $^2\mu'$.

Thus semi-Markov process of generation of transactions both from $^3\mu'$ to $^1\mu'$ and $^2\mu'$ is as follows

$$^3\gamma = \left\{ \{^3\alpha_0, {}^3\alpha_1, {}^3\alpha_2\}, \begin{bmatrix} 0 & \delta(t)\pi_U & \delta(t)\pi_S \\ f_U(t) & 0 & 0 \\ f_S(t) & 0 & 0 \end{bmatrix} \right\} \tag{14}$$

where $^3\alpha = \{^3\alpha_0, {}^3\alpha_1, {}^3\alpha_2\}$ – is set of states, when switching from which trans-action to $^1\gamma$ is generated; $^3\alpha_0$ – is the state, which define probability of the next switching; $\delta(t)$ – is the Diraq δ-function;

$$\pi_U = \sum_{m_3=1_3}^{U_3} \pi_{m_3}; \qquad \pi_S = \sum_{m_3=U_3+1}^{S_3} \pi_{m_3}, \tag{15}$$

$$f_U(t) = \frac{\sum\limits_{m_3=1_3}^{U_3} \pi_{m_3} \sum\limits_{n_3=1_3}^{S_3} h'_{m_3 n_3}(t)}{\sum\limits_{m_3=1_3}^{U_3} \pi_{m_3}};$$

$$f_S(t) = \frac{\sum\limits_{m_3=U_3+1}^{S_3} \pi_{m_3} \sum\limits_{n_3=1_3}^{S_3} h'_{m_3 n_3}(t)}{\sum\limits_{m_3=U_3+1}^{S_3} \pi_{m_3}}. \tag{16}$$

For evaluation of density of time between transactions from $^3\gamma$ to $^1\gamma$ one should to split the state $^3\alpha_1$ onto $^{3b}\alpha_1$ and $^{3e}\alpha_1$. Semi-Markov process $^3\gamma$ with divided state $^3\alpha_1$ is as follows:

$$^3\gamma_1 = \left\{ \{\alpha_1, {}^{3b}\alpha_1, {}^3\alpha_2, {}^{3e}\alpha_1\}, \begin{bmatrix} 0 & 0 & \delta(t)\pi_S & \delta(t)\pi_U \\ f_U(t) & 0 & 0 & 0 \\ f_S(t) & 0 & 0 & 0 \\ 0 & 0 & 0 & 0 \end{bmatrix} \right\}. \tag{17}$$

Density of time between transactions from $^3\gamma$ to $^1\gamma$ is equal to:

$$g_{31}(t) = L^{-1}\left[(0,1,0,0),\sum_{k=1}^{\infty}\left\{L\begin{bmatrix} 0 & 0\,\delta(t)\pi_S & \delta(t)\pi_U \\ f_U(t) & 0 & 0 & 0 \\ f_S(t) & 0 & 0 & 0 \\ 0 & 0 & 0 & 0 \end{bmatrix}\right\}^k\begin{pmatrix}0\\0\\0\\1\end{pmatrix}\right]. \tag{18}$$

Correspondingly, semi-Markov process $^3\gamma$ with divided state $^3\alpha_2$ and density of time between transactions from $^3\gamma$ to $^2\gamma$ are as follows:

$$^3\gamma_1 = \left\{\{\alpha_1,{}^3\alpha_1,{}^{3b}\alpha_2,{}^{3e}\alpha_2\},\begin{bmatrix} 0 & \delta(t)\pi_U & 0 & \delta(t)\pi_S \\ f_U(t) & 0 & 0 & 0 \\ f_S(t) & 0 & 0 & 0 \\ 0 & 0 & 0 & 0 \end{bmatrix}\right\}; \tag{19}$$

$$g_{32}(t) = L^{-1}\left[(0,0,1,0),\sum_{k=1}^{\infty}\left\{L\begin{bmatrix} 0 & \delta(t)\pi_U & 0 & \delta(t)\pi_S \\ f_U(t) & 0 & 0 & 0 \\ f_S(t) & 0 & 0 & 0 \\ 0 & 0 & 0 & 0 \end{bmatrix}\right\}^k\begin{pmatrix}0\\0\\0\\1\end{pmatrix}\right]. \tag{20}$$

Due to the fact, that transactions are generated as a result of random wandering through the states of semi-Markov processes, train of transactions, being generated when moving through every separate trajectory, may be considered as independent flow. So, united train of transactions may be considered as combination of transactions. In accordance theorem by Grigelionis B. [14] this united flow is a Poisson one. This is why next restrictions to densities of time between transactions may be accepted:

$$g_i(t) = \lambda_i \exp(-\lambda_i t); \tag{21}$$

$$g_{3j}(t) = \lambda_{3j} \exp(-\lambda_{3j} t); \tag{22}$$

where $\lambda_i, \lambda_{3j}, i, j = 1, 2$ – are the densities of flows of transactions

$$\lambda_i = \frac{1}{T_i}; \qquad \lambda_{3j} = \frac{1}{\int\limits_0^{\infty} t g_{3j}(t)dt}. \tag{23}$$

So processes $^1\gamma, ^2\gamma, ^3\gamma$ may be considered as the Markov processes with continual time.

3   "Competition" of Transactions

As it follows from models above, switching in every process $^1\gamma, ^2\gamma, ^3\gamma$ lead to generation a transaction into adjacent Markov process. When transaction comes restart of corresponding Markov process takes place. When restarting, transaction is not generated. In such a way in the states $^i\alpha, i = 1, 2, ^3\alpha_1, ^3\alpha_2$

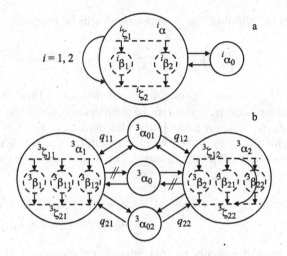

Fig. 2. "Competition" of transactions

"competitions" are evolved (Fig. 2a, b) [15,16]. Models of "competition", as fragments of Petri-Markov nets [17] are shown with dash lines within states $^i\alpha, i = 1, 2, ^3\alpha_1, ^3\alpha_2$ of processes $^1\gamma, ^2\gamma, ^3\gamma$ on Fig. 2a, b.

Into models additional states $^i\alpha_0, i = 1, 2, ^3\alpha_{01}, ^3\alpha_{02}$ are inserted for simulation of restart processes. After restarting of the processes $^1\gamma, ^2\gamma$, unconditional switching to state $^i\alpha, i = 1, 2$ takes place (Fig. 2a). Process $^3\gamma$ switches into states $^3\alpha_1, ^3\alpha_2$, with probabilities q_{11}, q_{12}, if transaction comes from $^1\gamma$, and with probabilities q_{21}, q_{22}, if transaction comes from $^2\gamma$ (Fig. 2b).

Petri-Markov net, evolved in state $^i\alpha$, has places $^i\beta_1, ^i\beta_2$, and transitions $^i\zeta_1, ^i\zeta_2, i = 1, 2$. Places simulate next processes: residence in state $^i\alpha$ with full-time completion and residence in state $^i\alpha$ till interrupt delivery from $^3\gamma$. Transitions $^i\zeta_1, ^i\zeta_2, i = 1, 2$ simulate start and finish of "competition", correspondingly.

Petri-Markov nets, evolved in states $^3\alpha_j, j = 1, 2$, have places $^3\beta_i, ^3\beta_{j1}, ^3\beta_{j2}, i = 1, 2$, and transitions $^3\zeta_{ij}, i, j = 1, 2$. Places simulate next processes: $^3\beta_j$ – residence in states $^3\alpha_j, j = 1, 2$ with full-time completion; $^3\beta_{ij}$ – residence transactions from $^i\gamma_i, i, j = 1, 2$. Transitions simulate $^3\beta_{1j}$ – start, $^3\beta_{2j}$ – finish of "concurrency".

Switching from transitions $^i\zeta_1, i = 1, 2$, or $^3\zeta_{1j}, j = 1, 2$ is executed simultaneously. "Winner" of "competition" is the place, which switches to the transition $^i\zeta_2, i = 1, 2$, or $^3\zeta_{2j}, j = 1, 2$ the first.

Taking into account (21), (22) density of at least one switch to $^i\zeta_2$, is as follows (15)

$$^i f_\zeta(t) = \lambda_i \exp\left[-t\left(\lambda_i + \lambda_{3i}\right)\right] + \lambda_{3i} \exp\left[-t\left(\lambda_i + \lambda_{3i}\right)\right], i = 1, 2. \qquad (24)$$

Densities of time of switches to $^i\zeta_2$, whatever the outcome of the "competition", are quite equal and as follows

$$^i f_{i\zeta}(t) = {}^i f_{3i\zeta}(t) = \left(\lambda_i + \lambda_{3i}\right) \exp\left[-t\left(\lambda_i + \lambda_{3i}\right)\right], i = 1, 2. \qquad (25)$$

Probabilities of "winning" in "competition" will be the next:

$$\begin{pmatrix} {}^i p_{i\zeta} \\ {}^i p_{3i\zeta} \end{pmatrix} = \frac{1}{\lambda_{3i} + \lambda_i} \begin{pmatrix} \lambda_i \\ \lambda_{3i} \end{pmatrix}, i = 1, 2, \tag{26}$$

where ${}^i p_{i\zeta}$ and ${}^i f_{i\zeta(t)}$ – are a probability and density of time of "winning" in "competition" the place ${}^i \beta_1$, accompanied with generation of the transaction to ${}^3\gamma$; ${}^i p_{3i\zeta}$ and ${}^i f_{3i\zeta(t)}$ – are a probability and density of time of "winning" in "competition" the place ${}^i \beta_2$, linked with receiving transaction from ${}^3\gamma$.

Taking into account (21), (22) density of at least one switch to ${}^3\zeta_{2j}, j = 1, 2$, is as follows (15)

$$f_{\gamma j}(t) = \lambda_{3j} \exp\lfloor -t\left(\lambda_{3j} + \lambda_1 + \lambda_2\right)\rfloor + \lambda_1 \exp\lfloor -t\left(\lambda_{3j} + \lambda_1 + \lambda_2\right)\rfloor + \\ + \lambda_2 \exp\lfloor -t\left(\lambda_{3j} + \lambda_1 + \lambda_2\right)\rfloor. \tag{27}$$

Densities of time of switches to ${}^3\zeta_{2j}$, whatever the outcome of the "competition", are quite equal and as follows

$${}^3 f_{3\zeta j}(t) = {}^3 f_{1\zeta j}(t) = {}^3 f_{2\zeta j}(t) = (\lambda_1 + \lambda_2 + \lambda_{3i}) \exp\left[-t\left(\lambda_1 + \lambda_2 + \lambda_{3i}\right)\right]. \tag{28}$$

$$\begin{pmatrix} {}^3 p_{3\zeta j} \\ {}^3 p_{1\zeta j} \\ {}^3 p_{2\zeta j} \end{pmatrix} = \frac{1}{\lambda_{3j} + \lambda_1 + \lambda_2} \begin{pmatrix} \lambda_{3j} \\ \lambda_1 \\ \lambda_2 \end{pmatrix}, j = 1, 2. \tag{29}$$

4   Iterative Procedure of Correction of Parameters of Flows of Transactions

It is obviously, that transactions, incoming from adjacent Markov processes, change parameters of flows of transactions (21), (22). Subsequent correction of parameters may be obtained with use of iterative procedure. For starting such a procedure one should nominate parameters of (21), (22) as follows:

$$g_i^0(t) = g_i(t); \quad \lambda_i^0 = \lambda_i; \quad g_{3j}^0(t) = g_{3j}(t); \\ \lambda_{3j}^0 = \lambda_{3j}; \quad i, j = 1, 2; \pi_1^0 = \pi_U; \pi_2^0 = \pi_S. \tag{30}$$

Parameters obtained on the l-th step of iteration one should nominate as $g_i^l(t), \lambda_i^l, g_{3j}^l(t), \lambda_{3j}^l, i, j = 1, 2; \pi_1^l, \pi_2^l$. Density of time between transactions from ${}^i\gamma$, to ${}^3\gamma$ is as follows

$$g_i^{l+1}(t) = \lambda_i^{l+1} \exp(-t\lambda_i^{l+1}), i = 1, 2 \tag{31}$$

where

$$\lambda_i^{l+1} = \cfrac{1}{\int\limits_0^\infty t L^{-1}\lceil (1,0,0) \sum\limits_{k=1}^\infty \{L\left[{}^i \boldsymbol{h}'(t)\right]\}^k \begin{pmatrix} 0 \\ 0 \\ 1 \end{pmatrix} \rceil dt}, \tag{32}$$

$$^i\boldsymbol{h}^l(t) = \begin{pmatrix} 0 & \lambda_i^l \exp\lfloor -t\left(\lambda_i^l + \lambda_{3i}^l\right)\rfloor & \lambda_{3i}^l \exp\lfloor -t\left(\lambda_i^l + \lambda_{3i}^l\right)\rfloor \\ \delta(t) & 0 & 0 \\ \lambda_i \exp\lfloor -t\left(\lambda_i^l + \lambda_{3i}^l\right)\rfloor & 0 & 0 \end{pmatrix}$$

$$(33)$$

For evaluation parameters of flow of transactions from $^3\gamma$ to $^1\gamma$ one should to change the arc $(^3\alpha_1, ^3\alpha_0)$ (marked with dash on Fig. 2b) to arc and absorbing state $^{3e}\alpha_1$. In this case set of states of Markov process will be the next: $\{^3\alpha_1, ^3\alpha_2, ^3\alpha_{01}, ^3\alpha_{01}, ^3\alpha_{02}, ^{3e}\alpha_1\}$.

For evaluation parameters of flow of transactions from $^3\gamma$ to $^2\gamma$ one should to change the arc $(^3\alpha_2, ^3\alpha_0)$ (marked with double dash on Fig. 2b) to arc and absorbing state $^{3e}\alpha_2$. In this case set of states of Markov process will be the next: $\{^3\alpha_1, ^3\alpha_2, ^3\alpha_{01}, ^3\alpha_{01}, ^3\alpha_{02}, ^{3e}\alpha_2\}$.

Values λ_{3j}^{l+1} are as follows

$$\lambda_{3j}^{l+1} = \frac{1}{\int\limits_0^\infty tL^{-1}\left[^r\boldsymbol{I}_j \sum\limits_{k=1}^\infty \left\{L\left[^3_j\boldsymbol{h}^l(t)\right]\right\}^k {}^c\boldsymbol{I}_6\right]dt}, j = 1, 2 \qquad (34)$$

where $^3_j\boldsymbol{h}^l(t)$ – semi-Markov matrix of size 6×6, j-th column of which is moved to sixth column, j-th column and sixth row are fulfilled with zeros; $^r\boldsymbol{I}_j$ – is the row vector, j-th element of which is equal to one, and all other elements are equal to zeros; $^r\boldsymbol{I}_j$ – is the column vector, sixth element of which is equal to one, and all other elements are equal to zeros;

$$^3_1h_{14}^l(t) = ^3_2h_{14}^l(t) = \lambda_1^l \exp\lfloor -\left(\lambda_{31}^l + \lambda_1^l + \lambda_2^l\right)\rfloor;$$

$$^3_1h_{15}^l(t) = ^3_2h_{15}^l(t) = \lambda_2^l \exp\lfloor -\left(\lambda_{31}^l + \lambda_1^l + \lambda_2^l\right)\rfloor;$$

$$^3_1h_{24}^l(t) = ^3_2h_{24}^l(t) = \lambda_1^l \exp\lfloor -\left(\lambda_{32}^l + \lambda_1^l + \lambda_2^l\right)\rfloor;$$

$$^3_1h_{25}^l(t) = ^3_2h_{25}^l(t) = \lambda_2^l \exp\lfloor -\left(\lambda_{32}^l + \lambda_1^l + \lambda_2^l\right)\rfloor;$$

$$^3_1h_{16}^l(t) = \lambda_{31}^l \exp\lfloor -\left(\lambda_{31}^l + \lambda_1^l + \lambda_2^l\right)\rfloor; \qquad ^3_2h_{26}^l(t) = \lambda_{32}^l \exp\lfloor -\left(\lambda_{32}^l + \lambda_1^l + \lambda_2^l\right)\rfloor;$$

$$^3_1h_{23}^l(t) = \lambda_{32}^l \exp\lfloor -\left(\lambda_{32}^l + \lambda_1^l + \lambda_2^l\right)\rfloor; \qquad ^3_2h_{13}^l(t) = \lambda_{31}^l \exp\lfloor -\left(\lambda_{31}^l + \lambda_1^l + \lambda_2^l\right)\rfloor;$$

$$^3_1h_{31}^l = ^3_2h_{31}^l = \pi_1^l\delta(t); \quad ^3_1h_{32}^l = ^3_2h_{32}^l = \pi_2^l\delta(t); \quad ^3_1h_{41}^l = ^3_2h_{41}^l = q_{11}\delta(t);$$

$$^3_1h_{42}^l = ^3_2h_{42}^l = q_{12}\delta(t); \quad ^3_1h_{51}^l = ^3_2h_{51}^l = q_{21}\delta(t); \quad ^3_1h_{52}^l = ^3_2h_{52}^l = q_{22}\delta(t).$$

Values π_1^{l+1}, π_2^{l+1}, one can define from analysis Markov process $^3\gamma$ without splitting of states. In steady regime for external observer

$$\pi_1^{l+1} = \frac{p_{21}\left(\lambda_{32}^l + \lambda_1^l + \lambda_2^l\right)}{p_{21}\left(\lambda_{32}^l + \lambda_1^l + \lambda_2^l\right) + p_{12}\left(\lambda_{31}^l + \lambda_1^l + \lambda_2^l\right)};$$

$$\pi_2^{l+1} = \frac{p_{12}\left(\lambda_{31}^l + \lambda_1^l + \lambda_2^l\right)}{p_{21}\left(\lambda_{32}^l + \lambda_1^l + \lambda_2^l\right) + p_{12}\left(\lambda_{31}^l + \lambda_1^l + \lambda_2^l\right)}; \tag{35}$$

$$p_{12} = \frac{\lambda_{31}^l \pi_2^l + \lambda_1^l q_{12} + \lambda_2^l q_{22}}{\lambda_{31}^l + \lambda_1^l + \lambda_2^l}; \quad p_{21} = \frac{\lambda_{32}^l \pi_2^l + \lambda_1^l q_{12} + \lambda_2^l q_{22}}{\lambda_{32}^l + \lambda_1^l + \lambda_2^l}. \tag{36}$$

So formulae (30)–(35) describe iteration for correction parameters $\lambda_{31}, \lambda_{32}$, λ_1, λ_2 of flow of transactions. Procedure may be finished on one of the next criteria;

$$\frac{|\lambda_{31}^l - \lambda_{31}^{l+1}|}{\lambda_{31}^l} < \varepsilon_{31}, \frac{|\lambda_{32}^l - \lambda_{32}^{l+1}|}{\lambda_{32}^l} < \varepsilon_{32}, \frac{|\lambda_1^l - \lambda_1^{l+1}|}{\lambda_1^l} < \varepsilon_1, \frac{|\lambda_2^l - \lambda_2^{l+1}|}{\lambda_2^l} < \varepsilon_2,$$
$$\tag{37}$$

or on the summing criterion

$$\frac{|\lambda_{31}^l - \lambda_{31}^{l+1}|}{\lambda_{31}^l} + \frac{|\lambda_{32}^l - \lambda_{32}^{l+1}|}{\lambda_{32}^l} + \frac{|\lambda_1^l - \lambda_1^{l+1}|}{\lambda_1^l} + \frac{|\lambda_2^l - \lambda_2^{l+1}|}{\lambda_2^l} < \varepsilon, \tag{38}$$

where $\varepsilon_{31}, \varepsilon_{32}, \varepsilon_1, \varepsilon_2, \varepsilon$ – are some small pre-determined threshold empirically selected.

5   Conclusion

In such a way, the analytical model of remote control of mobile robot is worked out. For construction of such a model both actions of human operator and functioning of dialogue and onboard computer are divided onto sequence of operations for which it is simple to determine time characteristics and probabilities of transfer to other operation. During execution of operations all subjects of process of control generate transactions to adjacent subject. So parameters of transactions were found with the aid of iterative procedure. Result obtained may be also used for working out other dialogue systems, for example for industry ergatic systems control.

Further continuation of investigations in this domain may be directed to an improvement of iteration procedure, to optimization of dialogue algorithms and adaptation it to characteristics of both human operator and mobile robot, to optimization the transaction flows in concrete systems etc.

The research was carried out within the state assignment of The Ministry of Education and Science of Russian Federation (No. 2.3121.2017/PCH).

References

1. Ivutin, A., Larkin, E., Kotov, V.: Established routine of swarm monitoring systems functioning. In: Tan, Y., Shi, Y., Buarque, F., Gelbukh, A., Das, S., Engelbrecht, A. (eds.) ICSI 2015. LNCS, vol. 9141, pp. 415–422. Springer, Cham (2015). doi:10. 1007/978-3-319-20472-7_45

2. Shneier, M., Bostelman, R.: Literature review of mobile robots for manufacturing. National Institute of Standards and Technology, US Department of Commerce (2015)

3. Angerer, S., Strassmair, C., Staehr, M., Roettenbacher, M., Robertson, N.M.: Give me a hand the potential of mobile assistive robots in automotive logistics and assembly applications. In: 2012 IEEE International Conference on Technologies for Practical Robot Applications (TePRA), pp. 111–116. IEEE (2012)

4. Angerer, S., Pooley, R., Aylett, R.: Mobcomm: Using BDI-agents for the reconfiguration of mobile commissioning robots. In: 2010 IEEE Conference on Automation Science and Engineering (CASE), pp. 822–827. IEEE (2010)

5. Graf, B., Reiser, U., Hägele, M., Mauz, K., Klein, P.: Robotic home assistant care-o-bot® 3-product vision and innovation platform. In: 2009 IEEE Workshop on Advanced Robotics and its Social Impacts (ARSO), pp. 139–144. IEEE (2009)

6. Tzafestas, S.G.: Introduction to Mobile Robot Control. Elsevier, New York (2013)

7. Hassan, M.A.A.: A review of wireless technology usage for mobile robot controller. In: Proceeding of the International Conference on System Engineering and Modeling (ICSEM 2012), pp. 7–12 (2012)

8. Argall, B.D., Browning, B., Veloso, M.M.: Teacher feedback to scaffold and refine demonstrated motion primitives on a mobile robot. Rob. Auton. Syst. 59(3), 243–255 (2011)

9. Sisbot, E.A., Marin-Urias, L.F., Alami, R., Simeon, T.: A human aware mobile robot motion planner. IEEE Trans. Rob. 23(5), 874–883 (2007)

10. Larkin, E., Ivutin, A., Kotov, V., Privalov, A.: Semi-markov modelling of commands execution by mobile robot. In: Ronzhin, A., Rigoll, G., Meshcheryakov, R. (eds.) ICR 2016. LNCS, vol. 9812, pp. 189–198. Springer, Cham (2016). doi:10.1007/978-3-319-43955-6_23

11. Korolyuk, V., Swishchuk, A.: Semi-markov random evolutions. In: Korolyuk, V., Swishchuk, A. (eds.) Semi-Markov Random Evolutions, vol. 308, pp. 59–91. Springer, Dordrecht (1995)

12. Ivutin, A., Larkin, E.: Estimation of latency in embedded real-time systems. In: 2014 3rd Mediterranean Conference on Embedded Computing (MECO), pp. 236–239. IEEE (2014)

13. Larkin, E., Ivutin, A., Esikov, D.: Recursive approach for evaluation of time intervals between transactions in polling procedure. In: MATEC Web of Conferences, vol. 56, p. 01004 (2016)

14. Grigelionis, B.: On the convergence of sums of random step processes to a poisson process. Theory of Probab. Appl. 8(2), 177–182 (1963)

15. Cleaveland, R., Smolka, S.A.: Strategic directions in concurrency research. ACM Comput. Surv. (CSUR) 28(4), 607–625 (1996)

16. Ivutin, A., Larkin, E.: Simulation of concurrent games. Bull. South Ural State Univ. Ser.: Math. Model. Program. Comput. Softw. 8(2), 43–54 (2015)

17. Ivutin, A.N., Larkin, E.V., Lutskov, Y.I., Novikov, A.S.: Simulation of concurrent process with petri-markov nets. Life Sci. J. 11(11), 506–511 (2014)

Desktop Gestures Recognition
for Human Computer Interaction

Qingjie Zhu[1,2,6(✉)], Hang Pan[3,4,6], Minghao Yang[6],
and Yongsong Zhan[5]

[1] Guangxi Key Laboratory of Trusted Software,
Guilin University of Electronic Technology, Guilin 541004, Guangxi, China
zhuqj21@foxmail.com
[2] Guangxi Cooperative Innovation Center of Cloud Computing and Big Data,
Guilin University of Electronic Technology, Guilin 541004, Guangxi, China
[3] Guangxi Key Laboratory of Cryptography and Information Security,
Guilin University of Electronic Technology, Guilin 541004, Guangxi, China
[4] Guangxi Colleges and Universities Key Laboratory of Intelligent Processing
of Computer Image and Graphics, Guilin University of Electronic Technology,
Guilin 541004, Guangxi, China
[5] Key Laboratory of Cloud Computing and Complex System,
Guilin University of Electronic Technology, Guilin 541004, Guangxi, China
[6] Institute of Automation, Chinese Academy of Sciences, Beijing 100190, China

Abstract. A dynamic gesture recognition and understanding method in natural human-computer interaction under desktop environment is proposed, including the "reach", "take up", "move", "put down", "return", "point" and other natural interactive gestures. In preprocess procedure of each frame of the video, the Gaussian background model and HSV skin-color model is employed to remove background and segment hand gestures. The temporal and spatial information of multi frame images is combined to construct temporal and spatial features of dynamic gestures images. Then a convolution neural network is built for recognize the dynamic characteristics of gesture image. Finally, the classification result is denoised to achieve the robust recognition and understanding of gestures. Experimental results show that the proposed method has a good ability of recognizing and understanding the dynamic gestures in the desktop environment.

Keywords: Gesture recognition · Human-computer interaction · Computer vision

1 Introduction

With the development of human-computer interaction technology, the interaction between human and computer becomes more and more important in people's daily life. Gesture interaction is a common way in human-computer interaction, and it is also a method which is natural, intuitive and easy to learn. For example, in interactions with computers through gestures, people make a series of gestures under the camera to give instructions to computers. In order to enable the computer to understand the user's behavior in preliminary, we must first make the computer understand the meaning of

© Springer International Publishing AG 2017
Y. Tan et al. (Eds.): ICSI 2017, Part II, LNCS 10386, pp. 578–585, 2017.
DOI: 10.1007/978-3-319-61833-3_61

each gesture. For example, the behavior, "pick up the cup on the table and drink", can be divided into "reach", "take up" and "move" gestures. In this paper, we mainly study the common understanding of human-computer interaction in the context of desktop environment.

Research on gesture recognition technology dates back to 1990s. In 1993, B. Thamas et al.'s free hand remote control system adopted a data glove as the input, to recognize gestures, but this requires the experimenter to wear a special equipment [1]. Later, people are committed to the study of marker gestures, by marking in the hand, such as pasting or painting dot in special color on the wrist or finger, gesture can be tracked through the markers motion [2]. Finally, people focus attention on the natural hand, with the development of deep learning and deep neural networks, people began to use deep learning method in gesture recognition. In 1999, Peter Vamplew used the recurrent neural network (RNN) to train 16 different dynamic gestures, and good classification results are achieved [3]. Compared with the ordinary convolution neural network (CNN), the RNN's ability of extracting the temporal information is much stronger, but its ability of processing images is worse. So some researchers began to try some new CNN structures, history motion information in adjacent frames is captured and achieved a good results [4]. Whether in static gesture recognition or dynamic gesture recognition, many works have been done and achieved good results. According to the specific circumstances of gestures, a new dynamic gesture recognition algorithm is proposed in this paper.

2 Approach

The architecture of the algorithm is shown in Fig. 1.

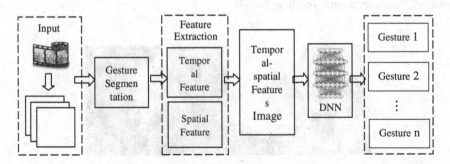

Fig. 1. The architecture of the algorithm

The input is a sequence of video frames. In this paper, the fused image of the spatial and temporal characteristics is a gray image, in which the degree of light and shadow of the gesture image represents the temporal information, the position of the gesture image represents the spatial information of the gesture. According to the temporal-spatial features image of the dynamic gesture, then enter the convolution neural network

training, and finally based on the training model for the understanding and classification of gestures.

3 Gesture Segmentation and Image Preprocessing

In order to make a better recognition of neural network with different gestures, eliminate other interference, this paper adopts gauss background modeling method combined with HSV color modeling to remove the background information, and abandon
the texture information of the hand, only retain gesture shape information.

In the context of the desktop background, gesture recognition background is relatively simple, this paper uses a single Gauss background modeling method to model
the background. According to the single Gauss background model, for a background
image, the distribution of the intensity of a specific pixel satisfies the Gauss distribution
[5]. That is, for the background image B, the brightness of point (x, y) satisfies:

$$B(x, y) \sim N\left(\mu, \sigma^2\right) \tag{1}$$

When using a single Gauss background model, use background subtraction method
to segment foreground. As formula (2), (x, y) is the gray value of the point (x, y) in the
image, use the current frame to subtract the background image to determine whether the
point is a foreground point.

$$(x, y) = \begin{cases} background, |F_t(x, y) - B_t(x, y)| < threshold \\ foreground, |F_t(x, y) - B_t(x, y)| > threshold \end{cases} \tag{2}$$

Using single Gauss background model to remove desktop environment background, the experimental result as Fig. 2:

Fig. 2. The left image is original image, and the right image is a remove background image that
using single Gauss background model

According to the experimental results, we can see that the single Gauss background
model is not ideal to remove the background. Due to the movement of the hand,
resulting in changes in the environment of local light, and thus have a certain impact on
the judgment of the background. As Fig. 2, the darker background area below the hand
has not been effectively removed.

In this paper, we uses the HSV(Hue, Saturation, Value) space model to solve this problem, and it also called Hexcone Model [6]. H channel determines the color of the object information, when skin color detecting, we only pay attention to hue H, saturation S and brightness V is not too concerned about [7], so we only use the H channel of HSV color space to quickly split the gesture area in this paper.

In order to build a good skin color model, we collected 88 hand images under different circumstances, cut out 113 non overlapping skin areas, a total of 6007466 skin pixels. Convert this result to HSV space, the histogram distribution of H channel as shown in Fig. 3(a):

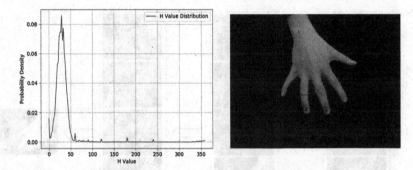

Fig. 3. (a) The histogram distribution of H channel in collected skin color image. (b) The gesture segmentation result using HSV color model

We can see from the chart, the vast majority H value of the histogram evenly and centrally distributed in the left about $0 \sim 60$ range, there is a small part of the H value in the vicinity of 80, 180, 360. In order to make the selected threshold has a good anti-noise ability, we selected H threshold is $6 \sim 48$ through the experimental measurement, using this range can detect most of the gesture area, and it also has a good anti-noise performance. The gesture segmentation experimental result using this color model as shown in Fig. 3(b).

4 The Training Method Based on Convolution Neural Network

CNN (Convolutional Neural Networks) can be used to classify the single image easily, while the dynamic hand gesture is a video clip, only one frame cannot represent the temporal information contained in dynamic gestures. The dynamic gestures contains multiple frames make it difficult for CNN to handle. In this paper, we use a dynamic gesture temporal information extraction method, which can use a temporal-spatial feature image to express dynamic gesture.

4.1   Extraction and Fusion of Temporal-Spatial Feature Images

Firstly, this method decompose many original images from a dynamic gesture video clip, then segment gesture based on Gauss background modeling and HSV skin color modeling in the previous section, because the different gestures are only related to the shape and motion information of different gestures, so in order to eliminate other interference, we make a binary processing for each frame. Each frame after processing is a binary image that retains only the gesture shape, then based on the continuous multi binary image in time domain space, the weight of each frame decay from near to distant in the temporal space, finally, all frames in the current image background region are fused based on each weight. The architecture of the algorithm is shown in Fig. 4.

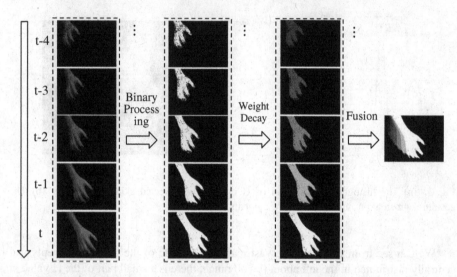

Fig. 4. The extraction process of dynamic gesture temporal-spatial feature image

In order to express the decay of historical gestures with time more accurately, the exponential decay model is used to simulate the decay of historical gestures, as follows:

$$w_x = e^{-\beta x}(x \geq 0). \tag{3}$$

Where w_x is the weight of the frame that distance from the current frame x frames ahead. In order to make the weight decay not too fast, let $\beta = 0.5$. If fused too many frames, then the feature map fused is larger in the temporal span, when temporal span is large, it is high possibility that the last frame is not the current gesture frame; If fused too few frames, then it will appear the problem that current temporal-spatial feature image cannot express the current dynamic gestures figuratively. Considering the above two problems, and experimental tests, this paper selects the fusion frame number is 6 frames, which can both describe dynamic gestures vividly and reduce the impact on the history frame to the current gesture. Based on formula (3), we can get the weights of the 6 frames: $w_0 = 1, w_1 = 0.60, w_2 = 0.37, w_3 = 0.22, w_4 = 0.14, w_5 = 0.08$.

In this paper, the gesture feature images fuse based on background region of current frame when fused this 6 frames, and first fuse the frame which weight is the highest. Firstly, we should select a frame which weight is the highest as the image to be fused, call it M, then select the highest weight frame in others, call it M1, traverse it by pixel, the fusion formula of two images in point $P = (x, y)$ as follows:

$$M(P) = \begin{cases} M1(P) & \begin{array}{c} P\ is\ foreground\ of\ M1 \\ and\ is\ background\ of\ M \end{array} \\ M(P) & others \end{cases} \qquad (4)$$

There, $M(P)$ is the pixel value of P in image M, $M1(P)$ is the pixel value of P in image $M1$. After fused this tow images, image M is replaced by the fused image, then select the highest weight image in the rest of the others as the new image M1. Repeat the above steps until all the images are fused.

4.2 Architecture of CNN

Convolution Neural Network has proven highly successful at static image recognition problems. The famous CNN network structure has LeNet, AlexNet and GoogLeNet, etc. the complexity of these three network structures increases in turn, considering the needs of this paper, we select the AlexNet network, and make a slight modification on the original network structure to meet our needs.

The network structure of AlexNet is divided into eight layers, it has five convolution layers and three full connection layers, the size of input image is 227*227*3. The final output results in a total of one thousand categories. In this paper, the input temporal-spatial feature image is gray image, the background is black, and it not has detailed texture information of hand, so the feature images not need a detailed description. For the input image, we select a size is 64*64*1, this will not lose the important details of the feature image, but also increase the speed of model training and real-time gesture recognition.

5 Experiments

In order to test the effect of gesture recognition, we take sample individually for the gesture "reach", "take up", "move", "put down", "return" and "point", we sampled 1800 images each gesture which set 1500 images as training set and 300 images as testing set, and sampled a total of 10800 images. Then, trained based on the Convolution Neural Network. Finally, using the trained model done the real-time gesture classification experiment. The results of experiment as follows (Table 1):

It can be seen from the experimental results, the recognition accuracy of gesture "move" is relatively low, the gesture "point" can get highest recognition accuracy because it is relatively simple. In the process of gesture recognition, there may be have a recognition error happened on one frame caused by external interference or the model itself is not accurate enough, For example, when the gestures is "reach", a series of

Table 1. Experimental results

Gesture	reach	take up	move	put down	return	point
Test amount	75	101	105	105	70	72
Hit@1	0.893	0.841	0.733	0.819	0.857	0.944
Hit@2	0.933	0.910	0.819	0.857	0.900	0.986

continuous recognition results as follows: {"reach"," reach", "reach", "reach", "move"," reach", "reach"}. This is a series of recognition results of a "reach" gesture, the correct result should be: {"reach"," reach", "reach", "reach", "reach ","reach", "reach"}. The fifth result "move" should be judged as noise. So, we adopted a denoising method to make the recognition results more smooth and improve the recognition results.

When smoothing processing the results, we set a one-dimensional filter kernel firstly, then set the gesture which has a highest occurrence number as the current result. Assume the length of the one-dimensional filter kernel is n, then count the historical results $\{t - n + 1, t - n + 2. \ldots .t\}$, calculate the highest occurrence number gesture, if it has many such results, then choice the result which is the nearest from the current time. In this paper, the length of one-dimensional filter kernel we selected is 5, the accuracy of real time gesture recognition has been improved after smoothing filtering. In order to compare the experimental results, we also test the method of "Raw Frames + CNN". The raw frames do not contain temporal information, but it contains rich texture information, color information and spatial information. The results of experiment as follows (Table 2):

Table 2. Comparison of experimental results

Method	Hit@1	Hit@2
Raw Frames + CNN	0.541	0.615
Temporal-Spatial Feature Images + CNN	0.839	0.894
Temporal-Spatial Feature Images + CNN + Denoising	0.899	0.931

6 Conclusion

In this paper, a method of understanding and recognizing the common gestures in the desktop environment is proposed. Through extraction and fusion the temporal information and spatial information of gesture video, an image with temporal-spatial features that is able to represent a dynamic gesture is obtained. There are great differences among the temporal-spatial features images of different gestures, so different temporal-spatial features' images can express different gestures. The accuracy of gesture segmentation has a great influence on generating the temporal-spatial features images. The performance of Gauss background modeling method for hand gesture segmentation is not good, and the skin color modeling will be affected by the background color if it's similar to the color of hand, so both methods are combined to get a good segmentation result. In this paper, the method of deep learning is employed to train different gestures, which learn the different representations between different gestures, it overcomes the shortcomings of the complexity in hand crafted features of describing and distinguishing the different gestures. The experimental results show that the proposed algorithm can achieve good classification results.

In this paper, these aspects could be improved in the following works: (1) the denoising procedure would delay the real time recognition results; (2) in the extreme environment, such as the circumstance is too bright or too dark, the result of gesture segmentation is not good enough. For the first problem, it is mainly caused by the one-dimensional filter kernel. It delays longer as the length of the filter kernel increase. If the filter kernel length is too short, the filtering effect is not obvious. A trade-off is made but there is still a delay of 2 frames. For the second problem, an adaptive algorithm should be discussed to solve this problem. These issues should be solved in the future's work.

Acknowledgments. This research work is supported by the grant of Guangxi science and technology development project (No: 1598018-6, AC16380124), the grant of Guangxi Key Laboratory of Trusted Software of Guilin University of Electronic Technology (No: KX201601), the grant of Guangxi Colleges and Universities Key Laboratory of Intelligent Processing of Computer Images and Graphics of Guilin University of Electronic Technology (No: GIIP201602), the grant of Guangxi Key Laboratory of Cryptography & Information Security of Guilin University of Electronic Technology (No: GCIS201603), the grant of Guangxi Cooperative Innovation Center of Cloud Computing and Big Data of Guilin University of Electronic Technology (No: YD16E11), the grant of Key Laboratory of Cloud Computing & Complex System of Guilin University of Electronic Technology (No: 15210).

References

1. Sturman, D.J.: Hand on interaction with virtual environments. In: ACM SIGGERAPH Symposium on User Interface and Software Technology (1989)
2. Sternberg, M.L.: A American Sign Language, A Comprehensive Dictionary. Harper and Row, New York (1981)
3. Vamplew, P., Adams, A.: Recognition and anticipation of hand motions using a recurrent neural network. In: Proceedings of the IEEE International Conference on Neural Networks, vol. 6(1), pp. 1226–1231 (1999)
4. Shuiwang, J., Wei, X.: 3D convolutional neural networks for human action recognition. IEEE Trans. Pattern Anal. Mach. Intell. **35**(1), 221–231 (2013)
5. Kang, X.J., Wu, J.: Object detecting technology based on gauss background modeling. Chin. J. Liquid Crystals Displays **25**(3), 454–459 (2010)
6. Garcia, C., Tziritas, G.: Face detection using quantized skin color regions merging and wavelet packet analysis. IEEE Trans. Multimedia **1**(3), 264–277 (1999)
7. Sural, S., Qian, G., Pramanik, S.: Segmentation and histogram generation using the HSV color space for image retrieval. In: Proceedings of the International Conference on Image Processing, vol. 2, pp. II-589–II-592. IEEE (2002)

Approach to the Diagnosis and Configuration of Servo Drives in Heterogeneous Machine Control Systems

Georgi M. Martinov, Sergey V. Sokolov, Lilija I. Martinova,
Anton S. Grigoryev, and Petr A. Nikishechkin[✉]

FSBEI HPE MSTU "STANKIN", Moscow, Russia
pnikishechkin@gmail.com

Abstract. During modernization of machines, development of prototypes, development or the retrofit of special modifications, a part of previously used components stay in the system. It is not rational to replace it. CNC system in this case should support the multiprotocol control and provide the possibility to configure mixed equipment. Modern servo drives have built-in diagnostic tools, they are designed for devices of the same manufacturer and cannot be used in heterogeneous environment. The article describes the approach of development and the use of special-purpose application tools for diagnostics, allowing to collect and analyze data simultaneously from the equipment of different manufacturers with various industrial communication protocols. The research of the application tools in the process of the startup of the milling machine Quaser with SERCOS III and EtherCAT servo drives is also presented. Solution for the milling machine with servo drives SERCOS III and EtherCAT is demonstrated; the results of machining the test part are analyzed.

Keywords: Multiprotocol CNC system · Servo drive diagnostic application tools · SERCOS III · EtherCAT

1 Introduction

In the field of industrial automation is often a need of modification or upgrading of individual machines or entire product lines of machine-tool plants [1, 2]. This need arise in the case of:

- transition to new equipment because of discontinuation of previously used components, such as servo drives;
- transition to other, cheaper components with similar specifications, but with other communication protocols;
- need for rapid prototype production of new machine modification to test technological solutions, when a part of equipment due to technical difficulties is advisable to leave unchanged; servo drivers in such systems can be controlled by different industrial communications protocols;
- application of ready-tested components and assemblies from previous models of machines in design of the new machine.

© Springer International Publishing AG 2017
Y. Tan et al. (Eds.): ICSI 2017, Part II, LNCS 10386, pp. 586–594, 2017.
DOI: 10.1007/978-3-319-61833-3_62

There is a need for diagnostic and configuration of the machine, composed of multi-vendor equipment with heterogeneous interaction protocols. The latter circumstance is typical for small-scale production. One of the most difficult tasks is to configure and synchronize the interpolation of servo drives, controlled by different fieldbuses, so as to ensure the technological accuracy of the machine.

Manufacturers of technological equipment supply their own software tools for diagnosing and tuning servo drives and PLC remote I/O modules, such as a digital oscilloscope or logic analyzer, which can only be used with the equipment of the manufacturer [3, 4]. On machines with multi-protocol management [5] and heterogeneous actuating devices, these software packages do not provide the possibility to set up joint work of multiple devices.

Using during commissioning of machines external instrumentation such as oscilloscopes, ballbars, laser interferometers and trackers, allows us to estimate the parameters of the individual parts of the technological system [6]. These devices are used for final adjustment of the machine to achieve the specified accuracy of the positioning and performance. These devices are expensive, and often require special conditions to ensure an environment for its operation. By providing the ability to measure the resulting trajectory of the executive bodies of the machine, at the same time, they do not provide the opportunity to compare these data with internal data of the control system, which is critical in the development and debugging of new algorithms and control approaches.

In view of the above factors, there is a need to support, from the side of the CNC system, of multiprotocol management and parametrization and debugging of heterogeneous hardware component in the machine. The paper considers a unique solution for configuring and synchronizing of the SERCOS III spindle and EtherCAT feed drives on a 4-axis milling machine.

2 Architectural Model of Data Collection Subsystem

NC system kernel getting all necessary information about the logical state and motion parameters of servo drive, which is required for diagnosis purposes and control loop parameters adjustment. The numeric values of these data in NC kernel are updated with a sufficient frequency, equal to fieldbus cycle time for most of parameters.

If additional signals from servo drive side not relative to drive control procedure should be analyzed, they can be temporary added to a cyclic telegram transmitted between drive and CNC system [7, 8]. Some of industrial fieldbuses, such as SERCOS III [9], provides special mechanisms of transmitting the lists of selected parameter values, which can be configured on the flight without passing the device into parametrization mode.

For a user-configured measurement points a special mechanism is presented, which allows real-time recording of values of almost any memory area of SoftPLC controller and NC kernel (Fig. 1). This provides the NC kernel developer with the powerful support tool for debugging and interaction processes adjustment in a modular NC kernel system.

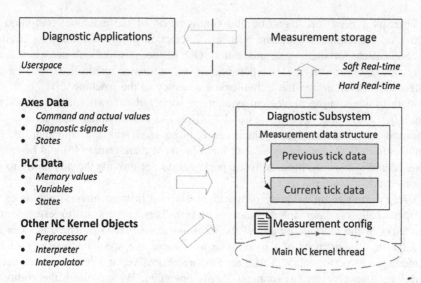

Fig. 1. Architectural model of special-purpose tools for diagnostics and configuration.

Measurement data collection in diagnostic subsystem is made in context of main NC kernel thread. Meeting the requirements of hard real-time functioning applies the corresponding restrictions on the software and architectural solutions can be used on this level.

In particular, the list of signals was defined, composed mostly of parameters relative to axis motion control process (actual and command position, velocity and acceleration for each drive), which is recording regardless to measurement configuration. Signal point for each of these parameters is created in special buffer any time when parameter value changes in CNC system course of work by the help of special inline wrapper functions, which is used in NC kernel code instead of assignment operators. User defined measurement points are recorded once at the end of main NC kernel cycle.

To prevent negative performance effects of cross-thread blocking recorded data are stored in switchable double buffer, each containing the array of measurement points. On each NC kernel cycle one of these arrays is used to store new signal point values, another one at the same time should be read and stored by less priority thread, which serves the measurement storage.

All the actions related to measurement recording are made in hard real-time thread regardless of the fact, whether the measurement is started or not. This allows avoiding the computation load bursts in main NC kernel thread. Measurement control functions, such as starting, stopping, collecting and sending data to user interface diagnostic application, are made in measurement storage software components, which work in soft real-time mode.

Measurement storage software components accumulate signal points in a buffer of configurable length of 10 MB to 100 MB, allowing to continuously record data for a long period of time (up to 10 min for a single axis movement) with a sampling time of 1 ms.

Around 30 signals for each NC axis and additionally up to 128 user configurable measurement points are stored.

Application of described technics in core software components of diagnostic subsystem result in increasing the computation load of main NC thread for just 3–5% depending on hardware platform used, which is quite acceptable.

3 Structure of Heterogeneous Technological System by the Example of Quaser Milling Machine

The prototype model of Quaser high performance milling machine center is the example of heterogeneous technological system. The machine is manufactured under a license, certain parts of it, primarily the control system, is being localized and replaced by domestically produced.

Linear and rotational feed drives where replaced with cheaper but powerful enough localized EtherCAT servo drives (Fig. 2). Some machine equipment, such as motor and servo drive of milling spindle under SERCOS III interface [10, 11], EnDat 2.2 linear measurement encoders and tool changing mechanism, remains unchanged at current stage of localization. Multiprotocol capable NC system AxiOMA Control is used to simultaneously operate the equipment under different industrial fieldbuses [12, 13]. The configuration of the machine parameters in the CNC kernel allowed controlling the spindle using the SERCOS III protocol and the servo drives with the EtharCAT protocol with 1 ms interpolation cycle. This ensures the execution of technological operations that use drive control on both protocols simultaneously. For example, threading with a tap (without a compensating chuck) or with threading tool, when the spindle needs to be switched to the rotary axis and interpolated together with the feed drives.

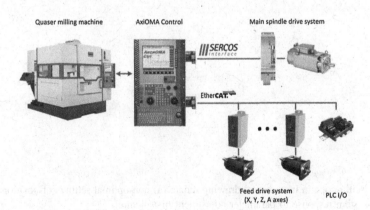

Fig. 2. Quaser machine heterogeneous machine control system.

4 Application of Software Tools for Diagnostics and Configuration for Quaser Machine

On the basis of developed software tools for diagnostics and configuration a set of user interface applications for machine and NC system commissioning was developed.

Digital oscilloscope application is intended for visualization of measured signals in time, allowing to analyze in details the behavior of transient processes in drive subsystem. Due to measurement recording process takes place in NC kernel, actual servo drive position and velocity signals are available to be analyzed along with internal NC kernel variables used in motion control process. The toolset supports the functionalities of scaling, using of cursors to obtain exact parameter values of measured signals, possibility of combining position signals from the set of axes to visualize path of motion in space according to kinematic configuration of machine.

Incremental drawing mode allows rating of different aspects of machine axis motion control process and observing the response of electromechanical drive system on control loop parameter changes. Signal points in this mode are collected at slower rates, than in regular measurement recording mode, number of simultaneously displayed signals is also limited. Only values of displayed on screen signals are being stored in this mode. Collected signal points are being combined into medium sized packets of total length up to 100 ms and then transferred to user interface application, optimizing the utilization of data transfer channel between NC kernel and terminal applications. For a diagnostic application such a delay is not critical and doesn't affect information perception by the operator.

Figure 3 illustrates the usage of incremental drawing oscilloscope application for commissioning of position loop regulator proportional coefficient of SERCOS III milling spindle drive for its usage in position control mode, in example for workpiece measurement cycles. The value of position error between command and actual axis position decreases as a result of parameter value tuning.

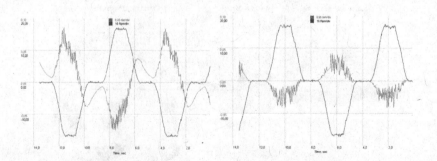

Fig. 3. Oscilloscope in incremental drawing mode: (a) non-optimal settings of position control loop; (b) system response to parameter adjustment first picture.

Circular test is defined by ISO 230-4 standard and being used in drive system commissioning for the purposes of minimization of machine motion contour errors by equalization of axes dynamic response.

Special purpose NC program is being executed in this mode, performing motions of contour entry and exit, circular motion, triggering the beginning and ending of measurement. To analyze contour diagrams diagnostic application provides a set of utility functions, such as changing the contour error scaling, source signal filtering for digital noise smoothing, automatic detection of movement direction (clockwise or counterclockwise), saving and loading measurement results, comparison of two diagrams to rate the effects caused by change of motion control parameters. It should be noted that in distinct from using of external measuring devices, the developed tool allows to analyze not only the actual positioning of the axes, but also visualize the trajectory generated by the interpolator that gives the developer of CNC system important feedback during the implementation of new motion control algorithms [14, 15].

Figure 4 shows the example of circle test tool application for setting up the X and Y linear axes of cross motion machine table, controlled through EtherCAT communication protocol. Two circular motions were made in XY plane, one clockwise one counterclockwise, with radius of 10 mm and 1000 mm/min feed. Y axis of this machine is loaded heavily than X one, so as it carries the weight of mechanical system of X axis in addition to weight of table itself. Initial measurement (drawn in chart with semi-transparent lines) reveals the dynamic error along the circle up to 10 μm, caused by the lack of dynamic responsiveness of Y-axis drive. After increasing of the coefficients of the speed and current control loop in the drive the dynamic characteristics of the axes were aligned and the error along the trajectory decreased to 2 μm. The developed toolkit will allow to preventively determine the necessity of setting up complex technological equipment.

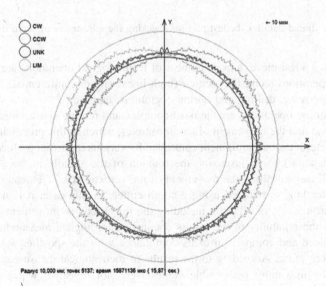

Радиус 10,000 мкм; точек 5137; время 15871136 мкс (15,87 сек)

Fig. 4. Circle test application – a tool for compensation drives' dynamic response mismatch.

5   Practical Application of Instrumentation for Diagnosis and Adjustment of Servo Drives for Milling Machining Center Quaser 184P

The check of the described approach and correctness of the adjustment of the servo drives for feed and spindle axes of the Quaser MV184P milling machining center was made by processing a test piece (Fig. 5).

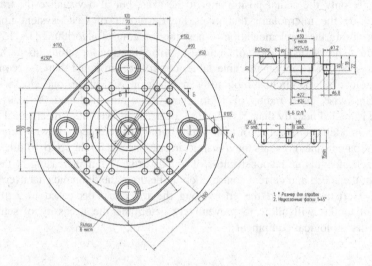

Fig. 5. Initial data for the test part for checking the adjustment of servo drives

To produce a test piece, a large number of technological operations are used, such as: milling operations along the contour (both linear and circumferential), flat surface machining, grooving, drilling and boring operations, tapping.

Thread cutting operations are the most complex and require synchronization of the spindle rotation and the movement of the linear axes, which in this processing center is realized by heterogeneous equipment controlled by various industrial protocols. Thread tapping is realized by synchronizing the rotation of the spindle in the axis control mode, as well as controlling the movement of the vertical axis Z. Thread cutting of a larger diameter hole is realized using a comb cutter. In this case, it is necessary to ensure synchronization of all axes, including the linear axis of movement X and Y.

To verify the reliability of the results obtained, metrological measurements of the treated test piece and comparison of its compliance with the specified standards and deviations were made. According to the results of metrological measurements, deviations from the maximum permissible values were not revealed, which proves the usefulness of the developed toolkit.

The presented results of measurements of the test part indicate the expediency of using the approach proposed in the work, which allows to collect and analyze data from various technological equipment and to adjust their synchronous operation, which allows achieving the necessary accuracy when manufacturing complex products on such machines.

6 Conclusion

The formalization of the basic concepts of tools for diagnostics and configuration of servo drivers, a properly chosen software framework and openness of the CNC architecture allow you to create software solutions that are not tied to specific manufacturers of technological equipment. Using such techniques allows configuring the machine components both individually and in their joint operation.

Application of tools for diagnosing and commissioning of servo drives in heterogeneous systems management does not rule out the use of external tools for machine fine tuning. The results of the research were checked and confirmed for the test part.

Acknowledgements. This research was supported by the Ministry of Education and Science of the Russian Federation as a public program in the sphere of scientific activity (N 2.1237.2017/PCH).

References

1. Martinov, G.M., Martinova, L.I.: Trends in the numerical control of machine-tool systems. Russian Eng. Res. **30**(10), 1041–1045 (2010)
2. Martinov, G.M., Obuhov, A.I., Martinova, L.I., Grigoriev, A.S.: An approach to building specialized CNC systems for non-traditional processes. Procedia CIRP **14**, 511–516 (2014)
3. Martinov, G.M., Grigoryev, A.S., Nikishechkin, P.A.: Real-time diagnosis and forecasting algorithms of the tool wear in the CNC systems. In: Tan, Y., Shi, Y., Buarque, F., Gelbukh, A., Das, S., Engelbrecht, A. (eds.) ICSI 2015. LNCS, vol. 9142, pp. 115–126. Springer, Cham (2015). doi:10.1007/978-3-319-20469-7_14
4. Grigoriev, S.N., Martinov, G.M.: Control and diagnosis of CNC machine tool digital drives. Kontrol'. Diagnostika **12**, 54–60 (2012)
5. Martinov, G.M., Lyubimov, A.B., Bondarenko, A.I., Sorokoumov, A.E., Kovalev, I.A.: An approach to building a multiprotocol CNC system. Autom. Remote Control **72**(10), 345–351 (2015)
6. Lei, W.T., Paung, I.M., Yu, C.C.: Total ballbar dynamic tests for five-axis CNC machine tools. Int. J. Mach. Tools Manuf. **49**(6), 488–499 (2009)
7. Grigoriev, S.N., Martinov, G.M.: Scalable open cross-platform kernel of PCNC system for multi-axis machine tool. Procedia CIRP **1**, 238–243 (2012)
8. Grigoriev, S.N., Martinov, G.M.: Research and development of a cross-platform CNC kernel for multi-axis machine tool. Procedia CIRP **14**, 517–522 (2014)
9. Wei, G., Zongyu, C., Congxin, L.: Investigation on full distribution CNC system based on SERCOS bus. J. Syst. Eng. Electron. **19**(1), 52–57 (2008)
10. Martinov, G.M., Ljubimov, A.B., Grigoriev, A.S., Martinova, L.I.: Multifunction numerical control solution for hybrid mechanic and laser machine tool. Procedia CIRP **1**, 260–264 (2012)

11. Grigoriev, S.N., Martinov, G.M.: The control platform for decomposition and synthesis of specialized CNC systems. Procedia CIRP **41**, 858–863 (2016)
12. Martinov, G.M., Kozak, N.V.: Numerical control of large precision machining centers by the AxiOMA contol system. Russ. Eng. Res. **35**(7), 534–538 (2015)
13. Martinov, G.M., Nezhmetdinov, R.A.: Modular design of specialized numerical control systems for inclined machining centers. Russ. Eng. Res. **35**(5), 389–393 (2015)
14. Martinova, L.I., Pushkov, R.L., Kozak, N.V., Trofimov, E.S.: Solution to the problems of axle synchronization and exact positioning in a numerical control system. Autom. Remote Control **75**(1), 129–138 (2014)
15. Martinova, L.I., Sokolov, S.S., Nikishechkin, P.A.: Tools for monitoring and parameter visualization in computer control systems of industrial robots. In: Tan, Y., Shi, Y., Buarque, F., Gelbukh, A., Das, S., Engelbrecht, A. (eds.) ICSI 2015. LNCS, vol. 9141, pp. 200–207. Springer, Cham (2015). doi:10.1007/978-3-319-20472-7_22

Other Applications

Gravitational Search Algorithm in Recommendation Systems

Vedant Choudhary[(⊠)], Dhruv Mullick, and Sushama Nagpal[(⊠)]

Netaji Subhas Institute of Technology,
Sector-3, Dwarka, Delhi 110078, India
{vedantc.ec,dhruvml.co}@nsit.net.in,
sushmapriyadarshi@yahoo.com

Abstract. Recommendation Systems have found extensive use in today's web environment as they improve the overall user experience by providing users with personalized suggestions. Along with the traditional techniques like Collaborative and Content-based filtering, researchers have explored computational intelligence techniques to improve the performance of recommendation systems. In this paper, a similar approach has been taken in the form of applying a heuristic based technique on recommendation systems. The paper proposes a recommendation system based on a less explored nature-inspired technique called Gravitational Search Algorithm. The performance of this system is compared with that of a system using Particle Swarm Optimisation, which is a similar optimisation technique. The results show that Gravitational Search Algorithm excels in improving the accuracy of the recommendation model and also surpasses the model using Particle Swarm Optimization.

Keywords: Recommendation systems · Computational intelligence · Gravitational Search Algorithm · Particle Swarm Optimisation · Collaborative filtering

1 Introduction

In recent times, due to the advancements in web technologies and popularization of internet, there has been a surge in the content present on the web which makes it difficult for the users to access the information of relevance to them. To tackle this information overload, recommendation systems (RS) have been widely deployed by companies to offer a better user experience to the customers. Many companies conduct surveys for users so that they can improve on the content delivering system or know more about the interests of the users. Amazon uses recommendations extensively in the form of "People who bought this, also buy" suggestions to users, while, Spotify takes a curated daily mix to users on their past music choices.

Generally, the most common filtering algorithms are categorized as: Content based filtering [1], Demographic based filtering (DF) [2], Collaborative filtering (CF) methods

---

The original version of this chapter was revised. A few errors in the equations on pages 600 and 601 were corrected. The erratum to this chapter is available at 10.1007/978-3-319-61833-3_67

© Springer International Publishing AG 2017
Y. Tan et al. (Eds.): ICSI 2017, Part II, LNCS 10386, pp. 597–607, 2017.
DOI: 10.1007/978-3-319-61833-3_63

[3] and Hybrid filtering methods [4]. Content-based filtering, often referred to as cognitive filtering, takes into account the similarities between the items or products for a user. Recommendations are provided by comparing the available items to the items bought by the user in the past. Demographic filtering employs demographic data such as gender, age, occupation, place of living etc. to infer recommendations. The third is the most conventional filtering method used, called Collaborative filtering. It is based on analyzing a cornucopia of data on user behavior, preferences etc. and then predicting what users might like based on their similarity to other users. It has been extensively deployed in areas where the recommendations should be independent of representation of the items. Many hybrid filtering methods have also been devised to offer the combined features of CF and DF methods [5], content-based and CF methods [6] and so on. to generate better recommendations to the users. Apart from the general filtering methods, a lot of research has also been done in applying computational intelligence (CI) to RS models. The algorithms discussed above lack the ability to provide customized recommendations since they consider equal weight for all the features used to generate recommendations. CI techniques like genetic algorithms [7], fuzzy logic [8, 9], swarm intelligence algorithms like ant colony [10], bee colony [11], particle swarm optimization (PSO) [5] etc. provide user-specific recommendations by optimization of feature weights, thereby improving upon the accuracies.

The motivation that stemmed the conceptualization of this paper is along the lines of fusing computational intelligence techniques with CF methods. This paper has explored a relatively new bio-inspired meta-heuristic algorithm named Gravitational Search Algorithm (GSA) [12] for the purpose of recommending jokes to users. Being an optimization algorithm, it helps in selection and fine tuning of the feature weights. One major reason for taking up GSA has been because of its similarity to PSO, in the sense that both are population-based algorithms. The experiments performed are compared to that of PSO as it is a well-researched algorithm in RS, while, GSA has not yet been applied to RS. Moreover, research between PSO and GSA has shown that GSA offers superior aspects like parameter optimization than PSO in non-linear systems [13].

The rest of the paper is organized as follows: Sect. 2 reviews the related work which has been done in recent past. Section 3 explains the working of GSA with mathematical equations, and compares its theoretical advantages and disadvantages against PSO. Section 4 shows the results of the experiments conducted. Section 5 concludes the work, and indicates the future work to be undertaken.

2   Related Work

Continuous research has been going on in the field of recommendation engines in recent times. Major work in this field has been done to generate better customized recommendations, tackle cold-start problems, sparsity of data etc. This paper targets in improving accuracies by working on top of CF method. CF methods have seen an unprecedented growth since its inception in recommendation systems. Its usage can date back to the work done by Badrul Sarwar et al. [14] in which they introduced item-based collaborative filtering algorithms and tested them on MovieLens dataset.

Others like David M. Penncock et al. evaluated an algorithm known as Personality Diagnosis with CF, which gave the probability of a user being of the same "personality type" as others, which in turn helped in making recommendations [15].

The research has seen advances in RS models by applying many hybrid techniques and algorithms along with CF methods. P. Bedi et al. [10] have used the behavior of ants for clustering user-item matrices based on communication between ants through pheromones to produced good quality recommendations. Apart from using ACO, PSO has also been used. Supiya Ujjin and Peter J Bentley's work on Particle Swarm Optimization [5] involves the use of PSO to fine tune the feature weights of MovieLens dataset and improve upon the accuracy of the recommendation model. From [5], it has been established that the prediction accuracy of a PSO model outperforms GA (Genetic algorithm) and PA (Pearson Algorithm) models. Moreover, to reduce the multidimensional data space, [16] adopts a type division method and classifies the movies to users. This has been done by providing initial parameters to PSO from K-means algorithm. PSO further optimizes fuzzy c-means for soft clustering of data.

Others have focused their work on tackling cold-start and data sparsity in RS. One way of doing so is using reclusive methods/fuzzy methods. Ronald R. Yager [9] has proposed an RS model, in which, recommendations have been derived from preferences of individuals. The recommendation rules have been represented and constructed with the help of fuzzy set methods. Some researchers have also integrated collaborative filtering with reclusive methods (RMs). The advantage RMs possess over CF is that they are less prone to sparsity and new item problems, however they suffer from over-specialization. The paper [8] integrates both by proposing a fuzzy naive Bayesian classifier for handling correlation-based similarity problems. Another improvement in RS has been by the inclusion of trust-based metrics to alleviate sparsity problems by enhancing inter user connectivity. Vibhor Kant et al. [17], have constructed fuzzy models for both trust and distrust through knowledge factors based on linguistic expressions.

The technique extensively explored in this paper, Gravitational Search Algorithm, is a population-based heuristic technique. Though it is similar to PSO, it offers superior optimization capabilities [13]. Its implementation can be seen in the domain of control systems where, it optimizes the control variables of the optimal power flow problem [18]. It has also been applied in filter modelling [19], where the unknown modelling parameters are optimized. In Sect. 3, GSA has been explored in detail and compared with PSO.

3 Techniques Used

The technicalities of the proposed algorithm have been discussed in this section. According to Rashedi et al. [12], GSA is similar to PSO, apart from the expressions for finding velocity and position of particles. The agents or particles in this optimization algorithm are objects (masses). They are governed by the gravity force, and as suggested in the laws, the force causes objects to move in the direction of heavier objects (greater fitness value). Heavy objects are considered to be good solutions as they are closer to the optimal solution, since they move slow (no sudden changes in direction) compared to lighter objects which change their motion rather swiftly.

In contrast to how GSA originates, PSO has originated from the motion of birds in flocks. In the swarm, individual particles do not know the objective, but just follow the nearby particles towards the local or global optima. Some vital similarities and differences between both the algorithms are as follows:

- Both are stochastic methods, since they never give the same results.
- Both termed as population-based algorithms, but differ in how the particles in the population work.
- In PSO, direction changes are characterized only by two parameters i.e. pbest (particle best) and gbest (global best) positions, whereas, in GSA, the total force due to all the particles updates the direction.
- Since PSO uses pbest and gbest values, it can be said that it uses memory. However, this is not the case in GSA as it depends only on the current position of the particles.

The equations for PSO and GSA have been provided in the paper with adequate information about the same. For more reference, one can go through the works of [5, 12]. The GSA equations [12] are:

The position of the ith agent is given by –

$$Pos_i = (pos_i^1, .., pos_i^d, ..pos_i^n) \tag{1}$$

where $i = 1, 2, \ldots, N$. The force between two masses and acceleration is given by –

$$force_{ij}^d(t) = G(t) \times \frac{mass_{pi}(t) \times mass_{aj}(t)}{dist_{ij}(t) + \epsilon} \times (pos_j^d(t) - pos_i^d(t)) \tag{2}$$

$$acc_i^d(t) = \frac{force_i^d(t)}{mass_{ii}(t)} \tag{3}$$

where, $pos_i^d = i^{th}$ particles position in the dth dimension, t = time, $mass_{ii}$ is the inertial mass of the ith particle, $mass_{pi}$ and $mass_{aj}$ = passive and active gravitational masses related to particle i and particle j respectively, $G(t)$ = gravitational constant, ϵ = a small constant, $dist_{ij}(t)$ = euclidean distance between the particles, $acc_i^d(t)$ = acceleration of agent i at time t and d^{th} dimension.

The new position and velocity expressions can be calculated as shown below:

$$vel_i^d(t') = random_i + vel_i^d(t) + acc_i^d(t) \tag{4}$$

$$pos_i^d(t') = pos_i^d(t) + vel_i^d(t') \tag{5}$$

where,

$t' = t + 1$ (subsequent time interval), $random_i$ = uniform random variable ranging from 0 to 1, $acc_i^d(t)$ = gravitational acceleration of particle i in the dth direction.

Authors define the fitness function as the average deviation in the predicted ratings from the true ratings [20].

$$fitness_i(t) = \frac{\sum_{i=1}^{n} |PredictedRating_i - ActualRating_i|}{n} \tag{6}$$

The gravitational masses are calculated by the fitness function value for the agent i, given below:

$$m_i(t) = \frac{fitness_i(t) - worst(t)}{best(t) - worst(t)} \tag{7}$$

$$mass_i(t) = \frac{m_i(t)}{\sum_{j=1}^{n} m_j(t)} \tag{8}$$

where the $fitness_i(t)$ represents the fitness value of agent i at time t, $best(t)$ and $worst(t)$ are represented by Eqs. (9) and (10).

$$best(t) = \min(fitness_j(t)) \tag{9}$$

$$worst(t) = \max(fitness_j(t)) \tag{10}$$

where, j goes from 1 to N (total number of agents).

The equations [5] guiding particle swarm optimization are (for agent i):

$$vel_i = w \times vel_i + c_1 \times r_1(pbest_i - pos_i) + c_2 \times r_2(gbest - pos_i) \tag{11}$$

$$pos_i = pos_i + vel_i \tag{12}$$

where, pos_i = current position, $pbest_i$ = particle's best position, $gbest$ = swarm's best position, vel_i = velocity, w = random weight in the range [0.5, 1], c_1 and c_2 = spring constants (set to 1.494), r_1 and r_2 = random variables in the range [0, 1].

While calculating the Euclidean distance between the users, weights are assigned to different features. These feature weights are calculated using the Gravitational Search Algorithm, as shown in Algorithm 1 below [12]. The GSA program is terminated after a fixed number of iterations. However, the termination can also be done using a convergence condition [12]. Algorithm 1 is as follows:

```
%GSA is applied to optimize specified parameters %
%This algorithm represents a single run of GSA algorithm%
Initialize ε, N-swarm size, t_max-iterations in each run;
Randomly initialize the set of N agents (masses) as Pos_i ∀
i Є N using Eq.(1);
for t = 1 to t_max do
  Evaluate fitness function for each agent, using Eq.(6);
  Update the gravitational constant G(t) [12];
  Find best(t) and worst(t) values using Eq.(9),(10);
  Calculate mass M_i for agent_i ∀ i ∈ N using Eq.(7),(8);
  Calculate force_ij acting between every pair of agents;
  Find the net force force_i acting on every agent;
  for i=1 to N do
    for d=1 to Number of dimensions do
      Calculate acceleration a^d_i using Eq.(3);
      Update the velocity vel^d_i(t+1) using Eq.(4);
      Update position pos^d_i(t+1) using Eq.(5);
    end for
  end for
end for
```

While generating recommendations for a user, all the users in the dataset are not considered. Instead, a set of closest neighbors for the user is defined and used for generating recommendations [10]. While finding the neighbors, the Euclidean Distance of the users is calculated on the feature vectors. For this purpose, it is not correct to assign equal importance to each feature of the user [5]. The GSA technique has thus been used for developing appropriate weights for the features, which are then used for finding neighbors of the user.

For the user under consideration, neighbors are used to generate a list of recommendations. Now, the Top-N items generated can be recommended to the user [21]. More details can be found in the Experiments section.

The technique for building a recommender system, as explained above is shown in Algorithm 2:

%GSA is applied to optimize the feature weights%
Input U - set of users, U_{active} - users to generate recommendations for, $U_{predict}$ - users to use for generating recommendations
Define feature vector $F_i \forall i \in U_{active} \cup U_{predict}$
for user $\in U_{active}$ **do**
    %Training phase %
    Generate feature weights using GSA as in Algorithm 1;
    Using the feature weights, find the Euclidean distance with the users in $U_{predict}$;
    %Testing phase %
    On the basis of the Euclidean distance, select the closest users to *user*. These users are *neighborhood*;
    Using the above, generate predictions for the user with the CF approach, and the feature weights;
    Recommend the Top-N recommendations generated;
end for

4 Experiments

The experimental study has been conducted on the Jester Dataset available on the website http://eigentaste.berkeley.edu/dataset/. Dataset-1 has been used which consists of over 4.1 million ratings of 100 jokes by 73,421 users. Each user in the dataset has rated 36 or more jokes. The experiment conducted has been restricted to the first 1000 users from the dataset. The ratings are continuous in nature, from −10.0 to +10.0. Unrated jokes being marked with a 99. Corresponding to every user in the dataset, the feature vector has been defined as a 100 dimensional vector. Each cell of the vector represents the rating given to a joke by the corresponding user. For every user, ratings are calculated by dividing into two sets: training set (67%) and testing set (33%) for process of finding the optimal feature weights using GSA.

Owing to the similarities between GSA and PSO, both have been applied as optimization techniques and further compared to see how well GSA performs over PSO.

Experiments are conducted by dividing the dataset into Active set and Predict set. The system uses Predict set to generate or train the prediction model, whereas predictions are made on the Active set according to the trained model. The top-N algorithm has been used for evaluating the recommendation system [21]. Corresponding to each of Top5, Top10 and Top15 predictions [21], four types of experiments are carried out.

- In the first experiment, the first 10 users are in the Active set and 50 random users are in the Predict set.
- In the second, randomly selected 10 users are in the Active set and 50 random users are in the Predict set.

- In the third, the first 50 users are in the Active set and 100 random users are in the Predict set.
- In the fourth, randomly selected 50 users are in the Active set and 100 random users are in the Predict set.

Making use of random users makes sure that the results are not affected by a bias in the dataset.

4.1 Evaluation Metrics

The most popular measures for evaluating the performance of a recommendation system are Mean Absolute Error (MAE), Precision (P), Recall (R) and F1 Score.

For an active user, MAE is defined as the average deviation in the predicted ratings from the true ratings. To measure the performance of a recommendation system, the value of MAE is averaged over the active users. Precision is defined as the fraction of the retrieved samples that are relevant, whereas recall is defined as the fraction of relevant samples that are retrieved. It has been observed that in many cases, when one of these measures increases, the other measure decreases which makes the evaluation of the system difficult. For this reason, F1 scores are also used to evaluate the overall performance of the system. It is defined as the harmonic mean of Precision and Recall.

The quality measures used in the paper are as follows:

$$MAE = \frac{\sum_{i=1}^{n} |PredictedRating_i - ActualRating_i|}{n} \tag{13}$$

$$P = \frac{Relevant\ Samples\ Selected}{Total\ Samples\ Selected} \tag{14}$$

$$R = \frac{Relevant\ Samples\ Selected}{Total\ Relevant\ Samples} \tag{15}$$

$$F1\ Score = \frac{2 \times P \times R}{P + R} \tag{16}$$

After calculating the predicted ratings for all the jokes, the TopN jokes are selected which shall be recommended to the user using the recommendation system. The values of N for the TopN algorithm are selected as 5, 10 and 15 while conducting the experiments [10].

4.2 Experimental Results

The experimental parameters taken in the paper are presented under Table 1. The values have been taken according to [20].

Corresponding each of TopN = 5, 10 and 15, we perform the four experiments on the dataset, evaluating the MAE and Precision (Table 2), Recall and F1 Score (Table 3) for PSO and GSA. Figure 1(a) and (b) show the precisions for the Top 10 predictions

Table 1. Parameter values for both PSO and GSA

Parameter name	Parameter	Description
Swarm size	10	Number of agents
Iterations in each run	30	Iterations before best solution
Number of runs	10	Iterations for each active user

Table 2. MAE and Precision values for all four experiments conducted (P – PSO, G – GSA)

Exp.	MAE						Precision					
	TopN = 5		TopN = 10		TopN = 15		TopN = 5		TopN = 10		TopN = 15	
	P	G	P	G	P	G	P	G	P	G	P	G
1	3.11	3.05	2.94	2.93	2.81	2.67	80.00	80.00	71.00	73.00	70.00	70.67
2	2.62	2.58	3.01	2.75	3.15	3.02	68.00	88.00	81.00	82.00	68.05	78.16
3	3.09	2.93	2.91	2.98	3.01	2.96	79.60	83.20	77.20	78.00	74.48	75.02
4	2.99	2.35	3.04	2.96	3.35	2.97	74.80	83.60	75.40	79.00	67.66	71.53

Table 3. Recall and F1 Score values for all four experiments conducted (P – PSO, G – GSA)

Exp.	Recall						F1 Score					
	TopN = 5		TopN = 10		TopN = 15		TopN = 5		TopN = 10		TopN = 15	
	P	G	P	G	P	G	P	G	P	G	P	G
1	31.40	26.69	45.16	46.42	72.34	69.30	45.09	40.02	55.20	56.75	71.15	70.13
2	23.31	28.40	42.20	53.72	72.06	73.87	34.71	42.94	55.49	64.91	69.99	75.95
3	26.51	27.75	51.01	51.00	72.37	72.73	39.77	41.61	61.43	61.67	73.40	73.85
4	27.33	31.12	57.12	56.60	75.93	73.69	40.03	45.35	64.99	65.94	71.56	72.59

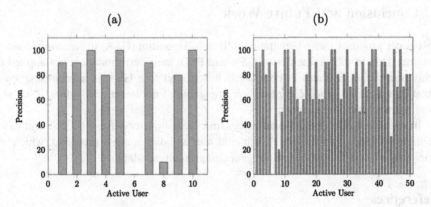

Fig. 1. (a) GSA Precision values of individual users for Top10 Exp. 1. (b) GSA Precision values of individual users for Top10 Exp. 3.

for every individual active user in Experiment 1 and Experiment 3 respectively. Since experiments 2 and 4 select active users on a random basis, their graphs have been omitted for clarity of representation.

4.3 Result Analysis

The results are analyzed with the evaluation metrics MAE, Precision, Recall and F1 Score, and the following conclusions can be made:

Mean Absolute Error values give the average variation of the predicted ratings from the actual ratings. This gives a measure of accuracy with which the ratings have been predicted. It can be observed that in terms of MAE, the recommendations generated using GSA are superior to those generated using PSO by 5.63%.

Precision values give a measure of how many items were correctly recommended out of the total items recommended by the algorithm. It can be observed that in terms of Precision, GSA is better than PSO by 6.41%.

Recall values give a measure of how many items which were to be recommended were actually recommended by our algorithm. It can be seen that in terms of Recall, GSA is better than PSO by 4.21%.

F1 Score values are a harmonic mean of the Precision and Recall values. F1 Score is a combined measure which gives an accurate representation of the accuracy of the system. In terms of F1 Score, GSA is better than PSO by 5.11%.

All the four experiments indicate that in most aspects, GSA fares better than PSO while in others, they both are comparable. Experiments 2, 3, and 4 show that GSA takes over PSO in optimizing the problem when the users are chosen randomly or are large in number.

As can be seen from the Tables 2 and 3, as the value of N is increased, the precision decreases and recall increases as was observed by the researchers in [10].

5 Conclusion and Future Work

This paper proposes using Gravitational Search Algorithm (GSA) in Recommendation Systems. Due to the similarities of GSA and PSO, their performance when applied to Recommendation Systems is compared. It is found that GSA is a promising optimization technique which can produce comparable, if not better results than PSO, in a more memory efficient manner.

In the future, the authors plan to construct a hybrid model using GSA and other techniques based on trust, to deal with cold start and data sparsity problems, which are other major issues associated while generating recommendations.

References

1. Aggarwal, C.C.: Content-based recommender systems. In: Recommender Systems, pp. 139–166. Springer International Publishing, Cham (2016)
2. Pazzani, M.J.: A framework for collaborative, content-based and demographic filtering. Artif. Intell. Rev. **13**(5), 393–408 (1999)

3. Ben Schafer, J., Frankowski, D., Herlocker, J., Sen, S.: Collaborative filtering recommender systems. In: Brusilovsky, P., Kobsa, A., Nejdl, W. (eds.) The Adaptive Web. LNCS, vol. 4321, pp. 291–324. Springer, Heidelberg (2007). doi:10.1007/978-3-540-72079-9_9
4. Burke, R.: Hybrid web recommender systems. In: Brusilovsky, P., Kobsa, A., Nejdl, W. (eds.) The Adaptive Web. LNCS, vol. 4321, pp. 377–408. Springer, Heidelberg (2007). doi:10.1007/978-3-540-72079-9_12
5. Ujjin, S., Bentley, P.J.: Particle swarm optimization recommender system. In: Proceedings of the 2003 IEEE Swarm Intelligence Symposium, SIS 2003, pp. 124–131, April 2003
6. Melville, P., Mooney, R.J., Nagarajan, R.: Content-boosted collaborative filtering for improved recommendations. In: AAAI/IAAI, pp. 187–192 (2002)
7. Kim, K.J., Ahn, H.: A recommender system using ga k-means clustering in an online shopping market. Expert Syst. Appl. **34**(2), 1200–1209 (2008)
8. Kant, V., Dwivedi, P.: A fuzzy bayesian approach to integrate user and item based collaborating filtering for enhanced recommendations. In: Proceedings of the 17th International Conference on Information Integration and Web-based Applications & Services, iiWAS 2015, pp. 75:1–75:7. ACM, New York (2015)
9. Yager, R.R.: Fuzzy logic methods in recommender systems. Fuzzy Sets Syst. **136**(2), 133–149 (2003)
10. Bedi, P., Sharma, R.: Trust based recommender system using ant colony for trust computation. Expert Syst. Appl. **39**(1), 1183–1190 (2012)
11. Ju, C., Xu, C.: A new collaborative recommendation approach based on users clustering using artificial bee colony algorithm. Sci. World J. **2013**, 1–9 (2008)
12. Rashedi, E., Nezamabadi-pour, H., Saryazdi, S.: GSA: a gravitational search algorithm. Inf. Sci. **179**(13), 2232–2248 (2009). Special Section on High Order Fuzzy Sets
13. Rozali, S.M., Rahmat, M.F., Husain, A.R.: Performance comparison of particle swarm optimization and gravitational search algorithm to the designed of controller for nonlinear system. J. Appl. Math. **2014** (2014)
14. Sarwar, B., Karypis, G., Konstan, J., Riedl, J.: Item-based collaborative filtering recommendation algorithms. In: Proceedings of the 10th International Conference on World Wide Web, pp. 285–295. ACM (2001)
15. Pennock, D.M., Horvitz, E., Lawrence, S., Giles, C.L.: Collaborative filtering by personality diagnosis: a hybrid memory-and model-based approach. In: Proceedings of the Sixteenth Conference on Uncertainty in Artificial Intelligence, pp. 473– 480. Morgan Kaufmann Publishers Inc. (2000)
16. Katarya, R., Verma, O.P.: A collaborative recommender system enhanced with particle swarm optimization technique. Multimedia Tools Appl. **75**(15), 9225–9239 (2016)
17. Kant, V., Bharadwaj, K.K.: Fuzzy computational models of trust and distrust for enhanced recommendations. Int. J. Intell. Syst. **28**, 332–365 (2013)
18. Duman, S., Gven, U., Snmez, Y., Yrkeren, N.: Optimal power flow using gravitational search algorithm. Energy Convers. Manag. **59**, 86–95 (2012)
19. Rashedi, E., Nezamabadi-pour, H., Saryazdi, S.: Filter modeling using gravitational search algorithm. Eng. Appl. Artif. Intell. **24**(1), 117–122 (2011)
20. Wasid, M., Kant, V.: A particle swarm approach to collaborative filtering based recommender systems through fuzzy features. Procedia Comput. Sci. **54**, 440–448 (2015)
21. Herlocker, J.L., Konstan, J.A., Terveen, L.G., Riedl, J.T.: Evaluating collaborative filtering recommender systems. ACM Trans. Inf. Syst. **22**(1), 5–53 (2004)

A Driver Model Based on Emotion

Qiong Xiao[1]([⊠]), Changzhen Hu[2], and Gangyi Ding[2]

[1] School of Computer Science, Beijing Institute of Technology, Beijing, China
qiongxiao@bit.edu.cn
[2] School of Software, Beijing Institute of Technology, Beijing, China
{chzhoo,dgy}@bit.edu.cn

Abstract. In this paper, according to the problem that existing driver model can't reflect some situations in actual driving, a driver model which considers the factors such as driver's emotion, personality and road condition is built, the emotion model is introduced when the driver is in the process of traffic jams, car following, overtaking, lane changing, emergency situation and so on, the influence of the emotion state, driving destination and so on to the driving decision is analyzed, in addition, the driver's states are detected by a camera, incentive or warning strategies will be given out according to the situation. At last, a virtual vehicle simulation is experimented based on this model.

Keywords: Driver model · Emotion · Vehicle simulation · Driving decision · Personality

1 Introduction

In the city traffic safety, besides the external factors such as weather, surrounding vehicles, the driver's personality, emotional states and so on also have a significant effect on traffic safety. In addition, in the vehicle assistant system, it is usually need to simulate the driving scene, thus to help the driver to make a better decision, building an appropriate driver model is an important content.

2 Related Works

The research about the driver model is as follows: Guo et al. [5] identified the driver model's parameter by identification algorithm of global evolution and local optimization in 2002. Shi et al. [3] researched the data fusion algorithm of the auxiliary driving system, Zhang [2] proposed a driver unified decision model, in 2008, Reichardt [7] generated believable and consistent behavior of simulated drivers by using driver models which imitated emotional influence on the human driver's decision, a specific driver's assistance system was presented. Xie et al. [6] proposed a driver emotion model under simplified traffic condition, and they discussed the influence of varying cognitive emotion on driving strategies. The model' predictions were then verified with real-world data in the virtual scene of overtaking experiment. Leu et al. [12] proposed a drivers' behavioral model for real-time action formation in traffic conditions and tested

© Springer International Publishing AG 2017
Y. Tan et al. (Eds.): ICSI 2017, Part II, LNCS 10386, pp. 608–614, 2017.
DOI: 10.1007/978-3-319-61833-3_64

it by embedding it in a conventional rule-based traffic model, drivers were modeled using human personality traits and emotions. Roidl et al. [13] created a multivariate model and revealed the influence of emotions, personal characteristics and other factors on driving by two studies, at last, predictors which may increase the risk of accidents in traffic were given out. Jeon et al. [14] introduced the effects of specific emotions on subjective judgment, driving performance, and perceived workload.

3 The Driver Model

3.1 The Emotion Model

In this study, the basic emotion model considers the cognitive and non-cognitive factors of emotion's generation; it is based on the Paul Ekman's theory of classifying emotion. The model which is represented by the emotion space is built; emotion's attenuation mechanism is analyzed, emotion's attenuation function is built, and many kinds of physiological needs' effects on the emotion's generation are considered.

3.1.1 Emotional Space
Ekman's six basic emotions: happy, sad, angry, surprise, fear and disgust constitutes six dimensional emotion spaces, in here, the six basic emotions of happy, sad, anger, surprise fear and anxious which is similar to the above ones are used, so emotion vector is represented as follows:

$$E(t) = (happy(t), sad(t), anger(t), surprise(t), fear(t), anxious(t)). \quad (1)$$

The emotion intensity can be represented as follows:

$$Ie(t) = (Ie1(t), Ie2(t), Ie3(t), Ie4(t), Ie5(t), Ie6(t)). \quad (2)$$

3.1.2 Mood Attenuation
In the general case, when the stimulus conditions don't exist, the mood will gradually decay as time goes on, this means mood intensity will tend to 0, which represents the calm state. Suppose that at time t0, the mood has the initial value, at time t, the mood intensity is calculated using following formula:

$$Em(t) = Em(t0) \times e^{-b(t-t0)}. \quad (3)$$

In the above formula, b is the coefficient of mood attenuation, which controls the mood attenuation's speed.

3.1.3 Inside Variables
According to the Maslow's hierarchy of needs, human's needs include physiological needs, safety needs, social needs and so on. In here, only physiological needs and safety needs are considered, so the inside variables include some physiological

variables and safety variable, in which the physiological variables include the basic physiological needs of fatigue, hunger and thirsty. Values of inside variables' demand intensity values range from 0 to 1. Normally, if the driver has many kinds of needs at the same time, when in the face of danger or threats, security needs will play a leading role, otherwise, needs of fatigue, hunger or thirsty will play a leading role if its intensity value is bigger than the threshold.

3.1.4 Personality Trait

In here, the drivers of different personality characteristics are classified using Eysenck' trait theory, in which personality structures are divided into three basic dimensions: extraversion, neuroticism and psychoticism, the high neurotic people is emotionally stable, while the low neurotic people lack emotional stability.

3.2 The Driver Model Based on Emotion

The driving status is relative to driving object, emotional status, personality, physiology, etc., for example, good emotional status can promote making the right decisions, and fatigue driving may cause accidents. Meanwhile, emotional status may be influenced by the factors such as outside stimulus events and physiology, the structure of driver model based on the above emotion model can be shown in Fig. 1. The flow chart of the driver's emotional interaction algorithm is shown in Fig. 2.

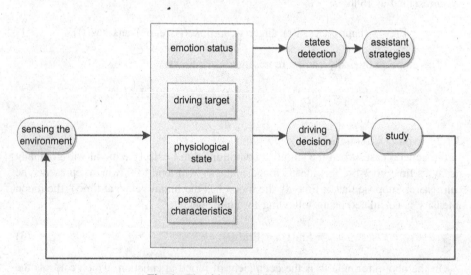

Fig. 1. The driver model's structure.

In traffic environment, outside stimulus events may have effect on the driver's emotional status; these events can be divided into the following categories: traffic jams, car following movement, lane changing, overtaking movement and other traffic

1. Initialize the driver's emotional state by e_0
2. While it is not at the stop time do
3. Update the driving state D_{i+1}
4. Update the emotional state e_{i+1}
5. While the driver is at the negative or dangerous emotional state
6. If the duration $t_d > t_{max}$
7. Give incentives or a warning to the diver
8. Update the iteration variable using $t = t + 1$.
9. End while.

Fig. 2. The driver's emotional interaction algorithm.

accidents. In the following, the driver model under some different stimulus is introduced in detail.

3.2.1 Traffic Jams

In the case of traffic jams, it is presumed that the change of the emotion value is relative to the driver's time to the nearest destination, waiting time, the driver's character, cognition and etc., if the driver waits for too long, he may be anxious or even angry. It is presumed that the driver's total waiting time is t; the threshold of the waiting time is T minutes, when t > T, the value of anxious or angry in the emotional space may change. T is relative to the driver's individual character, the remaining waiting time and the physiological states:

$$T = \alpha * \text{wait } (\mu, Iph). \tag{4}$$

α is the driver's individual factor, such as the personality' influence, μ represents the longest remaining waiting time of reaching the destination punctually, Iph is the vector of physiological variables including fatigue, hunger and thirsty.

The intensity value of anxiety can be calculated by the following formula:

$$anx(t) = \begin{cases} 0, & t < T \\ \alpha * e^{bt} + \beta * Iph, & T \leq t < T1 . \\ \alpha * maxa, & t \geq T1 \end{cases} \tag{5}$$

T1 is the time needed for reaching the maximum anxious value of maxa; b is the coefficient of mood increase. β is the coefficient of physiological variables, $\beta = [\beta1, \beta2, \beta3]$, Iph is the vectors of physiological variables, $Iph = [p1, p2, p3]^T$.

3.2.2 Car Following Movement

In car following status, presumed that at time of t, the driver's velocity is v, the front vehicle's velocity is v1, which is driving at the same direction to the driver and on the same road, when the distance of the driver and the front vehicle is smaller than the safe distance of d, rear-end accidents may happen easily, or the distance of the driver and the left or right vehicle is smaller than the safe distance, side crash may happened, the driver will feel scared, so that the fearful component of the emotion space will change, the safe distance of d is relative to the driver's personality factor of α, driving proficiency of s, driving velocity relative to the front vehicle, pavement condition of r, pavement condition is relative to the smooth degree of road surface, the width of the road, the road's steepness which can be represented by the angle between the road and the horizon, whether there are obstacles and so on. The intensity value of fear can be computed as follows:

$$Ie5(t) = \sum_{i=0}^{i=n} \lambda_i * fear(i) + k * f\ (\alpha, r, s) .\tag{6}$$

fear(i) is the fearful intensity of the ith surrounding vehicle in the field of the driver's view, λi is its influence coefficient, f (α,r,s) is a function relative to driver's individual factor, pavement condition and driving proficiency, k is its influence coefficient.

$$fear(i) = \begin{cases} 0, & dis \geq d \\ \frac{\alpha * |vi-v|}{dis}, & d0 \leq dis \leq d \\ F, & dis \leq d0 \end{cases}\tag{7}$$

vi is the velocity of the ith vehicle, dis is the distance between the driver and the surrounding vehicle, d0 is the distance threshold leading to the fear's biggest value of F.

3.2.3   Lane Changing

When the driver is in the status of waiting and so on, if the road is multilane and allows lane changing, the driver may change lane to move faster. Suppose that the distance between the driver and front vehicle is d, the safe distance to change lane is d0, the velocity of a vehicle behind the driver is v, if the driver is in the condition such as the value of d is too small, the value of v is too fast, there are too many of the vehicles, the fear component's value may increase, meanwhile, if the width of the road which can be represented as rw is narrow, it is also unfavorable for lane changing, the stimulus intensity of these factors to the driver can be computed as follows:

$$P = \alpha * change\ (d0, v, rw) = \alpha * (r1 * d0 + r2 * v + r3 * rw).\tag{8}$$

α is the driver's individual factor, r1 is the effective coefficient of the d0, r2 is the effective coefficient of the v, r3 is the effective coefficient of the rw.

Driver's appraisal to the possibility and the safety of the lane changing can be represented by the following function:

$$apc(t) = \alpha * m1 * m2 * r * d * n/v.\tag{9}$$

m1 is a bool value, when the road is multilane and the dividing line is not the solid line, the value of m1 is true, otherwise its value is false. m2 is the driving proficiency, even if it is in a unfavorable conditions, the proficient driver also may change lane, n is the general number of the vehicles near the objective position, r is the condition of the road,v1 is the general velocity of a vehicle behind the objective position, v is the velocity of a vehicle behind the driver, g is a threshold of lane changing' condition. If it is in the condition of apc(t) > g, the driver may change the lane.

In general, the time of overtaking is very short, if its starting time is t0, the intensity value of anxiety can be calculated by the following formula:

$$Ie5(t) = Ie5(t0). \tag{10}$$

3.2.4 Overtaking

In the processing of overtaking, the driver's emotional changes are relative to the factors such as the width of the road, the distance between the front vehicle, the velocity of the behind vehicle and so on, which is similar to processing of lane changing.

3.2.5 Detection and Solutions

The driver's states are detected from the image captured by a camera. If the driver is in the negative emotion, a light music may be played, if the driver is in the other abnormal conditions, such as napping, an alarm may be played.

4 Simulation Experiment

The simulation experiment is implemented in a vehicle simulation system, in the condition of car following and traffic jams, the driver model described previously which is applied in the three different drivers' emotional changes and driving decisions is demonstrated, the driver A is high neurotic, whose mood is not stable, driver B is low neurotic, whose mood is stable, driver C is low neurotic but has an urgent task and must reach destination in a short time. The driving processing is as follows: t = 0, there is a traffic jam, vehicles are in the status of waiting, t = 6 min, driver A is in the status of changing lane, t = 7 min, driver A is in the status of freely driving. Driver B is always in the status of waiting, t = 2 min, driver C is in the status of changing lane, t = 2.5 min, driver C is in the status of freely driving. Their anxious values can be displayed by the following table (Table 1):

Driver A has the unstable emotional personality, so after waiting a period of time, he changes the lane, driver B has the stable emotional personality, so he is always in the same traffic lane, although driver C has the stable emotional personality, he has an urgent task to reach the destination, so he changes the lane in the shortest time.

Table 1. The drivers' emotion values.

	The driver A	The driver B	The driver C
t=0	0	0	0
t=3	0	0	1
t=6	6	0.04	2.25
t=9	4.5	0.9	1.43

5 Conclusion

In this paper, a new driver model is presented, which considers some factors having effects to the driver's emotion, such as the driver's destination, personality, road condition and so on, some factors which have influence to the driver's tactics are also analyzed, if the driver is in some abnormal states, it may take according measures, at last, a simulation experiment is made. In this driver model, driving behaviors are mainly analyzed from the affective factors, which can be suitable for some complicated vehicle conditions.

References

1. Song, H., Chen, Z.J., Kong, F.E.: Study on aiding-level judgment of pilot assistant system. J. Syst. Simul. **21**(1), 208–212 (2009)
2. Zhang, L.: Researches on Driver Unified Decision Model for Vehicle Assistant Control [Ph. D. dissertation], Jilin University, China (2007)
3. Shi, W.Y., Zhao, M.H.: Application of information fuse technique in assistant-driving system. Modern Electron. Tech. **26**(17), 84–87 (2003)
4. Wang, G., Wan, J.: Review of information perception technology for driving safety assistance system. Comput. Commun. **26**(3), 50–54 (2008)
5. Guo, K.H., Ma, F.J., Kong, F.S.: Driver model parameter identification of the driver-vehicle-road closed-loop system. Autom. Eng. **24**(1), 20–24 (2002)
6. Xie, L., Wang, Z., Ren, D.C., Teng, S.D.: Research of driver emotion model under simplified traffic condition. Acta Automatica Sinica **36**(12), 1732–1743 (2010)
7. Reichardt, D.M.: Approaching driver models which integrate models of emotion and risk. In: Proceedings of the IEEE Intelligent Vehicles Symposium, Eindhoven, pp. 234–239. IEEE Press, Netherlands (2008)
8. Nass, C., Jonsson, I.M., Harris, H., Reaves, B., Endo, J., Brave, S.: Improving automotive safety by pairing driver emotion and car voice emotion. In: Proceedings of the CHI Extended Abstracts of Human Factors in Computing Systems, pp. 1973–1976. ACM Press, Portland, USA(2005)
9. Leng, H., Lin, Y., Zanzi, L.A.: An experimental study on physiological parameters toward driver emotion recognition. In: Dainoff, M.J. (ed.) EHAWC 2007. LNCS, vol. 4566, pp. 237–246. Springer, Heidelberg (2007). doi:10.1007/978-3-540-73333-1_30
10. Fukuda, S.: Detecting driver's emotion: a step toward emotion-based reliability engineering. In: Pham, H. (ed.) Recent Advances in Reliability and Quality in Design, pp. 491–507. Springer, Berlin (2008)
11. Yu, L., He, M.: Study of driving behavior model based on agent and ambient intelligence. Technol. Dev. Enterp. **28**(3), 19–21 (2009)
12. Leu, G., Curtis, N.J., Abbass, H.: Modeling and evolving human behaviors and emotions in road traffic networks. Procedia Soc. Behav. Sci. **54**(2290), 999–1009 (2012)
13. Roidl, E., Siebert, F.W., Oehl, M., Höger, R.: Introducing a multivariate model for predicting driving performance: the role of driving anger and personal characteristics. J. Safety Res. **47**, 47–56 (2013)
14. Jeon, M., Walker, B.N., Yim, J.B.: Effects of specific emotions on subjective judgment, driving performance, and perceived workload. Transp. Res. Part F: Traffic Psychol. Behav. **24**, 197–209 (2014)

A Binaural Signal Synthesis Approach for Fast Rendering of Moving Sound

Hui Zhou[1,2,3(✉)], Yi Ning[3,4], Jinlong Chen[3,5], Bin Liu[3],
Yongsong Zhan[6], and Minghao Yang[7]

[1] Guangxi Key Laboratory of Trusted Software, Guilin University of Electronic
Technology, Guilin 541004, Guangxi, China
zhouwork2017@foxmail.com
[2] Guangxi Colleges and Universities Key Laboratory of Intelligent Processing
of Computer Image and Graphics, Guilin University of Electronic Technology,
Guilin 541004, Guangxi, China
[3] Guangxi Key Laboratory of Cryptography and Information Security,
Guilin University of Electronic Technology, Guilin 541004, Guangxi, China
[4] Guangxi Cooperative Innovation Center of Cloud Computing and Big Data,
Guilin University of Electronic Technology, Guilin 541004, Guangxi, China
[5] Key Laboratory of Cloud Computing and Complex System,
Guilin University of Electronic Technology, Guilin 541004, Guangxi, China
[6] Guangxi Experiment Center of Information Science,
Guilin University of Electronic Technology, Guilin 541004, Guangxi, China
[7] Chinese Academy of Sciences, Beijing 100190, China

Abstract. Considering the difficulty of modeling the individual head related
transfer function (HRTF), a binaural moving sound source-oriented stereo audio
synthesis approach is proposed, which refers to the interpolation method of
HRTF. By means of spherical recording, the spatial audio of points from different directions can be obtained. In order to achieve more realistic effect, the
spatial bilinear interpolation method is employed to calculate the weight of the
relevant points which used to synthesize the moving sound source, and finally
the Doppler Effect is simulated with the interpolation and extraction method in
frequency domain. The experiment results show that, our method is capable of
replacing HRTF to synthesize moving sound sources approximatively, and the
generated performance of moving sound source is closer to the real recording.

Keywords: Moving sound source · Spatial interpolation · Stereo audio · Head
related transfer function · Frequency-domain interpolation and extraction ·
Doppler effect

1 Introduction

3D sound technology has great application potentials in high immersive virtual reality,
human-computer interaction, home entertainment and multimedia applications, which
makes virtual reality companies pay more and more attention to research in. Continuous movement of the sound source makes the synthetic parameters change constantly,
which makes it difficult to render 3D sound. Incessant switching of the signal may

© Springer International Publishing AG 2017
Y. Tan et al. (Eds.): ICSI 2017, Part II, LNCS 10386, pp. 615–623, 2017.
DOI: 10.1007/978-3-319-61833-3_65

create discontinuous phenomenon [1, 2]. Moreover, the rapid movement of the sound source will cause the Doppler Effect [3]. Therefore, it is significant to study how to generate realistic moving sound sources.

Audio spatialization is one of the key technologies in creating realistic 3D audio which has been a subject of many studies. Typically, the moving spatial audio is thought as containing two components: (1) the interpolation algorithm generates audio in arbitrary directions from a limited number of measured signal, (2) the pitch shift algorithm can be used to simulate the Doppler effect. In terms of interpolation, a straightforward way to perform HRTF interpolation is the bilinear method [4, 5], whose basic idea is to realize the target position's interpolation with its four nearest measured HRTFs according to their ubiety. In literature [6], a neural network interpolation algorithm based on radial basis function is proposed. In terms of Doppler Effect, the pitch synchronous overlap and add (PSOLA) method is proposed to realize the algorithm of pitch shifting [7].

In general, the current synthesis of moving sound source is focused on HRTF [8–11], a virtual source can be spatialized by convoluting its mono audio and the corresponding position of the head-related impulse response (HRIR). However, the accurate measurement of HRTF has a rigorous requirement on the devices, and the measurement process is complex and time-consuming, which makes it hard to model [12]. Therefore, a moving sound source synthesis algorithm is proposed.

2 Moving Sound Image Rendering Algorithm

Referring to the present problem of personalized HRTF that is hard to obtain, we propose a binaural moving sound source-oriented spatial audio synthesis method. Figure 1 shows an overview of our algorithm. Firstly, the loudspeaker was spoken in a bee voice and binaurally recorded at 44.1 kHz using an artificial head (Head Acoustics HMS IV [13]) in different spherical radius, azimuth, elevation. During the subset election step, to minimize computational load at run-time, some pre-processing is done to load correlated points into memory. Next, we can calculate the interpolation weights by using the measurement coordinates, and the 3D audio data is generated by spatial linear interpolation. Finally, we can use the technology of interpolation and extraction to simulate the Doppler Effect in the frequency domain, and now discuss each step in detail.

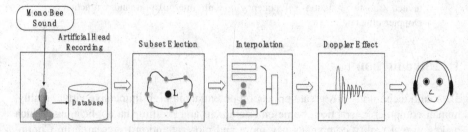

Fig. 1. Overview of our spatial sound pipeline for moving sound sources. The pipeline can be used to produce 3D sound in a dynamic scene.

2.1 Artificial Head Recording

In order to simplify the workload of the recording, this paper proposes a new approach to obtain the measurement audio. The audio can be recorded in radius, azimuth, elevation, and stored in the two-channel database. Figure 2(a) shows the recording scene.

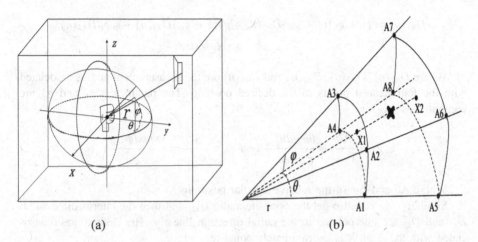

Fig. 2. (a) Head related coordinates r, θ, φ and room related coordinates x, y, z; (b) desired source position x and measurement positions A_i forming hexahedron used for interpolation.

2.2 Subset Selection Algorithm

In order to carry out spatial linear interpolation, we first need to find the necessary audio clips for interpolation. The traditional subset selection algorithm is mainly to find the nearest three or four points from the measured point [14]. However, the most difficult thing of the interpolation algorithm is how to locate the adjacent audio quickly. the binary search method can be applied to improve the search speed. First, the upper and lower bounds of the radius can be found by the radius of the target point, and then the upper and lower bounds of the horizontal and elevation angles are searched respectively. Finally, we can find out all 8 nearest measurement points. As shown in Fig. 2(b), the 8 measurement points are located at $A_i(i = 1, 2, 3 \cdots 8)$ in the spherical space.

2.3 Bilinear Interpolation

Once a hexahedron mesh of measurement points has been generated via subset election algorithm, an audio estimate for any point X lying inside the mesh can be obtained by interpolating the points of the hexahedron enclosing X. The entire interpolation process consists of a total of two parts, in the first place, bilinear interpolation is implemented in the intrados and extrados, and then two intermediate audio data from the previous step can be interpolated in the radius direction. Consider a hexahedron formed by the points $A_i(i = 1, 2, 3, \cdots, 8)$ as depicted in Fig. 2(b). In that method, if the related audios have been measured over a spherical grid with steps θ_{grid} and φ_{grid}, relative to azimuth and elevation, respectively, the estimate of the audios at an arbitrary coordinate (θ, φ), can be obtained by

$$
\begin{aligned}
D_{X1}(n) \approx (1 - g_\theta)(1 - g_\varphi)D_{A1}(n) + g_\theta(1 - g_\varphi)D_{A2}(n) + g_\theta g_\varphi D_{A3}(n) \\
+ (1 - g_\theta)g_\varphi D_{A4}(n).
\end{aligned}
\tag{1}
$$

Where $D_{A1}(n)$, $D_{A2}(n)$, $D_{A3}(n)$ and $D_{A4}(n)$ are the measurement audios associated with the four nearest points to the desired position. The parameters g_θ and g_φ are computed as

$$
g_\theta = \frac{\Delta\theta}{\theta_{grid}} = \frac{\theta \bmod \theta_{grid}}{\theta_{grid}} \ and \ g_\varphi = \frac{\Delta\varphi}{\varphi_{grid}} = \frac{\varphi \bmod \varphi_{grid}}{\varphi_{grid}}.
\tag{2}
$$

Where $\Delta\theta$ and $\Delta\varphi$ of the relative angular positions.

Similarly, we can also get the opposite audio D_{x2}, and then the intermediate audio D_{x1} and D_{x2} are interpolated in the radial direction linearly. The desired audio associated with the point X is approximately equal to

$$
D_x \approx g_2 D_{x1}(n) + g_1 D_{x2}(n).
\tag{3}
$$

Here, g_1 and g_2 are the interpolation weights and can be calculated by the equation set

$$
g_1 = \frac{r - r_1}{r_2 - r_1}, g_2 = 1 - \frac{r - r_1}{r_2 - r_1}.
\tag{4}
$$

For a desired point moving inside a hexahedral, the weights can change smoothly as a function of the vertex-distance. For a desired point lying on a vertex A, the interpolation weights are 1 at A and 0 otherwise; hence, the interpolation at A is exact. For a desired point lying on an edge or facet of the hexahedral, only the vertices forming that edge or facet have nonzero interpolation weights. Therefore, when a virtual source moves smoothly from one hexahedral to another across a shared vertex, edge, or facet, the sound source estimate changes smoothly, including at the crossing point.

2.4 Doppler Effect Implementation

In a stationary virtual scene, the reproduction of the sound field is only the generation of the time domain audio signal. When there is a relative movement between the sound source and the observer, the Doppler effect will be produced. Therefore, in order to simulate more realistic binaural signal of the moving sound source, the interpolation and extraction in frequency domain are used to simulate the changes of sound field caused by the moving sound source.

On the basis of physical research, when there is the relative movement between wave source and observer, the frequency received by the observer will be different from the wave source sent, which is the Doppler shift caused by the movement of the sound source. In a stationary medium, when the relative velocity exists on the line between the observer and the wave source, the frequency received by the observer will be related to the direction of motion of sound source. If the wave source is close to the observer, the received frequency f_0 is greater than the frequency f_s sent by the sound source; if it is far away, the received frequency f_0 is less than the frequency f_s. The received frequency f_0 can be written in the equation

$$f_0 = \left(\frac{c + v_0}{c - v_s}\right)f_s. \tag{5}$$

Where v_0 denotes the speed of the observer relative to the medium, v_s the velocity of the wave source relative to the medium, c denotes the propagation velocity of the waves in the stationary medium.

According to the above analysis, the Doppler effect of the moving sound source can be simulated by changing the frequency of the sound source as depicted in Eq. (5). In this research, the interpolation can expand the signal to realize the up-regulation and the extraction can compress the signal to realize the down-regulation. Suppose the original speech sequence can be denoted as $x(n)$, $n = 0, \ldots, N - 1$, and the adjusted sequence is denoted as $y(n)$. To ensure the length of the post-tone signal, $y(n)$ is approximately equal to

$$X(k) = FFT\{x(n)\}, k = [0, N - 1], k \in Z. \tag{6}$$

$$Y(k) = \begin{cases} \{X(\lfloor \rho k \rfloor + 1) - X(\lfloor \rho k \rfloor)\}(\rho k - \lfloor \rho k \rfloor) + X(\lfloor \rho k \rfloor), \\ k = [0, min\{\lfloor (N-1)/2 \rfloor, \lfloor (N-1)\rho/2 \rfloor\}], k \in Z. \\ 0, k = [\lfloor (N-1)\rho/2 \rfloor, (N-1)/2], k \in Z. \\ Y^*(N - 1 - k), k = [\lfloor (N-1)/2 \rfloor + 1, N - 1], k \in Z. \end{cases} \tag{7}$$

$$y(n) = IFFT\{Y(k)\}, n = [0, N - 1], n \in Z. \tag{8}$$

Where *FFT* and *IFFT* are the Fast Fourier Transform and Inverse Fast Fourier Transform respectively, ρ is the tonal coefficient, *min()* denotes the minimum function, * denotes the conjugation, when we turn down the signal frequency, the *Y(k)* execute the zero-setting operation.

3   Simulation and Evaluation

To evaluate the performance of the proposed method, we firstly compare the objective interpolation results of the HRTF and proposed method in the horizontal plane, and then the effect of moving sound source is evaluated by subjective way. The experiments above are conducted using one dataset: CIPIC HRTF.

3.1   Objective Test with Moving Sources

To evaluate the performance of interpolation, the signal distortion ratio of each reconstructed frequency point is defined as SDR

$$SDR(\theta_0) = 10log \frac{\sum_{n=0}^{T} D^2(\theta_{0,n})}{\sum_{n=0}^{T} (D(\theta_{0,n}) - \hat{D}(\theta_{0,n}))^2} (dB). \tag{9}$$

Where $D(\theta_0, n)$ and $\hat{D}(\theta_0, n)$ denote the magnitudes of measured and interpolated signal in the comparison subset respectively, and n is the number of sample. A large value of SDR means a good interpolation performance. For briefness, only the left-ear signals are presented here.

The average SDR for the left ear with $\Delta\theta = 2°$ are shown in Fig. 3(a). It can be seen that the moving sound source generated by mentioned method is closer to HRTF method. Moreover, there is an increasingly common view that the SDR is relatively low at azimuths $0° - 180°$ contralateral to the left ear, and high at azimuths $180° - 360°$ ipsilateral to the left ear. This phenomenon is due mainly to the head shadow effect, which has a complicated structure due to the interference of multi-path waves scattering across the head, as well as the low SDR of the measured contralateral signals caused by the head shadow effect.

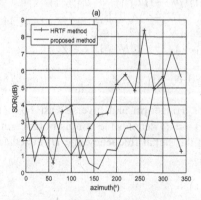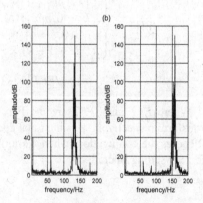

Fig. 3. (a) The SDR for the left ear with $\Delta\theta = 20°$; **(b)** magnitude spectra of moving sound source before pitch shifting (left) and after (right).

To fully validate the performance of interpolation and extraction in frequency domain, the simulation experiment is carried out on a bee signal. Figure 3(b) shows the amplitude spectrum before the change of the frequency and after. Taking the peak of 130 Hz for example, after the interpolation in frequency domain, the frequency has become 160 *Hz*. Therefore, the frequency becomes about 1.2 times the original. It is difficult for us to validate the performance from the spectrum. Thus, subjective listening test with moving sources can be put into effect.

3.2   Subjective Test with Moving Sources

To further evaluate the performance of virtual sources, we conducted a subjective experiment of moving sound source. The procedure was the following: the sound source was located randomly somewhere in the front for 4–6 s so that the subject had enough time to perceive the source. Then, the source began to move with the constant speed of $10°/s$ for a certain time, 16 individuals with normal hearing will choose which is the best one in naturalness. In the end, the source stood still again for 4 s, and noted down the direction of sound source. In total, there were 104 trials preceded by 4 training trials to familiarize subjects with the listening apparatus. The following conditions were used for these trials: (1) 2 directions of movement (clockwise /counterclockwise), (2) 3 acoustic stimulus, (3) 4 different trajectory lengths $(30°/60°/90°/120°)$, (4) each trial was repeated after 1 week.

To evaluate the performance of moving sound source, two parameters were used to judge the performance. (1) Naturalness, defined as the fluency of moving sound source, is the number of selected signals divided by the total number of signals. The total number of signals in this formula is 1600, which is equal to multiply 16 by 100. (2) The *RMS* error is the localization error of the moving sound source, which defined the gap between the direction angle of person and the real angle of the moving sound source.

The naturalness and localization error of moving sound source is shown in Fig. 4. From the histogram, there are no significant differences between real sources and simulated sources, and the overall hearing sense of the signal generated by the proposed method is better than that of HRTF method. In addition, Fig. 4(a) shows that, when the sound source is moving in the radial direction, the two interpolation signals in the naturalness are worse than the real signals, this phenomenon can be analyzed from the essence of algorithm, it due mainly to the interpolation effect in the frequency, which introduces some redundant frequency components, and make the virtual signal distorted. The average localization error of the sound source is depicted in Fig. 4(b), we can see that the localization error of the moving sound source generated by HRTF is higher than the proposed method. The most principal reason is that the HRTF of every individuals is different. The head effect, elevation confusion and others in HRTF method has increased the average localization error.

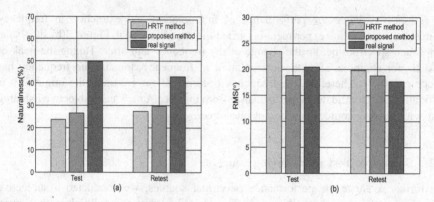

Fig. 4. (a) Naturalness of moving sound source via a subjective listening test; (b) mean RMS localization error of a moving source during the moving phase.

4 Conclusion

This paper takes moving sound source as the research object, which involves the synthesis and pitch shifting of speech. On the basis of HRTF, given the convenience of recording and interpolation, a spatial synthesis method for the binaural moving sound source is proposed. Firstly, spatial discrete audio is collected by spherical recording. Then, bilinear interpolation can be used to generate arbitrary spatial audio. Finally, interpolation and extraction can be used to simulate the Doppler effect. As seen from the results, the proposed method can approximatively substitute HRTF to generate moving sound source and performs better than HRTF in auditory effect. As for future work, other interpolation methods will be taken into consideration, and improving the performance of interpolation will be the most important future work.

Acknowledgments. This research work is supported by the grant of Guangxi science and technology development project (No: AC16380124, 1598018-6), the grant of Guangxi Key Laboratory of Trusted Software of Guilin University of Electronic Technology (No: KX201601), the grant of Guangxi Colleges and Universities Key Laboratory of Intelligent Processing of Computer Images and Graphics of Guilin University of Electronic Technology (No: GIIP201602), the grant of Guangxi Key Laboratory of Cryptography & Information Security of Guilin University of Electronic Technology (No: GCIS201601), the grant of Guangxi Cooperative Innovation Center of Cloud Computing and Big Data of Guilin University of Electronic Technology (No: YD16E11), the grant of Guangxi Experiment Center of Information Science of Guilin University of Electronic Technology (No: 20140208), the grant of Key Laboratory of Cloud Computing & Complex System of Guilin University of Electronic Technology (No: 14105).

References

1. Kudo, A., Hokari, H., Shimada, S.: A study on switching of the transfer functions focusing on sound quality. Acoust. Sci. Technol. **26**(3), 267–278 (2005)
2. Keyrouz, F., Diepold, K.: A new HRTF interpolation approach for fast synthesis of dynamic environmental interaction. J. Audio Eng. Soc. **56**(1), 28–35 (2008)
3. Ahrens, J., Spors, S.: Reproduction of moving virtual sound sources with special attention to the doppler effect. Conv. Aes. **114**(4), 63–73 (2008)
4. Biscainho, L.W.P., Freeland, F.P., Diniz, P.S.R.: Using inter-positional transfer functions in 3D-sound. In: IEEE International Conference on Acoustics, Speech, and Signal Processing. II-1961–II-1964 (2002)
5. Zhang, J., Wu, Z.: A piecewise interpolation method based on log-least square error criterion for HRTF. In: Workshop on Multimedia Signal Processing, pp. 1–4 (2005)
6. Zhong, X.: Interpolation of head-related transfer functions using neural network. In: International Conference on Intelligent Human-Machine Systems and Cybernetics, pp. 565–568 (2013)
7. Mousa, A.: Voice conversion using pitch shifting algorithm by time stretching with PSOLA and re-sampling. J. Electr. Eng. **61**(1), 57–61 (2013)
8. Algazi, V.R., Duda, R.O., Thompson, D.M.: The CIPIC HRTF database. In: Application of Signal Processing to Audio and Acoustics, pp. 99–102 (2001)
9. Gardner, W.G., Martin, K.D.: HRTF Measurements of a KEMAR. J. Acoust. Soc. Am. **97**(6), 3907–3908 (1995)
10. Marcelo, Q.: Efficient binaural rendering of moving sound sources using HRTF interpolation. J. New Music Res. **40**(3), 239–252 (2011)
11. Freeland, A.P., Wagner, L., Biscainho, P.: Efficient HRTF interpolation in 3D moving sound. In: International Aes Conference, pp. 1–9 (2002)
12. Zotkin, D.N., Hwang, J., Duraiswaini, R.: HRTF personalization using anthropometric measurements. In: Applications of Signal Processing To Audio and Acoustics, pp. 15–160 (2003)
13. Mninaar, P.: Localization with binaural recordings from artificial and human heads. J. Audio Eng. Soc. **49**(5), 323–336 (2012)
14. Gamper, H.: Selection and interpolation of head-related transfer function for rendering moving virtual sound sources. In: Conference on Digital Audio Effects, pp. 1–7 (2013)

Semantic Evolutionary Visualization

Marwa Keshk[✉]

University of New South Wales Canberra, Canberra, Australia
marwa.hassan@student.adfa.edu.au

Abstract. The Evolutionary optimization (EO) field has become an active area of research for handling complex optimization problems. However, EO techniques need tuning to obtain better results. Visualization is one approach used by EA researchers to identify early stagnation, loss of diversity, and other indicators that can help them to guide evolutionary search to better areas. In this paper, a Semantic Evolutionary Visualization framework (SEV) is proposed for analysing and exploring the potential EA dynamics. Empirical results have shown that the SEV can help to reveal and monitor information on evolutionary dynamics; thus, it can assist researchers in adapting the evolutionary parameters to obtain better performance.

Keywords: Semantic visualization · Fitness landscape analysis · EA

1 Introduction

Evolutionary Computation is broadly used in a wide range of problem solving contexts. It is simple to develop a very naive Evolutionary algorithm (EA) for solving a problem, but it is too hard to analyse and validate the outcomes of the algorithm to determine if the algorithm is efficient enough or if there is a need to improve it [1]. An EA is a generic population-based, metaheuristic algorithm developed for global optimization problems. It borrows concepts from Darwinian evolution of species to global optimization problems. A new solution is generated at each generation by merging two or more existing solutions based on some evolutionary strategy [2]. A careful setting of different parameters (e.g. population sizes, mutation and crossover rates, selection pressure etc.) is essential for EA to find high quality solutions [3].

Evolutionary dynamics should be understood by visualization, experience, statistical test of significance, and fitness landscape analysis methods. Visualization offers unprecedented information to the human when attempting to understand complex evolutionary dynamics. Many techniques have been suggested in the literature of evolutionary computation [4,5].

Visualization is a form of reasoning. In general, reasoning is the process of inferring conclusions (i.e. outcomes) from certain premises (i.e. propositions) assumed to be true. It is a sequential, directed process and is better modelled by various visual graphs. In analytical reasoning, the visualization of information describes the search space using different interactions to enable the user to

© Springer International Publishing AG 2017
Y. Tan et al. (Eds.): ICSI 2017, Part II, LNCS 10386, pp. 624–635, 2017.
DOI: 10.1007/978-3-319-61833-3_66

explore and analyze the findings of the visualization process [6]. We call a visualization that supports a user to reason about a problem as 'semantic' visualization as it offers analysts with knowledge about the problem at hand.

In this paper, a Semantic Evolutionary Visualization (SEV) framework is proposed. We propose a visualization framework that can keep track of information and offer a direct reflection of evolutionary dynamics. In classic genetics research, a common approach to capture the changes occurring in a population is through a pedigree graph visualization.

The framework consists of three primary components. First, an evolutionary algorithm; which is the Differential Evolution (DE) [7] algorithm for the purpose of this paper. Second, a hierarchical profiling algorithm to transform the data being generated by EA/DE into an appropriate format for visualization. SEV involves a variety of graphs. The main focus here is on two types of visualization: a hierarchy graph to track the solutions that led to the best found solution at the end of a run, and correlation graphs depicting the correlation coefficients between children and parents.

The rest of the paper is organised as follows. In Sect. 2, previous research on visualization techniques in evolutionary computation is illustrated. The proposed visualization framework is then presented in Sect. 3 followed by a case study in Sect. 4. Finally, Sect. 6 concludes the paper.

2   Related Work and Background

Visualisation of evolutionary algorithms assists in the analysis of the search space and evolutionary dynamics. It can explore convergence behaviours and fitness landscape dynamics, and assist in human-guided parameter settings (i.e., crossover and mutation) of evolutionary mechanisms. In evolutionary algorithms, visualisation techniques can be executed to study EAs both online – during evolution – and offline – after evolution. We categorise evolutionary visualization techniques into three types: (1) Population Based Feature (PBF), (2) Local Dynamics Based Features (LDBF), and (3) Global Dynamics Based Features (GDBF).

Figure 1 shows the previous research for visualising evolutionary algorithms and their potential functions based on the aforementioned types. The key terms that are provided in the figure are type, methodology, advantage, disadvantage and the relation to our study.

The first category, PBF, consists of frequency-based and projection-based approaches. An imaging matrix model [8] measures the frequency ratio of individual values occurring during evolution. Unfortunately, this approach does not truly capture population wide characteristics. Projection approaches attempt to reduce the high dimensionality of the data by projecting it to lower dimensions. These are not visualization techniques per se, as they act as preprocessing steps to enable visual representation of the information using a few number of dimensions.

Secondly, LDBF involves a wide variety of techniques to visualise the internal evolutionary dynamics. For instance, VIPER [9], which is conceptually similar

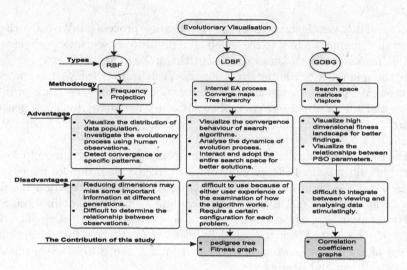

Fig. 1. Categorisation of evoluationay visualisation techniques

to one of the two visualisations we introduce in this paper, it is not suited for EO as it cannot handle, and it is not obvious how to even adapt it to handle, information like fitness.

The third group of visualisation techniques, GDBF, focuses on the overall search space itself. Examples include: Search Space Matrices (SSM) [10] that present a single mapping for the complete search space based on configuration space analysis.

The variety of visualization techniques discussed above can assist in studying evolutionary or swarm dynamics. However, their reasoning power is sitting at a high level of abstraction. The proposed visualisation framework in this paper attempts to offer a depth representation for the internal dynamics of EA. The proposed pedigree graph visualisation and fitness graphs are a type of LDBF, while the correlation coefficient graphs are a type of GDBF.

3   Semantic Evolutionary Visualization

In this section, Semantic Evolutionary Visualization framework (SEV) is introduced. The core motivation is to offer visualisation tools that can assist a user to reason - hence our choice of the word 'semantic' - about evolutionary dynamics at the right level of resolution. We present two new visualisation tools with a conventional one. The three together triangulate three different pieces of complementary information to the user.

We use DE as an example of an evolutionary optimisation heuristic that is fast and popular. However, any evolutionary optimisation method can fit the framework without any modifications.

Figure 2 shows the interconnectivity between the three major components of SEV: the evolutionary optimisation algorithm - DE in this paper, hierarchy

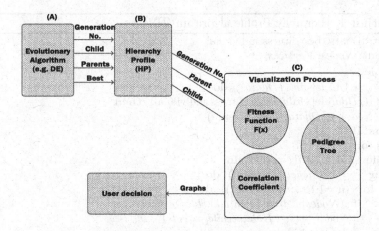

Fig. 2. Semantic Evolutionary Framework (SEV).

profiling as a preprocessing step to prepare the data for visualisation, and the visualisation process itself involving three different types of diagrams.

3.1 Differential Evolution Algorithm (DE)

DE, introduced by Storn and Price [11], is today one of the most commonly used EAs for solving global optimization problems [12], which is a fast and simple EA because it does not rely on sorting. The classic parameters in DE include: NP for the population size; CR for the crossover rate; and F for the scaling factor. CR can be seen as a heredity parameter (an approximation of the inbreeding coefficient in classic genetics), and F can be seen as the step length (a scaling factor) in classic optimisation.

This algorithm is slightly modified to explicitly store the data needed for visualisation. Especially, a data matrix with some parental information (i.e., each child with its parent identifiers) is stored to support the analysis of ancestry for the best solution found.

3.2 Hierarchy Profile Algorithm (HP)

Because of the enormous amount of data produced by DE, a hierarchy profile algorithm (HP) is presented in Algorithm 1. The algorithm recursively preprocesses the data generated by evolution and prepares it in a format suitable for visualisation. The algorithm identifies the best solution found then recursively generates a table of ancestry up to the first generation during evolution. The input of the HP algorithm is the output of the DE algorithm (i.e., IDdata) and the best individual found (i.e., best_fitness_individual). *NodeParts* stores the child and parents identifiers to keep track the hierarchy of each parent while *Vertices* saves the child identifier over run for establishing the relationship between the parents and their children to avoid the duplication during the pedigree tree representation.

Algorithm 1. Hierarchy Profile algorithm (HP)

Input: IDdata, best_fitness_individual
Output: *Nodes, Vertices*

1: Set $d=1$
2: **for** $(r = 1$ to *length* $(IDdata))$ **do**
3:   **if** $(IDdata[r].child = $best_fitness_individual$)$ **then**
4:     $Node1 \leftarrow IDdata(child, parents)$
5:   **end if**
6: **end for**
7: **while** $(d < length(Node1))$ **do**
8:   **for** $(k = 1$ to $length(IDdata))$ **do**
9:     **for** $(u = 1$ to $length(Node1))$ **do**
10:       **if** $(Nodes[u,d] = IDdata[k].child)$ **then**
11:         $NodeParts \leftarrow IDdata(child, parents)$
12:         $d = d + 1$
13:       **end if**
14:     **end for**
15:   **end for**
16: **end while**
17: $Nodes \leftarrow Node1$ *and* $NodeParts$
18: $Vertices \leftarrow IDdata.child$
19: **return** $(Nodes, Vertices)$

3.3   Visualization Methodology Process

The three visualisations we present in this section can work in online and offline mode. The framework involves three visualisations, a classic one, and two newly proposed one. The classic visualisation presents a line-graph of the best, average and worst fitness at each generation. This graph is classic, but essential as it shows an analyst critical information on convergence.

The two new visualisations are a pedigree graph of the best individual found so far and a child-parent correlation graph. The pedigree graph needs to be recalculated every time a new better solution is found. The correlation graph can incrementally be adjusted over time. Each visualisation type is explained below.

– **Pedigree Graph**
   A Pedigree Graph represents the parental hierarchy for the best solution. It simply traces the parents of the best solution and their parents back to the first generation. The primary objective of this graph is not to visualize the ancestor of the best solutions alone. Instead, this graph establishes the basis for verifying the evolutionary trajectories towards the best solution. We have two settings for this graph. One where each node has a unique id, thus, the graph shows ancestors. The second where each node is labelled with its fitness value, which offers a way to see the relationship between parents and children from a fitness perspective.

- **Fitness Graphs**
  These are classic visualisations that we can see in almost every paper published on EA. They describe population statistics related to the distribution of the fitness values. As a minimum, these graphs show the best, worst and average fitness in each generation. However, they can also include other statistics including mode, median and a specific percentile. The objective is to observe the changes occurring in the distribution of the fitness function at each generation.

- **Correlation Graphs**
  The correlation coefficient between two solution vectors in the continuous domain is a measure of inheritance equivalent to schema length in binary representation. However, when the fitness of children is correlated with the fitness of parents, the correlation coefficient acts as an indicator of the fitness landscape.
  For example, under the effect of local mutation alone, if the correlation coefficient between the fitness of the children and parents is high, it may imply a non-rugged fitness landscape. If it is low, it is an indication that the fitness value of individuals locally vary significantly; thus, a very rugged fitness landscape.
  We use Pearson's correlation coefficient (PCC) [13] as a measure of the strength of the association between the fitness values of parents and children.

$$PCC\,(Children, Parents) \leftarrow \frac{cov\,(Children, Parents)}{\delta\,(Children)\,.\delta\,(Parents)} \tag{1}$$

Equation 1 is the classic correlation coefficient calculated between children and parents at each generation. The relationship measures the rate of variations of fitness values for children relative to parents over generations.

4   Example

In this section, we present an example to demonstrate SEV and possible information that can be extracted from these graphs to guide the user to adjust evolutionary parameters. We use two different functions; a simple unimodal function (DeJong F1) and the multimodal function (Rastrigin F2), defined as Eqs. 2 and 3 in 2 dimensions (d) to explain the concept.

This case study runs DE twice with different parameters. In both trials, the classic DE/rand/1/bin strategy is used. The crossover rate is $CR = 0.1$ and $CR = 0.5$, respectively for each function. The F factor is 0.5. We use 10 generations for the sake of illustration. The output from both trials were then used by the hierarchy profiling algorithm to prepare the Pedigree Graph visualisation data.

$$f_1(x) = \sum_{i=1}^{d} x_i^2, i = 1.....d, -5.12 < x_i < 5.12 \tag{2}$$

$$f_2(x) = 10.d + \sum_{i=1}^{d} x_i^2 - 10.cos(2.\pi.x_i), i = 1...d, -5 < x_i < 5 \qquad (3)$$

The HP profiles of the two trials are used to generate the Pedigree graph. The Pajek [14] tool plots the two profiles to explore the best individuals for each trial. Figure 3 shows the F1 first and second trials of DE, while Fig. 4 displays the F2 two trials of DE . The two figures represent the best solution identifier from the last generation recursing backwards to the first generation in each trial.

In Fig. 3, the best solution identifier of the first trial is 101, whereas that of second trial is 80. We need to recall that the objective of this graph is not to see information on a screen per se. Instead, it establishes a verification network that can be queried. In addition, Fig. 4 of the multimodal Rastrigin function, the best solution identifiers are 78 and 59 for first and second trial, respectively.

Similar to any graph used for verification, it can get complex. Thankfully, we can use graph rewrite rules to skip intermediate parents and simplify the graph when needed. Leveraging the transitive nature of the clerestory relationship, if $p1$ is the parent of $p2$ and $p2$ is the parent of $p3$, then we can rewrite the two

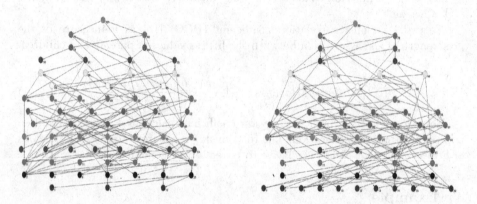

Fig. 3. Pedigree Graph Visualization of the DeJong's two trials using DE algorithm.

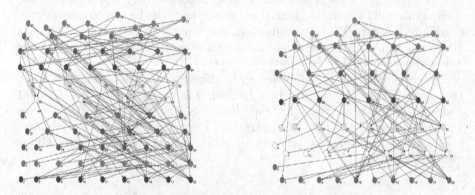

Fig. 4. Pedigree Graph Visualization of the Rastrigin's two trials using DE algorithm.

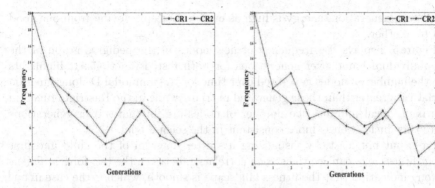

Fig. 5. Frequency of graph nodes over generations for DeJong and Rastrigin functions, respectively.

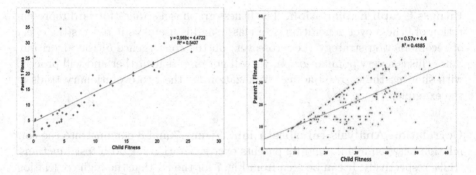

Fig. 6. A Scatter diagram showing the relationship between the child and first parent fitness for DeJong and Rastrigin functions, respectively.

connections and replace them with one from $p1$ to $p3$. In this way, we drop $p2$ (father) and maintain instead the child and grandfather. We can skip as many as needed to keep simplifying this graph.

We also noticed that almost all solutions in the first population appears as parents. Focusing only on parents from the second or third generations onwards can reduce the number of nodes significantly.

When the nodes are labelled with the fitness function, querying this graph can shed light on evolutionary dynamics. Examples of these queries is to estimate the manifold where the solutions associated with this graph are embedded.

Equally, we can use social network analysis to analyze the ruggedness of the fitness landscape to reach this best solution. A simple example of social network analysis is to visualize the energy associated with the evolution of the network. When evolution stagnates, the best solution does not change from one generation to another. Consequently, no new nodes get added at each generation after stagnation occurs; thus, the energy is zero. However, if new nodes are being added each generation, energy is high as evolution keeps moving from one good area to another.

Figure 5 displays the frequency of new nodes in the pedigree graph of the best individual found over generations for both tested functions. It highlights that the number of nodes generated over time for the unimodal DeJong function in trial 1 is greater than those generated in trial 2 while in the Rastrigin function with many local minima, the number of nodes are fluctuates over generations where this fluctuation is more consistent in the second trial.

Last, but not least, by visualizing a scatter diagram of the child and first parent fitness, one can see the strong correlation between the two for F1. This is a strong indication that the fitness landscape is smooth, which is the case in our unimodal function. On the other hand, the correlation is decreased for the F2 as its landscape is rugged somehow, which is the multimodal functions nature.

Fitness Graph Explanation. The fitness graph is a straightforward representation of fitness over generation. is a classic and is useful as it reflects the effect of recombination strategy (i.e. crossover) on the convergence of the algorithm. Since this is a very popular graph, we will not discuss it further and will proceed with showing the corresponding visualization for the two evolutionary trials in the example.

Correlation Analysis Interpretation. Figures 7 and 8 demonstrate the correlation between a child and its parents over generation for both test functions' trials, respectively. It can be seen from Fig. 7 for the F1 that the high correlation between the child and parent 1 is natural given the nature of recombination in DE. It is consistently high because of the fixed step and length and the non-rugged nature of the fitness landscape. This correlation is particularly useful in rugged landscapes and when the algorithm is self-adaptive as in Fig. 8 where the correlation between the child and its first parent is diminished compared with the F1.

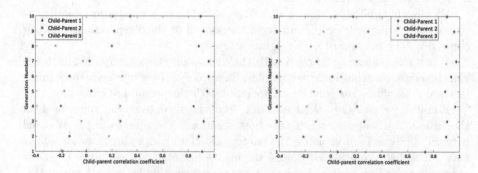

Fig. 7. Correlation coefficient of Child & Parents for DeJong trials 1 and 2, respectively.

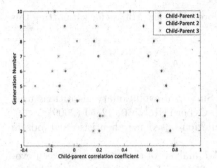 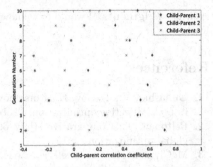

Fig. 8. Correlation coefficient of Child & Parents for Rastrigin trials 1 and 2, respectively.

5 Comparison with Existing Techniques

We compare our framework with recent visualisation techniques to position the contribution of the proposed SEV framework. To begin with, the techniques of PBF [8,15,16] generally represent the frequency of the population vectors. The use of this summary statistics hide the relationships among these vectors. In our study, we use the frequency of graph nodes (solutions) in Fig. 5 and the scatter diagram in Fig. 6 for showing the relationship between childrens and their parents as an indicator for pathways generated by evolution in the fitness landscape.

Secondly, the techniques of LDBF [9,17,18] show the evolutionary dynamics while we use the pedigree tree and fitness representations to examine the internal evolutionary process. However, LDBF assumes a more experienced user with an understanding of evolutionary dynamics, while pedigree trees are simple heredity concepts understood by less experienced and expert users equally.

Finally, the techniques of GDBF [10,19] represent the search space characteristics of the data matrices while we use second order statistics, in particular the correlation coefficient, as shown in Figs. 7 and 8 for demonstrating the relationship between the population elements in order to identify the actual changes of the vectors.

6 Conclusions and Future Work

This paper proposed the Semantic Evolutionary Visualization (SEV) framework to verify the dynamics of an evolutionary algorithm. In this paper, Differential Evolution was used as the evolutionary algorithm testbed. The framework consists of an evolutionary algorithm, a Hierarchy profiling (HP) technique for data preprocessing, and three visualizations including two newly proposed ones. We have demonstrated the use of the framework. In future work, we will extend the framework and design a suitable query engine to assist analysts in understanding the progress of an evolutionary algorithm.

Acknowledgments. I would to acknowledge prof. Hussein Abbass for supporting this work.

References

1. Guliashki, V., Toshev, H., Korsemov, C.: Survey of evolutionary algorithms used in multiobjective optimization. Prob. Eng. Cybern. Rob. **60**, 42–54 (2009)
2. Peltonen, T.: Comparative study of population-based metaheuristic methods in global optimization (2015)
3. McClymont, K., Keedwell, E., Savic, D.: An analysis of the interface between evolutionary algorithm operators and problem features for water resources problems. A case study in water distribution network design. Env. Model. Softw. (2015)
4. Tušar, T., Filipič, B.: Visualization of pareto front approximations in evolutionary multiobjective optimization: a critical review and the prosection method. IEEE Trans. Evol. Comput. **19**(2), 225–245 (2015)
5. He, Z., Yen, G.G.: Visualization and performance metric in many-objective optimization. IEEE Trans. Evol. Comput. **20**(3), 386–402 (2016)
6. Meyer, J., Thomas, J., Diehl, S., Fisher, B.D., Keim, D.A., Laidlaw, D.H., Miksch, S., Mueller, K., Ribarsky, W., Preim, B., et al.: From visualization to visually enabled reasoning. Sci. Vis. Adv. Concepts **1**, 227–245 (2010)
7. Qin, A.K., Huang, V.L., Suganthan, P.N.: Differential evolution algorithm with strategy adaptation for global numerical optimization. IEEE Trans. Evol. Comput. **13**(2), 398–417 (2009)
8. Collins, T.D.: Visualizing evolutionary computation. In: Advances in Evolutionary Computing, pp. 95–116. Springer, Heidelberg (2003)
9. Paterson, T., Graham, M., Kennedy, J., Law, A.: Evaluating the viper pedigree visualisation: detecting inheritance inconsistencies in genotyped pedigrees. In: 2011 IEEE Symposium on Biological Data Visualization (BioVis), pp. 119–126. IEEE (2011)
10. Collins, T.D.: Using software visualisation technology to help evolutionary algorithm users validate their solutions. In: ICGA, pp. 307–314. Citeseer (1997)
11. Storn, R., Price, K.: Differential evolution-a simple and efficient heuristic for global optimization over continuous spaces. J. Global Optim. **11**(4), 341–359 (1997)
12. Thangaraj, R., Pant, M., Abraham, A., Badr, Y.: Hybrid evolutionary algorithm for solving global optimization problems. In: Corchado, E., Wu, X., Oja, E., Herrero, Á., Baruque, B. (eds.) HAIS 2009. LNCS, vol. 5572, pp. 310–318. Springer, Heidelberg (2009). doi:10.1007/978-3-642-02319-4_37
13. Cohen, J., Cohen, P., West, S.G., Aiken, L.S.: Applied Multiple Regression/correlation Analysis for the Behavioral Sciences. Routledge, New York (2013)
14. De Nooy, W., Mrvar, A., Batagelj, V.: Exploratory Social Network Analysis with Pajek, vol. 27. Cambridge University Press, Cambridge (2011)
15. Romero, G., Merelo, J.J., Castillo, P.A., Castellano, J.G., Arenas, M.G.: Genetic algorithm visualization using self-organizing maps. In: Guervós, J.J.M., Adamidis, P., Beyer, H.-G., Schwefel, H.-P., Fernández-Villacañas, J.-L. (eds.) PPSN 2002. LNCS, vol. 2439, pp. 442–451. Springer, Heidelberg (2002). doi:10.1007/3-540-45712-7_43
16. Amir, E.-A.D., Davis, K.L., Tadmor, M.D., Simonds, E.F., Levine, J.H., Bendall, S.C., Shenfeld, D.K., Krishnaswamy, S., Nolan, G.P., Pe'er, D.: visne enables visualization of high dimensional single-cell data and reveals phenotypic heterogeneity of leukemia. Nat. Biotechnol. **31**(6), 545–552 (2013)

17. Jornod, G., Di Mario, E.L., Navarro Oiza, I., Martinoli, A.: Swarmviz: an open-source visualization tool for particle swarm optimization. In: IEEE Congress on Evolutionary Computation, no. EPFL-CONF-206841 (2015)
18. de Freitas, A.R., Fleming, P.J., Guimarães, F.G.: Aggregation trees for visualization and dimension reduction in many-objective optimization. Inf. Sci. **298**, 288–314 (2015)
19. Khemka, N., Jacob, C.: Visplore: a toolkit to explore particle swarms by visual inspection. In: Proceedings of the 11th Annual conference on Genetic and Evolutionary Computation, pp. 41–48. ACM (2009)

Erratum to: Gravitational Search Algorithm in Recommendation Systems

Vedant Choudhary[✉], Dhruv Mullick, and Sushama Nagpal[✉]

Netaji Subhas Institute of Technology,
Sector-3, Dwarka, Delhi 110078, India
{vedantc.ec,dhruvml.co}@nsit.net.in,
sushmapriyadarshi@yahoo.com

Erratum to:
Chapter "Gravitational Search Algorithm
in Recommendation Systems" in: Y. Tan et al. (Eds.),
Advances in Swarm Intelligence, LNCS 10386,
DOI: 10.1007/978-3-319-61833-3_63

The initial version of the paper contained a few errors in the equations on pages 600 and 601. These have now been rectified.

The updated online version of this chapter can be found at
http://dx.doi.org/10.1007/978-3-319-61833-3_63

© Springer International Publishing AG 2017
Y. Tan et al. (Eds.): ICSI 2017, Part II, LNCS 10386, p. E1, 2017.
DOI: 10.1007/978-3-319-61833-3_67

Author Index

Printed in the United States
By Bookmasters

Printed in the United States
By Bookmasters